DATE DUE			

Magnetic Atoms and Molecules

Magnetic Atoms and Molecules

By

William Weltner, Jr.
Chemistry Department
University of Florida

Dover Publications, Inc., New York

Published in Canada by General Publishing Company, Ltd., 30 Lesmill Road, Don Mills, Toronto, Ontario.
Published in the United Kingdom by Constable and Company, Ltd., 10 Orange Street, London WC2H 7EG.

This Dover edition, first published in 1989, is an unabridged, corrected, slightly enlarged republication of the work first published by Scientific and Academic Editions (division of Van Nostrand Reinhold Company Inc.), New York, 1983. For this edition, the author has corrected a number of errors, provided a new illustration for Fig. V-32, and added a new Appendix ("E"), "Fundamental Physical Constants; Units and Conversion Factors."

Manufactured in the United States of America
Dover Publications, Inc., 31 East 2nd Street, Mineola, N.Y. 11501

Library of Congress Cataloging-in-Publication Data

Weltner, William.
Magnetic atoms and molecules / by William Weltner, Jr.
p. cm.
Reprint. Originally published: New York : Scientific and Academic Editions, c1983.
Includes bibliographical references.
ISBN 0-486-66140-7
1. Electron magnetic resonance spectroscopy. 2. Molecular spectroscopy. I. Title.
QD96.E4W44 1989
543′.0877—dc20 89-17107
CIP

To Cam and the Boys

Preface

This book began as a "primer," written to acquaint new graduate students and postdoctoral fellows in my laboratory with the electron-spin-resonance theory of randomly-oriented molecules. The later chapters were added during periods of sabbatical from the University of Florida. This exposition of theory and experimental work owes much to my past students and associates who have contributed and taught me the subject. There are not too many that they cannot all be named here, so with the hope that no one has been inadvertently omitted, I gratefully thank the following:

J. H. Ammeter
C. A. Baumann
S. V. Bhat
B. R. Bicknell
W. D. Bird
J. M. Brom, Jr.
C. M. Brown
T. Chen
R. L. DeKock
A. Dendramis
T. C. DeVore
K. I. Dismuke
C. H. Durham
W. C. Easley
R. F. Ferrante
W. R. M. Graham
W. D. Hewett, Jr.
P. H. Kasai
L. B. Knight, Jr.
R. R. Lembke
K. C. Lin
J. V. Martinez
N. S. McIntyre
D. McLeod, Jr.
M. L. Seely
G. R. Smith
K. R. Thompson
R. J. Van Zee
P. N. Walsh
J. R. W. Warn
E. B. Whipple
J. L. Wilkerson
K. J. Zeringue

In recent years I have particularly enjoyed collaboration with Dr. Richard J. Van Zee.

Chapter IV involves gas phase studies, and in this less familiar area, I am especially grateful to Dr. John M. Brown of the University of Southampton for making some critical remarks and suggestions. Any remaining errors are, of course, mine.

I also thank all those authors who have given me permission to use their Figures published in journals and books, and who often sent me larger prints for better duplication: R. Aasa, F. Adrian, J. H. Ammeter, R. A. Bernheim, V. Beltrán–López, T. G. Castner, Jr., K. Evenson, J. H. Freed, J. F. Gibson, W. Gordy, E. A. Harris, J. T.

Hougen, M. T. Hutchings, D. J. E. Ingram, K. Itoh, C. K. Jen, P. H. Kasai, G. F. Kokoszka, J. Kommandeur, R. Lefebvre, D. M. Lindsay, G. W. Ludwig, Z. Luz, H. M. McConnell, J. A. McMillan, J. R. Morton, J. Owen, P. Raghunathan, M. T. Rogers, D. Schoemaker, J. H. van der Waals, E. Wasserman, M. Weissbluth.

Particular thanks goes to Mrs. Wanda Douglas, a superior person and secretary, who persevered through many years with me and typed much of the manuscript for this monograph. My wife, Cam, has provided the final typing effort and as usual did a superb job.

Gainesville, Florida WILLIAM WELTNER, JR.

Contents

Preface / vii

I. ATOMS IN MAGNETIC FIELDS / 1

1. Electron Spin Resonance (ESR) / 3
2. Nuclear Hyperfine Structure / 4
   (a) Breit-Rabi Equation / 6
   (b) Magnetic Dipole Transitions / 7
3. Nuclear Quadrupole Effects / 12
4. The Meaning of g_J and A / 13
   (a) Isotropic Hyperfine Constant, A_{iso} / 14
   (b) Anisotropic or Dipolar Hyperfine Constant, A_{dip} / 14
   (c) Polarization, Correlation, and Relativistic Effects / 16
   (d) $|\Psi(0)|^2$ and $\langle r^{-3}\rangle$ / 17
5. Atoms and Atomic Ions in Solid Matrices / 19
   (a) S-State Atoms / 19
      (α) Matrix effects / 19
      (β) Half-filled shells / 25
      (γ) Superhyperfine interaction / 26
      (δ) High-spin atoms, crystal-field effects / 28
   (b) Orbitally Degenerate Atoms and Quenching / 29

II. $^2\Sigma$ AND OTHER DOUBLET ($S = \frac{1}{2}$) MOLECULES IN MAGNETIC FIELDS / 39

1. $^2\Sigma$ Molecules / 39
   (a) Spin Hamiltonian / 40
   (b) g Tensor / 42
      (α) $\Delta g_\perp = g_\perp - g_e$ / 46
      (β) Spin-rotation constant; the Curl equation / 50
   (c) Nuclear Hyperfine Interaction (A Tensor) / 52
      (α) Spin density and hyperfine splitting / 57
      (β) A^X_{iso} (atom) and A^X_{dip} (atom) values / 59

(γ) Interatomic dipole–dipole interaction / 61
(δ) Signs of molecular A components / 62
(ε) Nuclear Zeeman and quadrupole terms / 63
(d) Energy Levels in an External Magnetic Field / 65
(e) ESR Spectroscopy / 70
(α) Selection rules and transition probabilities / 70
(β) Resonant magnetic fields for allowed transitions ($\Delta m_S = \pm 1, \Delta m_I = 0$) / 72
(γ) Forbidden transitions ($\Delta m_S = \pm 1, \Delta m_I = \pm 1$ or ± 2) / 74
2. Other Axial Doublet Molecules / 77
(a) Orbitally Degenerate Molecules and Quenching / 77
(b) Axial Nonlinear Molecules / 79
3. ESR Experimental Measurements in Crystals and Matrices / 81
(a) ESR of Molecules in Single Crystals / 81
(b) ESR of Matrix-Isolated Molecules / 86
(α) Randomly oriented molecules / 89
(β) Nuclear hyperfine effects / 93
(γ) Nonrandom molecular orientation / 95
$Cu(NO_3)_2$ and CuF_2 / 95
BO / 99
Other observations of orientation / 100
(δ) Off-principal-axis absorption; "extra" lines / 102
4. Nonaxial Doublet Molecules / 103
(a) g and A Having the Same Principal Axes; $^{14}NO_2$, $^{13}CO_2^-$ / 104
(b) Axes of g and A_i Not Coincident; $N^{17}O_2$, $^1H^{13}CO$ / 109
(c) Randomly Oriented Nonaxial Radicals / 117
(α) Transition probabilities / 119
5. Free and Hindered Rotation of Trapped Radicals / 120
6. Magnetic Properties of Small Doublet Molecules in Crystals and Matrices / 126
(a) Tabulated Magnetic Properties / 126
(b) Representative Doublet Radicals / 126
BeH, MgH, CaH, SrH, and BaH / 126
BeF, MgF, CaF, SrF, and BaF / 129
BeOH and MgOH / 135
CN / 136
CH_3 and Its Homologues / 141
CH_3 / 141
CF_3 / 142
CH_2F, CHF_2 / 144
SiH_3, GeH_3, SnH_3 / 145
HCO / 145

Na_3 / 147
Alkali Superoxide Molecules: NaO_2, KO_2, RbO_2, and CsO_2 / 149

III. $^3\Sigma$ AND OTHER TRIPLET ($S = 1$) MOLECULES IN MAGNETIC FIELDS / 156

1. Zero-Field Tensor, $\boldsymbol{D}$ / 157
   (a) $\boldsymbol{D}_{SO}$ / 157
   (b) $\boldsymbol{D}_{SS}$ / 158
2. $^3\Sigma$ Molecules / 160
   (a) Eigenvalues in a Magnetic Field / 160
   (b) Magnetic-Dipole Transitions / 165
   (c) Eigenfunctions / 170
   (d) Transition Probabilities and Relative Intensities of Transitions / 172
   (e) $\Delta m_S = 2$ Transitions / 175
   (f) $\boldsymbol{g}$ Tensor / 176
   (g) ESR Spectra of Randomly Oriented $^3\Sigma$ Molecules / 176
   (h) Hyperfine Interaction / 179
3. Nonlinear Triplet Molecules / 183
   (a) Randomly Oriented Molecules / 190
   (b) Noncollinear $\boldsymbol{g}$ and $\boldsymbol{D}$ / 192
   (c) Two-Quantum Transition / 192
4. Motional and Matrix Effects Upon the ZFS / 193
   (a) Rotational Diffusion / 193
   (b) Torsional Oscillation; Isotope Effects / 193
   (c) Matrix Effects Upon $\boldsymbol{D}$ / 197
5. Theoretical Calculations of ZFS Parameters / 198
   (a) Molecules Containing No Metal Atoms / 199
      (α) O_2 and analogues SO, S_2 / 199
      (β) NH and analogues PH, NX, PF / 200
      (γ) CH_2 / 200
      (δ) Naphthalene, benzene and other aromatic molecules / 202
   (b) Transition-Metal Molecules / 204
6. Magnetic Properties of Triplet Molecules / 206
   (a) Tabulated Magnetic Properties / 206
   (b) Other Representative Triplet Radicals / 206
      Propargylene and its homologues (alternant methylenes) / 206
      C_4 / 209
      CNN, NCN, CCO / 211
      SiCO and SiNN / 213
      TiF_2 (a nonlinear triatomic) / 218

IV. MAGNETIC RESONANCE OF GAS-PHASE MOLECULES / 219

1. Experimental Techniques / 219
   (a) Electron Paramagnetic Resonance (EPR) / 220
   (b) Laser Magnetic Resonance (LMR) / 221
2. Experimental Data for Gaseous Radicals / 224
   (a) Hyperfine Parameters / 225
   (b) *g* Values; the Curl Equation / 226
   (c) Spin-Rotation Interaction in High Spin Molecules / 234

V. HIGH SPIN MOLECULES / 235

1. $^4\Sigma$ and Other Quartet ($S = \frac{3}{2}$) Molecules in Magnetic Fields / 236
   A. The Axial Case, $^4\Sigma$ Molecules / 236
      (a) Eigenvalues and Magnetic-Dipole Transitions / 236
      (b) Effective *g* Values / 239
      (c) Randomly Oriented Molecules / 241
      (d) Hyperfine Interaction / 243
   B. Individual Molecules / 247
      VO and NbO / 247
      VF_2 and MnO_2 / 250
      Organic Radicals / 252
      Impurity Pairs in Silicon Crystals / 254
      R-Center in KCl Crystals / 256
2. $^5\Sigma$ and Other Quintet ($S = 2$) Molecules in Magnetic Fields / 256
   A. Theory / 256
      (a) General Spin Hamiltonian / 256
      (b) Eigenvalues and Magnetic-Dipole Transitions / 257
      (c) Randomly Oriented Molecules / 262
   B. Individual Molecules / 264
      Organic Radicals / 264
      Crystals and Inorganic Molecules / 265
3. $^6\Sigma$ and Other Sextet ($S = \frac{5}{2}$) Molecules in Magnetic Fields / 266
   A. The Axial Case, $^6\Sigma$ Molecules / 266
      (a) Eigenvalues and Magnetic-Dipole Transitions / 268
      (b) Effective *g* Values / 271
      (c) Randomly Oriented Molecules / 271
      (d) Hyperfine Interaction / 272
   B. Nonaxial Molecules; The $g' = 4.3$ Signal / 277
   C. Individual Molecules / 281
      Ferrihemoglobin / 281

FeF_3 / 281
MnF_2, MnH_2, and MnO / 282
4. $^7\Sigma$ Molecules ($S = 3$) in Magnetic Fields / 287
A. Theory / 288
(a) Randomly Oriented Molecules / 288
B. Individual Molecules / 288
MnH, MnF / 288
Organic Radicals / 294
5. $^8\Sigma$ ($S = \frac{7}{2}$) and $^9\Sigma$ ($S = 4$) Molecules / 295
A. Axial and $^8\Sigma$ ($S = \frac{7}{2}$) Molecules / 295
GdF_3 / 295
CrCu / 297
B. $^9\Sigma$ ($S = 4$) Molecules / 301
GdO / 302
6. Hyperfine Interaction in High-Spin Molecules / 302
7. Electron Exchange Coupling / 304
(a) Anisotropic Exchange / 306
(b) Dimers, Judd-Owen Equations / 309
(c) Hyperfine Interaction / 312
(d) Direct and Super Exchange / 314
(e) Exchange-Coupled Dimers / 316
Copper Acetate Monohydrate $[Cu_2(CH_3CO_2)_4(H_2O)_2]$ / 316
Isolated Pairs of Magnetic Ions in Insulators / 318
Mn_2 Molecules in the Solid Rare Gases / 319

REFERENCES / 323

I. Books / 323
II. Articles in Journals and in Books / 324

APPENDIX A. STABLE MAGNETIC NUCLEI; THEIR ABUNDANCES, SPINS, MOMENTS, AND MAGNETOGYRIC RATIOS / 341

APPENDIX B. ATOMIC HYPERFINE PARAMETERS FOR USE WITH ESR DATA / 345

APPENDIX C. MAGNETIC PARAMETERS AND SPIN DENSITIES OF SMALL MOLECULES STUDIED BY ESR IN MATRICES AND CRYSTALS / 349

APPENDIX D. MAGNETIC RESONANCE PARAMETERS FOR GAS-PHASE MOLECULES / 389

APPENDIX E. FUNDAMENTAL PHYSICAL CONSTANTS; UNITS AND CONVERSION FACTORS / 400

AUTHOR INDEX / 402

SUBJECT INDEX / 409

Magnetic Atoms and Molecules

Chapter I

Atoms in Magnetic Fields

An atom in the gas phase may have a magnetic dipole moment $\boldsymbol{\mu}$, and if so, it will interact with a magnetic field $\boldsymbol{H}$. This interaction is described by the Hamiltonian

$$\mathcal{H} = -\boldsymbol{\mu} \cdot \boldsymbol{H} \tag{I, 1}$$

Orbital electronic motion is one source of a magnetic moment in an atom that results directly from the angular momentum, $\boldsymbol{p}$, of the charge. For an electron moving in an orbit, $\boldsymbol{\mu}_L$ is found to have the classical value

$$\boldsymbol{\mu}_L = (-e/2mc)\,\boldsymbol{p}_L \tag{I, 2}$$

where $\boldsymbol{p}_L$ is its orbital angular momentum. Here e and m are the magnitudes of the charge and mass of the electron, and c is the velocity of light. This illustrates the generality that $\boldsymbol{\mu}$ is proportional to angular momentum, but oppositely directed, for a negatively charged electron. This proportionality is often expressed by defining the magnetogyric ratio γ as

$$\boldsymbol{\mu} = \gamma \boldsymbol{p}.$$

The magnetic moment generated by an electron moving in a circular orbit of radius r is given classically by the current multiplied by the cross-sectional area of the orbit. The current generated by charge $-e$ rotating at $v/2\pi r$ revolutions/second is $-ev/2\pi r$, where v is its linear velocity. Then

$$\mu_L = (-ev/2\pi r) \cdot (\pi r^2) = (-evr/2c)\ emu$$

But $p_L = mvr$ so that

$$\boldsymbol{\mu}_L = (-e/2mc)\,\boldsymbol{p}_L, \quad \text{as in (I,2).}$$

The electron also spins about its own axis, that is, has an intrinsic magnetic moment, but here γ is anomalously larger, by a factor of 2, than for $\boldsymbol{\mu}_L$,

$$\boldsymbol{\mu}_S = 2(-e/2mc)\,\boldsymbol{p}_S$$

These orbital and spin contributions are then added to give the total magnetic moment of the atom.

In classical terms the solution of (I, 1) yields the energy of the magnetic particle in the field as

$$E = -\mu H \cos\theta$$

where θ is the angle between the vectors of magnitude μ and H, and the energy can therefore vary continuously with the orientation of the magnetic moment. However, quantum mechanically this angle is space-quantized and can only take $2J + 1$ orientations, where J is the quantum number for total angular momentum, equal to $J\hbar$ (where $\hbar = h/2\pi$). The allowed projections of J (or $\boldsymbol{\mu}$) along the magnetic field direction are given by $m_J\hbar$, where m_J is the magnetic quantum number taking the values

$$m_J = J, J-1, \cdots, -J.$$

If only spin angular momentum arises (as in a ${}^2S_{1/2}$ atom), m_J becomes

$$m_S = S, S-1, \cdots, -S$$

where S is the total electron spin quantum number. Then μ_H (along the field direction) is, since $p_S = m_S\hbar$,

$$\mu_H = -2(eh/4\pi mc)\, m_S.$$

Defining the Bohr magneton as $(eh/4\pi mc) = \beta_e = 9.2741 \times 10^{-21}$ erg/gauss, one then finds that the allowed (Zeeman) energies for an atom in a magnetic field are

$$E_{m_S} = -\mu_H H = 2\beta_e m_S H$$

For a free electron a small but important quantum electrodynamics correction requires that this equation be written as

$$E_{m_S} = g_e \beta_e m_S H \tag{I, 3}$$

where $g_e = 2.0023$. Thus, in this pure spin case, the familiar equal spacing of $2S + 1$ energy levels occurs with a separation of $g_e \beta_e H$.

For an orbitally degenerate state of an atom ($L \neq 0$), where there is strong (Russell–Saunders) coupling between L and S, then $J = L + S$, $L + S - 1, \cdots, |L - S|$, and for each J, $m_J = J, J - 1, \cdots, -J$,

$$E_{m_J} = g_J \beta_e m_J H \tag{I, 4}$$

where

$$g_J = 1 + \frac{S(S+1) + J(J+1) - L(L+1)}{2J(J+1)}$$

(g_J is known as the Landé splitting factor.) For example, for a $^2P_{1/2}$ state, $S = \frac{1}{2}$, $L = 1$, $J = \frac{1}{2}$, and $g_J = \frac{2}{3}$. Of course, g_J must attain the free electron value at $L = 0$.

Thus loss of orbital angular momentum will cause g_J to ultimately approach g_e, which is referred to as "quenching" of the orbital angular momentum. In nonlinear molecules the orbital angular momenta are routinely quenched by the asymmetric internal electric fields, but in free gas-phase atoms they are not [Van Vleck (22)]. However, when an atom is trapped in a solid matrix, the crystal fields produced by the asymmetric environment can, at least partially, quench the orbital angular momentum of the atom. The magnetic properties of the atom are then being influenced by the surrounding electric field, and this may provide information about its site in the lattice or matrix.

1. ELECTRON SPIN RESONANCE (ESR)

For the simplest case of one free electron spin, $m_J = m_S = \pm\frac{1}{2}$, only two Zeeman levels are possible according to (I, 3) or (I, 4), as indicated in Fig. I-1. A transition ΔE can be induced between these levels by applying an oscillating magnetic field of frequency ν_0 at

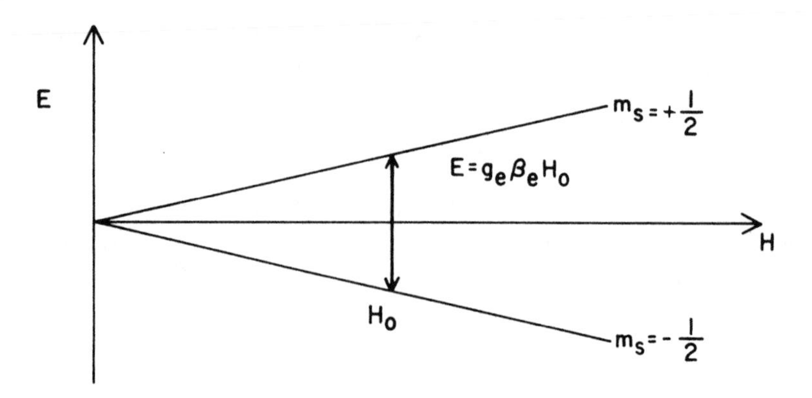

Fig. I-1. Zeeman effect and allowed magnetic-dipole transition of a free electron in a magnetic field.

right angles to $\boldsymbol{H}$ such as to cause the spin to flip at a particular resonance field H_0:

$$h\nu_0 = g_e \beta_e H_0 \tag{I, 5}$$

Putting in $g_e = 2.0023$, $\beta_e = 9.2741 \times 10^{-21}$ erg/gauss, $h = 6.6262 \times 10^{-27}$ erg sec, we find

$$\frac{\nu_0}{H_0} = 2.8025 \text{ MHz/G}$$

Most such electron spin resonance experiments utilize a fixed frequency $\nu_0 \cong 9.3$ GHz (so-called X band) which places the resonant magnetic field H_0 at approximately 3300 G when $g \cong 2$. One then expects to see one absorption line in the ESR spectrum of a ^2S atom at about this magnetic field, and similarly for an unquenched ^2P atom but at the magnetic field corresponding to its particular g_J value.

2. NUCLEAR HYPERFINE STRUCTURE

Almost all elements have at least one stable isotope with a nuclear magnetic moment. The natural abundance of some isotopes may be small, however, so that the influence of those magnetic nuclei may not be observed. Appendix A contains an up-to-date list of all the stable magnetic nuclei and their natural abundances, spins, nuclear magnetic moments, and electric quadrupole moments. Their derived magnetogyric ratios $\gamma_n = \mu_I/I\hbar$ are also tabulated. The magnetic moment due to the electrons in the atom can then interact with the nuclear moment and split the single ESR line to yield hyperfine structure (hfs). Thus if an unpaired electron interacts magnetically with a nucleus of magnetic moment μ_I and nuclear spin I, the one line previously observed in ESR will be split into $2I + 1$ equally (almost) spaced lines by the hyperfine interaction. For example, the proton H has a nuclear spin $I = \frac{1}{2}$ and $\mu_I = 2.7926$ n.m. (nuclear magnetons) and the deuteron D has $I = 1$ and $\mu_I = +0.85735$ n.m. H atoms then exhibit an ESR absorption spectrum containing two lines split by 508 G (see Fig. I-2), and D atoms yield three lines separated by 78 G. Also, since the splitting is proportional to γ_n, the hfs for one isotope can be calculated from that of the other.

One recognizes the analogy here between electronic and nuclear magnetic properties. μ_I is related to the nuclear g_I value by

$$\mu_I = g_I I \beta_n \tag{I, 6}$$

and the sign of g_I then depends upon the sign of the nuclear moment. Here β_n is the nuclear magneton, $e\hbar/2Mc$, where M is the proton mass, so that $\beta_n = 5.0510 \times 10^{-24}$ erg/gauss. The projection along the magnetic field is determined now by the values of m_I which vary over the range $I, \cdots, -I$. Thus, as Fig. I-2 shows, for each value of m_S, $m_I = \pm\frac{1}{2}$ for an $I = \frac{1}{2}$ nucleus in a strong magnetic field. However, for a positive nuclear moment, $\boldsymbol{\mu}_J$ and $\boldsymbol{\mu}_I$ oppose each other if $\boldsymbol{I}$ and $\boldsymbol{J}$ are parallel, so that the signs of the interactions of the electron and the nucleus with the magnetic field will also be opposed.

The effect of a magnetic nucleus on the Zeeman levels can be accounted for by including in the Hamiltonian considered above terms involving nuclear moment interactions. Concisely,

$$\mathcal{H} = g_J \beta_e \boldsymbol{H} \cdot \boldsymbol{J} + hA\boldsymbol{I} \cdot \boldsymbol{J} - g_I \beta_n \boldsymbol{H} \cdot \boldsymbol{I} \tag{I, 7}$$

where $\boldsymbol{I}$ and $\boldsymbol{J}$ are operators. (For a nucleus having an electric quadrupole moment an additional term should be added to the right-hand side; it will be considered at the end of this hyperfine structure dis-

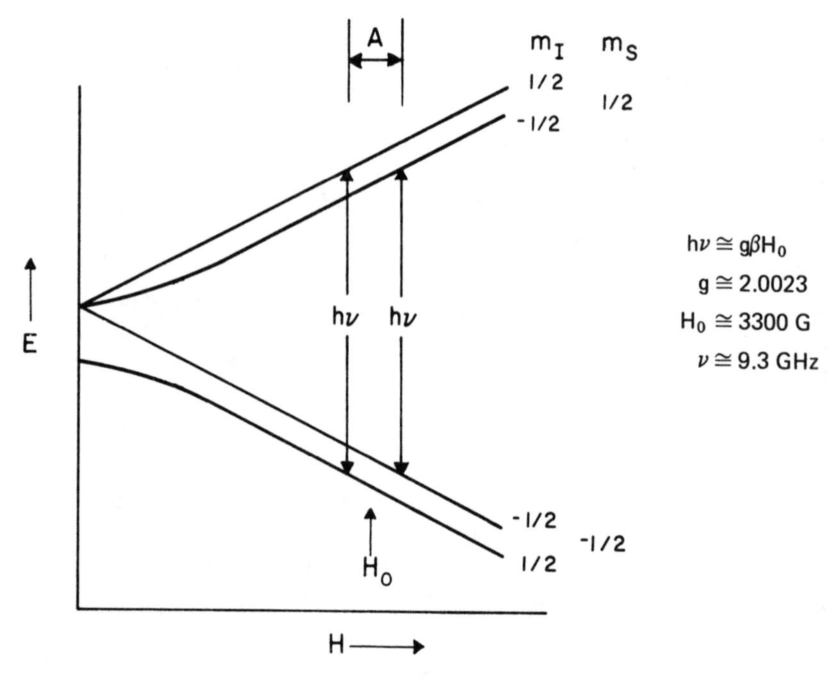

Fig. I-2. Hyperfine splitting (A) due to an electron interacting with a nucleus with spin $\frac{1}{2}$.

cussion.) The first term on the right is the electronic Zeeman effect already discussed. The last term is the so-called nuclear Zeeman effect resulting from the direct interaction of the magnetic field with the nuclear moment; it is usually negligible because β_n is ~1/2000 of β_e. Hyperfine splitting is produced by the remaining term, and its magnitude is proportional to the hf constant A (see Fig. I-2). Note that even when the magnetic field is zero, the hyperfine term involving $\boldsymbol{I} \cdot \boldsymbol{J}$ will produce a splitting of the energy levels. Measurement of the line positions in an ESR spectrum of an atom allows the calculation of g_J and A from the solution of the energy level problem, which is due to Breit and Rabi (84).

(a) Breit-Rabi Equation

In a weak external magnetic field the $hA\boldsymbol{I} \cdot \boldsymbol{J}$ term in equation (I, 7) is largest, and the angular momenta indicated by $\boldsymbol{I}$ and $\boldsymbol{J}$ are coupled to form a resultant $\boldsymbol{F} = \boldsymbol{I} + \boldsymbol{J}$. F, the good quantum number in this limit, takes the values $I+J, I+J-1, \cdots, |I-J|$. For each value of F there are $2F+1$ projections of $\boldsymbol{F}$ in the field direction indicated by $m_F = F, F-1, \cdots, -F$. In a very strong field, corresponding to the Paschen–Back effect, $\boldsymbol{I}$ and $\boldsymbol{J}$ are effectively acted upon separately by H, i.e., they are decoupled. F and m_F are no longer good quantum numbers and are replaced by m_J and m_I as defined previously. The number of levels is the same as in the weak field but is now $(2J+1)(2I+1)$. In intermediate fields m_F is replaced by $m = m_I + m_J$, and the Breit–Rabi equation, which gives the energy levels, $W(F, m)$ over all fields, is, for $J = \frac{1}{2}$ [Kopfermann (14), Ramsey (19)]:

$$W(F, m) = \frac{-\Delta W}{2(2I+1)} - \left(\frac{\mu_I}{I}\right) Hm \pm \frac{\Delta W}{2} \left(1 + \frac{4m}{2I+1} x + x^2\right)^{1/2} \tag{I, 8}$$

where $\Delta W = h(A/2)(2I+1)$, $x = [(g_J \beta_e - (\mu_I/I))/\Delta W]H$. The plus sign in (I, 8) applies for $F = I + \frac{1}{2}$ and $m = +(I + \frac{1}{2}), \cdots, -(I + \frac{1}{2})$, and the minus sign applies for $F = I - \frac{1}{2}$ when $m = (I - \frac{1}{2}), \cdots, -(I - \frac{1}{2})$. ΔW is the energy level spacing at zero magnetic field. The weak-field case corresponds to $x^2 \ll 1$ and the strong-field case to $x^2 \gg 1$.

The application of this equation to a $^2S_{1/2}$ atom with a nuclear spin $I = \frac{7}{2}$ gives the variation of the energy levels shown in Fig. I-3. The nuclear moment is assumed positive here; a negative moment would change the sign of A and thereby cause the levels to be inverted and also change the signs of the quantum numbers [Ramsey (19)]. Both sets of quantum numbers (F, m_F) and (m_J, m_I) are indicated for each level in Fig. I-3. $(\mu_I/I) = g_I \beta_n$ is usually negligible in (I, 8) so that x can be taken as $g_J \beta_e H/\Delta W$, where for the specific case of $I = \frac{7}{2}$, $\Delta W = 4hA$.

(b) Magnetic Dipole Transitions

Since transitions between levels involve changes in magnetic moments, we are interested in magnetic dipole (as opposed to electric dipole) transitions and the selection rules pertaining to them. In strong fields, these selection rules require that only one spin (i.e., electronic or nuclear) flip at a time: $\Delta m_J = \pm 1$, $\Delta m_I = 0$ or $\Delta m_J = 0$,

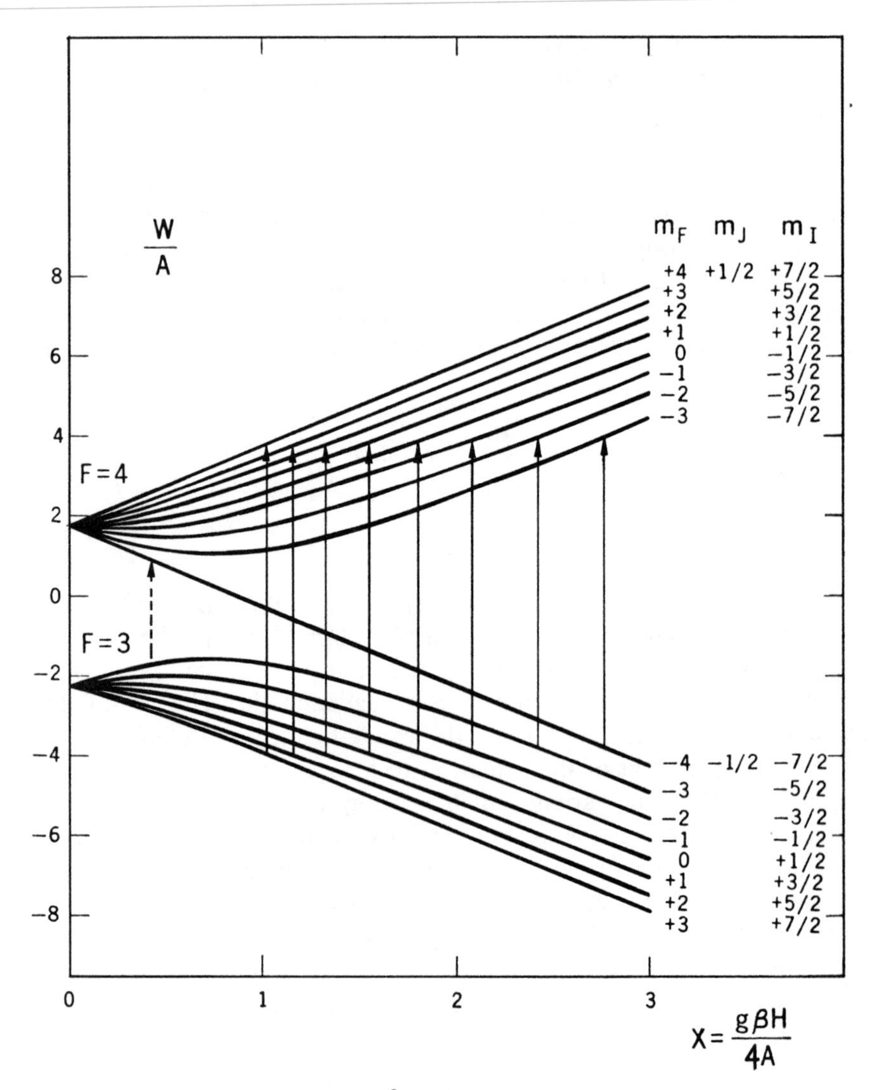

Fig. I-3. Zeeman energy levels (W) of a $^2S_{1/2}$ atom with a nuclear spin $I = \frac{7}{2}$ [from Weltner, McLeod, and Kasai (477)]. A is the hyperfine coupling constant, allowed "ESR" transitions indicated by solid arrows, and "NMR" transition by dashed arrow.

$\Delta m_I = \pm 1$. In ESR work where it is most convenient to vary the dc magnetic field and keep the inducing (microwave) field fixed, one is usually automatically limited to observing only the electron spin transitions $\Delta m_J = \pm 1$, $\Delta m_I = 0$ because of their larger energy difference. These are illustrated by the $2I + 1$ solid vertical lines shown in Fig. I-3, which, if they occurred in stronger fields, would be equally spaced (in gauss) as the magnetic field is varied. Here they occur in the intermediate-field region, and the spacing becomes greater at higher fields.

In weak fields, an inducing alternating magnetic field (H_1) which is perpendicular to the "static" magnetic field can produce transitions $\Delta m_F = \pm 1$ for $\Delta F = \pm 1$ or 0. (This is the usual orientation of H_1 versus H in ESR cavities in commercial instruments; if $H_1 \| H$, the selection rules are $\Delta m_F = 0$, $\Delta F = \pm 1$.) Such perpendicular (π) transitions are also in accord with the solid lines in Fig. I-3 in the intermediate-field region.

If the zero-field splitting (proportional to the hyperfine splitting constant A) were large enough relative to the microwave frequency, one can see that the nuclear spin (NMR) transition shown by the dashed vertical arrow in Fig. I-3 might be observed. Here the transition $\Delta m_J = 0$, $m_I = -\frac{7}{2} \leftrightarrow -\frac{5}{2}$ is indicated, and it is evident that it is the only such transition with a large enough energy difference to possibly be observed in the same spectrum with the electron spin transitions. If it were observed, it is clear that only a few of the other hyperfine transitions could also appear in the spectrum.

The magnetic fields at which resonant transitions occur may be derived from (I, 8). If terms involving μ_I/I are considered negligible, then the energy levels are given by

$$W = -\frac{A}{4} \pm \frac{1}{2}\left[A^2\left(I + \frac{1}{2}\right)^2 + 2Amg\beta H + g^2\beta^2H^2\right]^{1/2} \qquad \text{(I, 9)}$$

For the transition $\Delta F = \pm 1, 0$ and $\Delta m = \Delta m_F = \pm 1$ (i.e., $\Delta m_S = \pm 1$, $\Delta m_I = 0$) where m_F changes from $m - 1$ to m, transitions occur for

$$\begin{aligned} h\nu_0 = &\tfrac{1}{2}\,[A^2(I + \tfrac{1}{2})^2 + 2Ag\beta H(m - 1) + g^2\beta^2H^2]^{1/2} \\ &+ \tfrac{1}{2}\,[A^2(I + \tfrac{1}{2})^2 + 2Ag\beta Hm + g^2\beta^2H^2]^{1/2} \end{aligned}$$

Dividing through by $h\nu_0$ and letting

$$u = \frac{A}{h\nu_0} \quad \text{and} \quad v = \frac{g\beta H}{h\nu_0}$$

this reduces to

$$v = \frac{-u\left(m - \frac{1}{2}\right)}{1 - \frac{u^2}{4}} \pm \left[\frac{u^2\left(m - \frac{1}{2}\right)^2}{\left(1 - \frac{u^2}{4}\right)^2} + \frac{1 - u^2\left(I + \frac{1}{2}\right)^2}{\left(1 - \frac{u^2}{4}\right)}\right]^{1/2} \tag{I, 10}$$

If $S = \frac{1}{2}$, then $m - \frac{1}{2} = m_I$, so that

$$v = \frac{-um_I}{1 - \frac{u^2}{4}} \pm \left[\frac{u^2 m_I{}^2}{\left(1 - \frac{u^2}{4}\right)^2} + \frac{1 - u^2\left(I + \frac{1}{2}\right)^2}{\left(1 - \frac{u^2}{4}\right)}\right]^{1/2} \tag{I, 11}$$

A plot of u versus v can then be made for each of the $2I + 1$ transitions where $m_I = +I, \cdots, -I$. Such a plot for $S = \frac{1}{2}, I = \frac{7}{2}$ is shown in Fig. I-4. For a given ν_0 ($\cong$9300 MHz) and a given A value, a horizontal line at the corresponding value of $u = A/h\nu_0$ will then determine the values of v at which resonance will occur. Values of A and g are then obtained by trial such that the calculated line positions fit the experimental ESR lines. This is usually not done graphically but is done more expeditiously and accurately by a computer.

The variation of the nuclear transition between the levels $F = 3$, $m_F = -3$ and $F = 4$, $m_F = -4$ ($\Delta m_I = \pm 1$, $\Delta m_S = 0$) is also plotted in Fig. I-4. It may be shown from equation (I, 9) to follow the equation

$$v = (8u - 2)/(2 - u) \tag{I, 12}$$

Figure I-4 shows clearly what was expressed earlier, that for a sufficiently large A value only two lines would be observed, the NMR transition and ESR transition for $m_I = -I$.

The horizontal lines in Fig. I-4 may be considered as two possible atomic cases differing in A values. They are, in fact, actual values found for $^2\Sigma$ molecules LaO and ScO trapped in argon matrices where the hf and g components were isotropic and $I = \frac{7}{2}$ for both metal atoms (see section II-1) [Weltner, McLeod, and Kasai (477)]. Hence the Breit–Rabi equation could be used in the interpretation of these molecular spectra.

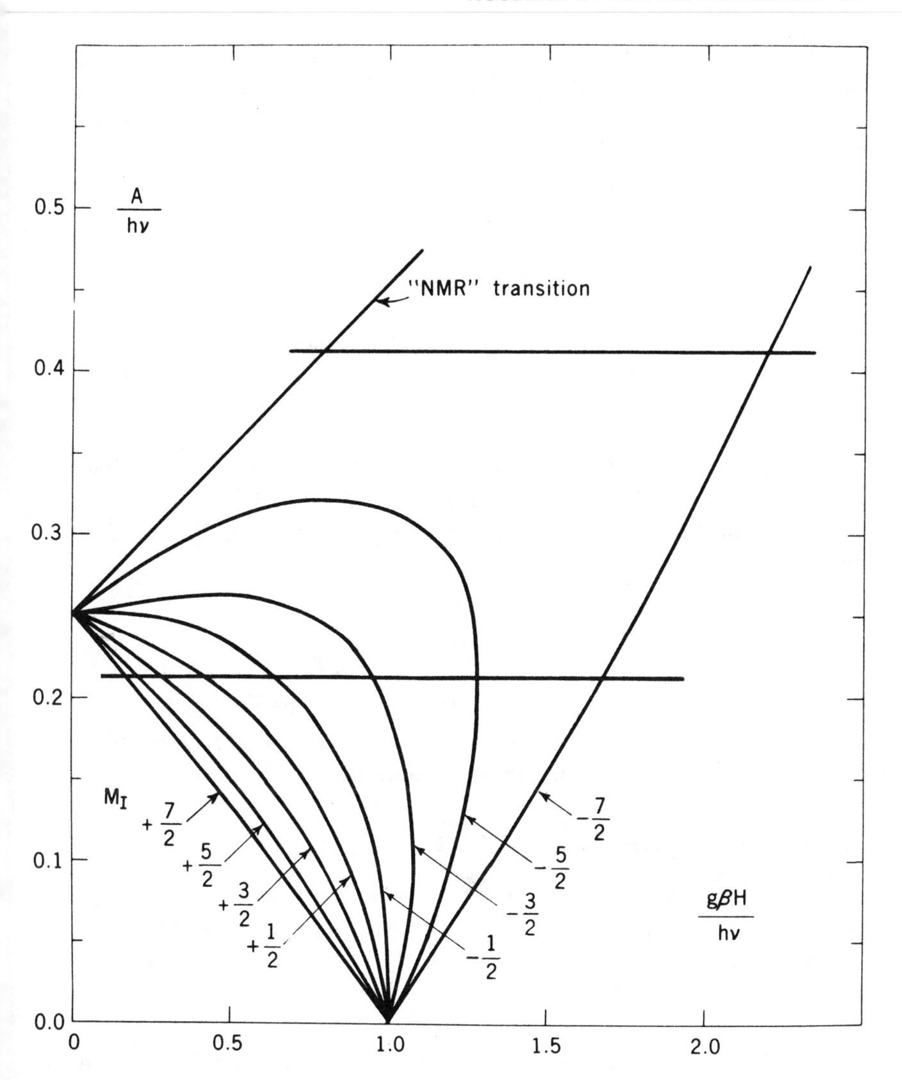

Fig. I-4. Resonant magnetic fields ($g\beta H/h\nu$) as a function of the hyperfine splitting constant ($A/h\nu$) for an $S = \frac{1}{2}$, $I = \frac{7}{2}$ atom [from Weltner, McLeod, and Kasai (477)]. ESR transitions are labeled by their m_I values assuming $A > 0$. The horizontal lines represent two extreme possibilities.

For the simplest case of an H atom, as shown in Fig. I-2, the positions of the two ESR lines can be obtained to a good approximation directly from (I, 11), since $u^2 \cong (500/3300)^2 \ll 1$. Equation (I, 11) becomes

$$v = -um_I \pm 1$$

or

$$H = \frac{h\nu_0}{g\beta} - \frac{Am_I}{g\beta} = H_0 - \frac{Am_I}{g\beta} \qquad \text{(I, 13)}$$

(where the positive sign must be chosen, since H must be positive). Here $m_I = \pm\frac{1}{2}$ so that the hyperfine splitting is just $|A|/g\beta$ in gauss. To this approximation H_0 and then g may be obtained directly from the positions of the two ESR lines,

$$H_0 = \tfrac{1}{2}\,(H_{+1/2} + H_{-1/2})$$

Note that the sign of A is not determined from the ESR spectrum so that the assignment of observed lines to particular values of m_I cannot be made without further information.

3. NUCLEAR QUADRUPOLE EFFECTS

Nuclei with spins $I \geqslant 1$ may have an electric quadrupole moment (departure from sphericity of charge) with which the atomic electrons may also interact. Then to the Hamiltonian in equation (I, 7) must be added the term [Abragham and Bleaney (1)]

$$\mathcal{H}_{\text{quad}} = q'[3(\mathbf{L}\cdot\mathbf{I})^2 + \tfrac{3}{2}(\mathbf{L}\cdot\mathbf{I}) - L(L+1)I(I+1)] \qquad \text{(I, 14)}$$

where q', analogous to the hyperfine coupling constant A, is the coefficient to be determined experimentally. For a free atom the corresponding energy contribution may be written [Kopfermann (14)]

$$E_{\text{quad}} = B\left[\frac{\frac{3}{4}C(C+1) - I(I+1)J(J+1)}{2I(2I-1)J(2J-1)}\right] \qquad \text{(I, 15)}$$

where $C = F(F+1) - I(I+1) - J(J+1)$, and as before

$$F = I + J, \cdots, |I - J|.$$

B is here equal to $eQq = eQ(\partial^2 V/\partial z^2)$ where Q is the nuclear quadrupole moment and q is the gradient of the electric field acting at the nucleus that is tending to orient the nuclei. The term represented by (I, 15) is usually very small and, at least in studies of atoms trapped in solid matrices, it can often be neglected.

The principal interest here is in the inclusion of this term in quenched atom analysis (see section I-5b) and particularly in dealing with linear molecules (section II-1c), where an axial crystal field or symmetry is effective. Then (I, 14) reduces to [Abragham and Bleaney (1)]

$$\mathcal{H}_{\text{quad}}\,(\text{axial}) = Q'[I_z{}^2 - \tfrac{1}{3}I(I+1)], \qquad \text{(I, 16)}$$

where

$$Q' = \frac{3eQq}{4I(2I-1)} \qquad \text{(I, 17)}$$

is commonly referred to as the quadrupole coupling constant.

4. THE MEANING OF g_J AND A

g_J and A depend upon the wavefunction of the unpaired electrons in the atom, or more precisely, upon the difference in magnetic properties and interactions of all electrons with spin α and all electrons of spin β. As indicated above, and as will be shown in more detail for molecules, the departure of g_J from g_e is determined by the contribution of orbital angular momentum (i.e., p, d electrons) to the wavefunction. A somewhat analogous separation of effects also may be made when considering the hyperfine coupling constant, A, determined by the magnetic interactions of the electrons with the nucleus. These may be described as isotropic (A_{iso}), as occurs for s electrons, and anisotropic or dipolar (A_{dip}), as occurs for p, d electrons. The hfs constant A is the sum of these two contributions, $A_{\text{iso}} + A_{\text{dip}}$. A_{iso} has no classical counterpart, whereas A_{dip} involves just the interaction between two magnetic dipoles.

(a) Isotropic Hyperfine Constant, A_{iso}

The isotropic interaction (A_{iso} or a_s) is also called the contact or Fermi interaction and depends upon the electron density at the nucleus, $|\Psi(0)|^2$, to which only *s* electrons significantly contribute. If only this interaction is involved, then the hyperfine structure constant in (I, 7) and (I, 8) is given by [Kopfermann (14), Ramsey (19)]:

$$a_s = A_{iso} = \frac{8\pi}{3h} g_e \beta_e g_I \beta_n |\Psi(0)|^2 \qquad \text{(I, 18)}$$

Thus, A for the hydrogen atom in its $(1s^1)\,^2S_{1/2}$ ground state in (I, 13) is accurately given by the isotropic interaction in (I, 18). Values of a_s, and thereby $|\Psi(0)|^2$, have been determined for many atoms in the gas phase by optical and atomic beam spectroscopy. $|\Psi(0)|^2$ can also be obtained from *ab initio* theoretical calculations [see, for example, Larsson, Brown, and Smith (306)]. This interaction is isotropic because there is obviously no angular dependence. The magnetic field at the nucleus can be very large (10^6 G), and *s* character in the wavefunction therefore makes a relatively large contribution to the hyperfine interaction.

$|\Psi(0)|^2$ is usually expressed in units of cm^{-3} or a.u.^{-3}, yielding A_{iso} in (I, 18) in Hz (1 a.u. = 5.29167×10^{-9} cm). A_{iso} is usually expressed in MHz or gauss (G), and if $g = g_e$ then $A\ (\text{MHz}) = [g_e\beta_e/10^6 h]\, A\ (\text{G}) = 2.8025\, A\ (\text{G})$. (I, 18) yields $A_{iso}\ (\text{MHz}) = 800.29\ (\mu_I/I)|\Psi(0)|^2$, where μ_I is in n.m. and $|\Psi(0)|^2$ is in a.u.^{-3}

(b) Anisotropic or Dipolar Hyperfine Constant, A_{dip}

The dipolar interaction energy arises classically from

$$E_{dip} = \frac{\boldsymbol{\mu}_J \cdot \boldsymbol{\mu}_I}{r^3} - \frac{3(\boldsymbol{\mu}_J \cdot \boldsymbol{r})(\boldsymbol{\mu}_I \cdot \boldsymbol{r})}{r^5}$$

where r is the distance between the magnetic dipoles of the electrons ($\boldsymbol{\mu}_J$) and the nucleus ($\boldsymbol{\mu}_I$). The corresponding quantum mechanical Hamiltonian is

$$\mathcal{H}_{dip} = g_e \beta_e g_I \beta_n \left[\frac{\boldsymbol{I} \cdot (\boldsymbol{L} - \boldsymbol{S})}{r^3} + \frac{3(\boldsymbol{I} \cdot \boldsymbol{r})(\boldsymbol{S} \cdot \boldsymbol{r})}{r^5} \right] \qquad \text{(I, 19)}$$

If the atom were truly hydrogenic one would find that the hyperfine interaction constant is [Kopfermann (14)]:

$$A_{\text{dip}}\,(\text{hydrogenic}) = a_J = \frac{g_e \beta_e g_I \beta_n L(L+1)}{hJ(J+1)} \langle r^{-3} \rangle \qquad \text{(I, 20)}$$

Then for the $^{27}Al\,(^2P_{1/2})$ atom, for example, where the observed value of $a_{1/2} = 502.0$ MHz [Lew and Wessel (308)], one finds from (I, 20) that $\langle r^{-3} \rangle_{3p} = 9.14 \times 10^{24}\ \text{cm}^{-3}$. This may be compared with the value of $\langle r^{-3} \rangle = 8.92 \times 10^{24}\ \text{cm}^{-3}$ (adjusted for the most recent $\mu_I = 3.6385$ n.m.) found by those authors where small corrections were made for relativistic effects and for the mixing of higher configurations into the ground state $1s^2\,2s^2\,2p^6\,3p^1$ configuration.

If the spin density of the electrons has spherical symmetry about the nucleus (synonymous with $L = 0$), then this dipole–dipole interaction averages to zero. Thus for filled and also half-filled shells of electrons, A_{dip} is expected to be zero. Examples are the $(2p^3)^4S\ ^{14}N$ atom and the $(3d^5)^6S\ ^{55}Mn$ atom. However, owing to higher-order effects, such as core polarization (see below), these two atoms do exhibit small hf splittings.

(I, 18) and (I, 20) for hydrogenic atoms have been modified by Goudsmit (192) to include approximate shielding effects and written in the forms:

$$a_S = \frac{8hcR_y \alpha^2 Z_i Z_a^{\,2} g_I}{3n_a^{\,3}(M/m)}$$

and

$$a_J = \frac{hcR_y \alpha^2 Z_i Z_a^{\,3} g_I}{n_a^{\,3}(L+\frac{1}{2})\,J(J+1)(M/m)}$$

These formulas can be used for predictions of these constants for atoms with one unpaired electron or hole outside of closed shells. Here R_y is the Rydberg constant ($me^4/4\pi\hbar^3 c\ \text{cm}^{-1}$), α is the fine structure constant ($e^2/\hbar c$), M/m is the ratio of the mass of the proton to the electron. Z_a is the effective nuclear charge acting on electrons outside the core, Z_i the effective nuclear charge acting on electrons inside the core ($\cong$ real nuclear charge), and n_a the effective principal quantum number outside the core. These are useful formulas especially when extended by correction factors for the finite volume of the nucleus and relativistic effects as given by Kopfermann (14).

(c) Polarization, Correlation, and Relativistic Effects

Departure from the central field approximation (polarization and correlation effects) and inclusion of relativistic effects make small alterations in this picture of A_{iso} and A_{dip}; however, for experimental purposes it appears to be adequate to parameterize the hyperfine interaction to account for these effects. If we exclude the nuclear quadrupole interaction given in (I, 14), the complete Hamiltonian for the hfs not considering these effects can be rewritten [Abragham and Pryce (27)]:

$$\mathcal{H}_{\text{hf}} = P\{(\boldsymbol{L}\cdot\boldsymbol{I}) + \xi L(L+1)(\boldsymbol{S}\cdot\boldsymbol{I}) - \tfrac{3}{2}\,\xi(\boldsymbol{L}\cdot\boldsymbol{S})(\boldsymbol{L}\cdot\boldsymbol{I}) - \tfrac{3}{2}\,\xi(\boldsymbol{L}\cdot\boldsymbol{I})(\boldsymbol{L}\cdot\boldsymbol{S}) - \kappa(\boldsymbol{S}\cdot\boldsymbol{I})\}, \tag{I, 21}$$

where κ accounts for the Fermi contact s electron interaction,

$$P = g_e g_I \beta_e \beta_n \langle r^{-3}\rangle,$$

and

$$\xi = \frac{(2l+1) - 4S}{S(2l-1)(2l+3)(2L-1)} \tag{I, 22}$$

($\xi = \frac{2}{5}$ for a single p electron, $\frac{2}{21}$ for a single d electron, etc.). This expression may now be "parameterized" by defining two values of $\langle r^{-3}\rangle$ entering into the anisotropic hf terms [Ammeter and Schlosnagle (39)]:

$$\begin{aligned} P_L &= g_e \beta_e g_I \beta_n \langle r_l{}^{-3}\rangle \\ P_S &= g_e \beta_e g_I \beta_n \langle r_{sC}{}^{-3}\rangle \end{aligned} \tag{I, 23}$$

and letting $-P\kappa$, the isotropic hf coefficient, be written as

$$A_0 = -P_S\,\kappa = g_e \beta_e g_I \beta_n \,\frac{8\pi}{3}\,|\Psi(0)|^2 \tag{I, 24}$$

$\langle r_l{}^{-3}\rangle$ is identified with the magnetic interaction of the orbital motion of the electrons with the nuclear dipole moment, and $\langle r_{sC}{}^{-3}\rangle$ with the spin magnetization "outside" the nucleus (as opposed to the contact A_0 term). [This symbolism comes from the early studies of relativity and configuration interaction in atomic hfs; see Sandars Beck (398) and Woodgate (480).] Then (I, 21) becomes

$$\mathcal{H}_{\text{hf}} = P_L(\boldsymbol{L} \cdot \boldsymbol{I}) + P_S\,\xi\{L(L+1)(\boldsymbol{S} \cdot \boldsymbol{I}) - \tfrac{3}{2}[(\boldsymbol{L} \cdot \boldsymbol{S})(\boldsymbol{L} \cdot \boldsymbol{I}) + (\boldsymbol{L} \cdot \boldsymbol{I})(\boldsymbol{L} \cdot \boldsymbol{S})]\} + A_0(\boldsymbol{S} \cdot \boldsymbol{I}) \tag{I, 25}$$

For a configuration of an atom involving only p or d electrons, with vanishing polarization, correlation, and relativistic effects, $P_L = P_S$ and $\kappa = 0$. The departure of $P_S/P_L = \langle r_{sC}^{-3}\rangle/\langle r_l^{-3}\rangle$ from unity is a measure of the spatial separation of spin and charge densities in the valence shell of the atom and is usually in the neighborhood of 10%. For example, from atomic beam spectroscopy this ratio is found to be 1.083 for $2p^1$ $(^2P_{1/2})$ ^{11}B, 1.058 for $3p^1$ $(^2P_{1/2})$ ^{27}Al [Harvey, Evans, and Lew (212)], and 1.108 for $3p^5$ $(^2P_{3/2})$ ^{35}Cl [Uslu, Code, and Harvey (434)].

The departure of κ from zero in such configurations ostensibly exhibiting no s character is attributed to a finite electron density at the nucleus due to a core-polarization effect where closed shells of electrons are spin-polarized by exchange interaction with unfilled shells. As might be expected, such polarization, and therefore A_0, is found to increase for $(p)^N$ ground configuration atoms as N increases. κ is small for such atoms; for example, for $^2P_{1/2}$ ^{27}Al it is found to be -0.0237 [Harvey, Evans, and Lew (212)] and for $^2P_{3/2}$ ^{35}Cl it is -0.0831 [Uslu, Code, and Harvey (434)].

The notation traditionally used in gas-phase studies involves the above parameters $\langle r_l^{-3}\rangle$ and $\langle r_{sC}^{-3}\rangle$, but $\langle r_s^{-3}\rangle$ is used as an analogous coefficient for the s electron interaction, i.e., $A_0 = -\kappa g_e g_I \beta_e \beta_n \langle r_s^{-3}\rangle$. Then κ as used here is related to it by $-\kappa\langle r_s^{-3}\rangle = (8\pi/3)|\Psi(0)|^2$. A discussion of the various $\langle r^{-3}\rangle$ values is given by Abragham and Bleaney (1).

(d) $|\Psi(0)|^2$ and $\langle r^{-3}\rangle$

The basic atomic properties obtained from the observation of hyperfine structure and the determination of A_{iso} and A_{dip} are $|\Psi(0)|^2$ and $\langle r^{-3}\rangle$. Although $\langle r_l^{-3}\rangle$ and $\langle r_{sC}^{-3}\rangle$ were introduced in the previous section, they can only be determined separately from accurate gas-phase atomic-beam spectroscopy, and their relatively small difference is not usually important in solid-state studies. However, as will be seen in the next section, they have been included by some authors [Vannotti and Morton (440); Ammeter and Schlosnagle (39)] in treatments of quenched atoms in solids, so that it was necessary to discuss their origin. $\langle r^{-3} \rangle$ calculated from an observed value of A_{dip} using (I, 20) will be some average value of these two quantities.

The experimentally determined values of $|\Psi(0)|^2$ and $\langle r^{-3} \rangle$ are characteristic of the electronic wavefunctions of atoms and therefore can provide a test of any atomic wavefunction derived theoretically. Correlation, relativistic effects, and polarization effects must be taken into account in accurate calculations, and for the heavier elements this is difficult at the present time [Descleaux and Bessis (138); Larsson, Brown, and Smith (306)]. For half-filled-shell atoms, such as nitrogen, where zero hfs is expected in first approximation, detailed configuration-interaction (CI) calculations have been made to account for the small hfs (10.5 MHz) actually observed [Bessis, Lefebvre–Brion, and Moser (64)].

Knowledge of these atomic parameters becomes very useful when hyperfine splittings in molecules are observed, since it permits an estimate to be made of the electronic spin distribution in the molecule. As will be discussed in Chapter II, section 1c, this can be done if hf interactions with several (preferably all) nuclei in the molecule are observed. Then with knowledge of A_{iso} and A_{dip} for the atoms, and assuming that the hf interaction at a particular nucleus is proportional to the electron spin density at that nucleus, one can obtain the fraction of the unpaired electron spin at each nucleus in the molecule.

Morton and co-workers, and others [see the references in Appendix B], have calculated $|\Psi(0)|^2$ and $\langle r^{-3} \rangle$ values, the former for a valence s electron, the latter for a valence p, d, or f electron, for all of the atomic elements with nuclear spin. As time has progressed, improved wavefunctions have been used, and the latest values of these calculated parameters (in a.u.$^{-3}$) are given in Appendix B. Also given there are A_{iso} (MHz) values calculated via (I, 18) with $g_e = 2.0023$ and $P\,(\text{MHz}) = g_e \beta_e g_I \beta_n \langle r^{-3} \rangle$. P may be taken as an approximate measure of A_{dip} so that A_{iso}/P for a given nucleus may be taken as a measure of the relative effectiveness of an s electron to a p (or d, f) electron in hyperfine interaction. This ratio can be seen from Table B1 to be in the range of 15 to 25, demonstrating clearly that the contact interaction of s electrons is most effective.

The accuracy of the parameters listed in Appendix B can be judged by comparing them with available gas-phase data. Thus, the appendix gives A_{iso} for ^{23}Na to be 927.1 MHz compared to the experimental value of 885.8 MHz [Kusch and Hughes (298)], and for the heavy element ^{197}Au, 2876 MHz versus 3053 MHz in the gas [Wessel and Lew (478)]. Correspondingly, comparison of $\langle r^{-3} \rangle$ values may be made for ^{27}Al as 1.493 a.u.$^{-3}$ from Appendix B versus the experimental value of 1.32 a.u.$^{-3}$ [corrected value of Lew and Wessel (308)]; see section (I-4b)]. For ^{69}Ga $P = 509.6$ MHz versus $\bar{P} = 478$ MHz [Ammeter and Schlosnagle (39), Table VIII]. The agreement between calculated and experimental parameters is then within about 10%, which is quite satisfactory for the purpose of estimating spin densities in molecules.

5. ATOMS AND ATOMIC IONS IN SOLID MATRICES

(a) *S*-State Atoms

For atoms and atomic ions possessing no orbital angular momentum, the values of g_J and A determined from their ESR spectra when isolated in the solid inert gases (neon, argon, etc.) are generally quite gaslike.

(α) Matrix Effects. However, it is often found that each expected ESR line observed in a matrix is split into two or more lines, as illustrated in Fig. I-5 for Na atoms trapped in an argon matrix at 4°K [Jen, Bowers, Cochran, and Foner (249)]. Since sodium is a $^2S_{1/2}$ ground-state atom with 100% natural abundance of the isotope ^{23}Na with $I = \frac{3}{2}$, one expects to observe just four hf lines centered about $g = 2.00$. Here six sets of lines were actually observed having slightly different g and A values. This is attributed to the trapping of sodium atoms in various sites in solid argon where they experience slightly different perturbations by the surrounding matrix. Often this site structure in spectra can be partially, or even completely, removed by careful annealing of the initial matrix by warming. For example, for the $m_I = \frac{1}{2}$ line shown in Fig. I-5 for Na, all components except a and b disappeared at $T \cong 36$°K, b disappeared at 38°K, and a at 40°K. Because of the ubiquity of this phenomenon in matrices, annealing procedures and effects are usually reported when discussing experimental results. Thus, solid argon matrices (melting point 84°K) are usually warmed to 30–35°K after the matrix is prepared and then

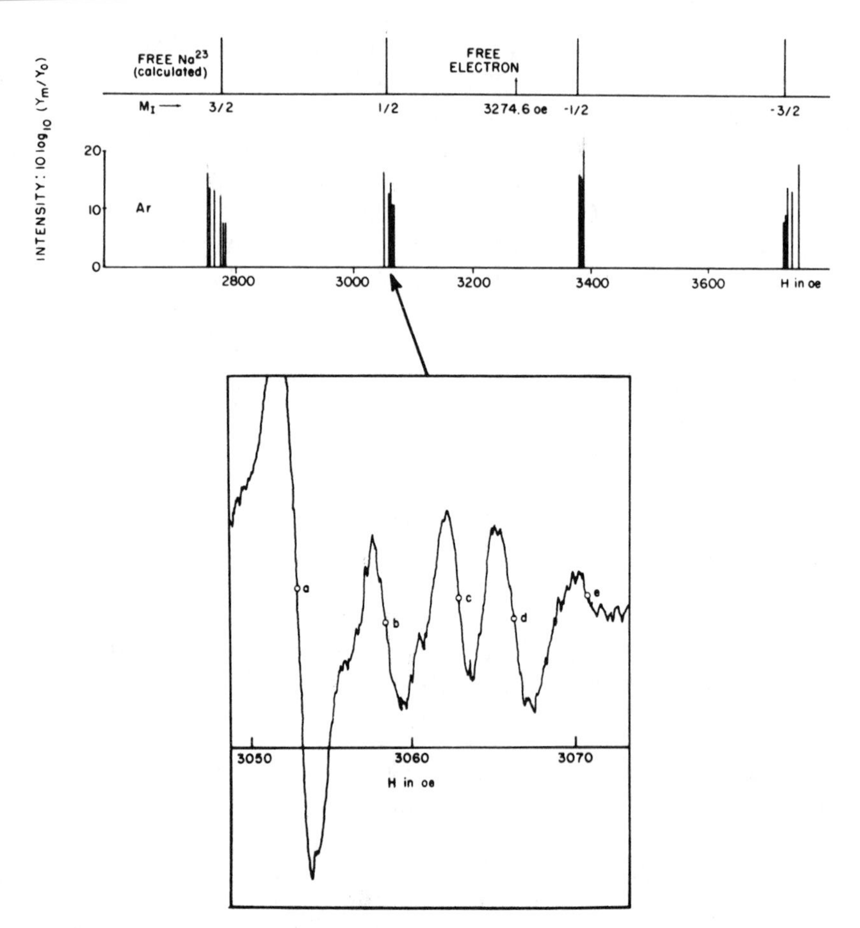

Fig. I-5. Electron-spin-resonance spectrum of sodium atoms trapped in a solid argon matrix at 4°K [from Jen, Bowers, Cochran, and Foner (249)]. (Top) Calculated spectrum for the free atoms; (bottom) multiple-site effect upon the spectrum.

immediately quenched again to 4–10°K. Some matrices, particularly those containing labile H atoms or where neon is the host, are difficult to anneal because the temperature difference between annealing and solid-state diffusion is very small.

In general, the exact nature of the sites in matrices is not known, and, of course, it will vary greatly with the properties of the trapped species. One speaks of substitutional and interstitial sites based upon a face-centered-cubic or hexagonal-close-packed structure for the crystalline rare-gas solid [Pollack (378); Meyer (17)], but this is an idealized picture, since packing around the isolated atom will depend upon its size relative to that of the matrix atom. From overlap calculations, Ammeter and Schlosnagle (39) conclude that all alkali, alkaline-earth, and transition-metal atoms are "too large" to fit into substitutional sites.

Rare-gas matrices are usually considered as polycrystalline based on X-ray diffraction evidence [Peiser (368)]. There is then also the possibility of intercrystalline (microcrystal surface or grain boundary) sites occurring in some cases. Kasai and McLeod (262) have suggested that the relatively large Cu, Ag, and Au atoms may occupy such sites in solid argon, krypton, and xenon. Kupferman and Pipkin (297) correlated their ESR signals of Rb atoms with optical absorption signals using a circular polarizer and concluded from the stability and reproducibility of the sites that the alkali atoms are surrounded by rare-gas atoms rather than being located on an interface or dislocation.

Table I-1 is a compilation of the shifts of g_J and A, usually relative to gas-phase values, for some atoms trapped in rare-gas matrices. The data given there are meant to be representative but may not encompass all observations. The number of sites and the shifts observed often appear to be a function of the method of preparation of the matrix, i.e., whether by condensation of the mixture from the gas, γ irradiation or photolysis of a trapped molecule, or nuclear reaction.

In general, g_J is lowered in the matrix relative to the gas phase, while the hyperfine interaction constant may increase or decrease. Understandably, most theoretical efforts to account for these shifts, both relative to the gas and among the various sites, have been made on the simplest case, hydrogen atoms trapped in various matrices. Table I-2 gives the data obtained by Foner, Cochran, Bowers, and Jen (1968) for H atoms produced by the photolytic dissociation of HI molecules in solid argon and studied by Adrian (28) and Smith (418, 419). H atoms were observed in three sites, but site B is apparently dependent upon the species being photolyzed, since it did not appear when H_2O was the precursor. One attributes such B sites, basically, to nonisolation, since apparently such a site is adjacent to or near the molecule being photolyzed. Sites A and C are attributed by Adrian to substitutional and interstitial sites, respectively, in the argon lattice on the basis of the difference in the signs of $\Delta A = A_{\text{matrix}} - A_{\text{gas}}$ in the two cases. Qualitatively, van der Waals

Table I-1. Matrix-Isolated S-State Atoms; Matrix Effects Upon g_J and A in the Solid Rare Gases

Atom or Ion	Gr. State	I	Free Atom		Matrix Effect[a]							
					$-100\,\Delta g_J = -100\,(g_J - g_J^*)$				$\Delta A/A = (A - A^*)/A^*$ (%)			
			A (MHz)*	g_J^*	Ne	Ar	Kr	Xe	Ne	Ar	Kr	Xe
^{1}H[b]	$^2S_{1/2}$	1/2	1420.41	2.00226	0.019 (1)	0.006, 0.065 (2)	0.047, 0.26 (2)	0.169 (u)	0.43	−0.46, 1.15 (2)	−0.59, 0.47 (2)	−1.04 (u)
^{1}H[c]			–	–		0.118 (1)	0.332 (1)	0.227 (1)	–	1.15 (1)	0.53 (1)	−1.02 (1)
^{7}Li[d]	$^2S_{1/2}$	3/2	401.76	2.00231	–	0.05, 0.13 (2)	0.36, 0.57 (2)	1.09 (u)	–	−1.6, 3.1 (2)	−1.7, 2.2 (2)	−1.2 (u)
^{23}Na[d]	$^2S_{1/2}$	3/2	885.81	2.00231	–	0.05–0.21 (6)	0.45, 0.93 (2)	0.98 (u)	–	−0.9 to 4.9 (6)	−1.4, 2.0 (2)	−1.3 (u)
^{39}K[d]	$^2S_{1/2}$	3/2	230.86	2.00231	–	0.08–0.37 (6)	0.59, 1.74 (2)	1.66 (u)	–	−0.4 to 11.8 (6)	−1.2, 6.6 (2)	1.7 (u)
^{85}Rb[d]	$^2S_{1/2}$	5/2	1011.91	2.00241	–	−0.85 to +0.89 (7)	0.65–1.07 (3)	2.02 (u)	–	2.6–8.4 (7)	−0.56 to 6.9 (3)	−1.6 (u)
^{63}Cu[e]	$^2S_{1/2}$	3/2	5867	(2.0023)	0.73 to −1.17 (5)	0.29 (1)	0.68 (1)	0.81 (1)	2.3 to −32 (5)	4.8 (1)	3.0 (1)	0.5 (1)
^{107}Ag[e]	$^2S_{1/2}$	1/2	−1713	(2.0023)	0.00 (1)	0.25 (1)	0.81 (1)	1.01 (1)	1.3 (1)	5.7 (1)	3.8 (1)	1.0 (1)
^{197}Au[e]	$^2S_{1/2}$	3/2	3053	(2.0023)	0.06–0.07 (2)	0.11 (1)	0.61 (1)	0.53 (1)	1.3–1.2 (2)	2.8 (1)	1.4 (1)	−0.9 (1)

$^{14}N^{f,g}$	$^4S_{3/2}$	1	10.45	2.0022	–	–	0.01^g	0.03^g	9.6^f	20.6^f	17.7^g	18.8^g
$^{14}N^h$			–	–	–	0.10	0.11	–	–	13.68	19.90	~31.1
$^{31}P^i$	$^4S_{3/2}$	1/2	55.06	2.0019	–	0.07	0.18	–	–	46.3	51.4	57.2
$^{75}As^i$	$^4S_{3/2}$	3/2	–66.20	1.9965	–	0.05	0.14	0.22	–	-47.2	-52.7	–
$^{111}Cd^{+j}$	$^2S_{1/2}$	1/2	–	(2.0023)	–	0.17	–	–	–	–	–	–
$^{53}Cr^j$	$^7S_3$	3/2	82.60	2.00135	–	0.17	–	–	–	14.5	–	–
$^{53}Cr^{+j}$	$^6S_{5/2}$	3/2	–	(2.0023)	–	0.12	–	–	–	–	–	–
$^{55}Mn^j$	$^6S_{5/2}$	5/2	72.42	(2.0023)	–	0.10	–	–	–	8.0	–	–
$^{55}Mn^{+j}$	$^7S_3$	5/2	–	(2.0023)	–	–0.01	–	–	–	–	–	–

[a] Values in parentheses under each matrix indicate number of observed sites; (u) indicates "unresolved."
[b] Foner, Cochran, Bowers, and Jen (168). Other research on H atoms by these authors is contained in references (250, 251, 117). H Atoms were deposited by electrical discharge of H_2 or by photolysis of HI, H_2O, etc. In Xe the spectrum was complex owing to hfs involving magnetic Xe nuclei. Recent superhyperfine studies are in reference (351).
[c] Jackel, Nelson, and Gordy (247). H atoms were produced by γ irradiation of 2% concentrations of NH_3, PH_3, etc. in the rare-gas solids. ESR was in the K band region ($\nu_0 \simeq 24$ GHz).
[d] Jen, Bowers, Cochran, and Foner (249). Other research on matrix-isolated alkalis is contained in references (487) and (188). ESR signals of Rb atoms in multiple sites in Ar have been correlated with optical absorption bands (297). A similar correlation (but with opposite optical shift behavior) was observed for ^{41}K [produced by the nuclear reaction $^{40}Ar\,(n, \beta^-)\,^{41}K$] trapped in Ar (124, 125).
[e] Kasai and McLeod (262). Neon spectra exhibited four "oriented" axial sites [see section II-3b(γ)] and one "amorphous" site. Hfs due to Xe nuclei, and also Kr nuclei in the case of Ag, was observed for all three metal atoms. Other research on trapped Cu atoms is contained in reference (486); on Ag in references (488), (489), and (485); on Au in reference (485).
[f] ESR of N atoms trapped in solid H_2, N_2, and CH_4 has been observed by Jen, Foner, Cochran, and Bowers (251, 169); in rare gases by Adrian (30).
[g] Jackel, Nelson, and Gordy (247).
[h] Fischer, Charles, and McDowell (165). N atoms produced by photolysis of HN_3. Other research on matrix-isolated N atoms in references (119) and (118).
[i] Jackel, Nelson, and Gordy (247). Earlier research in reference (347).
[j] Kasai (259). Ions were formed by photolysis of a matrix containing metal atom and HI: $M + HI \xrightarrow{h\nu} M^+ + I^- + H$.

Table I-2. g_J and A Values for H Atoms in an Argon Matrix (Produced by the Photolysis of HI)[a]

Site	g_J	$-100\ \Delta g_J$[b]	A (MHz)	$\Delta A/A$ (%)[c]	Disappearance Temperature (°K)
A	2.00220	0.006	1413.82	−0.46	39
B	2.00224	0.002	1416.31	−0.29	23
C	2.00161	0.065	1436.24	1.15	12
Free atom	2.00226	–	1420.41	–	–

[a] Taken from Adrian (31).
[b] $\Delta g_J = g_J - g_J^*$ (free atom).
[c] $\Delta A = A - A^*$ (free atom).

attractive forces between the H atom and its surrounding matrix atoms cause a small expansion of their charge clouds, thereby decreasing the hfs of the H atom. Repulsive (Pauli exclusion) forces also arise when the charge clouds of the H and matrix atoms overlap, causing the opposite effect and increasing the hfs. One expects the latter to dominate in the more crowded interstitial site and attractive forces to prevail in the substitutional site. For large atoms, the hfs should always be increased in the matrix, since even a substitutional site may be relatively small; this is evident from Table I-1. Because of the overlap there is also a small mixing of the matrix atom properties into the isolated-atom wavefunction. The spin-orbit interaction of the matrix then has an effect, which theory predicts [Adrian (28)] will produce a negative g shift, i.e., $\Delta g = g_{\text{matrix}} - g_{\text{gas}}$ is negative, which will be true regardless of the nature of the site.

Adrian (28, 31) derived for the shift of the isotropic hfs constant for hydrogen atoms in a matrix

$$A' \cong A\left(1 + \frac{3E_{\text{vdW}}}{\overline{E}_H}\right) \tag{I, 26}$$

and since the van der Waals energy E_{vdW} for interaction of H and a matrix atom is negative, $A' < A$. Here $\overline{E}_H$ is the ionization energy of the H atom, taken as an average excitation energy. Correspondingly, for the repulsive forces, he finds

$$A' \cong A\left[1 + \sum_i \langle \Psi_H | \Psi_i \rangle^2\right] \tag{I, 27}$$

where the last sum of overlap integrals between H and matrix atom i is small but positive. Hence here $A' > A$. Δg arises only from overlap with matrix atoms with spin-orbit coupling constant λ. It is given by

$$\Delta g = (4N/3\overline{E}_H)\langle\Psi_H|\Psi_{p\sigma}\rangle^2\lambda_p \qquad \text{(I, 28)}$$

where all quantities on the right-hand side except $\overline{E}_H$ are positive. The quantity in the bra-ket is the overlap between the s orbital of the unpaired electron and the σp orbital of the nearest matrix atom, N is the number of nearest-neighbor matrix atoms, and λ_p is the atomic spin-orbit splitting constant of the valence shell p orbital of these atoms. Smith (418) has made a more detailed "first-principles" variational calculation also including crystal field effects which, along with exchange effects contribute significantly in calculating shifts in A in interstitial sites. g shifts appear to be most reliably calculated, since they are independent of these latter effects.

The calculations of Δg have been extended by Smith (419) to alkali atoms trapped in solid Ar and Kr; all g shifts are negative. Goldsborough and Koehler (188) have observed other sites for the alkali atoms trapped in the solid rare gases when the sample was exposed to room temperature infrared radiation during deposition on the liquid helium cooled finger. [These may be due to ESR signals from triatomic molecules, such as Na_3, recently observed by Lindsay, Herschbach, and Kwiram (312).]

(β) Half-Filled Shells. For N, P, and As atoms having half-filled np shells and ^4S ground states, one expects no hfs, since there is no s electron contribution, and the spherical symmetry of the atom causes the anisotropic hfs to also vanish. The small hfs that is observed in these cases is attributed to spin polarization, as mentioned earlier. Arsenic is seen from Table I-1 to have polarization of the opposite sign to nitrogen and phosphorous; this has been explained by CI calculations by Pendlebury (369) [see Jackel, Nelson, and Gordy (247)].

The three large lines in the ESR spectrum of ^{14}N ($I = 1$) atoms trapped in solid molecular nitrogen shown in Fig. I-6 are due to this small hf interaction and are split by only about 12 MHz. For this atom, polarization effects are clearly very sensitive to the environment, as may be seen from the large values of $\Delta A/A^*$ in Table I-1. Following his earlier treatment of the trapped H atom, Adrian (30) has accounted for this shift in $A(^{14}N)$ in matrices as largely due to van der Waals interactions that effectively introduce $(2s)(2p^4)$ excited states into the N atom wavefunction. Since the $2p$ shell was half-filled with α spin electrons, the $2s$ electron with β spin had to be

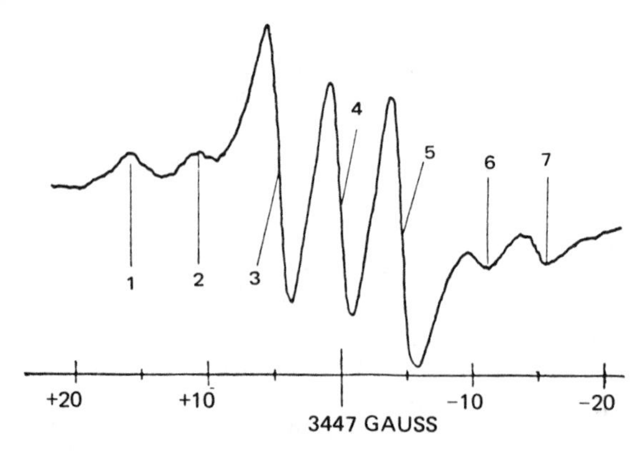

Fig. I-6. Electron-spin-resonance spectrum of atomic nitrogen trapped in solid N_2 at 4°K [from Cole and McConnell (119)]. The four weak lines are due to crystal-field effects; see section I-5a(δ) (ν = 9.6550(5) GHz).

excited into the $2p$ orbital, leaving a $2s$ electron with α spin providing increased hfs. These calculations have been repeated with improved wavefunctions by Jackel, Nelson, and Gordy (247) to provide better agreement with the experiment. P and As atoms in the solid rare gases also exhibit large matrix effects, and the theory of Adrian was generalized to account for those perturbations. Those authors indicate that the Group V atoms occupy substitutional lattice sites.

The half-filled d-shell high-spin atom ^{55}Mn also has a relatively small A value of 72 MHz (see Table I-1), but does not exhibit this extreme sensitivity to matrix effects. This may be taken as support for Adrian's theory, since an analogous $s \rightarrow d$ transition is not allowed.

(γ) Superhyperfine Interaction. The overlap of the isolated-atom and matrix-atom charge clouds referred to above also has the effect of allowing hf interaction between the isolated-atom unpaired electrons and those matrix nuclei that have magnetic moments and are present in relatively high natural abundance. This "superhyperfine" interaction can occur in krypton matrices where ^{83}Kr ($I = \frac{9}{2}$) is present in 11.6% natural abundance and in solid xenon where ^{129}Xe ($I = \frac{1}{2}$) and ^{131}Xe ($I = \frac{3}{2}$) are present in 26.24 and 21.24% abundance, respectively. The observed superhyperfine patterns can be rather complex, particularly for xenon matrices, since the trapped impurity is usually surrounded by at least six nearest-neighbor matrix atoms. Matrices enriched in ^{129}Xe have been used to simplify the shf structure [Morton, Preston, Strach, Adrian, and Jette (351)], and Fig. I-7

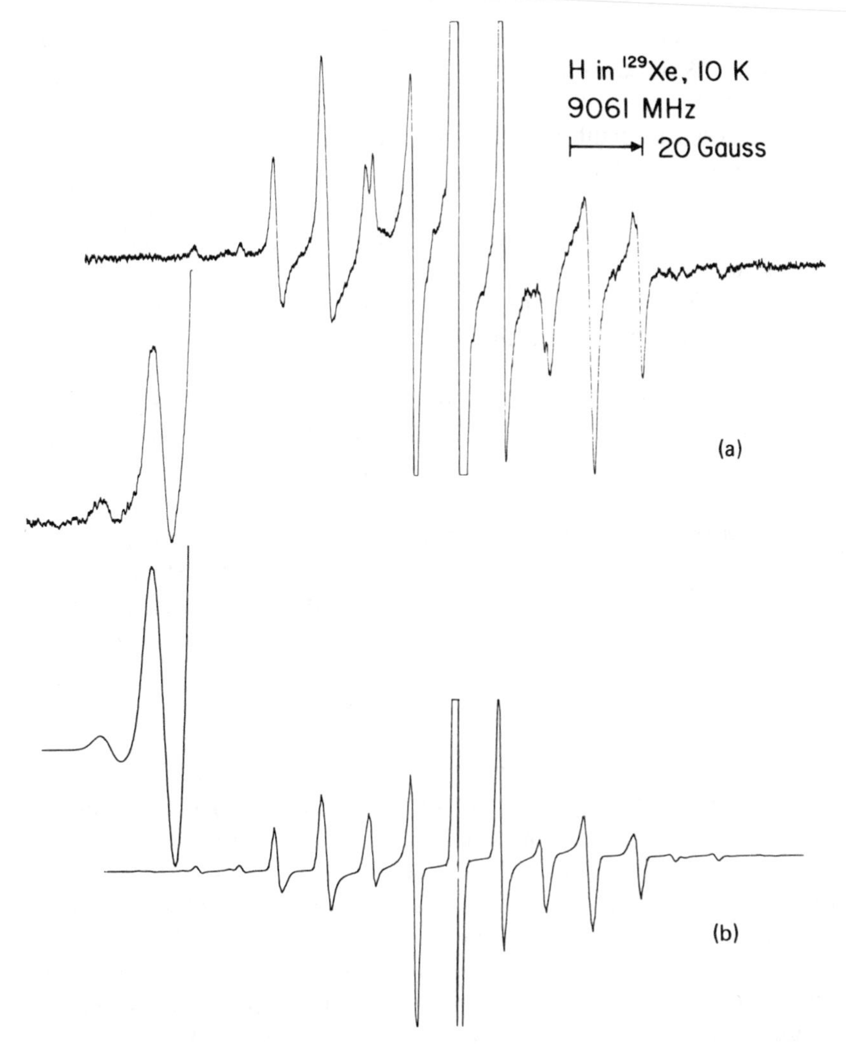

Fig. I-7. ESR of hydrogen atoms trapped at 10°K in a xenon matrix enriched to 37% in ^{129}Xe [from Morton, Preston, Strach, Adrian, and Jette (351)]. (a) The high field ($m_I = -\frac{1}{2}$) transition. (b) Computer simulation of (a) assuming an octahedral distribution of six xenon nuclei around the proton. The appropriate probabilities for $^{129}Xe_nXe_{6-n}$ are incorporated; $A_{\parallel}^{129}$ = 136 MHz, $A_{\perp}^{129}$ = 67 MHz.

shows that structure for what would ordinarily be one line in the ESR spectrum of matrix-isolated H atoms.

H atoms trapped in Kr and Xe were originally reported to exhibit isotropic superhyperfine structure of the host [Foner, Cochran,

Bowers, and Jen (168); Adrian (28)], but the reinvestigation by Morton et al. (351) gives $|A_{\parallel}| \simeq 2|A_{\perp}|$ shown in Fig. I-7. Theory then indicates that the atom is in an octahedrally symmetric interstitial site (six equivalent neighbors) in both matrices as opposed to a tetrahedrally arranged interstitial or a substitutional site. It is interesting to note that it was necessary to include a small amount of electron charge transfer in xenon ($Xe^{+} \cdots H^{-}$) in order to account for the hfs there.

Kr and Xe superhyperfine structure has been seen in other trapped atom cases such as Cu, Ag, and Au [Kasai and McLeod (262)].

(δ) High-Spin Atoms, Crystal-Field Effects. In atoms with $S \geqslant 1$, that is, with more than one unpaired electron, electronic spin–spin interaction must occur, but in the free atom, because of its spherical symmetry, this interaction averages to zero. However, the crystal fields in the matrix may destroy this symmetry, and then additional terms accounting for this interaction must be added to the Hamiltonian. The cases observed can be accounted for satisfactorily by assuming an axial crystal field (along z) so that the terms reduce to the form

$$D[S_z^2 - \tfrac{1}{3} S(S+1)] \qquad \text{(I,.29)}$$

where D is called the spin–spin interaction or zero-field-splitting parameter. This treatment is analogous to that of Abragham and Pryce (27) for the ^6S Mn^{2+} ion in a tetragonal or trigonal crystal field. It was first applied to matrices by Cole and McConnell (119) in order to explain the ESR spectrum of N atoms trapped in solid N_2 at 4°K (Fig. I-6). They proposed that the four visible weak lines in that spectrum were part of three sets of triplets (due to ^{14}N hfs) where the remaining two weak lines lie under the strong lines 3 and 5. These three fine structure transitions correspond to S_z transitions of $\frac{3}{2} \leftrightarrow \frac{1}{2}$, $\frac{1}{2} \leftrightarrow -\frac{1}{2}$, and $-\frac{1}{2} \leftrightarrow -\frac{3}{2}$ and are separated by $|D|$. $|D|$ is found to be only 31 ± 6 MHz $\cong 10^{-3}$ cm^{-1} for this $S = \frac{3}{2}$ case and $A(^{14}\text{N}) \cong 12$ MHz.

A similar kind of background spectrum, but with many more lines, has been observed when ^{55}Mn atoms, having a ^6S ground state and $I = \frac{5}{2}$, are trapped in relatively high concentrations in solid neon at 4°K [Bhat, Van Zee, and Weltner (66)]. Again presumably in some sites a predominately axial crystal field acts upon the Mn atoms

stal-Field Analysis of Representative[a] ESR Data for Partially nched Atoms in Solid Rare Gas Matrices at 4°K

Gas-Phase Parameters[b]				
ζ (cm^{-1})	P_L (10^{-4} cm^{-1})	P_S (10^{-4} cm^{-1})	A_0 (10^{-4} cm^{-1})	References (Gas & Matrix)
10.67	44.40	48.15	3.87	197
74.93	60.77	64.35	−1.528	39, 284
550.8	(153.7)[c]	(169.1)[c]	(−48.8)[c]	39
2456	592	807	58.2	65
5060	516.4[d]		(80)[d]	245

Matrix Data[e]						
rix	$g_{\parallel}$	$g_{\perp}$	$A_{\parallel}$ (10^{-4} cm^{-1})	$A_{\perp}$ (10^{-4} cm^{-1})	A_{iso}[f] (10^{-4} cm^{-1})	A_{dip}[f] (10^{-4} cm^{-1})
	2.001	1.965	42	−11	6.4	18
e	2.000	1.925	46.3	−35.3	−8.1	27.2
r	2.000	1.952	47.7	−34.0	−6.8	27.2
r	2.001	1.989	45.3	−30.0	−4.9	25.1
e	2.001	2.02	46.7	<−25	~0	<24
r	1.940	1.581	145.0	−200.6	−85.4	115.2
r	1.952	1.675	131.9	−180.8	−76.6	104.2
e	1.968	–	114.0	–	–	–
	1.55	2.646	141	645.7	478	−168
	1.400	2.532	297	535.3	455.7	−79.7

Static Crystal Field Parameters[g]					
$\Delta_{\perp}$ (cm^{-1})	P_S^M/P_S (%)	A_F (10^{-4} cm^{-1})	$A_F - A_0$ (10^{-4} cm^{-1})	$A_{iso}{}^{Mh}$ (10^{-4} cm^{-1})	$A_{dip}{}^{Mh}$ (10^{-4} cm^{-1})
380	91	7.4	3.6	–	–
1320	94.5	−5.1	−3.6	–	–
1990	98.0	−4.8	−3.3	–	–
7540	95.5	−4.4	−2.9	–	–
∞	–	–	–	–	–
2020	–	–	–	−84	109
2540	–	–	–	−76	99
3460	–	–	–	–	–

leading to the observation of five sets of six hf lines separated by $|D| = 0.0071$ cm^{-1} and $|A(^{55}Mn)| = 72$ MHz.

The zero-field-splitting parameter D will be discussed in more detail when triplet and higher multiplicity molecules are considered in later chapters. These matrix-perturbed atoms N and Mn are analogous to $^4\Sigma$ and $^6\Sigma$ ground-state molecules with very small D values.

(b) Orbitally Degenerate Atoms and Quenching

Atoms with orbitally degenerate ground states, as opposed to S-state atoms, are considerably perturbed in the matrix environment; in fact, to such an extent that they are either not observed at all via ESR or are strongly quenched. An excellent but rather extreme illustration of this is provided by the observations on matrix-isolated halogen atoms.

The ESR spectra of $^2P_{3/2}$ atoms, such as Br and I, had not been observed in inert-gas matrices until recently, although a thorough search had been made for them [Jen (248)], presumably because the atoms were trapped in sites in which anisotropic crystal fields interact with the orbital angular momenta of the atoms. The resonance lines were thereby broadened beyond detection either because of the partial quenching of orbital angular momentum, which will be discussed here, or possibly because of a shortening of the spin-lattice relaxation time due to the strong coupling of the spin system and the matrix. The latter mechanism has been briefly discussed by Adrian (31). However, in recent work in matrices signals due to partially quenched bromine [Bhat and Weltner (65)] and iodine [Iwasaki, Toriyama, and Muto (245)] atoms have been observed. {Quenched Br atoms have also been observed in X-irradiated N-bromosuccinimide single crystals [Muto and Kispert (356)]}. In solid argon the ESR spectrum of Br atoms appeared when they were formed by the reaction of H atoms with Br_2 (see Fig. I-8). This is so apparently because the majority of the halogen atoms were trapped in the vicinity of an HX molecule, and in each site a similar axial crystal field was produced. Perhaps for the same reasons Iwasaki et al. detected quenched I atoms in Xe matrices where they were produced by ultraviolet photolysis of HI, present in about 1% concentration. However, it is interesting that the same experiment in argon matrices gave no spectra of I atoms [Jen (248)].

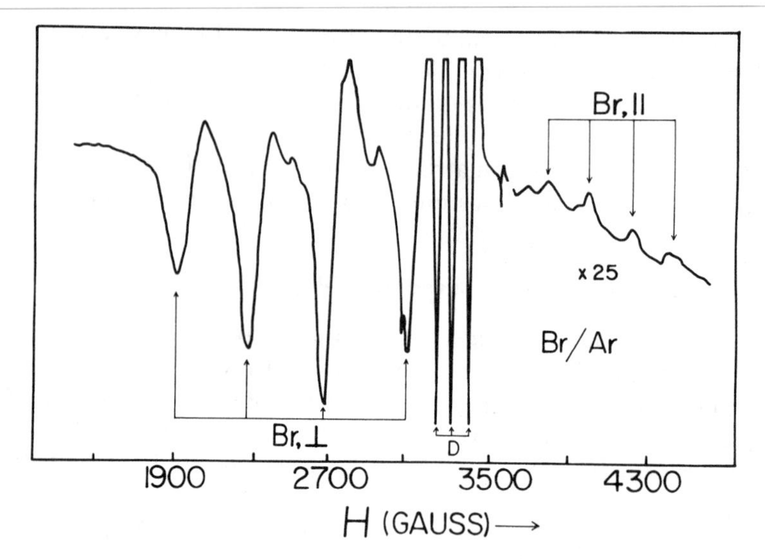

Fig. I-8. ESR spectrum of Br atoms electronically quenched in an argon matrix at 4°K [from Bhat and Weltner (65)]. Br atoms are formed during the reaction of deuterium atoms with Br_2.

Prior to the observation of the halogen atoms, $^2P_{1/2}$ aluminum atoms were observed via ESR [Knight and Weltner (284)], but the spectra were first recognized and interpreted as quenched atoms by Ammeter and Schlosnagle (39), who made a thorough and definitive study of this atom and also gallium. Boron atoms, in that same Group III series, were later observed by Graham and Weltner (197) and interpreted using the simple static crystal-field model (39).

The experimental matrix data, and gas-phase parameters, for these partially quenched atoms are given in Table I-3. To a fair approximation all of these cases can be treated by a static crystal-field model wherein the spherical symmetry of the atom is lost owing to an environment that effectively maintains the equivalence of the p_x and p_y orbitals but alters that of the p_z orbital. The site of the trapped atom is then such as to produce an axial electric field along the z direction of strength determined by a crystal-field parameter Δ [Van Vleck (22)]. The orbital angular momentum quantum states for a ^2P_J atom then no longer are degenerate and in axial symmetry are split into $^2\Sigma$, $^2\Pi_{1/2}$, and $^2\Pi_{3/2}$ states depending upon the degree of quenching, i.e., upon the magnitude of Δ, and also upon the atomic spin-orbit coupling constant ζ. This is shown in

Fig. I-9 for a $^2P_{1/2}$ $(np)^1$ atom
nagle (39).

In the cylindrically symm
become $\sigma_z(l = 0)$ and $\pi_{x,y}(l =$
will become clearer in the ne
sidered, the atomic g and A p

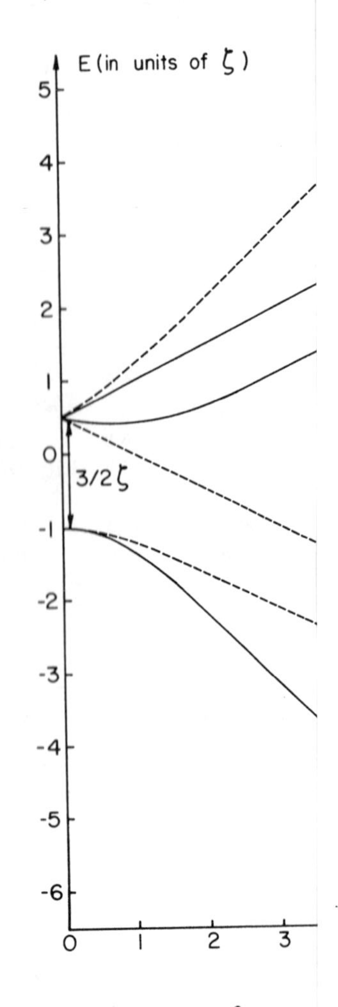

Fig. I-9. Energy levels of a $^2P_{1/2}$ (np
meter and Schlosnagle (39)]. Solid li
field parameter and ζ is the spin-orb
inverted.

Table I-3. Cr
Qu

Atom
$^{11}B\ (^2P_{1/2})$
$^{27}Al\ (^2P_{1/2})$
$^{69}Ga\ (^2P_{1/2})$
$^{79,81}Br\ (^2P_{3/2})$
$^{127}I\ (^2P_{3/2})$

Atom	M
$^{11}B\ (^2P_{1/2})$	
$^{27}Al\ (^2P_{1/2})$	
$^{69}Ga\ (^2P_{1/2})$	
$^{79,81}Br\ (^2P_{3/2})$	
$^{127}I\ (^2P_{3/2})$	

Atom	Δ (cm
$^{11}B\ (^2P_{1/2})$	
$^{27}Al\ (^2P_{1/2})$	
$^{69}Ga\ (^2P_{1/2})$	15 17 21

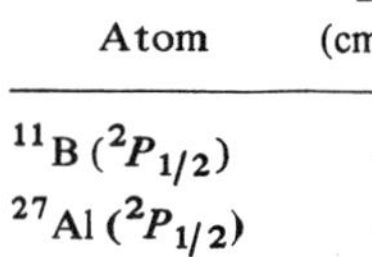

Table I-3. (Continued)

	Static Crystal Field Parameters[g]						
Atom	$\Delta_{\parallel}$ (cm^{-1})	$\Delta_{\perp}$ (cm^{-1})	P_S^M/P_S (%)	A_F (10^{-4} cm^{-1})	$A_F - A_0$ (10^{-4} cm^{-1})	A_{iso}^{Mh} (10^{-4} cm^{-1})	A_{dip}^{Mh} (10^{-4} cm^{-1})
$^{79,81}Br$ ($^2P_{3/2}$	710	2320	–	–	–	440	-49
^{127}I ($^2P_{3/2}$)	392	1076	–	–	–	375[d]	-24[d]

[a]Site 1 was arbitrarily chosen here whenever the atom was observed in several sites.
[b]One gas-phase parameter for Ga is unknown, $A_{3/2,1/2}$. $A_{3/2,1/2}$ for Br was approximated as 43×10^{-4} cm^{-1}, guided by theory (65) in order to calculate P_L, P_S, and A_0. For Br, P_L, P_S, and A_0 are natural-abundance weighted averages for the two isotopes.
[c]Calculated from known gas-phase values of $A_{1/2}$ and $A_{3/2}$, and assuming $P_S/P_L = 1.1$ in Eq. 11 of Ammeter and Schlosnagle (39).
[d]A_0 was approximated as an average of $[A_{iso}(I)/A_{iso}(X)] \cdot A_0(X)$, where X is either Br or Cl. The ratios in square brackets were obtained from Table A1, and $A_0(X)$ are the gas-phase values (10.80 for Cl). A_{iso}^M and A_{dip}^M were calculated from Eq. 7b of Ammeter and Schlosnagle (39) assuming either P_L or $P_S = 516.4 \times 10^{-4}$ cm^{-1}, and $P_S/P_L = 1.25$, and an average taken in each case.
[e]Experimental, except for signs of $A_{\parallel}$ and $A_{\perp}$, which determine the values of A_{iso} and A_{dip}.
[f]$A_{iso} = (A_{\parallel} + 2A_{\perp})/3$; $A_{dip} = (A_{\parallel} - A_{\perp})/3$.
[g]Utilizing the theory of Ammeter and Schlosnagle (39). $\Delta_{\parallel}$ and $\Delta_{\perp}$ are obtained from $\Delta g_{\parallel}$ and $\Delta g_{\perp}$, respectively.
[h]Calculated from (I, 33), or Eq. 7b of Ammeter and Schlosnagle (39), using gas-phase parameters and matrix Δg values.

different values along the z axis and perpendicular to it; that is, $g_{\parallel}$, $g_{\perp}$, $A_{\parallel}$, and $A_{\perp}$ are now required for a magnetic description of the quenched atom. If the odd electron were confined to the σ_z orbital and Δ were very large, the $^2P_{1/2}$ atom would be completely quenched to a $^2\Sigma^+$ ground state. However, mixing can, in general, occur with the upper $^2\Pi$ state, which itself will be split into $^2\Pi_{1/2}$ and $^2\Pi_{3/2}$ states by ζ. In heavy atoms one can, in fact, expect Δ and ζ to be of comparable magnitudes.

Explicitly, for a single p electron, in the lowest state [Ammeter and Schlosnagle (39)],

$$\Psi_\beta{}^\alpha(^2\Sigma^+) = a\sigma_\beta{}^\alpha - b\pi_\alpha{}^\beta \qquad \text{(I, 30)}$$

where a and b are normalized coefficients,

$$a = \{1 + \tfrac{1}{4}[5 - 6x + 9x^2 + 3(1 - 3x)(x^2 - \tfrac{2}{3}x + 1)^{1/2}\}^{-1/2}$$
$$b = (1 - a)^{1/2}, \qquad x = \Delta/\zeta \qquad \text{(I, 31)}$$

$$g_{\parallel} \cong 2 - 2b^2$$
$$g_{\perp} \cong 2 - 2b^2 - 2\sqrt{2}ab \qquad (\text{I}, 32)$$

and

$$A_{\parallel} \cong P_L\, 2b^2 + P_S\,[(\tfrac{4}{5} - \kappa)(1 - 2b^2) + \tfrac{3}{5}(2\sqrt{2}ab + b^2)]$$
$$A_{\perp} \cong -P_L\, 2\sqrt{2}ab - P_S\,[(\tfrac{2}{5} + \kappa)(1 - b^2) + \tfrac{3}{10}(2\sqrt{2}ab + 4b^2)]$$
$$(\text{I}, 33)$$

where P_L, P_S, and κ are the hyperfine anisotropic and isotropic parameters discussed above for gas-phase atoms. For strong quenching, where b^2 is small,

$$\Delta g_{\parallel} = g_{\parallel} - g_e = 0$$
$$\Delta g_{\perp} = g_{\perp} - g_e = -2\sqrt{2}b = -4\zeta/3\Delta \qquad (\text{I}, 34)$$

$$A_{\text{iso}} = \frac{A_{\parallel} + 2A_{\perp}}{3} = \frac{2}{3}P_L\,\Delta g_{\perp} - P_S\,\kappa = \frac{2}{3}P_L\,\Delta g_{\perp} + A_F$$

$$A_{\text{dip}} = \frac{A_{\parallel} - A_{\perp}}{3} = -P_L\,\Delta g_{\perp} + P_S\left(\frac{2}{5} - \frac{3}{10}\Delta g_{\perp}\right) \qquad (\text{I}, 35)$$

These latter equations apply satisfactorily to the $^2P_{1/2}$ atoms ^{11}B and ^{27}Al where $g_{\parallel} \cong 2.00$, but not to ^{69}Ga where $\Delta g_{\parallel} \neq 0$. The ESR spectra of ^{11}B ($I = \frac{3}{2}$) and ^{10}B ($I = 3$) atoms in argon matrices at 4°K are shown in Fig. I-10 [Graham and Weltner (197)], and one can see that the hyperfine structures of the parallel lines are in accord with the nuclear spins of the two isotopes, i.e., $2I + 1$ lines. The perpendicular hfs is unresolved. These ESR patterns are typical of those for randomly oriented nonrotating $^2\Sigma$ molecules where the hfs ($A_{\parallel}$) spacing of the lines associated with $g_{\parallel}$ is large relative to the hfs ($A_{\perp}$) associated with $g_{\perp}$ (see Chapter II).

For Ga atoms, separate values of Δ (indicated as $\Delta_{\parallel}$ and $\Delta_{\perp}$) can be derived from $\Delta g_{\parallel}$ and $\Delta g_{\perp}$; these would be the same if this simple theoretical model were correct. Assuming that the gas-phase hf parameters P_S, P_L, and κ for gallium remain essentially the same in the matrix, as was the case for B and Al (see Table I-3), one can calculate $A_{\text{iso}}{}^M$ and $A_{\text{dip}}{}^M$ from (I, 33) for comparison with the ex-

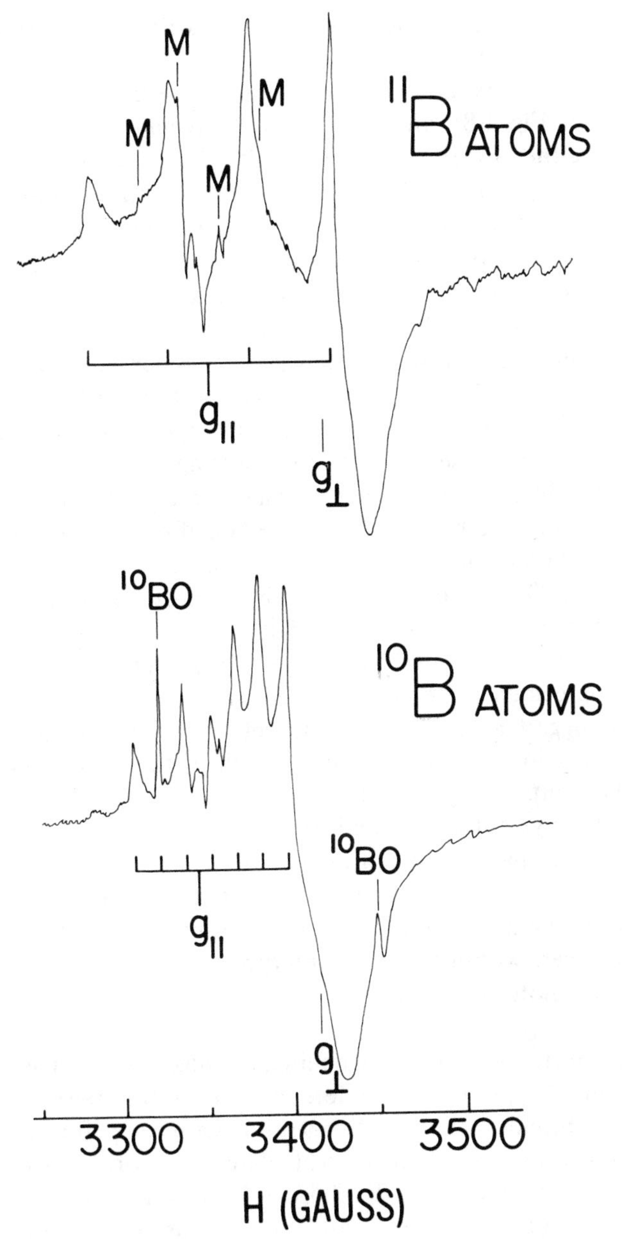

Fig. I-10. ESR spectra of electronically quenched ^{11}B ($I = \frac{3}{2}$) and ^{10}B ($I = 3$) atoms in argon matrices at 4°K [from Graham and Weltner (197)].

perimental A_{iso} and A_{dip} measured in the matrix. The agreement is excellent.

An analogous crystal-field treatment can be made for the $^2P_{3/2}$ atoms Br (see Fig. I-8), and I [Iwasaki, Toriyama, and Muto (245); Bhat and Weltner (65)]. Since here also $\Delta g_{\parallel} \neq 0$, one can derive two values of Δ, and $A_{iso}{}^M$, $A_{dip}{}^M$, as described for gallium. Agreement between calculated and observed hf parameters is not as good as for gallium but appears satisfactory.

An earlier study of a partially quenched $^2D_{5/2}$ atom is that of the $Ag^{2+}(4d^9)$ ion in acidic solutions at 77°K [McMillan and Smaller (341)]. Figure I-11 shows the observed first-derivative ESR spectrum, with $g_{\parallel} \simeq 2.27$ and $g_{\perp} \simeq 2.07$, and the corresponding absorption spectrum. Since silver has two isotopes with nuclear spin $I = \frac{1}{2}$, hfs was also observed, but the splittings due to each isotope were not resolved because their magnetic moments differ by less than 20%. From the hfs $A_{\parallel}$ (Ag) = −48 G and $A_{\perp}$ (Ag) = +31.6 G so that A_{dip} = −26.5 G and A_{iso} = +5.1 G, indicating a very small degree of s character. Quenching is produced by a linear (tetragonal) crystal field that the g values [with $\lambda = -1840\ \text{cm}^{-1}$; see Bowers (80)] indicate has $\Delta_{\parallel} = -53{,}500\ \text{cm}^{-1}$ and $\Delta_{\perp} = -48{,}750\ \text{cm}^{-1}$. These values are not in agreement with the observed optical absorption at 25,560 cm^{-1} and suggest an error in the model. It is inferred that the assumption of a localized hole in the d shell of the paramagnetic ion may not be completely justified.

The static crystal-field model is only approximate, and other considerations resulting from specific interactions with the matrix atoms and vibronic effects were found to be necessary to give good agreement with the data for Al and Ga atoms [Ammeter and Schlosnagle (39)]. However, within the data, interpreted by this simple model, there are some noteworthy trends:

1. For a given trapped atom, quenching, as judged by the magnitudes of $\Delta g_{\parallel}$ and $\Delta g_{\perp}$, is more complete in the heavier rare-gas matrices.
2. The crystal-field parameter $\Delta_{\perp}$ in solid argon varies for B, Al, and Ga atoms in proportion to the calculated $\langle r \rangle_{np}$ of the atom: 380, 1990, 2020 cm^{-1} versus 2.20, 3.43, 3.42 Å, respectively.
3. The agreement between the isotropic hf constant, i.e., the s electron contribution, in the matrix and in the gas, for B, Al, and Ga

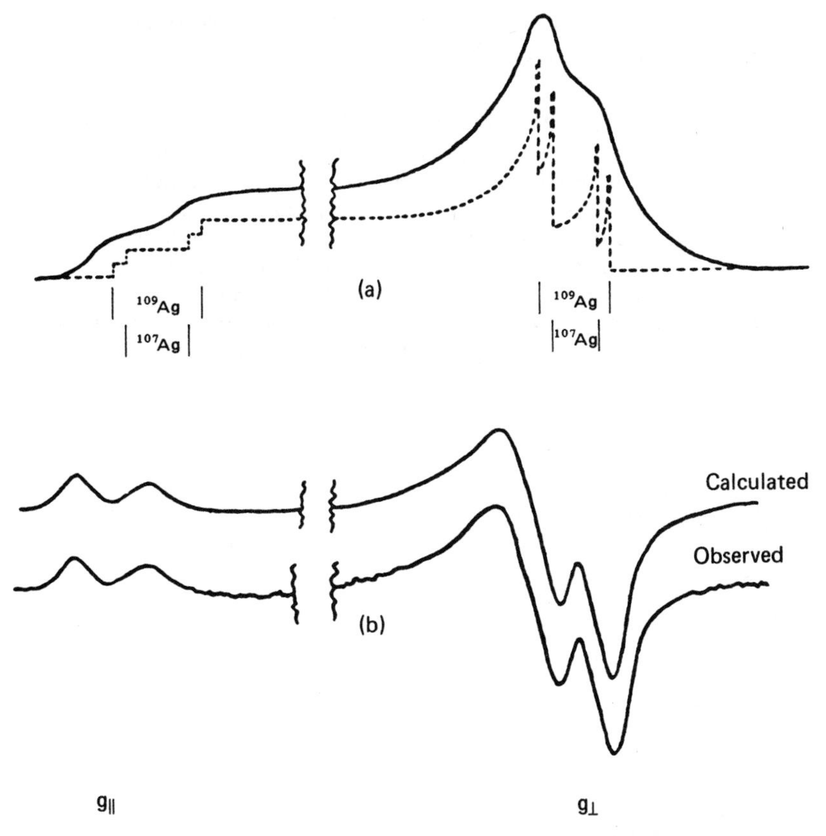

Fig. I-11. ESR spectrum of Ag^{2+} in frozen acid solution [from McMillan and Smaller (341)]. (a) Absorption spectrum; (b) first-derivative ESR spectrum. Ag has two isotopes each with spin $I = \frac{1}{2}$ but differing in μ_I by less than 10% so that they are not resolved in the spectrum.

atoms indicates that these atoms occupy centrosymmetric sites in all matrices.

4. The variations in environment necessary for observation of trapped $^2P_{3/2}$ atoms suggest that their ground-state wavefunctions are especially sensitive to small perturbations by the matrix. From the history of the halogen atom studies, there are indications that the detection of other orbitally degenerate atoms may require the presence in the matrix of another neighboring species to provide sites with the necessary homogeneous crystal fields. This, of

course, implies that the observed atoms can no longer be considered as truly matrix-isolated.

The recognition that the splitting in the excited *P* state in the absorption spectra of sodium and mercury atoms trapped in rare-gas matrices resulted from removal of orbital degeneracy was first made by McCarty and Robinson (328). Schnepp made analogous observations on magnesium and manganese atoms (405) and explained the quenching in terms of asymmetric sites in the matrix by considering covalent and van der Waals effects [Brith and Schnepp (86)].

Early ESR work on the partially quenched $^2P_{3/2}$ S^- and O^- ions in crystals was done by Vannotti and Morton (440) and Brailsford, Morton, and Vannotti (81), who used a static crystal-field model in the interpretation of their data.

Chapter II

$^2\Sigma$ and Other Doublet ($S = \frac{1}{2}$) Molecules in Magnetic Fields

It is intuitively understandable that the magnetic properties of a linear molecule may vary, depending upon the direction that the magnetic field makes with the axis of the molecule. In general, the g factor, introduced when discussing atoms, will have two extreme values, $g_\parallel$ and $g_\perp$, depending upon whether the magnetic field is parallel or perpendicular to the axis of the molecule. For intermediate angles, θ, g will be given by

$$g^2 = g_\parallel{}^2 \cos^2 \theta + g_\perp{}^2 \sin^2 \theta$$

Then if all of the molecules were lined up, as in a hypothetical single crystal, one could observe resonance first at a magnetic field $H_\parallel = h\nu_0/g_\parallel\beta$ and then finally at $H_\perp = h\nu_0/g_\perp\beta$ as the angle θ varied from 0° to 90° (ν_0 being, as usual, at some fixed frequency). Any molecule with one distinct axis of trigonal or higher symmetry will also have these two extreme perpendicular and parallel orientations. Such a molecule, e.g., CH_3, will be called axial. For a nonaxial molecule, a third g component is necessary with its axis orthogonal to the principal axes of the other two g components. Thus, a g tensor will be necessary to characterize the most general molecule, and it will have three components g_1, g_2, and g_3 when described in terms of the principal axes of the tensor, i.e., when it is diagonal.

1. $^2\Sigma$ MOLECULES

$^2\Sigma$ molecules are the simplest that can be investigated by electron-spin-resonance spectroscopy, since nominally $L = 0$, and essentially the entire magnetic moment is due to the one unpaired electron spin.

Any orbital angular momentum in the ground state effectively arises from the mixing-in of excited states with $L \neq 0$ ($^2\Pi$ states) via the molecular spin-orbit coupling constant ξ. If such states lie relatively high ($>\sim 10{,}000\ \text{cm}^{-1}$) above the ground state, as they usually do (or if ξ is small), this mixing will be small, so that one expects both $g_\parallel$ and $g_\perp$ to be near the pure spin value, g_e. For reasons to be discussed later, $g_\parallel$ will usually be closer to g_e than $g_\perp$ is; that is $g_\perp$ will be more indicative of the orbital angular momentum in the ground state. Thus, in general, the two g components will be different (the $\mathbf{g}$ tensor will be anisotropic) so that indeed the position of the resonance line will vary with the orientation of the molecule in the magnetic field.

In this description of the magnetic properties of these molecules, contributions of orbital angular momentum are absorbed into the g components; that is, $\Delta g = g - g_e$ is a measure of the divergence from pure spin. This is a simplification provided by the spin Hamiltonian.

(a) Spin Hamiltonian

The Hamiltonian for a molecule with $S = \frac{1}{2}$ in a magnetic field may be written

$$\mathcal{H} = \mathcal{H}_F + \mathcal{H}_{Ze} + \mathcal{H}_{LS} + \mathcal{H}_{hf} + \mathcal{H}_{Zn} \qquad \text{(II, 1)}$$

where the magnitude of the terms on the right-hand side generally decreases toward the right. Here $\mathcal{H}_F$ is the Hamiltonian of the free molecule in the absence of spin-orbit coupling, hyperfine interaction, and an external magnetic field. It therefore includes only coulombic terms involving electron–nuclear and electron–electron interactions. (Translational, rotational, and vibrational motions of the molecule are not of concern here; Born–Oppenheimer separation of electron and nuclear motions is assumed, and the nuclear framework fixed at the equilibrium bond lengths.) $\mathcal{H}_{Ze}$ and $\mathcal{H}_{Zn}$ are Zeeman terms indicating the interaction of the applied magnetic field directly with the electron and nuclear magnetic dipoles, respectively. $\mathcal{H}_{LS}$ is the coupling of the spin and orbital angular momenta of the electrons in the axial electric field of the molecule. $\mathcal{H}_{hf}$ is the hyperfine interaction arising from magnetic interactions with all nuclei having $I \neq 0$ in the molecule. Other terms that might have been included in this

Hamiltonian involve any effects of the crystal field produced by the surrounding matrix, and the weak interaction of the electron magnetic moment with any nucleus having an electric quadrupole moment Q.

Initially, the relatively small terms $\mathcal{H}_{hf}$ and $\mathcal{H}_{Zn}$ in (II, 1) may be neglected, since they are usually less than 10^{-2} cm^{-1}. The remaining two interaction terms on the right side are

$$\mathcal{H}_{Ze} = \beta_e \boldsymbol{H} \cdot (\boldsymbol{L} + g_e \boldsymbol{S}) \qquad \text{(II, 2)}$$

and

$$\mathcal{H}_{LS} = \xi \boldsymbol{L} \cdot \boldsymbol{S} \qquad \text{(II, 3)}$$

The Zeeman term includes an $\boldsymbol{H} \cdot \boldsymbol{L}$ interaction because, as mentioned above, the orbital angular momentum in the ground state of the molecule is zero only in the first approximation. $\mathcal{H}_{LS}$ in (II, 3) is given in the standard form for an atom but where ξ is now an effective spin-orbit constant for the molecule. Atomic spin-orbit constants (to be indicated here by ζ) vary approximately as Z^4, where Z is the atomic number, so that, for example, $\zeta = 26$ cm^{-1} for the C atom, whereas $\zeta = 4270$ cm^{-1} for Hg. ξ will reflect these atomic values, as is evident when the simple linear-combination-of-atomic-orbitals (LCAO) approximation is used to describe the molecular orbitals. In a molecule, as in an atom, the deviation of the g components from a pure spin value of 2 depends upon the strength of this coupling.

The Hamiltonian

$$\mathcal{H} = \mathcal{H}_F + \mathcal{H}_{Ze} + \mathcal{H}_{LS} \qquad \text{(II, 4)}$$

is still very complicated, but may be greatly simplified for our purposes by putting it in the form of a spin Hamiltonian first introduced by Abragham and Pryce (27). In principle, it amounts to retention only of operators in (II, 4) that act on electron and nuclear spin functions, since others will only provide constant terms as far as magnetic interactions are concerned. Thus $\mathcal{H}_F$ only involves spatial operators and can be neglected. Further simplification can be made if the spin-orbit term, as in our case, is relatively small so that it may be treated as a perturbation of the Zeeman term. Then a so-called spin Hamiltonian,

$$\mathcal{H}_{\text{spin}} = \beta_e \boldsymbol{H} \cdot \boldsymbol{g} \cdot \hat{\boldsymbol{S}} \tag{II, 5}$$

can be formally written in place of

$$\mathcal{H} = \beta_e \boldsymbol{H} \cdot (\boldsymbol{L} + g_e \boldsymbol{S}) + \xi \boldsymbol{L} \cdot \boldsymbol{S} \tag{II, 6}$$

where $\boldsymbol{g}$ is now an *effective* $\boldsymbol{g}$ tensor (no longer a g "factor" because of its molecular anisotropy), and $\hat{\boldsymbol{S}}$ is an operator corresponding to an *effective* spin.

(b) *g* Tensor†

In order to calculate $\boldsymbol{g}$ in (II, 5) for an $S = \frac{1}{2}$ molecule it is instructive, if not essential, to proceed through the first-order wavefunction obtained when the $\boldsymbol{L} \cdot \boldsymbol{S}$ term may be considered as a small perturbation [Carrington and McLachlan (6), page 134; Wertz and Bolton, (23), page 278)]. If the zeroth-order wavefunctions are the pure spin $|0, \alpha\rangle$ and $|0, \beta\rangle$ functions, where 0 indicates the ground state and α and β designate $m_S = +\frac{1}{2}$ and $-\frac{1}{2}$, the first-order ones are

$$|+\rangle = |0, \alpha\rangle - \sum_n \frac{\langle n|\xi \boldsymbol{L} \cdot \boldsymbol{S}|0, \alpha\rangle}{E_n - E_0} |n\rangle$$

and

$$|-\rangle = |0, \beta\rangle - \sum_n \frac{\langle n|\xi \boldsymbol{L} \cdot \boldsymbol{S}|0, \beta\rangle}{E_n - E_0} |n\rangle \tag{II, 7}$$

where n numbers the excited states. These wavefunctions are then used to calculate the matrix elements of $\beta \boldsymbol{H}(\boldsymbol{L} + g_e \boldsymbol{S})$ in the true Hamiltonian in (II, 6), and they are compared with the correspond-

---

†The $\boldsymbol{g}$ tensor and also the so-called hyperfine tensor $\boldsymbol{A}$ are not mathematically true tensors, although it has become habitual to refer to them as such. One of the consequences of this is that it may be impossible to diagonalize them both simultaneously; that is, they may not have the same principal axes. Usually it is assumed that they do have the same principal axes until experimental knowledge makes that position untenable. However, the fact they are not tensors makes it possible to diagonalize $\boldsymbol{A}$ relative to both $\boldsymbol{S}$ and $\boldsymbol{I}$ by using different sets of axes, as in section II-1d. [See Abragham and Bleaney (1), pages 170, 651.]

ing matrix elements of the spin Hamiltonian (II, 5). In the latter, the effective spin operator $\hat{\boldsymbol{S}}$ is defined such that

$$\hat{S}_z|+\rangle = \frac{1}{2}|+\rangle, \qquad \hat{S}_z|-\rangle = -\frac{1}{2}|-\rangle,$$

$$\hat{S}_x|+\rangle = \frac{1}{2}|-\rangle, \qquad \hat{S}_y|+\rangle = \frac{i}{2}|-\rangle, \text{ etc.} \tag{II, 8}$$

in analogy to $\boldsymbol{S}$ acting on $|\alpha\rangle$ and $|\beta\rangle$. For a linear molecule where z is along the internuclear axis, the above comparison yields the principal components of the effective $\boldsymbol{g}$ as

$$g_\parallel = 2\langle +|L_z + g_e S_z|+\rangle = 2\langle -|L_z + g_e S_z|-\rangle$$

and

$$g_\perp = 2\langle -|L_x + g_e S_x|+\rangle = 2\langle -|L_y + g_e S_y|+\rangle \tag{II, 9}$$

Putting in $|+\rangle$ and $|-\rangle$ then gives

$$g_\parallel = g_e - 2\xi \sum_n \frac{\langle 0|L_z|n\rangle\langle n|L_z|0\rangle}{E_n - E_0} \tag{II, 10}$$

and

$$g_\perp = g_e - 2\xi \sum_n \frac{\langle 0|L_x|n\rangle\langle n|L_x|0\rangle}{E_n - E_0} \tag{II, 11}$$

where ξ has been taken as some average constant value. However, $g_\parallel$ is simplified if the ground-state wavefunction, $|0\rangle$, is orbitally nondegenerate, since then $\langle n|L_z|0\rangle = 0$, and (II, 10) yields $g_\parallel = g_e$. In second order of perturbation a nonvanishing term will appear on the right side of (II, 10) for a Σ ground-state molecule [Tippins (430); Atkins and Jamieson (42); Stone (426)], so that

$$g_\parallel = g_e - 2\xi^2 \sum_n{}' \frac{|\langle n|L_x|0\rangle|^2}{(E_n - E_0)^2} \tag{II, 12}$$

This term, proportional to ξ^2, can be significant in molecules containing heavy atoms.

Then, in general, from (II, 11) and (II, 12), $g_\parallel \neq g_\perp$ for a Σ ground-state molecule. If the electronic properties of the ground and lowest-lying excited states are known, which is rarely true with accuracy, the $\boldsymbol{g}$ tensor components can be calculated from these equations. (Orbitally degenerate molecules exhibit much larger $\boldsymbol{g}$ tensor anisotropies, as will be discussed in section 2 of this chapter.)

It is not really necessary to deal with the eigenfunctions $|+\rangle$, $|-\rangle$ in deriving (II, 10) and (II, 11). An alternative, more direct, derivation also puts the final result in a more concise and general form [Abragham and Bleaney (1), page 745; Wertz and Bolton, (23), page 278]. Using the general functions $|0, m_S\rangle$, $|n, m_S\rangle$, perturbation theory applied to the true Hamiltonian in (II, 6) yields

$$\begin{aligned}\mathcal{H}(m_S, m_S') = \langle 0, m_S | g_e \beta H_z S_z | 0, m_S \rangle \\ - \sum_n{}' [\langle m_S | \beta \boldsymbol{H} + \xi \boldsymbol{S} | m_S' \rangle \langle 0 | \boldsymbol{L} | n \rangle] \\ \cdot [\langle n | \boldsymbol{L} | 0 \rangle \langle m_S' | \beta \boldsymbol{H} + \xi \boldsymbol{S} | m_S \rangle] / (E_n - E_0)\end{aligned} \quad \text{(II, 13)}$$

The orbital terms may be factored out and written as elements of a tensor,

$$\Lambda_{ij} = -\sum_n{}' \frac{\langle 0 | \boldsymbol{L}_i | n \rangle \langle n | \boldsymbol{L}_j | 0 \rangle}{E_n - E_0} \quad \text{(II, 14)}$$

where i and j are any of the cartesian coordinates. Then (II, 13) becomes

$$\mathcal{H}(m_S, m_S') = \langle m_S | \beta^2 \boldsymbol{H} \cdot \Lambda \cdot \boldsymbol{H} + 2\xi \boldsymbol{H} \cdot \Lambda \cdot \boldsymbol{S} + \xi^2 \boldsymbol{S} \cdot \Lambda \cdot \boldsymbol{S} | m_S' \rangle$$

The first term on the right-hand side is a constant not involving spin variables and can be neglected, so that

$$\mathcal{H}(m_S, m_S') = \beta \boldsymbol{H}(g_e \mathbf{1} + 2\xi \Lambda) \cdot \boldsymbol{S} + \xi^2 \boldsymbol{S} \cdot \Lambda \cdot \boldsymbol{S}$$

or more concisely,

$$\mathcal{H}(m_S, m_S') = \beta \boldsymbol{H} \cdot \boldsymbol{g} \cdot \boldsymbol{S} + \boldsymbol{S} \cdot \boldsymbol{D} \cdot \boldsymbol{S} \qquad \text{(II, 15)}$$

Here

$$\boldsymbol{g} = g_e \mathbf{1} + 2\xi\Lambda, \qquad \text{(II, 16)}$$

$$\boldsymbol{D} = \xi^2 \Lambda \qquad \text{(II, 17)}$$

and **1** is the unit tensor. $\boldsymbol{D}$ is only nonzero for molecules containing more than one unpaired electron, i.e., $S \geqslant 1$, and will be considered in more detail in future chapters. Then, as expected for a $^2\Sigma$ molecule, the spin Hamiltonians in (II, 15), with $\boldsymbol{D} = 0$, and in (II, 5) are equivalent, as are also the components of $\boldsymbol{g}$ derived from (II, 16) and those given in (II, 10), (II, 11).

For a linear molecule (or any molecule with an axis of trigonal or higher symmetry) the $\boldsymbol{g}$ tensor may be written in diagonal form if the principal axes are chosen along the molecular axis and perpendicular to it:

$$g = \begin{vmatrix} g_\perp & 0 & 0 \\ 0 & g_\perp & 0 \\ 0 & 0 & g_\parallel \end{vmatrix} \qquad \text{(II, 18)}$$

The spin Hamiltonian will then be

$$\mathcal{H}_{\text{spin}} = g_\parallel \beta H_z \hat{S}_z + g_\perp \beta (H_x \hat{S}_x + H_y \hat{S}_y) \qquad \text{(II, 19)}$$

$$\boldsymbol{L} \cdot \boldsymbol{S} = (L_x S_x + L_y S_y + L_z S_z) = L_z S_z + \tfrac{1}{2}(L_+ S_- + L_- S_+) \qquad \text{(II, 20)}$$

and the raising and lowering operators, S_+, S_-, cause the mixing of α and β excited states with the ground β and α states, respectively, when (II, 20), is inserted into (II, 7). Then

$$|+\rangle = |0, \alpha\rangle - \frac{1}{2}\xi \sum_n \left[\frac{\langle n|L_z|0\rangle}{E_n - E_0} |n, \alpha\rangle + \frac{\langle n|L_x + iL_y|0\rangle}{E_n - E_0} |n, \beta\rangle \right]$$

and

$$|-\rangle = |0, \beta\rangle + \frac{1}{2}\xi \sum_n \left[\frac{\langle n|L_z|0\rangle}{E_n - E_0} |n, \beta\rangle - \frac{\langle n|L_x - iL_y|0\rangle}{E_n - E_0} |n, \alpha\rangle \right] \quad \text{(II, 21)}$$

where $L_x + iL_y = L_+$ and $L_x - iL_y = L_-$. Note that, because of this mixing, $|+\rangle$ and $|-\rangle$, unlike $|\alpha\rangle$ and $|\beta\rangle$, are not eigenstates of the true spin S_z, i.e., $S_z|+\rangle \neq \frac{1}{2}|+\rangle$, etc. $|+\rangle$ and $|-\rangle$ may then be used to calculate the matrix elements of the true Hamiltonian $\beta \boldsymbol{H}(\boldsymbol{L} + g_e \boldsymbol{S})$. For example, for $H \| z$ in a linear molecule, one obtains [Carrington and McLachlan (6), pages 134, 135]:

	$\|+\rangle$	$\|-\rangle$
$\|+\rangle$	$\langle +\|L_z + g_e S_z\|+\rangle \beta H$	0
$\|-\rangle$	0	$\langle -\|L_z + g_e S_z\|-\rangle \beta H$

But, according to the spin Hamiltonian (II, 5), with the $\hat{S}$ operators defined as in (II, 8), these matrix elements should be

	$\|+\rangle$	$\|-\rangle$
$\|+\rangle$	$\frac{1}{2} g_\| \beta H$	0
$\|-\rangle$	0	$-\frac{1}{2} g_\| \beta H$

Thus, one derives the first of equations (II, 9) and $g_\perp$ follows in an analogous way.

(II, 13) is obtained from (II, 6) by first- and second-order perturbation to originally yield

$$\mathcal{H}(m_S, m_S') = \langle 0, m_S | g_e \beta H_z S_z | 0, m_S\rangle + \langle m_S | \beta H_z + \xi S_z | m_S\rangle \langle 0 | L_z | 0\rangle$$
$$- \sum_n{}' \frac{|\langle 0, m_S | (\beta \boldsymbol{H} + \xi \boldsymbol{S}) \cdot L + g_e \beta \boldsymbol{H} \cdot \boldsymbol{S} | n, m_S' \rangle|^2}{E_n - E_0}$$

But $\langle 0|L_z|0\rangle$ vanishes for an orbitally nondegenerate ground state, and $g_e \beta \boldsymbol{H} \cdot \boldsymbol{S}$ will also not appear because of the orthogonality $\langle 0|n\rangle = 0$. $\mathcal{H}(m_S, m_S')$ then reduces to (II, 13). Λ in (II, 16) and (II, 17) is a symmetric tensor:

$$\begin{vmatrix} \Lambda_{xx} & \Lambda_{xy} & \Lambda_{xz} \\ \Lambda_{xy} & \Lambda_{yy} & \Lambda_{yz} \\ \Lambda_{xz} & \Lambda_{yz} & \Lambda_{zz} \end{vmatrix}$$

(α) $\Delta g_\perp = g_\perp - g_e$. According to the earlier discussion, for a Σ ground-state molecule $g_\| = g_e$, or if higher-order perturbation is considered, according to (II, 12) $g_\|$ could be slightly less than g_e in particular cases. However, $g_\perp$ is obtained from (II, 11) and can vary considerably from g_e owing to the ability of $L_x(=L_y)$ to couple excited states with the ground state.

The matrix element $\langle\Sigma|L_x|n\rangle$ will be nonvanishing only if n corresponds to an excited Π state. Often the lowest-lying Π state will dominate the summation in (II, 11) so that $g_\perp$ may be written:

$$g_\perp = g_e - \frac{2}{E_\Pi - E_\Sigma}\langle\Pi|\xi L_x|\Sigma\rangle\langle\Sigma|L_x|\Pi\rangle \qquad \text{(II, 22)}$$

The sign of $\Delta g_\perp$ will depend upon whether that excited state has the properties of an electron in a π orbital or of a "hole" in a π orbital. Another way of stating this is to say that the spin-orbit coupling constant, ξ, is positive in the first case and negative in the second. In molecular orbital notation, the excited state would be obtained by excitation of the ground $^2\Sigma$ state with a configuration $\cdots\pi_1{}^4\sigma^1$ to $\cdots\pi_1{}^4\pi_2{}^1$, $^2\Pi_r$ (r for regular) or to $\cdots\pi_1{}^3\sigma^2$, $^2\Pi_i$ (i for inverted). Mixing a $^2\Pi_r$ state with the ground state will yield a negative $\Delta g_\perp$, whereas a $^2\Pi_i$ state will yield a positive $\Delta g_\perp$. If several excited states must be considered, then their individual effects on $\Delta g_\perp$ must be summed. However, in the simplest case, one can say that if the experimental value of $\Delta g_\perp$ is negative, usually a $^2\Pi_r$ excited state lies lowest, and if $\Delta g_\perp$ is positive, a $^2\Pi_i$ state is lowest. In this way qualitative information about the disposition of excited states of a $^2\Sigma$ molecule may be obtained immediately from the value of $g_\perp$.

The theory and application of (II, 11) to molecules has been considered in detail by Stone (426). Discussed there is the form that ξ, the molecular spin-orbit constant, should take, and the necessity for gauge-invariance of g; that is, for the calculated value to be invariant to where the origin is chosen. Assume for simplicity that $g_\perp$ is given by (II, 22) and that Σ and Π can be approximated by the LCAO wavefunctions

$$\Psi_\Sigma = \sum_k a_k \chi_k{}^\sigma \quad \text{and} \quad \Psi_\Pi = \sum_k b_k \chi_k{}^\pi \qquad \text{(II, 23)}$$

where $\chi_k{}^\sigma$ and $\chi_k{}^\pi$ are AO's in the ground and excited states, respectively, and the summation is over the k atoms. ξL_x and L_x may be written as sums of atomic values, i.e.,

$$\xi L_x = \sum_k \xi_k l_k{}^x \qquad \text{and} \qquad L_x = \sum_k l_k{}^x$$

where k indicates a particular atom in the molecule. Also ξ_k decreases rapidly for large r_k ($\sim$ as $r_k{}^{-3}$) so that ξ_k is essentially zero except near atom k, where it may be assumed to have the fixed atomic value ζ_k. Then (II, 22) becomes

$$\Delta g_\perp = -\frac{2}{E_\Pi - E_\Sigma} \sum_{k,k'} \langle \Pi | \zeta_{k'} l_{k'}{}^x | \Sigma \rangle \langle \Sigma | l_k{}^x | \Pi \rangle \qquad \text{(II, 24)}$$

The second matrix element on the right-hand side of (II, 24) will reduce to a sum of terms involving integrals of the type

(i) $\langle \chi_k{}^\sigma | l_k{}^x | \chi_k{}^\pi \rangle$ (only on atom k)

(ii) $\langle \chi_k{}^\sigma | l_{k'}{}^x | \chi_{k'}{}^\pi \rangle$ (both atoms involved)

(iii) $\langle \chi_{k'}{}^\sigma | l_{k'}{}^x | \chi_k{}^\pi \rangle$ (both atoms involved)

(iv) $\langle \chi_{k'}{}^\sigma | l_k{}^x | \chi_{k'}{}^\pi \rangle$ (both atoms involved) (II, 25)

The first matrix element in (II, 24) can be simplified, since ζ_k is assumed constant near atom k and zero elsewhere. Then it involves only integrals of type (i). Integrals of type (iv) require that the origin of operator $l_k{}^x$ be moved from atom k to atom k' along the internuclear axis by the relationship

$$l_k{}^x = l_{k'}{}^x + \hbar^{-1} R p_y \qquad \text{(II, 25a)}$$

where R is the interatomic distance and p_y is the *linear* momentum along the y axes. This introduces matrix elements involving the operator p_y into the evaluation of these integrals, but those elements are often zero or small so that they can be neglected [Stone (426); Atkins and Jamieson (42); Knight and Weltner (283)]. Then evaluation of all integrals involves only the orbital angular momentum operator $l_k{}^x$ operating on an AO on atom k, where

$$l^x = \frac{\hbar}{i}\left(y\frac{\partial}{\partial z} - z\frac{\partial}{\partial y}\right)$$

l^x is known to have the property of rotating σ AO functions into π functions, and π into δ and σ, and vice versa. In the case of integrals (ii) and (iii) in (II, 25), because two AO's on different atoms are involved after operating with l^x, an overlap integral will result,

$$S(i_k, j_{k'}) = \langle \chi_k{}^i | \chi_{k'}{}^j \rangle$$

These overlap integrals are usually small and in first approximation can be neglected, but their inclusion in calculations of $\Delta g_\perp$ for some molecules, such as CN [Easley and Weltner (153)], may be essential.

In the LCAO approximation, the orthogonal AO's in a linear molecule, where the internuclear axis is z, may be indicated as

$$p_x = p\pi', \quad p_y = p\pi, \quad p_z = p\sigma, \quad d_{xz} = d\pi', \quad d_{yz} = d\pi,$$
$$d_{xy} = d\delta', \quad d_{x^2-y^2} = d\delta, \quad \text{and} \quad d_{z^2} = d\sigma$$

Then if operated upon by l^x, the nonzero results are

$$l^x(d\pi') = -i\,d\delta', \quad l^x(d\pi) = i\,d\delta + i\sqrt{3}\,d\sigma, \quad l^x(d\delta') = i\,d\pi',$$
$$l^x(d\sigma) = -i\sqrt{3}\,d\pi, \quad l^x(d\delta) = -i\,d\pi$$

so that the only nonzero matrix elements are

$$\langle p\sigma|l^x|p\pi\rangle = i, \qquad \langle d\sigma|l^x|d\pi\rangle = \sqrt{3}i,$$
$$\langle d\pi'|l^x|d\delta'\rangle = i, \qquad \langle d\delta|l^x|d\pi\rangle = i, \quad \text{and the inverse of these,}$$

where $\langle i|l^x|j\rangle = \langle j|l^x|i\rangle^*$. These matrix elements justify the statement made above that L_x couples Σ states only with excited Π states.

Accurate calculation of $\Delta g_\perp$ for a $^2\Sigma$ molecule requires that good wavefunctions be known for the ground state and at least one or two excited $^2\Pi$ states. Lack of such information leads to the general use of LCAO wavefunctions, and those in the excited states are particularly doubtful. However, approximate spin densities at each atom in the molecule can be obtained from observed hyperfine structure in the ESR spectrum, as will be discussed in section II-1c, and thereby provide information about the ground-state wavefunction. Then the crude LCAO calculations can often show that the two pieces of experimental data, $\Delta g_\perp$ and hfs, are compatible.

In ESR investigations of the HgH and HgD molecules trapped in solid argon at 4°K, an isotope effect upon $\Delta g_\perp$ was observed [Knight and Weltner (283)]. In a particular site it appeared that $g_\perp$, which was of the order of 1.82, decreased by about 1.6% from HgH to HgD. This is explained by a small shift in spin density toward the metal atom in the molecule upon deuteration because of a slight shortening of the bond (0.0026 Å). This shift effectively leads to an increase in the matrix element $\langle 6p\sigma | l^x | 6p\pi \rangle$ between the ground $^2\Sigma$ and the excited $^2\Pi$ state and to a larger negative value of $\Delta g_\perp$. The findings are also in accord with an observed increase in $A_\perp$(Hg) when deuterium was substituted.

(β) Spin-Rotation Constant; the Curl Equation. Curl (127) first derived the approximate relationship between the molecular $\boldsymbol{g}$-tensor shifts and the electron spin rotational coupling constants, which for a linear molecule is

$$\gamma = -2B\Delta g_\perp. \qquad \text{(II, 26)}$$

B is here the rotational constant, $\hbar^2/2I$, for the molecule, which to the accuracy of (II, 26) can be taken as the ground-state value. This relationship was also obtained independently by Knight and Weltner (280) from Van Vleck's (442) derivation of γ.

In general, $\boldsymbol{\gamma}$ is an electron spin-rotation tensor contributing a term $\boldsymbol{S} \cdot \boldsymbol{\gamma} \cdot \boldsymbol{N}$, where $\boldsymbol{N}$ is the rotational angular momentum, to the Hamiltonian of the rotating molecule in the gas phase. Molecules in $^2\Sigma$ electronic states exhibit a splitting of their rotational levels given by $\gamma(K + \frac{1}{2})$. γ is very small relative to B and was earlier referred to as the spin-doubling constant. Gas-phase spectroscopic studies on Σ molecules have provided a large number of γ values, particularly for diatomics [Mulliken and Christy (354)].

Table II-1 gives a comparison of γ values observed in the gas-phase with those calculated via (II, 26) from $\Delta g_\perp$ observed in rare-gas matrices at 4°K [Weltner (475)]. The Curl equation appears to be correct to within about 10%. One should also note that there is a variation in sign and that the range in γ is quite large, varying from $+2.1\ cm^{-1}$ (HgH) to $-2.3\ cm^{-1}$ (PdH). [Curl (127) has also obtained good agreement of his more general version of (II, 26) when applied to the bent NO_2 molecule.]

This equation, in the same form, can also be applied to triplet and higher-spin Σ molecules [Weltner (475)], and they have been included in Table II-1 where data are available. We will return to a discussion of these molecules in later chapters.

It is clear from Table II-1 that $\Delta g_\perp$ values can provide reliable values of γ, and in fact, in the case of the C_2H radical, the matrix data have aided in its detection in interstellar space [Tucker, Kutner, and Thaddeus (433)].

Whether from Curl's (127) equation (6a) or from Van Vleck's (280, 442) equation (66), one finds

$$\gamma = 4 \sum_n [\langle 0|\xi l_x|n\rangle\langle n|Bl_x|0\rangle/(E_n - E_0)]$$

Table II-1. Comparison of the Spin-Rotation Constant γ (in cm^{-1}) Obtained from $\Delta g_\perp$ and Directly Observed

Molecule	$\Delta g_\perp$	γ (calc.)	γ (obs.)
		$^2\Sigma$ Molecules	
PdD[a]	+0.290(1)	-2.10(1)	-2.283(3)
RhC[b]	+0.052(1)	-0.062(1)	-0.065(3)
C_2H[c]	+0.0006(1)	-0.0016(2)	-0.00209(3)
CaF[d]	-0.002(1)	+0.0013(1)	+0.001223
AlO[e]	-0.0019(3)	+0.0014(4)	+0.00497(1)
CN[f]	-0.0020(1)	+0.0077(3)	+0.007250(1)
BS[g]	-0.0073(1)	+0.010(1)	+0.013(2)
SrF[h]	-0.004(1)	+0.0020(5)	+0.002503
BaF[i]	-0.006(1)	+0.0026(4)	+0.0028
MgH[j]	-0.002(1)	+0.024(12)	+0.020(5)
CaH[k]	-0.004(1)	+0.034(9)	+0.0424(10)
SrH[l]	-0.015(1)	+0.11(1)	+0.122
BaH[m]	-0.26(1)	+0.17(1)	+0.1927(6)
ZnH[n]	-0.015(1)	+0.21(1)	+0.248
YbH[o]	-0.0621(2)	+0.51(1)	+0.5688(8)
CdH[p]	-0.048(1)	+0.56(1)	+0.59
HgH[q]	-0.178(5)	+2.18(6)	+2.08

Table II-1. (Continued)

Molecule	$\Delta g_\perp$	γ (calc.)	γ (obs.)
		$^3\Sigma$ Molecules	
CCO[r]	+0.0006(4)	−0.0005(4)	−0.0007(3)
C_4[s]	+0.0018(5)	−0.0006(2)	- - - - -
		$^4\Sigma$ Molecules	
VO[t]	−0.0219(10)	+0.0231(10)	+0.02252(7)
NbO[u]	−0.0446(10)	+0.038(1)	- - - - -

[a]Knight and Weltner (281); Malmberg, Scullman, and Nylén (324).
[b]Brom, Graham, and Weltner (88); Kaving and Scullman (267).
[c]Cochran, Adrian, and Bowers (115); Graham, Dismuke, and Weltner (195); Adrian and Bowers (32); Tucker, Kutner, and Thaddeus (433).
[d]Knight, Easley, Weltner, and Wilson (278); Childs, Goodman, and Goodman (109).
[e]Knight and Weltner (284); Mahieu, Jacquinot, Schamps, and Hall (323).
[f]Cochran, Adrian, and Bowers (114); Easley and Weltner (153); Adrian and Bowers (32); Dixon and Woods (145).
[g]Brom and Weltner (89).
[h]Knight, Easley, Weltner, and Wilson (278); Domaille, Steimle, and Harris (147); Childs, Goodman, and Renhorn (110).
[i]Knight, Easley, Weltner, and Wilson (278); Barrow, Bastin, and Longborough (49).
[j]Knight and Weltner (282); Huber and Herzberg (12).
[k]Knight and Weltner (282); Berg and Klynning (57).
[l]Knight and Weltner (282); Huber and Herzberg (12).
[m]Knight and Weltner (282); Kopp, Kronekvist, and Guntsch (291).
[n]Knight and Weltner (283); Mulliken and Christy (354).
[o]Van Zee, Seely, and Weltner (451); Hagland, Kopp, and Åslund (202).
[p]Knight and Weltner (283); Deile (134).
[q]Knight and Weltner (283); Fujioka and Tanaka (175); Mulliken and Christy (354).
[r]Smith and Weltner (421); Devillers and Ramsay (139).
[s]Graham, Dismuke, and Weltner (196).
[t]Kasai (258); Cheung, Hansen, and Merer (108).
[u]Brom, Durham, and Weltner (87).

in the notation of $\Delta g_\perp$ in section (α) above. When B is assumed constant and removed from this summation and a comparison made with (II, 11), the Curl equation in (II, 26) results. The approximate nature of this expression is not due essentially to that assumption but to the neglect of other factors discussed by Curl (127) and Knight and Weltner (280).

(c) Nuclear Hyperfine Interaction (A Tensor)

If the molecule contains one or more nuclei with a magnetic dipole moment, hyperfine splittings of the energy levels may result (even in the absence of a magnetic field) due to the interaction of each such nucleus with the electronic magnetic moment. The strength of the interaction with a particular nucleus then depends not only upon its value of μ_I/I, but also upon the distribution of unpaired electron spin in the molecule. Thus, in the $^{13}C^{14}N\ ^2\Sigma$ molecule, hfs can be observed due to the interaction of the unpaired electron with both the $^{13}C\ (I = \frac{1}{2})$ nucleus and the $^{14}N\ (I = 1)$ nucleus, from which one can deduce that the spin density is about 65% near C and 35% near N [Easley and Weltner (153); Adrian and Bowers (32)]. It is then also understandable, particularly in covalently bounded molecules, that the spin distribution parallel and perpendicular to the axis of the molecule will be different, resulting in the hf constant for a given nucleus having the extreme values $A_\parallel$ and $A_\perp$. Thus, as with the $\boldsymbol{g}$ tensor, each $\boldsymbol{A}_i$ tensor is usually anisotropic, where i indicates the particular magnetic nucleus.

The hyperfine interaction contribution to the Hamiltonian, $\mathcal{H}_{hf}$, is the sum of electron-nuclear magnetic-dipole interactions and Fermi contact terms, as given in (I, 18) to (I, 20) for the atomic case. Then for interaction with one magnetic nucleus,

$$\mathcal{H}_{hf} = g_e \beta_e g_I \beta_n \left\{ \frac{(\boldsymbol{L} - \boldsymbol{S}) \cdot \boldsymbol{I}}{r^3} + \frac{3(\boldsymbol{S} \cdot \boldsymbol{r})(\boldsymbol{r} \cdot \boldsymbol{I})}{r^5} + \frac{8\pi}{3} \delta(r) \boldsymbol{S} \cdot \boldsymbol{I} \right\} \tag{II, 27}$$

The third term is the contact interaction written with a Dirac δ-function to indicate that it has a nonzero value only at the nucleus itself. Frosch and Foley (174), in an important paper, have reduced this Hamiltonian for a diatomic molecule to

$$\mathcal{H}_{hf} = aL_z S_z + b\boldsymbol{I} \cdot \boldsymbol{S} + cI_z S_z \tag{II, 28}$$

where z is the molecular axis. a, b, and c† depend only upon the electron distribution in the molecule relative to the particular nucleus

†Unfortunately, these hf parameters are indicated by the same symbols, a and b, as were used in Chapter I for coefficients in the wavefunctions of quenched atoms. There should, however, be no confusion, since the topics are quite distinct.

under consideration and are given by

$$a = g_e \beta_e g_I \beta_n \langle 1/r^3 \rangle$$

$$b = g_e \beta_e g_I \beta_n \left\{ \frac{8\pi}{3} |\Psi(0)|^2 - \langle (3\cos^2\alpha - 1)/2r^3 \rangle \right\}$$

$$c = 3 g_e \beta_e g_I \beta_n \langle (3\cos^2\alpha - 1)/2r^3 \rangle \qquad \text{(II, 29)}$$

r is the distance of the electron from the nucleus, and α is the angle between $\boldsymbol{r}$ and the axis of the molecule. The bra-kets indicate averages over the coordinates of the electron, for fixed nuclei. Thus the basic quantities determining the hf interaction at the ith nucleus are $\langle r^{-3} \rangle_i$, $|\Psi(0)|_i^2$ and $\langle (3\cos^2\alpha - 1)/r^3 \rangle_i$, which can be calculated if an accurate molecular electronic wavefunction is available. In general the $\boldsymbol{L} \cdot \boldsymbol{I}$ iteraction is small in a Σ molecule and can be neglected. Then only the contact term, $|\Psi(0)|_i^2$, and the dipolar term, $\langle (3\cos^2\alpha - 1)/r^3 \rangle_i$, are required to characterize the hfs, and we write

$$A_{\text{iso}}(i) = \frac{8\pi}{3} g_e \beta_e g_I \beta_n |\Psi(0)|_i^2 = \left(b + \frac{c}{3}\right)_i$$

$$A_{\text{dip}}(i) = g_e \beta_e g_I \beta_n \langle (3\cos^2\alpha - 1)/2r^3 \rangle_i = \left(\frac{c}{3}\right)_i \qquad \text{(II, 30)}$$

at the ith magnetic nucleus.

With the $\boldsymbol{L} \cdot \boldsymbol{I}$ term neglected, the Hamiltonian can also be written, for one magnetic nucleus, as

$$\mathcal{H}_{hf} = \boldsymbol{I} \cdot \boldsymbol{A} \cdot \hat{\boldsymbol{S}} \qquad \text{(II, 31)}$$

and in a linear molecule as

$$\mathcal{H}_{hf} = A_{\parallel} S_z I_z + A_{\perp}(S_x I_x + S_y I_y) \qquad \text{(II, 32)}$$

Comparison of this expression with (II, 28) yields

$$A_{\parallel} = b + c$$

and

$$A_\perp = b \tag{II, 33}$$

so that, from (II, 30), for the ith nucleus

$$A_{\text{iso}}(i) = \left(\frac{A_\parallel + 2A_\perp}{3}\right)_i$$

$$A_{\text{dip}}(i) = \left(\frac{A_\parallel - A_\perp}{3}\right)_i \tag{II, 34}$$

Inverting these equations (and dropping the i) gives

$$A_\parallel = A_{\text{iso}} + 2A_{\text{dip}}$$

$$A_\perp = A_{\text{iso}} - A_{\text{dip}} \tag{II, 35}$$

Then the $\boldsymbol{A}$ tensor, when referred to its principal axes, is

$$\boldsymbol{A} = \begin{vmatrix} A_\perp & 0 & 0 \\ 0 & A_\perp & 0 \\ 0 & 0 & A_\parallel \end{vmatrix} = \begin{vmatrix} A_{\text{iso}} & 0 & 0 \\ 0 & A_{\text{iso}} & 0 \\ 0 & 0 & A_{\text{iso}} \end{vmatrix} + \begin{vmatrix} -A_{\text{dip}} & 0 & 0 \\ 0 & -A_{\text{dip}} & 0 \\ 0 & 0 & +2A_{\text{dip}} \end{vmatrix}$$

or

$$\boldsymbol{A} = A_{\text{iso}}\mathbf{1} + \boldsymbol{A}_{\text{dip}} \tag{II, 36}$$

where $\mathbf{1}$ is the unit tensor, and $\boldsymbol{A}_{\text{dip}}$ is a traceless tensor. The importance of the traceless character of the dipolar contribution is that it implies that when spherical symmetry prevails, as when the molecule is rotating rapidly, or (see section 5a(β) of Chapter I) if the unpaired electron distribution is spherical, $\boldsymbol{A}_{\text{dip}}$ vanishes, and one can observe only isotropic hfs.

(II, 31) and therefore (II, 32), should include small terms arising from $\boldsymbol{L} \cdot \boldsymbol{I}$ interaction, and if these are added [Knight and Weltner (283)],

$$A_\parallel = b + c - \Delta g_\parallel a$$

and

$$A_\perp = b + \Delta g_\perp a \qquad \text{(II, 37)}$$

$\Delta g_\parallel$ is almost always <0.005, but $\Delta g_\perp$ may not be negligible, particularly where heavy atoms are involved. These corrections can be made accurately enough by using the theoretical value of $\langle r^{-3}\rangle$ for the atom to calculate the a parameter from (II, 29). They can lead to small alterations in the values of $|\Psi(0)|^2$ and $\langle(3\cos^2\alpha - 1/r^3\rangle$ derived from (II, 30) when $A_\parallel$ and $A_\perp$ are observed in ESR spectra. For example, they were found to be worth including when analyzing the spectrum of the HgH molecule, where $-\Delta g_\perp \cong 0.18$ (283).

Note that $A_\parallel$ and $A_\perp$ in MHz are connected via $g_\parallel$ and $g_\perp$, respectively, with the measured A values in gauss by

$$A_i\,(\text{MHz}) = (g_e\beta_e/h)(g_i/g_e)A_i\,(\text{G}) = 2.8025(g_i/g_e)A\,(\text{G})$$

It is clear that for more than one magnetic nucleus in the molecule (II, 31) must be written

$$\mathcal{H}_{hf} = \sum_i^N (\boldsymbol{I}_i \cdot \boldsymbol{A}_i \cdot \hat{\boldsymbol{S}}) \qquad \text{(II, 38)}$$

where the index i runs over the N nuclei. Then if the nuclei are inequivalent (as in $^{13}C^{14}N$) the number of possible hfs lines produced from each line in the ESR spectrum, when hf interaction did not occur, would be $(2I_1 + 1)(2I_2 + 1) \cdots (2I_N + 1)$. If the nuclei with spin are equivalent (as in $^{19}F_2^-$), then there will be overlap of some lines in the hfs that will produce a characteristic intensity pattern. Thus for $^{19}F_2^-$ ($I = \frac{1}{2}$), each line in the spectrum would be split by hfs into the following pattern:

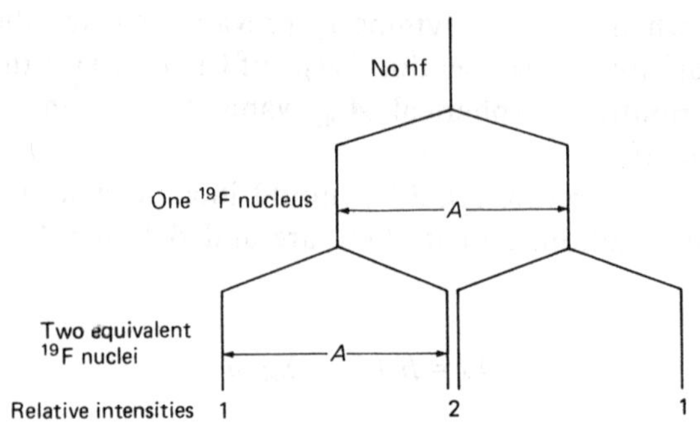

The parameters a, b, and c are those used by gas-phase spectroscopists when their spectra provide hf information [see Barrow (48) and Townes and Schawlow (21)], but they are not commonly used in solid-state work. Equations (II, 30) and (II, 33) are then necessary for interconversion of the two sets of parameters.

(α) Spin Density and Hyperfine Splitting. The electron spin density at nucleus X in a molecule is the unpaired-electron probability density at nucleus X; e.g., for $S = \frac{1}{2}$, it is the fraction of one unpaired electron/cm^3 at nucleus X. More correctly, it is the difference between the total probability densities of all electrons with spin α and with spin β at nucleus X [see, for example, Pople, Beveridge, and Dobosh (382)]. However, in first approximation (where spin polarization effects are ignored) the α and β spins in closed shells cancel, and only the spin density of the unpaired electrons contribute. The spin density can then be divided into the contributions from s, p, d, etc., electrons. For s electrons, the spin density at nucleus X is

$$\rho_{sX} |\chi_{sX}(0)|^2$$

and for $p\sigma$ electrons

$$\rho_{p\sigma X} \langle \chi_{p\sigma X} | (3 \cos^2 \alpha - 1)/r^3 | \chi_{p\sigma X} \rangle$$

with a similar expression for $p\pi$, $d\sigma$, etc. Here ρ_{sX} and $\rho_{p\sigma X}$ signify the unpaired spin population of the valence s and $p\sigma$ orbitals of atom X. The ground state of the molecule is assumed to be represented by a normalized LCAO wavefunction,

$$\Psi(^2\Sigma) = \sum_i a_i \chi_i \qquad \text{(II, 39)}$$

so that

$$\rho_i = a_i^2 \qquad \text{and} \qquad \sum_i \rho_i = 1 \qquad \text{(II, 40)}$$

It is usual to assume when calculating $|\Psi(0)|_X{}^2$ and $\langle (3 \cos^2 \alpha - 1)/r^3 \rangle_X$, as they appear in (II, 29) and (II, 30), that contributions from AO's not at atom X can be neglected, so that

$$|\Psi(0)|_X{}^2 = \rho_{sX} |\chi_{sX}(0)|^2$$

and

$$\langle(3\cos^2\alpha - 1)/r^3\rangle_X = \rho_{p\sigma X}\langle\chi_{p\sigma X}|(3\cos^2\alpha - 1)/r^3|\chi_{2p\sigma X}\rangle \tag{II, 41}$$

Then the isotropic and anisotropic hyperfine interaction parameters at a particular nucleus X in the molecule can be written

$$A_{\text{iso}}{}^X \text{ (molecule)} = \frac{8\pi}{3} g_e\beta_e g_I\beta_n \rho_{sX}|\chi_{sX}(0)|^2$$

and

$$A_{\text{dip}}{}^X \text{ (molecule)} = g_e\beta_e g_I\beta_n \rho_{p\sigma X}\langle\chi_{p\sigma X}|(3\cos^2\alpha - 1)/2r^3|\chi_{2p\sigma X}\rangle$$

But the integrals involved in these expressions are characteristic of atom X so that

$$A_{\text{iso}}{}^X \text{ (molecule)} = \rho_{sX} \cdot A_{\text{iso}}{}^X \text{ (atom)}$$

and

$$A_{\text{dip}}{}^X \text{ (molecule)} = \rho_{p\sigma_X} \cdot A_{\text{dip}}{}^X \text{ (atom)} \tag{II, 42}$$

or, conversely, the unpaired electron spin population can be obtained from

$$\rho_{sX} = \frac{A_{\text{iso}}{}^X \text{ (molecule)}}{A_{\text{iso}}{}^X \text{ (atom)}}$$

and

$$\rho_{2p\sigma X} = \frac{A_{\text{dip}}{}^X \text{ (molecule)}}{A_{\text{dip}}{}^X \text{ (atom)}} \tag{II, 43}$$

Assuming that the A^X (atom) values are known or can be calculated, then the ρ_i will be given by (II, 43) using A^X (molecule) parameters obtained from the observation of hfs in molecular spectra.

(β) $A_{iso}{}^X$ (Atom) and $A_{dip}{}^X$ (Atom) Values. The atomic hf parameters for one s, p, d, or f electron interacting with the nucleus have already been discussed, in Chapter I, section 4, and are tabulated in Appendix B. $A_{iso}{}^X$ (atom) for a single ns electron is given there in column 5, Table B1. However, since A_{dip} is given by (II, 30), $A_{dip}{}^X$ (atom) is obtained from the free atom value of $P = g_e \beta_e g_I \beta_n \langle r^{-3} \rangle$, given in column 7, by multiplying by an angular factor $\bar{\alpha}/2 = \langle (3\cos^2\alpha - 1)/2 \rangle$. This factor is $\frac{2}{5}$ for a p electron, $\frac{2}{7}$ for a d electron, and $\frac{4}{15}$ for an f electron. [Note that if A_{iso} or A_{dip} is desired for an isotope missing in the table (such as ^{10}B) it may be found by correcting the A value for the isotope listed there (^{11}B) by use of the ratio of γ_n values for the two isotopes given in Appendix A.]

The angular part of A_{dip} may be separated from $\langle r^{-3} \rangle$ and evaluated for a single p, d, or f electron. For example, for a $p\sigma$ electron ($l = 1$, $m_l = 0$),

$$\frac{\bar{\alpha}}{2} = \langle (3\cos^2\alpha - 1)/2 \rangle = \frac{1}{2} \langle P_l^{m_l} | 3\cos^2\alpha - 1 | P_l^{m_l} \rangle$$

where $P_l^{m_l}$ is the angular part of the atomic wavefunction [see, for example, Kopfermann (14), page 153]. Then

$$\frac{\bar{\alpha}}{2} = -\frac{1}{2}\left[\frac{3m_l^2 - l(l+1)}{l(2l-1)} \cdot \frac{2l}{2l+3}\right] \tag{II, 44}$$

which for $l = 1$ is

$$\frac{\bar{\alpha}}{2} = -\frac{1}{5}(3m_l^2 - 2)$$

and for $m_l = 0$ is

$$\frac{\bar{\alpha}}{2} = \frac{2}{5}$$

For a $2p\pi$ electron, $m_l = \pm 1$ and $\bar{\alpha}/2 = -\frac{1}{5}$ [see, for example, Dousmanis (148)].

These atomic hf parameters are calculated for the free atoms, and their use in deriving spin densities in molecules is fraught with problems. This is particularly true if the bonding in the molecule is highly ionic, since it is obvious in such a case that A_{iso} and A_{dip} values for the ions and not the atoms should be used in calculating the spin density distribution. For example, the $^2\Sigma$ molecule SrF is an ion-pair Sr^+F^- with the odd electron essentially localized on the Sr^+ ion, and by substitution of the magnetic nucleus ^{87}Sr the metal hfs can be measured (see section II-6). In calculating $\rho_s(Sr)$ from (II, 43) it is then correct to use $A_{iso}(^{87}Sr^+) = 992$ MHz, known from gas-phase measurements, rather than the value of $A_{iso}(^{87}Sr) = 853.6$ MHz from Appendix B. The difference between these A_{iso} values is due to the increased coulombic attraction at the nucleus in the positive ion; the opposite effect occurs when an anion is formed.

In general, partial ionic character presents the real problem in the use of (II, 43). This can be estimated for a given bond as one-half the electronegativity difference between the atoms [Gordy (8)] or from a graph given by Townes and Schawlow [(21), page 236; a table of electronegativities is also given there on page 237]. Thus, formal charges on the two bonded atoms can be approximated by setting them equal to the ionic character derived in this way. If the hf parameters for the atomic ions were known, one might then estimate more realistically the $A_{iso}{}^X$ and $A_{dip}{}^X$ values to be used in the denominator of (II, 43).

For p orbitals Gordy [(8), page 256] proposes an approximate correction to $A_{dip}{}^X$ (atom) for this ionic character. He uses the formal charge concept introduced by Townes and Schawlow for nuclear quadrupole coupling [(21), pages 234ff.], since in both parameters effective $\langle 1/r^3 \rangle$ values are involved. Then, in our symbolism, he writes for a positively charged or negatively charged atom, respectively,

$$\rho_{2p\sigma_X}{}^+ = \frac{A_{dip}{}^X\ \text{(molecule)}}{A_{dip}{}^X\ \text{(atom)} \cdot (1 + c^+\epsilon)}$$

$$\rho_{2p\sigma_X}{}^- = \frac{A_{dip}{}^X\ \text{(molecule)} \cdot (1 + c^-\epsilon)}{A_{dip}{}^X\ \text{(atom)}} \qquad \text{(II, 43a)}$$

instead of the relationship in (II, 43). ϵ values for various atoms are taken from the tabulation of Townes and Schawlow [(21), Table 9-5, page 239], and the net charge c^+ or c^- for cation or anion, respectively, is equated to the ionic character of the bond, as discussed above.

One then sees that the use of (II, 43) to find the unpaired spin distribution in a molecule is very approximate. It is most useful in organic molecules where empirical $A_{iso}{}^X$ (atom) values are derived by fitting many hfs observations to theoretical spin densities (382).

(γ) Interatomic Dipole–Dipole Interaction. There is an additional small contribution to the anisotropic hyperfine splittings in molecules from magnetic dipole–dipole interactions between a nuclear moment and unpaired electron spin on a *neighboring* atom. Since the interaction decreases as $1/r^3$, this contribution is never very large, but it can be relatively significant when other hyperfine effects are small. For example, the hyperfine interaction between a $p\pi$ electron on a carbon atom and an adjacent proton is isotropic and is caused by spin polarization [McConnell and Chesnut (335)]. It produces a small hfs of about 23 G [see page 141]. Dipole–dipole interaction provides an anisotropic contribution that has a maximum of about one-half of that value [McConnell and Strathdee (337); Gordy (8), pages 206–216; Goodman and Raynor (190), pages 170–173]. Also, in very ionic fluoride molecules where the electron spin is on an adjacent metal ion, this could be a significant contribution to $A(^{19}F)$, since, to a first approximation, the spin density on the F^- ion is zero. However, even here it turns out that the dipole–dipole contribution is small relative to that due to spin polarization and covalency. Thus in the CuF_2 molecule were $r \simeq 2$ Å the contribution to $A(^{19}F) = 115$ G from dipole–dipole interaction is calculated to be only about 7 G [Kasai, Whipple, and Weltner (266)]. On the other hand, for the diatomic alkaline-earth hydrides, BeH and MgH [see section II (6)], Gordy [(8), p. 383)] has calculated the opposite interaction, i.e., the dipole–dipole contribution to the metal anisotropic hfs. Thus with a spin density of about 0.14 on H, the dipole–dipole interaction at ^{9}Be is found to be -0.5 MHz. This amounts to about 10% of $A_{dip}(^9Be)$, the remainder being due to $2p\sigma$ electron character on that atom.

The contribution to A_{dip} from this interatomic dipole–dipole coupling is simply $A_{dip}{}' = g_e\beta_e g_I\beta_n \langle(3\cos^2\theta - 1)/2r^3\rangle$ from (II, 30), where r is the "effective" distance from the electron magnetic dipole to that of the adjacent nucleus. If the electron is in an $s\sigma$ or $p\sigma$ orbital, then the interaction will have cylindrical symmetry and $(A_{dip}{}')_\perp = -\rho g_e\beta_e g_I\beta_n(1/R^3) = -(\frac{1}{2})(A_{dip}{}')_\parallel$, where R is the bond length and ρ the spin density. For $p\pi$ (or $d\pi$, $d\delta$) the spin density is not on the axis, so that Gordy [(8), page 207] takes the effective off-axis distance as R_p and then derives the anisotropic components as approximately:

$$(A_{\text{dip}}')_x = +\rho_\pi g_e \beta_e g_I \beta_n (2R_p^2 - R^2)/(R^2 + R_p^2)^{5/2}$$

in the $p\pi$ orbital direction,

$$(A_{\text{dip}}')_y = -\rho_\pi g_e \beta_e g_I \beta_n/(R^2 + R_p^2)^{3/2}$$

$\perp$ to the bond and to the $p\pi$ orbital,

$$(A_{\text{dip}}')_z = +\rho_\pi g_e \beta_e g_I \beta_n (2R^2 - R_p^2)/(R^2 + R_p^2)^{5/2}$$

along the bond,

so that $(A_{\text{dip}}')_x + (A_{\text{dip}}')_y + (A_{\text{dip}}')_z = 0$. R_p may be estimated from measurements of anisotropic hfs (as in CH groups containing $p\pi$ electrons) or perhaps from electron radial distributions.

(δ) Signs of Molecular A Components. In ESR experiments in matrices only the absolute values of $A_\parallel$ and $A_\perp$ are determined. (Signs have been obtained in other media from other effects, such as linewidth variation [Hudson and Luckhurst (234)] or nuclear electric quadrupole perturbations [Bleaney (70)].) However, unless the hf splitting is small, the signs of the $\boldsymbol{A}$ components will usually be positive, *for positive* μ_I. This rather too general statement is derived from (II, 35) and the fact noted in Chapter I that A_{iso} (atom) is always more than an order of magnitude larger than A_{dip} (atom) (see Appendix B). Some verification of the latter may be made by referring to Appendix C where the data for most $^2\Sigma$ molecules are collected. It is true in spite of the fact that the hfs is, of course, modified by how the spin is distributed among the AO's.

Exceptions to this are the B, Al, and Ga atoms quenched in the axial crystal fields provided by the solid rare-gas sites, where their ESR spectra are analogous to those of $^2\Sigma$ molecules. In these three atoms, the odd electron is largely in a $p\sigma$ orbital so that $A_{\text{iso}} \cong 0$ and A_{dip} is positive; consequently, from (II, 36), $A_\perp$ is negative and $A_\parallel$ is positive, as was shown in Table I-3.

The hf interaction at ^{14}N in the CN radical is small, only 5 to 10 G, and from the observation that $|A_\parallel| < |A_\perp|$ and a direct measurement of $|A_{\text{iso}}| = 4.5$ G, it can be deduced that $|A_\parallel| = 6.5$ G and $|A_{\text{dip}}| = 5.5$ G [Easley and Weltner (153)]. Then since the major portion of the nitrogen spin density is in the $2p\sigma$ orbital, it follows that A_{dip} is positive, and therefore A_{iso} has to be negative to account for the observed value of $A_\perp$ (10 G). A_{iso} is then attributed predominantly to spin polarization of the inner s orbitals on the N atom.

A borderline case is in the ion-pair $M^+Cl_2^-$, where the unpaired electron is in an antibonding σ_u orbital on Cl_2^-. This orbital is an sp hybrid in which the s character is only about 2% [Schoemaker (406, 407, 408); Martinez and Weltner (326)]. This amount of s character makes $A_{iso}(^{35}Cl)$ and $A_{dip}(^{35}Cl)$ in the molecule comparable in size, so that $A_\parallel >> A_\perp$ (see also section II-3a). The small value of $A_\perp$ (~3 MHz) could be negative, but the overall evidence indicates that it is probably positive [Schoemaker (406, 408)]. However, generally if an $\boldsymbol{A}$ component is quite small, spin polarization may be large enough to affect the sign, and it is difficult to predict whether it will be negative or positive.

(ϵ) Nuclear Zeeman and Quadrupole Terms. The interaction between the external magnetic field and the nuclear magnetic moment is small relative to the electronic Zeeman energy, or usually even to the hf interaction, essentially because the nuclear magneton, β_n, is ~1/2000 of the Bohr magneton. (For example, if $\mu_I = 1\ \beta_n$ and $H = 3$ kg, then $\mu_I H < 10^{-4}$ cm^{-1}.) Its contribution to the Hamiltonian in (II, 1) is given by

$$\mathcal{H}_{Zn} = -\beta \boldsymbol{H} \cdot \boldsymbol{g}^{(I)} \cdot \boldsymbol{I} \qquad \text{(II, 45)}$$

where $\boldsymbol{g}^{(I)}$ is, like $\boldsymbol{g}$, a tensor due to the external field affecting anisotropically the electronic field at the nucleus [Abragham and Bleaney (1), p. 167; Low (15), p. 60]. If $\boldsymbol{g}^{(I)}$ may also be assumed axial, then

$$\mathcal{H}_{Zn} = -\beta[g_\parallel{}^{(I)} H_z I_z + g_\perp{}^{(I)}(H_x I_x + H_y I_y)]$$

However, because it is a small term, $\boldsymbol{g}^{(I)}$ may be taken to an adequate approximation as an isotropic $g^{(I)}$, and only the diagonal matrix elements need be considered. Then, with Abragham and Pryce (27)

$$\mathcal{H}_{Zn} = -\gamma_n \beta_n \boldsymbol{H} \cdot \boldsymbol{I} = -\frac{\mu_I}{I} \beta_n H I_z \qquad \text{(II, 46)}$$

for each magnetic nucleus in the molecule.

When $I \geqslant 1$, a nucleus has an electric quadrupole moment, Q, that can interact with the atomic electrons according to the expression given in (I, 14). In anticipation of the interest here in linear molecules, it was noted in (I, 16) that, to a sufficient approximation for such a small term,

$$\mathcal{H}_{\text{quad}} = Q'[I_z{}^2 - \tfrac{1}{3} I(I+1)] \qquad \text{(II, 47)}$$

where $Q' = 3eQq/4I(2I - 1)$. Q' is called the nuclear electric quadrupole coupling constant, and q is the gradient of the electric field in the molecule acting at the particular nucleus, $\partial^2 V/\partial z^2$. The unknown in this expression is usually q, which is difficult to calculate but can be estimated in some cases [Townes and Dailey (431)].

$\mathcal{H}_{Zn}$ and $\mathcal{H}_{\text{quad}}$ are to be added to the Hamiltonian in (II, 32), which with (II, 19) completes the spin Hamiltonian for the linear molecule.

The electronic potential at the nucleus due to charge e at a distance r is

$$V = \frac{e}{r} = \frac{e}{x^2 + y^2 + z^2}$$

Then

$$q = \frac{\partial^2 V}{\partial z^2} = e\left(\frac{3z^2 - r^2}{r^5}\right) = e\left(\frac{3\cos^2\alpha - 1}{r^3}\right)$$

where α is the angle between the z axis and the radius vector $\boldsymbol{r}$. For an atom with one valence electron (spherical shells do not contribute)

$$q = e\langle(3\cos^2\alpha - 1)/r^3\rangle \qquad \text{(II, 48)}$$

and if the electron is in a central field

$$q_J = -\frac{2le}{2l+3}\left\langle\frac{1}{r^3}\right\rangle$$

where l is the orbital angular momentum. This neglects polarization of inner shells by the electron, so that it may be in error by about 10% for actual atoms.

In molecules [see Townes and Dailey (431)], an LCAO wavefunction can be used in (II, 48), and then q will contain terms $q_{nlm,nlm}$ and $q_{nlm,n'l'm'}$ with nlm indicating the quantum numbers of a particular AO. Because of the $\langle r^{-3}\rangle$ factor, contributions to q from neighboring atoms are usually negligible. Furthermore, s electrons do not contribute, so that in most cases the major contributors to q are the lowest-energy atomic p wavefunctions, since q decreases rapidly with n and l because the electron then spends less time close to the nucleus. However, in the calculation of q (molecule) from (II, 48) all valence electrons can contribute, and not just unpaired electrons as in the hf A_{dip} parameter obtained from (II, 30). Hence q at each nucleus in a $^1\Sigma$ molecule is generally nonzero. Note that in a completely ionic molecule such as Na^+Cl^-, the closed shells about the two nuclei imply that, to a first approximation, $q_{\text{Na}} = q_{\text{Cl}} = 0$. This has formed a basis for estimating the ionic character in diatomic molecules from the measurements of quadrupole coupling constants of their nuclei. An excellent discussion of this subject is given by Townes and Schawlow [(21), Chapter 9].

(d) Energy Levels in an External Magnetic Field

So far in this chapter we have been concerned with the spin Hamiltonian and with the parameters $\mathbf{g}, \boldsymbol{A}, g_I$, and Q' contributing to that Hamiltonian. We still must calculate how the energy levels of a linear $S = \frac{1}{2}$ molecule are affected by the strength and orientation of a homogeneous magnetic field.

For a linear molecule the spin Hamiltonian including hyperfine interactions, but temporarily neglecting nuclear quadrupole and nuclear Zeeman terms, is given by the sum of (II, 19) and (II, 32):

$$\mathcal{H} = g_{\parallel} \beta H_z S_z + g_{\perp} \beta (H_x S_x + H_y S_y) + A_{\parallel} I_z S_z + A_{\perp} (I_x S_x + I_y S_y) \tag{II, 49}$$

Here the x, y, and z axes are fixed in the molecule with z being the axis of symmetry. If the external magnetic field is relatively strong, then the axis of that field will be the desired axis of quantization for the Zeeman term, so that we wish to change to another set of axes x', y', z', such that z' is parallel to H and

$$\mathcal{H}_{Ze} = g\beta H S_z' \tag{II, 50}$$

The direction of H can be taken as the polar axis and θ the angle between z and H. y can be arbitrarily chosen perpendicular to H so that $y = y'$ and only x and z need be rotated [Abragham and Bleaney (1), p. 173]. Since $H_x = H \sin\theta$, $H_z = H \cos\theta$, and $H_y = 0$, then

$$\mathcal{H}_{Ze} = \beta [g_{\parallel} \cos\theta S_z + g_{\perp} \sin\theta S_x] H \tag{II, 50a}$$

If new direction cosines are chosen,

$$g_z = g_\parallel \cos\theta/g, \qquad g_x = g_\perp \sin\theta/g$$

where

$$g = (g_\parallel{}^2 \cos^2\theta + g_\perp{}^2 \sin^2\theta)^{1/2} \tag{II, 51}$$

then

$$\mathcal{H}_{Ze} = g\beta[q_z S_z + q_x S_x]H$$

This rotation of the electron coordinates allows us to write

$$\mathcal{H}_{Ze} = g\beta H S_z{}'$$

where

$$\begin{aligned} S_z{}' &= q_z S_z + q_x S_x \\ S_x{}' &= -q_x S_z + q_z S_x \\ S_y &= S_y \end{aligned} \tag{II, 52}$$

When the hyperfine terms in (II, 49) are included, then the nuclear coordinate system (assumed axial) must, correspondingly, be rotated relative to the original coordinates in order to eliminate the term involving $I_x S_z{}'$ [just as (II, 52) were chosen to eliminate S_x in (II, 50a)]. Then, choosing direction cosines $q_i{}^n$ such that

$$\begin{aligned} I_z &= q_z{}^n I_z{}' - q_x{}^n I_x{}' \\ I_x &= q_x{}^n I_z{}' + q_z{}^n I_x{}' \\ I_y &= I_y{}' \end{aligned} \tag{II, 53}$$

and defining

$$\begin{aligned} q_z{}^n &= A_\parallel g_\parallel \cos\theta/Kg \\ q_x{}^n &= A_\perp g_\perp \sin\theta/Kg \end{aligned}$$

where

$$K^2 g^2 = A_\parallel{}^2 g_\parallel{}^2 \cos^2\theta + A_\perp{}^2 g_\perp{}^2 \sin^2\theta \tag{II, 54}$$

(II, 49) becomes

$$\mathcal{H} = g\beta H S_z' + K I_z' S_z' + A_{\parallel} A_{\perp} I_x' S_x' + \frac{A_{\perp}^2 - A_{\parallel}^2}{K} \frac{g_{\parallel} g_{\perp}}{g^2} \sin\theta \cos\theta I_z' S_x' + A_{\perp} I_y' S_y' \quad \text{(II, 55)}$$

Dropping the primes and substituting the raising and lowering operators for S_x, I_x, etc., (II, 55) finally becomes

$$\mathcal{H} = g\beta H S_z + K S_z I_z + \frac{A_{\perp}^2 - A_{\parallel}^2}{K} \cdot \frac{g_{\parallel} g_{\perp}}{g^2} \cdot \cos\theta \sin\theta \left(\frac{S_+ + S_-}{2}\right) I_z$$
$$+ \left(\frac{A_{\parallel} A_{\perp}}{4K} - \frac{A_{\perp}}{4}\right)(S_+ I_+ + S_- I_-) + \left(\frac{A_{\parallel} A_{\perp}}{4K} + \frac{A_{\perp}}{4}\right)(S_+ I_- + S_- I_+) \quad \text{(II, 56)}$$

For convenience, let

$$X_+ = \frac{A_{\parallel} A_{\perp}}{4K} + \frac{A_{\perp}}{4}$$

$$X_- = \frac{A_{\parallel} A_{\perp}}{4K} - \frac{A_{\perp}}{4}$$

Note that g and K are given by (II, 51) and (II, 54). If the electron spin states are indicated by the quantum number m_S, where $m_S = -S, \cdots, +S$, and the nuclear spin states by $m_I = -I, \cdots, +I$, then the only nonvanishing matrix elements are

$$\langle m_S | S_z | m_S \rangle = m_S, \qquad \langle m_I | I_z | m_I \rangle = m_I$$

$$\langle m_S \pm 1, m_I | S_{\pm} I_z | m_S, m_I \rangle = m_I [S(S+1) - m_S(m_S \pm 1)]^{1/2}$$

$$\langle m_S \pm 1, m_I \pm 1 | S_{\pm} I_{\pm} | m_S, m_I \rangle = [S(S+1) - m_S(m_S \pm 1)]^{1/2} \cdot [I(I+1) - m_I(m_I \pm 1)]^{1/2}$$

$$\langle m_S \pm 1, m_I \mp 1 | S_{\pm} I_{\mp} | m_S, m_I \rangle = [S(S+1) - m_S(m_S \pm 1)]^{1/2} \cdot [I(I+1) - m_I(m_S \mp 1)]^{1/2}$$

We are often only interested in either $\theta = 0°$ or $90°$, and for those two angles the $\cos\theta \sin\theta$ term in (II, 56) vanishes. Let us consider a specific case $S = \frac{1}{2}$, $I = 1$; then the general matrix for the spin states, excluding $\cos\theta \sin\theta$ terms, with $G = \frac{1}{2} g\beta H$, is

	$\vert\frac{1}{2}, 1\rangle$	$\vert\frac{1}{2}, 0\rangle$	$\vert-\frac{1}{2}, 1\rangle$	$\vert\frac{1}{2}, -1\rangle$	$\vert-\frac{1}{2}, 0\rangle$	$\vert-\frac{1}{2}, -1\rangle$
$\vert\frac{1}{2}, 1\rangle$	$G + \frac{K}{2}$				$\sqrt{2}X_-$	
$\vert\frac{1}{2}, 0\rangle$		G	$\sqrt{2}X_+$			$\sqrt{2}X_-$
$\vert-\frac{1}{2}, 1\rangle$		$\sqrt{2}X_+$	$-G - \frac{K}{2}$			
$\vert\frac{1}{2}, -1\rangle$				$G - \frac{K}{2}$	$\sqrt{2}X_+$	
$\vert-\frac{1}{2}, 0\rangle$	$\sqrt{2}X_-$			$\sqrt{2}X_+$	$-G$	
$\vert-\frac{1}{2}, -1\rangle$		$\sqrt{2}X_-$				$-G + \frac{K}{2}$

Notice that rows and columns with the same values of $m_S + m_I$ have been placed adjacently in this matrix. The eigenvalues are then the roots of this secular determinant.

If the molecule is oriented such that its axis is parallel to the magnetic field H, this matrix becomes even simpler. Then $\theta = 0°$, $g = g_{\parallel}$, $K = A_{\parallel}$, $X_+ = A_{\perp}/2$, and $X_- = 0$, so that the determinant can be solved exactly, since it breaks down into two 1×1 and two 2×2 matrices. For H perpendicular to $z(\theta = 90°)$, $g = g_{\perp}$, $K = A_{\perp}$, $X_+ = (A_{\parallel} + A_{\perp})/4$ and $X_- = (A_{\parallel} - A_{\perp})/4$, so that an exact solution is more difficult. However, sometimes X_- is small and may be neglected, and the eigenvalues are then again easily found.

Usually $g\beta H >> A_{\parallel}$ and $A_{\perp}$, so that second-order perturbation will provide the desired accuracy in the solution of such secular determinants. In a classic paper, Bleaney (70) [see also Low (15)] has worked out the perturbation-theory solution to (II, 49) but including also the nuclear Zeeman and quadrupole terms. The derived energy is, for $S = \frac{1}{2}$ (since D would also enter for $S \geqslant 1$):

$$E(m_S, m_I) = g\beta H m_S + K m_I m_S + \frac{A_\parallel A_\perp^2}{2Kg\beta H}[m_S^2 - S(S+1)]\, m_I$$
$$+ \frac{(A_\parallel^2 + K^2)}{4K^2 g\beta H} A_\perp^2 m_S [I(I+1) - m_I^2]$$
$$- \frac{(A_\parallel^2 - A_\perp^2)^2}{2K^2 g\beta H}\left(\frac{g_\parallel g_\perp}{g^2}\right)^2 \sin^2\theta \cos^2\theta\, m_S m_I^2$$
$$+ Q'\left[m_I^2 - \frac{1}{3} I(I+1)\right]$$
$$- \frac{Q'^2 \cos^2\theta \sin^2\theta}{2Km_S}\left(\frac{A_\parallel A_\perp g_\parallel g_\perp}{K^2 g^2}\right)^2$$
$$\cdot\, m_I[4I(I+1) - 8m_I^2 - 1]$$
$$+ \frac{Q'^2 \sin^4\theta}{8Km_S}\left(\frac{A_\perp g_\perp}{Kg}\right)^4 m_I[2I(I+1) - 2m_I^2 - 1]$$
$$- \frac{\mu_I}{I}\beta_n H m_I \qquad \text{(II, 57)}$$

Note that, except for $g\beta H m_S$, the first-order hyperfine term $K m_I m_S$ is dominant here. It is orientation-dependent because K is a function of θ.

Zeldes, Trammell, Livingston, and Holmberg (484) and Blinder (75) have derived the first-order contribution to (II, 57), assuming an isotropic $\boldsymbol{g}$ tensor, in the form:

$$E(m_S, m_I) = g\beta H m_S + [(A_{\text{iso}} - A_{\text{dip}})^2 + 3A_{\text{dip}}(2A_{\text{iso}} + A_{\text{dip}})\cos^2\theta]^{1/2} m_S m_I \qquad \text{(II, 58)}$$

Their symbols have been changed to those used here, i.e., $A_0 = A_{\text{iso}}$ and $B = A_{\text{dip}}$. [That this form corresponds to the first two terms in (II, 57) may be seen by recognizing through (II, 54) that the quantity in square brackets is K^2.] (II, 58) shows explicitly the approximate angular variation of the energy levels as a function of the s and/or p character of the unpaired electron. In particular if $A_{\text{iso}} = 0$, as when the electron is in a pure p orbital, the hfs varies as $A_{\text{dip}} \cdot (1 + 3\cos^2\theta)^{1/2} m_S m_I$. This corresponds to the case where the hyperfine magnetic field at the nucleus is larger than the applied field. When $A_{\text{iso}} \gg A_{\text{dip}}$, i.e., large reinforcement of the applied field occurs due to the isotropic field at the nucleus, then the square root in (II, 58) can be expanded to obtain the hf energy as $[A_{\text{iso}} - A_{\text{dip}}(1 - 3\cos^2\theta)]\, m_S m_I$. The difference between these two variations of the anisotropic hf with θ then reflects the relative magnitudes of the applied field and the hyperfine magnetic field at the nucleus. [See Goodman and Raynor (190), pages 150–152; Wertz and Bolton (23), pages 140–144; Atkins and Symons (2), pages 17-18.]

(e) ESR Spectroscopy

The heart of the ESR spectrometer is the microwave cavity shown in Fig. II-1 [see Poole (18)]. In it, standing microwaves of fixed frequency ν provide the oscillating magnetic field H_1 that induces magnetic dipole transitions in the sample. It is essential that the slowly varying field H (called the static or dc field) which produces the Zeeman effect, is at a right angle to H_1. This may be shown by calculating the transition probability between spin states as a function of the angle between $\boldsymbol{H}_1$ and $\boldsymbol{H}$.

(α) Selection Rules and Transition Probabilities. For a magnetic dipole transition the probability is proportional to

$$P = |\langle m_S, m_I | \boldsymbol{\mu} \cdot \boldsymbol{H}_1 | m_S', m_I' \rangle|^2 \qquad \text{(II, 59)}$$

where $\boldsymbol{\mu}$ is essentially $-g\beta\boldsymbol{S}$ if we assume an isotropic $\boldsymbol{g}$ tensor. If H_1 were parallel to H in Fig. II-1, then

$$\boldsymbol{\mu} = -g\beta S_z$$

and since $S_z|m_S\rangle = m_S|m_S\rangle$, P will be nonvanishing only if $m_S = m_S'$, $m_I = m_I'$, which is no transition at all. At the other extreme, if H_1 is perpendicular to H (as in Fig. II-1), then

$$\boldsymbol{\mu} = -g\beta S_x$$

and since $S_X|m_S\rangle = \frac{1}{2}\,[\,|m_S + 1\rangle + |m_S - 1\rangle]$, P will be nonvanishing only if $m_S' = m_S \pm 1$, $m_I = m_I'$ or $\Delta m_S = \pm 1$, $\Delta m_I = 0$. These the reader will recognize as the customary selection rules for allowed transitions in ESR. Since only one or the other of these transitions can occur, the relative intensities of transitions in the rectangular cavity oriented as in Fig. II-1 are found from

$$P = \tfrac{1}{4}\, g_1{}^2 \beta^2 H_1{}^2 |\langle m_S \mp 1, m_I | S_\mp | m_S, m_I \rangle|^2 \qquad \text{(II, 60)}$$

where $g_1{}^2 = g^2$ in the isotropic case.

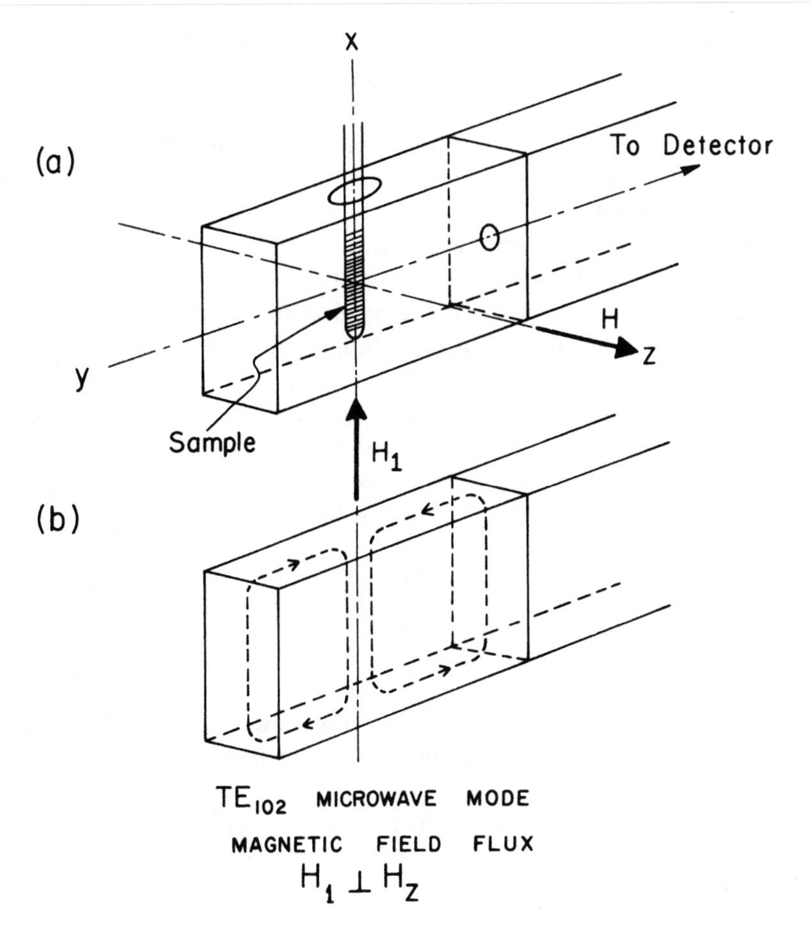

Fig. II-1. Rectangular ESR cavity and orientation of static (H) and microwave (H_1) magnetic fields. (a) Configuration of sample and usual $H_1 \perp H$ orientation. (b) Magnetic flux in the TE_{102} microwave mode.

If the $\mathbf{g}$ tensor is anisotropic, then the g_1 factor is a function of the orientation of the molecule in the oscillating field H_1 [Bleaney (71)]. If H_1 is perpendicular to H, as in Fig. II-1, and the $\mathbf{g}$ tensor has axial symmetry,

$$g_1{}^2 = \left(\frac{g_x g_z}{g}\right)^2 \sin^2 \eta + g_y{}^2 \cos^2 \eta \qquad \text{(II, 61)}$$

where $g^2 = g_z{}^2 \cos^2\theta + g_x{}^2 \sin^2\theta$ as in (II, 51). η is the angle between $\boldsymbol{H}_1$ and the principal y axis of $\boldsymbol{g}$, where z is the axis of symmetry. If now the angle η is changed without changing the other conditions [i.e., keeping $H_1 \perp H$ and H at a fixed angle θ to z, turning the linear molecule about its z axis such that the orthogonal $\boldsymbol{g}$ tensor axes x and y rotate], a maximum value of g_1 can be found. In such a single crystal rotation the value of η at which g_1 will be a maximum will depend upon whether $|g_y|$ is greater than or less than $|g_x g_z/g|$, as seen from (II, 61). Since for an axial $\boldsymbol{g}$ tensor $g_x = g_y = g_\perp$ and $g_z = g_\parallel$, if the molecule has $|g_\perp| > |g_\parallel|$, g_1 will be a maximum when $\eta = 0$. However, if $|g_\perp| < |g_\parallel|$, g_1 will be a maximum when $\eta = \pi/2$.

Of more importance to us here is the case of an assembly of randomly oriented molecules, as in a matrix or powder. Then all values of η are equally probable and when averaged $\langle\sin^2\eta\rangle = \langle\cos^2\eta\rangle = \frac{1}{2}$, so that an average value of $g_1{}^2$ from (II, 61) is

$$g_1{}^2 = \frac{1}{2} g_\perp{}^2 \left[\left(\frac{g_\parallel}{g}\right)^2 + 1\right] \qquad \text{(II, 62)}$$

for $H_1 \perp H$.

(β) Resonant Magnetic Fields for Allowed Transitions ($\Delta m_S = \pm 1$, $\Delta m_I = 0$). For the allowed transitions $|m_S, m_I\rangle \leftrightarrow |m_S - 1, m_I\rangle$,

$$h\nu = E(m_S, m_I) - E(m_S - 1, m_I) = g\beta H_0;$$

then from (II, 57)

$$\begin{aligned} H = H_0 &- Km_I - \frac{A_\parallel A_\perp{}^2}{2Kg\beta H_0}(2m_S - 1)\,m_I \\ &- \frac{(A_\parallel{}^2 + K^2)}{4K^2 g\beta H_0} A_\perp{}^2\,[I(I+1) - m_I{}^2] \\ &- \frac{(A_\parallel{}^2 - A_\perp{}^2)^2}{2K^2 g\beta H_0}\left(\frac{g_\parallel g_\perp}{g^2}\right)^2 \sin^2\theta\cos^2\theta m_I{}^2 \\ &- \frac{Q'^2\cos^2\theta\sin^2\theta}{2Km_S(m_S - 1)}\left(\frac{A_\parallel A_\perp g_\parallel g_\perp}{K^2 g^2}\right)^2 m_I[4I(I+1) - 8m_I{}^2 - 1] \\ &+ \frac{Q'^2\sin^4\theta}{8Km_S(m_S - 1)}\left(\frac{A_\perp g_\perp}{Kg}\right)^4 m_I[2I(I+1) - 2m_I{}^2 - 1] \qquad \text{(II, 63)} \end{aligned}$$

Or, if all coupling constants are expressed in gauss (G),

$$A_\parallel\,(\mathrm{G}) = \frac{A_\parallel\,(\mathrm{ergs})}{g_\parallel \beta}, \quad A_\perp\,(\mathrm{G}) = \frac{A_\perp\,(\mathrm{ergs})}{g_\perp \beta}, \quad K\,(\mathrm{G}) = \frac{K\,(\mathrm{ergs})}{g\beta},$$

$$Q'\,(\mathrm{G}) = \frac{Q'\,(\mathrm{ergs}),}{g_\perp \beta}, \quad \text{and} \quad K^2 g^4 = A_\parallel^{\,2} g_\parallel^{\,4} \cos^2\theta + A_\perp^{\,2} g_\perp^{\,4} \sin^2\theta$$

with g given by (II, 51)

$$\begin{aligned} H = H_0 &- Km_I - \frac{A_\parallel g_\parallel A_\perp^{\,2} g_\perp^{\,2}}{2Kg^2 \cdot gH_0}(2m_S - 1)\, m_I \\ &- \left(\frac{A_\parallel^{\,2} g_\parallel^{\,2} + K^2 g^2}{K^2 g^2}\right) \frac{A_\perp^{\,2} g_\perp^{\,2}}{4g^2 H_0} [I(I+1) - m_I^{\,2}] \\ &- \frac{1}{2H_0} \left(\frac{A_\parallel^{\,2} g_\parallel^{\,2} - A_\perp^{\,2} g_\perp^{\,2}}{Kg^2}\right)^2 \left(\frac{g_\parallel g_\perp}{g^2}\right)^2 \sin^2\theta \cos^2\theta m_I^{\,2} \\ &- \frac{Q'^2 g_\perp^{\,2} \cos^2\theta \sin^2\theta}{2Kg^2 m_S(m_S - 1)} \left(\frac{A_\parallel A_\perp g_\parallel^{\,2} g_\perp^{\,2}}{K^2 g^4}\right)^2 \\ &\cdot m_I[4I(I+1) - 8m_I^{\,2} - 1] \\ &+ \frac{Q'^2 g_\perp^{\,2} \sin^4\theta}{8Kg^2 m_S(m_S - 1)} \left(\frac{A_\perp g_\perp^{\,2}}{Kg^2}\right)^4 m_I[2I(I+1) - 2m_I^{\,2} - 1] \end{aligned} \qquad \text{(II, 64)}$$

[This is the general form of equation (7) of Rollmann and Chan (394), which is for $S = \frac{1}{2}$.] One sees that the first two terms on the right-hand side result from diagonal matrix elements and yield equidistant hf lines; the third and fourth terms cause increasing spacing at higher fields. (II, 63) and/or (II, 64) is generally applicable because the hfs is usually not comparable with the electronic Zeeman energy.

It may be necessary to include third-order perturbation terms in the above equations if the hyperfine splitting is not small enough to obtain sufficient accuracy from second-order analysis. Such terms can be found in the third-order expressions of Friedman and Low (172), Bleaney and Rubins (74), de Wijn and van Balderen (143), and Golding and Tennant (187). Of course, an exact expansion of the secular determinant is necessary if the hfs is comparable to the electronic Zeeman energy, $g\beta H$.

(γ) Forbidden Transitions ($\Delta m_S = \pm 1, \Delta m_I = \pm 1$ or ± 2). The usual selection rules in ESR can be broken down when either the nuclear Zeeman or the nuclear quadrupole interaction is comparable with the hyperfine splitting parameters [Abragham and Bleaney (1) pages 181–186, 195–200]. This leads to so-called forbidden transitions being also observed, i.e., those where $\Delta m_I = \pm 1$ or ± 2. The physical basis for these transitions is that the nuclear states become mixed by opposing and interacting effects; for example, by the quadrupole interaction tending to align the nucleus along the symmetry axis while the electronic magnetic field tries to align the nucleus perpendicular to the axis via $A_\perp$. This amounts to off-diagonal terms becoming important relative to the situation where $A \gg Q'$ or $\gamma_n \beta_n H$.

If higher-order energy terms, proportional to $(Q'/A)^2$, are neglected, Bleaney (70) has shown that the field at which resonance occurs in (II, 64) is altered by the addition of (in gauss)

$$\pm \left\{ \frac{kQ'' g_\perp}{g} - \frac{\gamma'}{g\beta} \right\} \tag{II, 65}$$

where

$$Q'' = Q'[(3A_\parallel{}^2 g_\parallel{}^4 / K^2 g^4) \cos^2 \theta - 1]$$

and

$$\gamma' = (g_I \beta_n H_0 / K g^2)(A_\parallel g_\parallel{}^2 \cos^2 \theta + A_\perp g_\perp{}^2 \sin^2 \theta)$$

In (II, 65) $k = (I - \frac{1}{2}), (I - \frac{3}{2}), \cdots, -(I - \frac{1}{2})$ is used instead of m_I so that in this notation the $\Delta m_S = -1, \Delta m_I = \pm 1$ transition is

$$(m_S, k \pm \tfrac{1}{2}) \longrightarrow (m_S - 1, k \mp \tfrac{1}{2}).$$

Corresponding expressions have also been derived for $\Delta m_S = -1$, $\Delta m_I = \pm 2$ transitions (70).

An example of forbidden transitions arising because of only the nuclear Zeeman term is shown in Fig. II-2. There the ESR spectrum

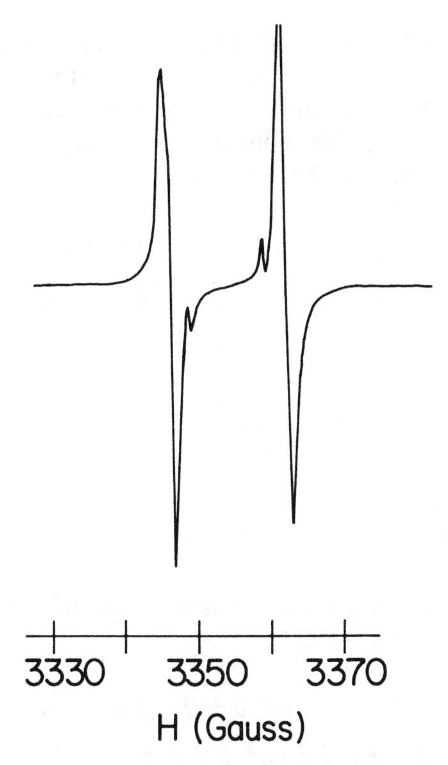

Fig. II-2. Forbidden ($\Delta m_I = \pm 1$) transitions in the ESR spectrum of the C_2H radical [from Graham, Dismuke, and Weltner (195)]. C_2H is trapped in an argon matrix at 4°K ($\nu = 9.398$ GHz).

of the trapped C_2H radical [Graham, Dismuke, and Weltner (195)] exhibits a strong doublet due to the hf interaction of the odd electron with the proton nuclear magnetic moment ($I = \frac{1}{2}$), but also a weaker inner doublet due to the forbidden $\Delta m_I = \pm 1$ transition. Here A(H) is small because the electronic spin density is largely localized at the other end of the linear molecule, so that the nuclear Zeeman term is not negligible compared to the hfs. Miyagawa and Gordy (345), McConnell, Heller, Cole, and Fessenden (336), and Weil and Anderson (469) have considered the theory in detail for the particular case of one electron and one nucleus.

The Hamiltonian, neglecting the nuclear quadrupole term, is

$$\mathcal{H} = \beta \boldsymbol{S} \cdot \boldsymbol{g} \cdot \boldsymbol{H} + \boldsymbol{S} \cdot \boldsymbol{A} \cdot \boldsymbol{I} + g_I \beta_n \boldsymbol{I} \cdot \boldsymbol{H} \qquad \text{(II, 66)}$$

Assuming, as usual, that the electronic Zeeman term is dominant, proper rotation of the magnetic field axis relative to those fixed in the molecule, in a manner similar to that described earlier, eliminates the off-diagonal hyperfine contributions and yields four energy levels [using the notation of Kasai (260)]:

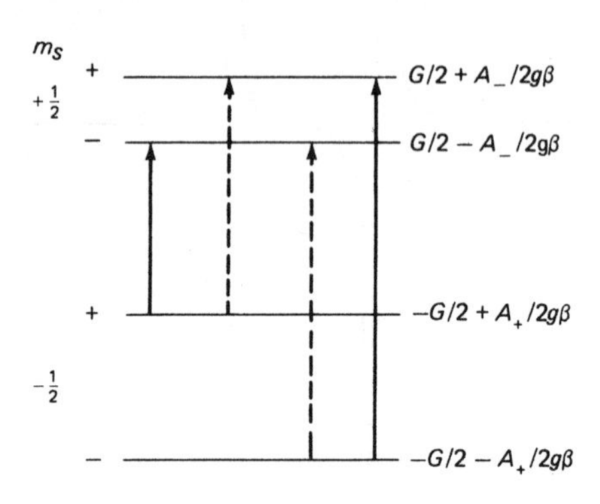

Here $G = g\beta H$, $H_0 = h\nu/g\beta$ is the center of the hf pattern, and if the nucleus is a proton, $H_p = (g_p \beta_n / g\beta) H_0$,

$$(A_\pm)^2 = (H_p \pm A/2)^2 \cos^2 \theta + (H_p \pm B/2)^2 \sin^2 \theta \cos^2 \phi + (H_p \pm C/2)^2 \sin^2 \theta \sin^2 \phi \qquad \text{(II, 67)}$$

A, B, C are the principal hf tensor components of the molecule, and θ and ϕ define the direction of the magnetic field relative to the principal axes. $m_I = +\frac{1}{2}$ and $m_I = -\frac{1}{2}$ states are mixed so that the usual selection rules do not apply, and four transitions can be observed. For $h\nu$ fixed, as in an ESR experiment, the positions (in gauss) of these two sets of doublets (inner and outer sets in Fig. II-2) are

$$H_{\pm\,\text{out}} = H_0 \pm \tfrac{1}{2}(A_+ + A_-)$$
$$H_{\pm\,\text{in}} = H_0 \pm \tfrac{1}{2}(A_+ - A_-) \qquad \text{(II, 68)}$$

The relative intensity of the two doublets is given by

$$\frac{I_{\pm\,\text{out}}}{I_{\pm\,\text{in}}} = \frac{1 - \cos \tau}{1 + \cos \tau} \qquad \text{(II, 69)}$$

where $\cos \tau = (4H_p{}^2 - A_0{}^2)/4A_+A_-$ and

$$A_0{}^2 = A^2 \cos^2 \theta + B^2 \sin^2 \theta \cos^2 \phi + C^2 \sin^2 \theta \sin^2 \phi$$

For C_2H trapped in solid argon, the hf pattern in Fig. II-2 arises largely from molecules aligned perpendicular to the fixed magnetic field so that $(A_\pm)^2 = (H_p \pm A_\perp/2)^2$ and $A_0 \cong A_\perp$. Experimentally the outer and inner doublet spacings are 14.5 G and ~9.7 G, corresponding approximately to $A_\perp$ and $2H_p$, respectively (H_p = 5.014 G at H_0 = 3300 G). The inner doublet appears on the wings of the outer doublet, and its exact position is difficult to estimate; however, the observed relative intensity of the two doublets is in approximate agreement with that calculated from (II, 69).

Poole and Farach (380) have given a more detailed discussion of the theory of the spacing and intensities of the two doublets as a function of $(2H_p/A)$. Kasai (260) has considered more than one proton when including these transitions in the calculation of the spectra expected for randomly oriented hydrocarbon radicals.

A relevent study is that of Trammell, Zeldes, and Livingston (432) who observed satellite lines on atomic hydrogen absorptions produced in glassy solids by γ irradiation. These weak transitions were shown to be produced by simultaneous spin flips of an electron and of a neighboring nucleus. The satellite-line to main-line (atomic H hf line) intensity ratio varied approximately with the inverse square of the applied field strength.

2. OTHER AXIAL DOUBLET MOLECULES

(a) Orbitally Degenerate Molecules and Quenching

$^2\Pi$, $^2\Delta$, etc., radicals are distinct from $^2\Sigma$ molecules in generally having very anisotropic $\boldsymbol{g}$ tensors; that is, their magnetic moments are very different along directions parallel and perpendicular to the molecular axis. As will be demonstrated in section 3b, this means that if they are randomly oriented, as in a matrix, the ESR spectrum will be so spread out as to be undetectably weak. On the other hand, their orbital angular momenta are frequently sensitive to an unsymmetrical environment, so that in matrices, and particularly in crystals, they may be observed as quenched species.

First it is to be recalled that, in linear doublet molecules with electronic orbital angular momentum, spin-orbit coupling causes multiplet splitting into two substates. Each of these substates essentially has its own magnetic properties. [However, in the gas phase these substates may be mixed by rotational distortion; see Chapter IV.] The magnetic moment due to orbital and spin angular momentum is proportional to $\Lambda + g_e \Sigma$. For a $^2\Pi$ molecule $\Lambda = \pm 1$ and $\Sigma = \pm\frac{1}{2}$ so that $\Lambda + g_e \Sigma = \pm 2$ in a $^2\Pi_{3/2}$ and ± 0.001 in a $^2\Pi_{1/2}$ state [Townes and Schawlow (21), page 284]. [This near vanishing of the magnetic moment in the $^2\Pi_{1/2}$ state arises infrequently among other multiplet states, but it is clear that it will also occur in $^3\Delta_1$ and $^3\Pi_0$ states.] Thus, the NO molecule is usually considered as strongly paramagnetic, which is only justified at temperatures where the $^2\Pi_{3/2}$ multiplet, 123 cm^{-1} about the $^2\Pi_{1/2}$, is occupied; at 4°K, the nonrotating molecule has no magnetic moment.

Orbital angular momentum causes the $\boldsymbol{g}$ tensor of linear molecules to be extremely anisotropic, or if not, it balances the spin angular momentum, as in a $^2\Pi_{1/2}$ state. For example, a molecule in a $^2\Pi_{3/2}$ state will have, from (II, 9),

$$\begin{aligned} g_\parallel &= 2\langle +1, +\tfrac{1}{2}\, | L_z + 2S_z | + 1, +\tfrac{1}{2}\rangle \\ &= 2\{\langle +1 | L_z | + 1\rangle + 2\langle +\tfrac{1}{2} | S_z | + \tfrac{1}{2}\rangle\} = 4 \\ g_\perp &= 2\langle +1, -\tfrac{1}{2}\, | L_x + 2S_x | - 1, +\tfrac{1}{2}\rangle = 0 \end{aligned}$$

This $g_\perp$ component implies that a resonant transition in the ESR spectrum occurs at infinite magnetic field when the static field is perpendicular to the molecular axis. Hence the resonant magnetic field is very sensitive to the orientation of the molecule, and as will be seen in section 3b, if the molecules are randomly oriented in a solid matrix, the spectrum of a $^2\Pi_{3/2}$ radical will extend over a large range of magnetic fields and therefore be too weak to be detected.

A similar calculation for a $^2\Pi_{1/2}$ state leads, as expected, to $g_\parallel = g_\perp = 0$. Thus, regardless of whether the ground state is $^2\Pi_r$ or $^2\Pi_i$, the ESR spectrum would not be observable for matrix-isolated randomly oriented molecules. Here r stands for regular and means that the $^2\Pi_{1/2}$ multiplet lies lowest in energy, while i stands for inverted, so that the $^2\Pi_{3/2}$ is lowest. One can also speak of regular as being an "electron" state versus inverted as a "hole" state relative to a closed π shell [see section II-1b (α)]. $^2\Delta$, $^2\Phi$, etc., molecules also have $g_\perp = 0$ and are not observable in matrices via ESR.

However, if the orbital angular momentum is partially quenched by the asymmetrical effects of the environment, as discussed for ^2P atoms in Chapter I, then the g values approach g_e, and the spectrum approaches that of a $^2\Sigma$ molecule. Whereas in a ^2P atom the *spherical* symmetry is destroyed by the interaction with the environment, in a $^2\Pi$ molecule the *cylindrical* symmetry is destroyed by the interaction. This readily occurs in crystals where the neighboring ions can provide the strong interactions, but in rare-gas matrices the fields are not usually adequate. Quenching, and the appearance of an ESR spectrum, can occur in rare-gas matrices if an ion-pair is trapped. For example, O_2^- is a $^2\Pi_i$ molecule that has been observed as a quenched radical in crystals [Bennett, Ingram, Symons, George, and Griffith (56); Känzig and Cohen (257)] and in an ion-pair, $Na^+O_2^-$, in argon and krypton matrices [Adrian, Cochran, and Bowers (36); Lindsay, Herschbach, and Kwiram (310)]. In fact, in these two kinds of environments the most effective quenching of O_2^- occurs in NaO_2 in solid argon because the single cation lying perpendicular to the molecular axis is the most unsymmetrical perturber of axial symmetry. This results in $g_\parallel = 2.112$, $g_\perp$ (to plane) $= 2.0029$, and $g_\perp$ (in plane) $=$ 2.0063. The unpaired electron is largely on O_2^- (although small ^{23}Na hfs was observed) in an antibonding π_g orbital with the occupied O $2p\pi$ orbitals oriented perpendicular to the plane of the triangular molecule. The spectrum of this and other alkali-metal dioxides will be discussed in more detail in section II-6. Data and references for other quenched $^2\Pi$ radicals, such as N_2^-, ClO, and OH, are given in Appendix C.

(b) Axial Nonlinear Molecules

Molecules with $S = \frac{1}{2}$ and an axis of at least trigonal symmetry can be expected to exhibit an ESR spectrum similar to a $^2\Sigma$ molecule. Examples are planar (D_{3h}) NO_3, pyramidal (C_{3v}) SiH_3, and the effectively square planar (D_{4h}) molecule Cu $(NO_3)_2$, where the symmetry is determined by the essential equality of the four nearest and most influential oxygen atoms surrounding the Cu^{+2}. In each of these cases only parallel and perpendicular $\boldsymbol{g}$ tensor components need be specified in the free molecule, although in some cases the radicals may be distorted slightly in the crystals in which they are trapped [Morton (349)].

Among the most interesting tetra-atomic radicals are the planar radicals $^{14}NO_3$, $^{13}CO_3^-$, and particularly $^{13}CH_3$. In the simplest theory, in all three of these molecules the odd electron should not interact with the nuclear moment of the central atom. In the first two isoelectronic oxides, the electron occupies an (a_2') orbital which is an out-of-phase combination of in-plane oxygen lone-pair orbitals. In contrast, the odd electron on CH_3 is in a $p\pi$ orbital on the carbon, perpendicular to the plane and with a node in the plane. Actually, small hfs due to electron density at the central nucleus is observed in all three cases, and in CH_3 also at the protons. All of these hf splittings must be attributed largely to spin polarization or configuration interaction (see section II-6).

Although the $\boldsymbol{g}$ tensor in each of these molecules has axial symmetry, that is only rigorously true for a hyperfine tensor if the nucleus is located on the symmetry axis. For nuclei located off-axis the principal axes of the $\boldsymbol{g}$ and $\boldsymbol{A}$ tensors may not be parallel, which can considerably complicate the ESR spectrum and its analysis. The only axial molecule where this appears to have been worked out in some detail is pyramidal CF_3 [Maruani, McDowell, Nakajima, and Raghunathan (327)]. There the $\boldsymbol{A}(^{19}F)$ tensor is found to be essentially axial but, surprisingly, with its symmetry axis (z) almost parallel to the molecular symmetry axis but tilted by 18° so that it is perpendicular to the C–F bond. The actual hf components given are $A_x \simeq 87$ G, $A_y \simeq 80$ G, and $A_z = 263.5$ G. Rogers and Kispert (393) found the $\boldsymbol{g}$ components to be $g_\parallel = 2.0024$, $g_\perp = 2.0042$. It should be noted that the hf coupling constants are still quite small compared to those for the ^{19}F atom (see Appendix B), but that they are very anisotropic, indicating mainly $2p_z$ spin density on the fluorine atoms. These findings are not in agreement with what one might expect from theory, but there seems to be no doubt that the rather complex spectrum is that of a randomly oriented, nonrotating, noninverting CF_3 radical.

Since in most radicals with low symmetry there will be noncoincidence between the principal axes of the $\boldsymbol{g}$ and $\boldsymbol{A}_i$ tensors, that is an important general consideration, which is discussed in more detail in section II-4b. Further attention will be given to CF_3 when representative doublet molecules are discussed in section II-6. First, however, the experimental methods and ESR spectra in crystals and matrices will be briefly considered.

3. ESR EXPERIMENTAL MEASUREMENTS IN CRYSTALS AND MATRICES

The experimental problems involved in the investigation of radicals are their formation (and also subsequent trapping if in the solid state) and the extraction of their magnetic parameters from the measured ESR spectra. In the gas phase the preparation of a particular reactive, short-lived, radical in sufficient concentration for magnetic resonance investigation is often a primary obstacle. This can be less difficult in the solid state, since the radical population can usually be increased by accumulation of the trapped species. Consideration will be given specifically to the gas phase in Chapter IV, and here only to crystal and matrix investigations.

(a) ESR of Molecules in Single Crystals

This is a field of extensive research activity ranging from production of the radical, defect, or center by irradiation of the crystals with γ, X-ray or ultraviolet radiation to the formation of mixed crystals containing a paramagnetic impurity. The virtue of single crystal studies is that if the crystal structure of the precursor is known and the radical has a preferential orientation in the lattice, then the ESR spectrum can be measured with the dc magnetic field along particular lattice directions. One can then deduce not only the complete hf and $\mathbf{g}$ tensor components of the radical but also its orientation in the lattice [Morton (349); Wertz and Bolton (23), pages 131, 164; Gordy (8), page 150].

A good example of this is the classic work of Castner and Känzig (105) on the V_K center formed in alkali halide crystals by X-ray irradiation at liquid nitrogen temperature. This center contains a Cl_2^- molecule with two cations as closest neighbors. Although it is perturbed somewhat by the crystalline environment, it is a σ radical and a $^2\Sigma$ ground-state molecule. Thus it is different from such π radicals as O_2^-, which is a $^2\Pi$ molecule with g values near 2 in crystals due solely to its quenched orbital angular momentum. [An informative discussion of the V_K center is given by Slichter (20)].

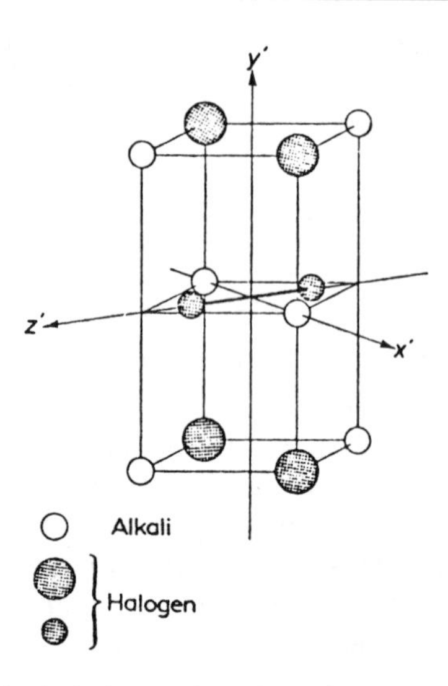

Fig. II-3. Orientation and principal axes of the halogen molecule-ion (X_2^-) in the V_K center in alkali halide crystals [from Castner and Känzig (105)].

If KCl is the single crystal, then the Cl_2^- molecule is formed and "frozen in" along the [110] axis in the crystal, as indicated in Fig. II-3 by z'. The internuclear distance in the molecule is between the $Cl^- \cdots Cl^-$ distance of 4.44 Å in the crystal and 2.00 Å in Cl_2 gas. However, the environment of the Cl_2^- is not the same in the x' and y' directions, since there are two cations closer to its axis in the x' direction, and this then distorts its axial magnetic symmetry. Hence, the $\mathbf{g}$ tensor components are found to be $g_{x'} = 2.0428$, $g_{y'} = 2.0447$, and $g_{z'} = 2.0010$. Hyperfine splitting is also observed since there are two magnetic nuclei ^{35}Cl (75% abundant) and ^{37}Cl (25% abundant) in natural KCl. There are then three kinds of isotopically distinguishable Cl_2^- molecules with different statistical weights and different hf splitting patterns, since $\gamma_{37}/\gamma_{35} = 0.83$ (although $I_{35} = I_{37} = \frac{3}{2}$).

Figure II-4 illustrates the variation of the ESR pattern as the crystal is rotated relative to the static magnetic field. θ indicates the angle between H and molecule axis, and the arrows distinguish the hf structure specifically of the $^{35}Cl_2^-$ species. It must be remembered that there are six equivalent [110]-axes in the crystal, all of which are equally populated by Cl_2^- radicals. Thus, if the magnetic field is parallel to [100], the axes of two thirds of the molecules form an angle $\theta = 45°$ with the field and one third an angle of 90°.

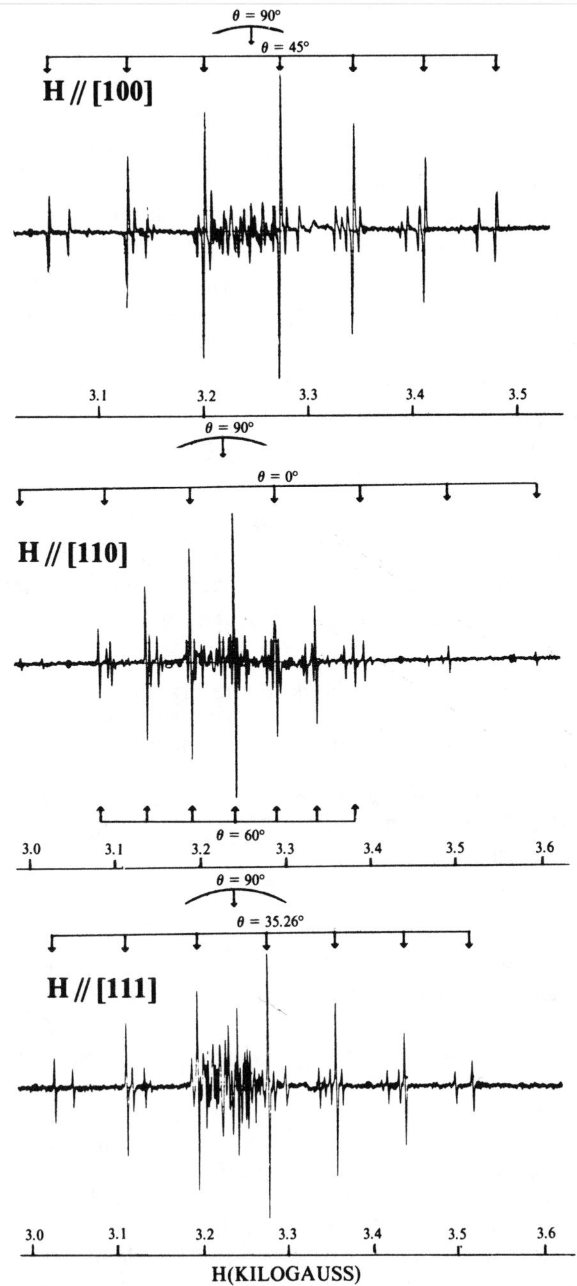

Fig. II-4. Orientation effects in the ESR spectrum of the V_K center in the KCl crystal [from Castner and Känzig (105)]. θ is the angle between the magnetic field and the molecular axis. The arrows indicate the $^{35}Cl_2^-$ spectra. ($\nu \simeq 9.3$ GHz).

Also, because the electron in Cl_2^- is essentially in a $3p\sigma$ antibonding orbital, $A_{z'}$ is very large relative to $A_{x'}$ and $A_{y'}$ so that the hfs for $\theta = 0°$ is also much larger than that for $\theta = 90°$. The latter spacing is, in fact, so small that it appears almost as "noise" in Fig. II-4 from the contributions of the three Cl_2^- species present. Careful analysis of the hfs yielded for $^{35}Cl_2^-$, $A_{\parallel} = 101.39$ G, $|A_{\perp}| = 9$ G at $\nu = 9.26$ GHz [in our symbolism, or $(a + b)$ and $|a|$ in Castner and Känzig's], where any effects of distortion from axial symmetry were negligible. Thus, a complete picture of the $Cl_2^- V_K$ center and its orientation in the KCl crystal was obtained. Such a V_K center in NaCl gave slightly different magnetic parameters, again oriented along the [110]-axes in the crystal (105). Similar studies were made of X-irradiated LiF and KBr crystals, yielding V_K centers containing F_2^- and Br_2^- radicals.

This original work by Känzig and coworkers led to much further research on V centers in crystals, also involving other halogens. Schoemaker (408) has most recently reviewed the extensive ESR and theoretical studies and carried out a general analysis based upon a crystal-field model.

In a V_K center the Cl_2^- ion can be considered to be perturbed by the two nearest positively charged M^+ ions lying on a perpendicular axis, x, to the molecular axis, z, as shown in Fig. II-3 [Castner and Känzig (105)]. The effect is to cause a splitting of the degenerate π orbitals in Cl_2^-, since the x and y directions are now distinct. The free Cl_2^- molecule with a $^2\Sigma_u^+$ ground state and a configuration

$$\cdots (\sigma_g)^2 (\pi_u)^4 (\pi_g)^4 (\sigma_u)^1$$

now must be considered as part of a molecule of D_{2h} symmetry with a $^2B_{1u}$ ground state and with the electrons in a configuration (see Fig. II-5)

$$\cdots (a_g)^2 (b_{3u})^2 (b_{2u})^2 (b_{2g})^2 (b_{3g})^2 (b_{1u})^1$$

The fact that the observed g_x and g_y are different attests to the position of the M^+ ions, i.e., that they are not situated along the molecular axis. The positive shift of $g_x \cong g_y = g_{\perp}$ from g_e is due to the coupling of the excited Π_u states (in the linear symbolism) resulting when an electron is excited into the half-empty b_{1u} orbital from the b_{2u} and b_{3u} orbitals as indicated in Fig. II-5. Then in the ground state the odd electron (or hole) occupies the molecular orbital

$$|\sigma_u(b_{1u})\rangle = \alpha_u(s_1 - s_2) + \beta_u(p_{z1} + p_{z2})$$

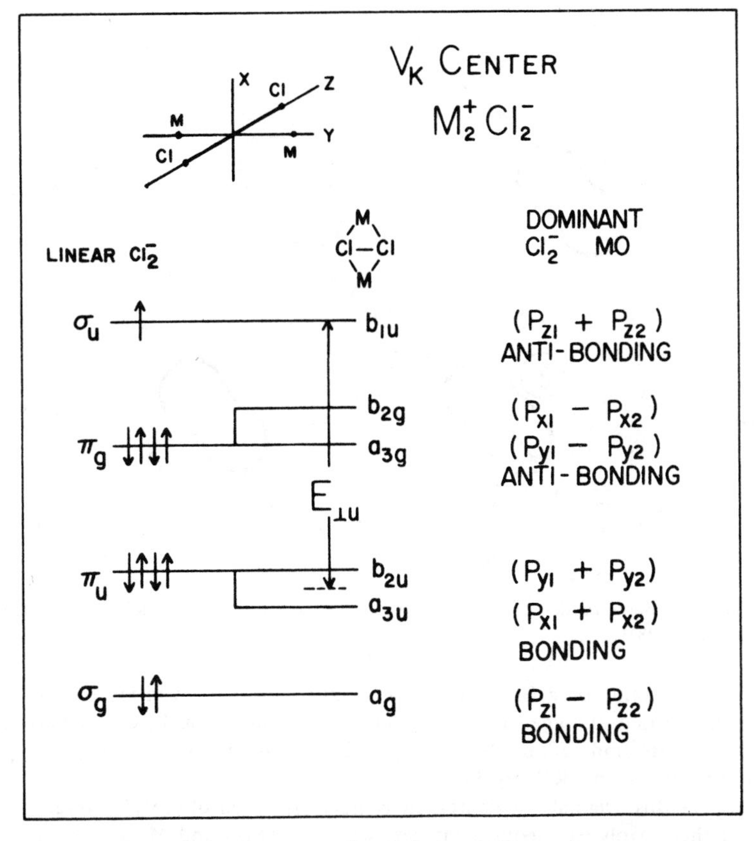

Fig. II-5. Molecular orbitals of $M^+Cl_2^-M^+$ in the V_K center in KCl crystals.

and in the excited Π state split by the crystal field it occupies either

$$|\pi_u(b_{3u})\rangle = \mu_u(p_{x1} + p_{x2})$$

or

$$|\pi_u(b_{2u})\rangle = \mu_u(p_{y1} + p_{y2})$$

Then

$$g_\perp - g_e \cong \frac{8\beta_u^2\mu_u^2\lambda}{E_{\perp u}} \qquad \left(\text{to first order in } \frac{\lambda}{E_\perp}\right)$$

from (II, 11). Here β_u and μ_u are coefficients in the above wavefunctions, $E_{\perp u} = \frac{1}{2}(E_{1u} + E_{2u})$ as shown in Fig. II-5, and $\lambda = 536\ \text{cm}^{-1}$ is the spin-orbit

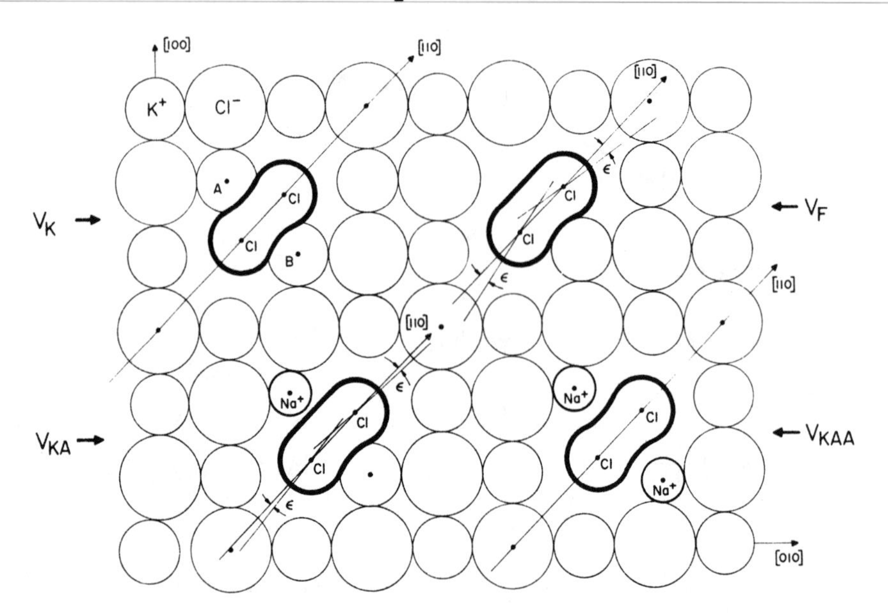

Fig. II-6. Schematic models of the V_K, V_{KA}, V_{KAA} and V_F centers in KCl crystals [from Schoemaker (408)].

coupling constant for a Cl atom. Schoemaker (408) has given more exact expressions (carried to higher order in $\lambda/E_\perp$) that account for the difference between g_x and g_y. He finds $2\beta_u \mu_u \cong 0.78$, $E_{\perp u} = 2.4$ eV, and the splitting $E_{2u} - E_{1u} = 0.135$ eV for Cl_2^- in a KCl crystal.

Besides this original V_K center, there have also been observed several variations either involving inequivalent cations, V_{KA} [Bass and Mieher (51)] and V_{KAA} [Schoemaker (408)], or only one cation, V_F [Känzig (256); Schoemaker (407)]. These centers are illustrated in Fig. II-6.

(b) ESR of Matrix-Isolated Molecules

Whether the radicals of interest are trapped in a frozen inert gas at low temperatures or are formed by photolysis of a stable precursor in a frozen solvent, they can be expected to be randomly oriented, if they are unable to rotate. There are, in fact, very few molecules, e.g., H_2O, HCl, NH_3, and NH_2, exhibiting any degree of rotational freedom when isolated in rare-gas matrices at 4–10°K. In a few cases of bent triatomics, such as CH_2 and NO_2, there is evidence of rotation in the solid about the axis of smallest moment of inertia. However, in general, even for molecules that have a rather spherical electron distribution, such as BaH, there is no indication of rotation from the ESR spectrum, and the matrix contains randomly oriented isolated radicals. [It should be mentioned here that there is often evidence that, in the preparation of a matrix from the gas phase, some preferential orientation of the trapped species relative to the cooled substrate is occurring. In a few cases this can be quite extreme. This phenomenon and its effect upon the observed ESR spectrum will be discussed below in section 3b (γ).]

Figure II-7 gives a cross-sectional view of a cryostat apparatus used to prepare rare-gas matrices containing radicals by deposition of discharge products [Jen, Foner, Cochran, and Bowers (251)]. Condensation occurs on a single-crystal sapphire rod cooled to 4°K that is then lowered into the microwave cavity by means of a mechanism at the top of the dewar (not shown on the figure) described by Duerig and Mador (151). The microwave cavity is in contact with the liquid nitrogen shield so that the matrix is still in a cooled environment when ESR measurements are made. A variation of this apparatus places the sapphire rod permanently inside the cavity with both in contact with the liquid helium container [Farmer, Gerry, and McDowell (161)]. To minimize the evaporation of liquid helium due

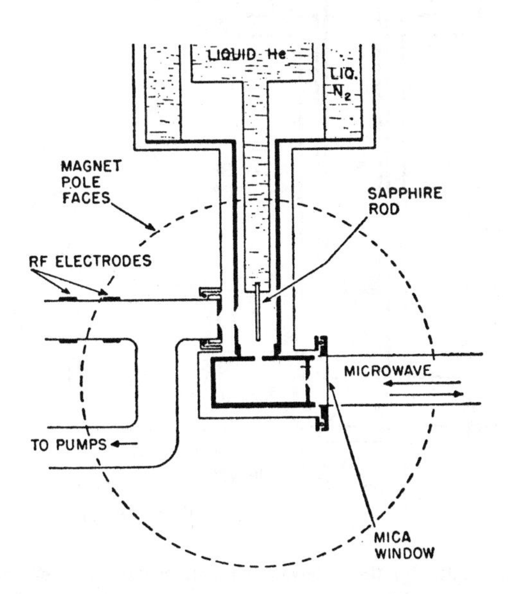

Fig. II-7. ESR apparatus for the observation of free radicals trapped in solidified gases at 4°K [from Jen, Foner, Cochran, and Bowers (251)]. The radicals are here prepared as discharge products, but ultraviolet irradiation can also be used.

to the entry of heat along the waveguide, the latter was connected to the cavity by a length of thin-walled stainless-steel waveguide.

Figure II-8 shows an apparatus, similar to that of Jen et al. used to prepare rare-gas matrices containing radicals vaporized from solids at high temperatures [Kasai, Whipple, and Weltner (266)]. The solid sample is vaporized (e.g., solid CuF_2 at 900°C) from the resistance-heated cell and mixes with neon (or argon) just before condensation on the sapphire rod. The important innovations here are the use of a 2-mm-diameter rod that had been ground flat over about a 3-cm length and the ability to rotate the rod about its vertical axis. When one is preparing the matrix, the rod is turned by 180° at intervals so that a flat face is always toward the incoming matrix gas. The dc magnetic field is, of course, perpendicular to the paper with its axis at the center of the cavity, as in Fig. II-1. Although this rotatable

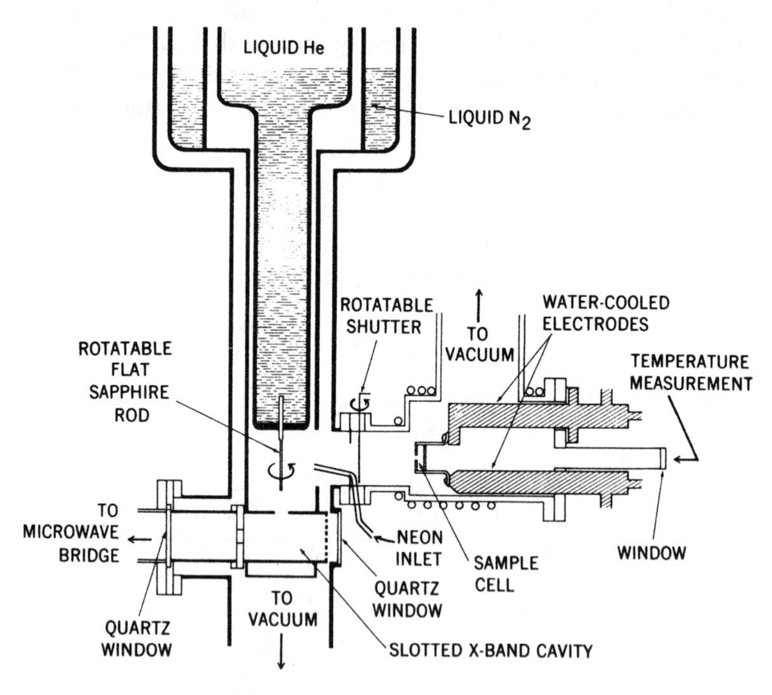

Fig. II-8. ESR apparatus for the observation of molecules vaporized from solids at high temperatures by trapping them in the solid rare gases at 4°K [from Kasai, Whipple and Weltner (266)].

flattened rod was originally used in order to make possible both optical absorption and ESR measurements on the same trapped radicals, it led to the discovery of preferential orientation of radicals in rare-gas matrices [see section 3b (γ)]. Optical measurements were made through two windows (not shown in Fig. II-8) in the dewar above the cavity by use of proper focusing lenses.

(α) Randomly Oriented Molecules. The question that arises is what sort of a spectrum will be observed if the molecules are randomly oriented but rigidly held, as in a solid matrix formed in this way? Here obviously the spectrum will be independent of the angle that the magnetic field makes with the solid sample. This problem was treated first by Bleaney (69, 70) and by Sands (399), and since then by many other authors [see ref. 1 of Weil and Hecht (470) for an extensive bibliography].

If the N_0 axial molecules are randomly oriented with respect to the direction of the applied magnetic field, then the number within an increment of angle $d\theta$, where θ is the angle measured from the field direction, is

$$dN = \frac{N_0}{2} \sin\theta \, d\theta \tag{II, 70}$$

Thus dN is proportional to the area of the surface of a sphere included within an angle variation of $d\theta$; the factor two enters because it is only necessary to cover a hemisphere. Note that the $\sin\theta$ factor is indicative of the predominance of molecules with axes nearly perpendicular to the field direction relative to those with axes more nearly aligned along the field direction. The absorption intensity in the ESR spectrum as a function of angle is proportional to the number dN of molecules lying between θ and $\theta + d\theta$, assuming the transition probability is independent of orientation. But the g value is a function of θ given by (II, 51), and therefore, for fixed frequency ν, the resonant magnetic field is given by

$$H = \frac{h\nu}{\beta} (g_{\parallel}^2 \cos^2\theta + g_{\perp}^2 \sin^2\theta)^{-1/2} \tag{II, 71}$$

and from this

$$\sin^2\theta = \frac{(g^0H^0/H)^2 - g_\parallel{}^2}{g_\perp{}^2 - g_\parallel{}^2} \tag{II, 72}$$

where

$$g^0 = \tfrac{1}{3}(g_\parallel + 2g_\perp) \tag{II, 73}$$

and

$$H^0 = h\nu/g^0\beta \tag{II, 74}$$

One then finds from (II, 72)

$$\sin\theta\, d\theta = -\frac{(g^0H^0)^2}{H^3}\{(g_\parallel{}^2 - g_\perp{}^2)[(g^0H^0/H)^2 - g_\perp{}^2]\}^{-1/2}\, dH \tag{II, 75}$$

The intensity of absorption in a range of magnetic field dH is proportional to

$$|dN/dH| = (dN/d\theta)\cdot|d\theta/dH| \tag{II, 76}$$

where $dN/d\theta = (N_0/2)\sin\theta$ from (II, 70) and $|d\theta/dH|$ is obtained from (II, 75). Note that from (II, 71) and (II, 74)

$$H = h\nu/g_\parallel\beta = g^0H^0/g_\parallel \qquad \text{at } \theta = 0^\circ$$

and

$$H = h\nu/g_\perp\beta = g^0H^0/g_\perp \qquad \text{at } \theta = 90^\circ \tag{II, 77}$$

Then, at these two extremes, the intensity of absorption varies from

$$|dN/dH| = N_0 g_\parallel{}^3/2g^0H^0(g_\parallel{}^2 - g_\perp{}^2) \quad \text{at } \theta = 0^\circ$$

to

$$|dN/dH| = \infty \quad \text{at } \theta = 90^\circ \tag{II, 78}$$

If plotted versus magnetic field, this absorption would have the following shape (for $g_\parallel > g_\perp$):

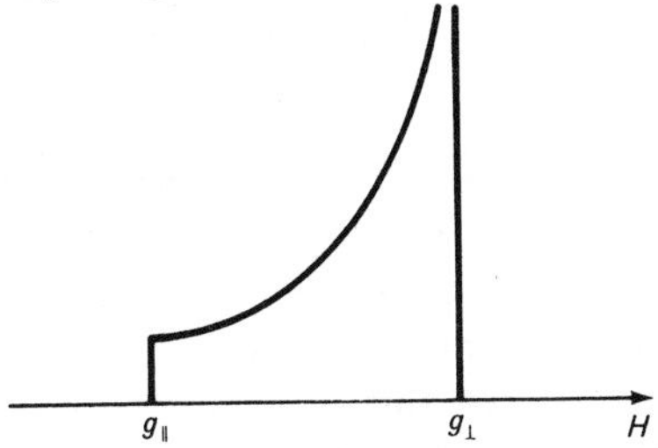

If the natural width of the lines of the individual molecules contributing to this absorption is considered, then the sharp angles in the above curve become rounded, and it takes more of the appearance of Fig. II-9a. Since in ESR the first derivative of this absorption curve is usually observed, the spectrum would then appear as in Fig. II-9b. If the $\mathbf{g}$ tensor is not too anisotropic, $g_\parallel$ and $g_\perp$ can be readily deter-

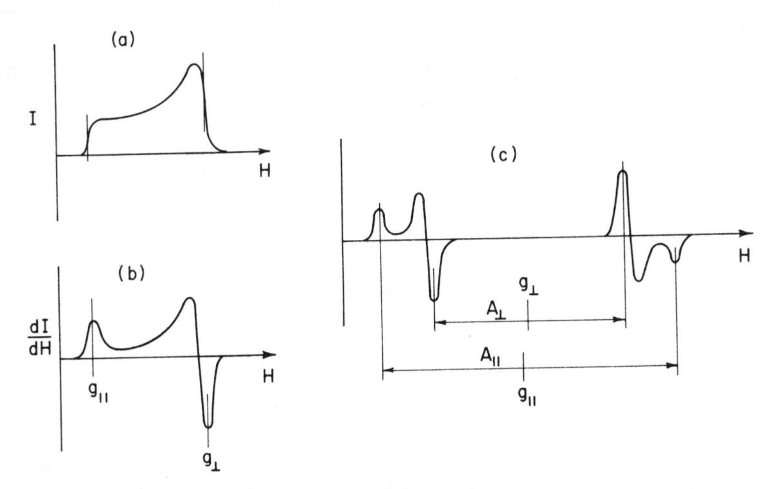

Fig. II-9. ESR spectrum of randomly oriented $^2\Sigma$ molecules [from Weltner (474)]. (a) Absorption spectrum; (b) ESR spectrum [first derivative of (a)]; (c) ESR spectrum of a $^2\Sigma$ molecule exhibiting hyperfine splitting due to a nucleus of spin $\frac{1}{2}$.

mined as shown there. In general, the perpendicular component can be easily distinguished in such a "powder pattern"; the weaker parallel peak is often more difficult to detect.

In the above derivation, it was assumed that the transition probability is independent of orientation, which we know from section II-1e is incorrect. g_1 in (II, 60) is a function of the angle between the molecular axis and the inducing field H_1 (taken here as always perpendicular to H, as in Fig. II-1.) For randomly oriented $S = \frac{1}{2}$ molecules, g_1^2 is given by equation (II, 62), as derived by Bleaney (71), and the angular dependence arises from the dependence of g in the denominator upon θ, as given in (II, 51). (For an isotropic $\mathbf{g}$ tensor we note that $g_1^2 = g^2$.) This expression for g_1^2 would then be multiplied by dN/dH to give the true absorption spectrum of the molecule and then the first derivative taken to obtain the expected pattern in the ESR.

This, however, is only correct for a frequency-swept spectrometer but not for the usual field-swept ESR instrument [Aasa and Vänngard (25)]. In the latter case g_1^2 should be replaced by g_1^2/g, which means that for an isotropic $\mathbf{g}$ tensor the multiplicative factor is not $g_1^2 = g^2$ but just g. For the anisotropic axial case it would be g_1^2/g with g_1^2 given by (II, 62). Since the anisotropy is not large in $^2\Sigma$ molecules, this difference is not important here, but it will be significant in high-spin molecules where $D \neq 0$ and the effective $\mathbf{g}$ can be very anisotropic (see later chapters).

g_1 is a function of the frequency of oscillation of H_1, so that $g_1^2 f(\nu)$ should express the transition probability, where $f(\nu)$ is a normalized function such that $\int f(\nu)\, d\nu = 1$. The intensity of the signal integrated over the magnetic field is then

$$\int g_1^2 s(H)\, dH = \int g_1^2 f(\nu) \left(\frac{dH}{d\nu}\right) d\nu$$

For $S = \frac{1}{2}$ the resonance condition is $h\nu = g\beta H$ and $dH/d\nu = h/g\beta$, so that $g_1^2(dH/d\nu)$ is essentially a constant which can be removed from the integral. Then the integral is proportional to g_1^2/g [Aasa and Vänngard (25)]. Pilbrow (373) has more recently discussed this g^{-1} factor for high-spin molecules where the effective g values are very anisotropic at low magnetic fields. Van Veen (441) has presented an efficient method for simulating complete powder spectra that reduces the considerable computation time involved, particularly in high-spin cases.

(β) Nuclear Hyperfine Effects. If now there is also hyperfine interaction in the randomly oriented molecules, the pattern in Fig. II-9b will be split into $2I+1$ patterns, if one nucleus of spin I is involved. Such a spectrum for $I=\frac{1}{2}$ is shown in Fig. II-9c where because $g_{\parallel} \cong g_{\perp}$ and $A_{\perp} < A_{\parallel}$, the lines for $m_I = +\frac{1}{2}$ and $-\frac{1}{2}$ point in opposite directions. If, instead, $g_{\perp}$ were shifted up-field relative to $g_{\parallel}$ and $A_{\perp} \cong A_{\parallel}$, one could obtain two patterns similar to that in Fig. II-9a but separated by the hyperfine splitting. Such a spectrum is shown in Fig. II-10. This is the spectrum of the randomly oriented BaF molecule (trapped in solid neon at 4°K) where the one unpaired electron is largely on the nonmagnetic Ba atom but is interacting weakly with the nuclear moment ($I=\frac{1}{2}$) of the fluorine atom to produce a splitting of about 20 G for both $A_{\parallel}$ and $A_{\perp}$ (see section II-6).

The Hamiltonian for a linear $S=\frac{1}{2}$ molecule in a magnetic field and containing a nucleus with spin was given in (II, 49). If the hfs is small relative to the Zeeman energy, second-order perturbation yielded the energy levels given in (II, 57) and the resonant magnetic fields for allowed transitions in (II, 64). Here we cannot solve explicitly for $\sin\theta$ as in (II, 72) so that $|dN/dH|$ cannot be written as

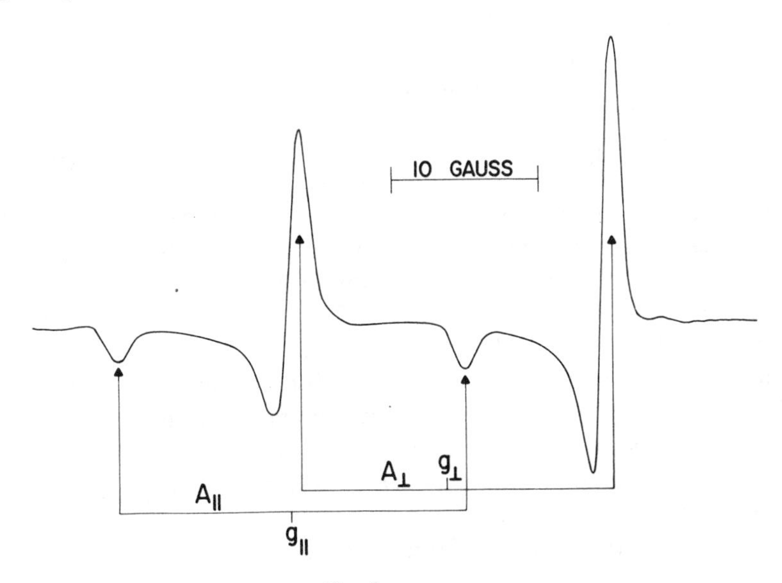

Fig. II-10. ESR spectrum of the $Ba^{19}F$ ($^2\Sigma$) molecule [from Knight, Easley, Weltner, and Wilson (278)].

a function of only the magnetic parameters. From (II, 64) for $S = \frac{1}{2}$, one finds $d\theta/dH$, and therefore, from (II, 76) [Rollmann and Chan (394)],

$$\frac{dN}{dH} = \frac{N_0}{2}\left(\frac{g^2}{\cos\theta}\right)\Bigg\{(g_\parallel{}^2 - g_\perp{}^2)H_0 + m_I\left[\frac{(A_\parallel{}^2 g_\parallel{}^4 - A_\perp{}^2 g_\perp{}^4)}{Kg^2}\right.$$
$$\left. + 2K(g_\perp{}^2 - g_\parallel{}^2)\right] - [I(I+1) - m_I{}^2]\left(\frac{A_\parallel{}^2 A_\perp{}^2 g_\parallel{}^2 g_\perp{}^2}{4H_0 g^4}\right)$$
$$\times\left[\frac{2(A_\parallel{}^2 g_\parallel{}^4 - A_\perp{}^2 g_\perp{}^4)}{K^4 g^2} + \frac{(g_\perp{}^2 - g_\parallel{}^2)}{K^2}\right]$$
$$+ [I(I+1) - m_I{}^2]\left(\frac{A_\perp{}^2 g_\perp{}^2}{4H_0 g^2}\right)(g_\perp{}^2 - g_\parallel{}^2)$$
$$- \left[\frac{(A_\parallel{}^2 g_\parallel{}^2 - A_\perp{}^2 g_\perp{}^2)^2 g_\parallel{}^2 g_\perp{}^2 m_I{}^2}{H_0 g^4}\right]\frac{(\cos^2\theta - \sin^2\theta)}{K^2 g^2}$$
$$- \left[\frac{(A_\parallel{}^2 g_\parallel{}^2 - A_\perp{}^2 g_\perp{}^2)^2 g_\parallel{}^2 g_\perp{}^2 m_I{}^2}{2H_0 g^4}\right]\left[\frac{2(A_\parallel{}^2 g_\parallel{}^4 - A_\perp{}^2 g_\perp{}^4)}{K^4 g^6}\right.$$
$$\left. - \frac{3(g_\perp{}^2 - g_\parallel{}^2)}{K^2 g^4}\right]\sin^2\theta\cos^2\theta\Bigg\}^{-1}$$
$$+ \text{terms involving } Q' \qquad \text{(II, 79)}$$

This expression then yields the intensities of transitions corresponding to each m_I value (since $\Delta m_I = 0$) if it is also multiplied by $g_1{}^2/g$ to account for the change of transition probability with orientation. One must now solve (II, 64) and (II, 79) (multiplied by $g_1{}^2/g$) for a series of values of θ to obtain the resonant fields and intensities as a function of orientation. As noted by Rollman and Chan, although (II, 79) is lengthy it can be readily programmed for a computer. Those authors have given the Q' terms and also the corresponding expressions for the "forbidden" $\Delta m_I = \pm 1, \pm 2$ transitions, which are nevertheless often observed.

With hyperfine splitting, from (II, 64),

$$H \cong H_0 - A_\parallel m_I \qquad \text{at } \theta = 0°$$

$$H \cong H_0 - A_\perp m_I \qquad \text{at } \theta = 90°$$

and again $|dN/dH| \rightarrow \infty$ at $\theta = 90°$ from (II, 79). The absorption pattern for a randomly oriented molecule in which there is also hf interaction is then a superposition of $2I + 1$ patterns of the type shown in Fig. II-9a. The first derivative of that superposition is the observed spectrum. In Figs. II-9c and II-10 where $I = \frac{1}{2}$, the two patterns did not overlap, so that the spectra are relatively simple (see Fig. I-11). An example of a more complex nuclear hyperfine pattern was actually given in Fig. I-10 for quenched ^{10}B and ^{11}B atoms in argon at 4°K. Because it sits in a site of essentially cylindrical symmetry, the perturbed B atom has the properties of a $^2\Sigma$ molecule with disparate g values along and perpendicular to the site axis. Also, the hf splitting parallel to the axis is much larger than that perpendicular to the axis, which is unresolved in that figure. The absorption pattern (a more complex version of Fig. II-9a) is then made up of the sum of four for ^{11}B ($I = \frac{3}{2}$), or seven for ^{10}B ($I = 3$), overlapping curves. Figure II-11 shows an idealized ESR spectrum of a $^2\Sigma$ molecule containing a nucleus with $I = \frac{3}{2}$, anisotropic $\boldsymbol{g}$ tensor, and $A_{\parallel}$ slightly larger than $A_{\perp}$. It is similar to the ESR spectrum for ^{11}B but with resolved perpendicular hfs.

The copper nitrate and copper fluoride molecules, to be discussed in the next section, also have anisotropic $\boldsymbol{g}$ and $\boldsymbol{A}$ tensors, and, as will be seen, they have ESR spectra in matrices similar to those of quenched boron atoms. In some matrices they also exhibit the interesting phenomenon of preferential orientation.

(γ) Nonrandom Molecular Orientation. It has been observed that during the condensation of a beam of radicals in solid neon or argon matrices at temperatures near 4°K; preferential orientation of the isolated molecules may occur relative to the flat condensing surface. Nonrandom orientation can easily be detected by turning the matrix in the magnetic field; if the ESR spectrum changes at all at any position of the sample, then the detected radicals have been trapped with some preferential alignment.

$Cu(NO_3)_2$ and CuF_2. This phenomenon was first observed when the $Cu(NO_3)_2$ molecule vaporizing at 150°C was isolated in a neon matrix at 4°K [Kasai, Whipple, and Weltner (266)]. The copper nitrate molecule is planar with the four nitrate oxygen atoms in an approximately square array about the Cu^{+2} ion:

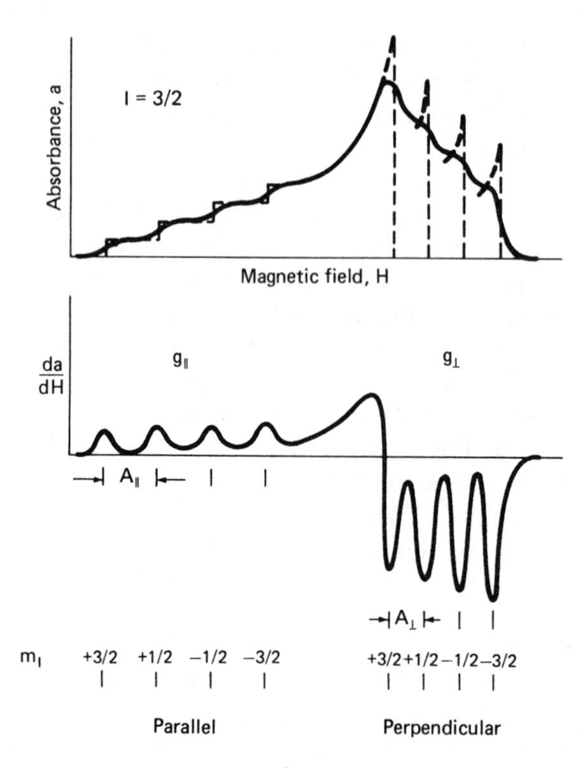

Fig. II-11. Idealized ESR spectrum of a $^2\Sigma$ molecule containing an $I = \frac{3}{2}$ nucleus and with anisotropic $\boldsymbol{g}$ and $\boldsymbol{A}$ tensors. (Top) Absorption spectrum; (bottom) ESR spectrum, first derivative of the absorption.

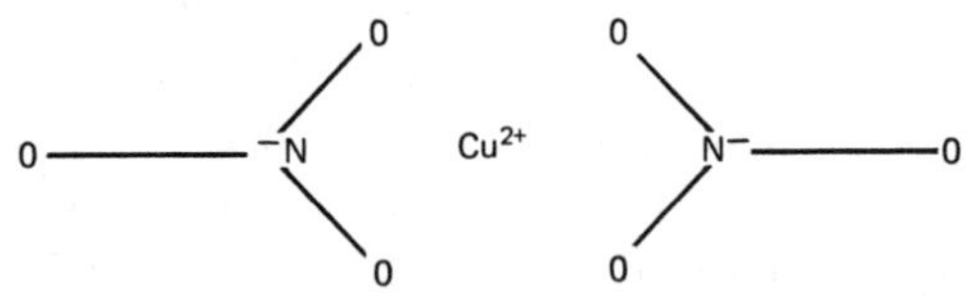

The magnetic properties of a cupric compound with D_{4h} symmetry are characterized by a $g_\parallel$ and $g_\perp$, relative to its symmetry axis normal to the plane of the $S = \frac{1}{2}$ molecule. This then leads to an ESR spectrum for randomly oriented radicals like that in Fig. II-9c, except that the $I = \frac{3}{2}$ nuclear spin of the Cu isotopes produces hf splitting into four lines at each g value. The striking orientation effects that occur are shown in Fig. II-12, where the flat face of the cold sapphire

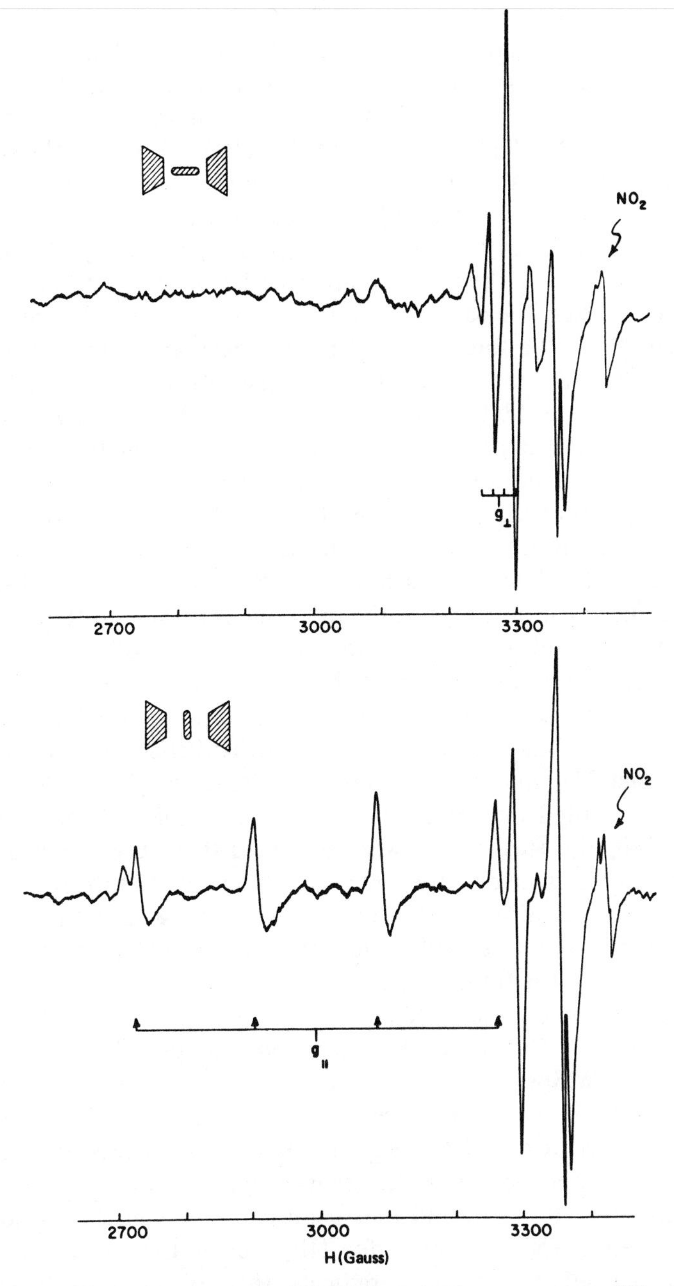

Fig. II-12. ESR spectrum of the $Cu(NO_3)_2$ molecule in a neon matrix at 4°K exhibiting orientation effects [from Kasai, Whipple, and Weltner (266)]. (a) Observed with the flat face of the rod perpendicular to the magnetic field; (b) with the flat face of the rod parallel to the magnetic field.

rod is turned perpendicular and parallel to the applied magnetic field. It is not possible to eliminate NO_2 from the matrix, since it is a decomposition product of $Cu(NO_3)_2$, and unfortunately its spectrum overlaps the high-field lines of the latter. However, the spectra demonstrate the almost complete disappearance of the low-field ($g_\parallel$) lines when the flat sapphire faces are parallel to the magnetic field. This indicates that the $Cu(NO_3)_2$ molecules are preferentially oriented such that their planes are parallel to the flat faces of the sapphire. When the sapphire flats are turned perpendicular to the magnetic field, the four low-field lines appear, and a decrease occurs in the high-field lines. Two of the strong signals near 3290 and 3360 G in Fig. II-12 are "extra" lines (to be discussed in the next section) that have their maximum intensities at some intermediate angle and should therefore not disappear at either orientation of the rod.

Whipple (266) has considered the orientation distribution function of the nitrate molecules necessary to produce the observed spectrum in solid neon. It is concluded that the peak of the distribution occurs when the plane of the molecule is parallel to the substrate surface, and that the distribution drops off rapidly as the angle increases. The best fit indicates that no molecules are turned at 90° to the surface. It is also significant that no preferential orientation was observed in argon matrices. On the other hand, the same workers found that the linear CuF_2 molecule (vaporizing at 925°C) was strongly oriented in argon matrices with the molecular axis parallel to the sapphire surface. It was concluded that "the mechanism of orientation must involve the large thermal gradient at the rod surface (during condensation) as the driving force in causing the neon or argon atoms to preferentially adjust themselves to form an atomically smooth site for the trapped species."

Preferential orientation in rare gas matrices also occurred when the nonlinear NO_2 [Kasai, Weltner, and Whipple (264)] and NF_2 [Kasai and Whipple (265)] molecules were isolated. A thorough study of the orientation of these molecules in several matrices was made by McDowell, Nakajima, and Raghunathan (337a), where a thin, flat, copper target, rather than sapphire, was the surface for condensation. It was found that neon is the best host for orienting NO_2, whereas argon is better for NF_2. Even though they are differently classified as σ and π radicals, this was presumably not an important factor since both molecules were found to have their molecular planes preferentially oriented parallel to the surface of deposition.

BO. Another case of extreme preferential orientation occurred when the BO molecule was trapped in neon matrices at 4°K on a flattened single-crystal sapphire rod [Knight, Easley, and Weltner (277)]. The BO molecule was vaporized from a mixture of solid boron plus alkaline-earth oxide at 1800°K. The overall ESR spectrum of ^{10,11}BO in a neon matrix is presented in Fig. II-13, showing clearly the strong perpendicular lines, four due to the 80% natural

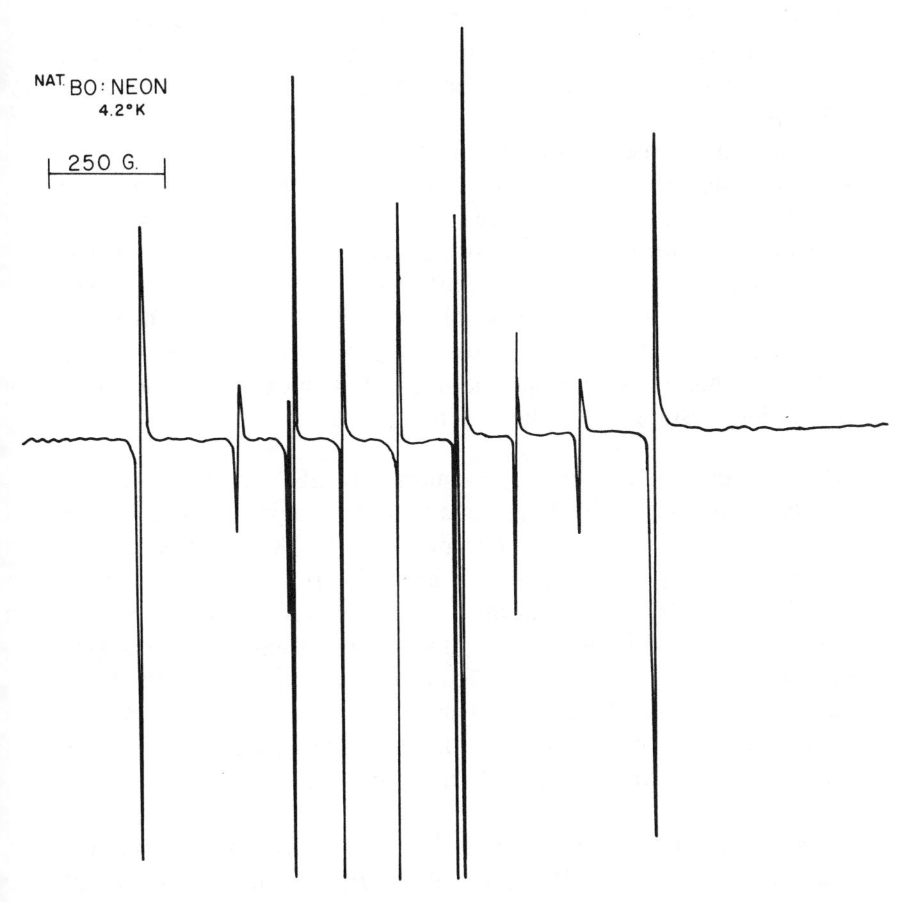

Fig. II-13. ESR spectrum of the BO molecule in a neon matrix at 4°K [from Knight, Easley, and Weltner (277)]. X-band, center of spectrum at ~3300 G.

abundance of ^{11}B ($I = \frac{3}{2}$) and seven due to the 20% natural abundance of ^{10}B ($I = 3$) in the molecule. An expanded view of the individual lines in that spectrum with the matrix oriented at various angles to the magnetic field indicated that strong preferential orientation of the trapped BO was occurring. Recent more thorough studies have distinguished several matrix sites and have clarified the earlier observations [Knight, Wise, Davidson, and McMurchie (286)]. Two major sites in neon were observed, one in which motional averaging was occurring (isotropic, ISO, site) and the other exhibiting strong orientation effects (OR site). Figure II-14 shows the variation of the $m_I = +3/2$ line of ^{11}BO in the two sites as the sapphire rod is turned about its vertical axis in the magnetic field. In an argon matrix prepared in the same manner only randomly oriented molecules were observed.

The high degree of orientation of BO in neon is probably chiefly a consequence of its size and its ability to fit into a substitutional or interstitial site in the neon lattice (277). Rough considerations of the sizes of the molecule and of possible neon lattice sites indicate that it could possibly occupy a substitutional plus octahedral-interstitial site. Since the matrix is probably polycrystalline, it is then also necessary that the crystallites be strongly oriented along the rod surface. This is all very speculative, but it does indicate a rather fascinating crystal growth phenomenon.

Other Observations of Orientation. From matrix ESR spectra of many, chiefly linear, molecules it appears that some degree of orientation is the rule, rather than the exception. VO [Kasai (258)], MgF [Knight, Easley, Weltner, and Wilson (278)], and ZnF [Knight, Mouchet, Beaudry, and Duncan (279)] all exhibit strong orientation effects. It should be noted that almost all of these molecules were also prepared by vaporization from a relatively high-temperature source. Generally, the degree of preferential orientation in rare-gas matrices appears to depend upon the size and shape of the molecule, the properties of the matrix, and other factors that are not completely understood.

Williams and co-workers (342, 352) have also observed extreme orientation effects in ESR spectra of "matrix-isolated" radicals, but they were produced in a completely different manner. Slow cooling of a 5 mol % solution of C_2F_4, for example, in methylcyclohexane-

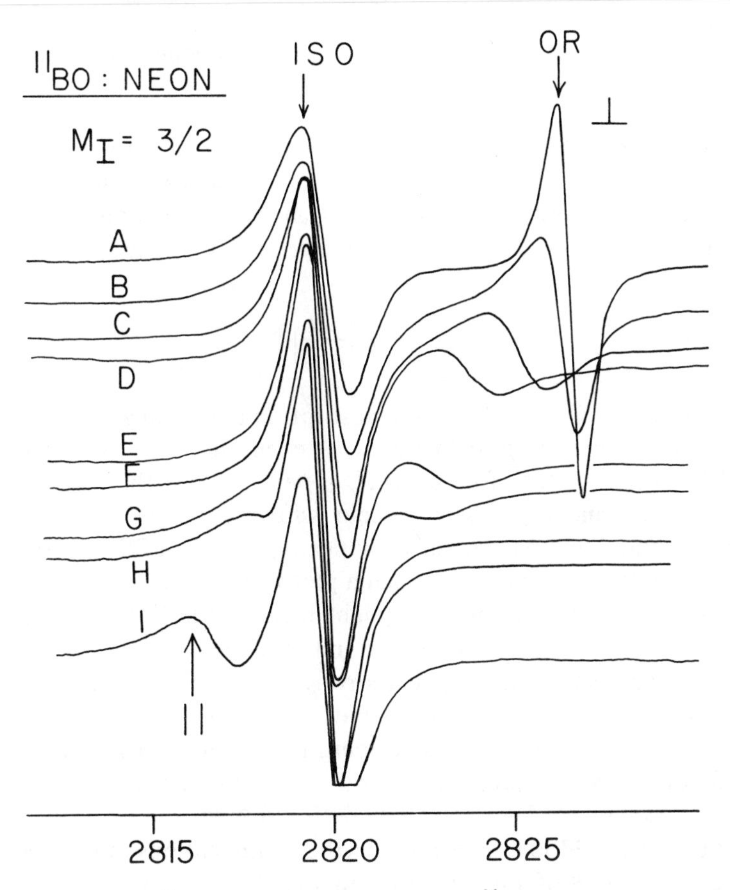

Fig. II-14. The $m_I = +\frac{3}{2}$ lines in the ESR spectrum of the ^{11}BO molecule in a neon matrix at 4°K [from Knight, Wise, Davidson, and McMurchie (286)]. Two sites are indicated: ISO, in which rotation is occurring, and OR, in which the molecule is strongly oriented relative to the sapphire rod surface. $A(\alpha = 0°)$ to $I(\alpha = 90°)$ indicate various angles of rotation of the rod surface relative to the magnetic field, where α is the angle between the normal to the surface and the magnetic field direction.

d_{14} (MCHD) (at ~1°C per minute to 77°K) produces a pseudocrystalline matrix with strong alignment of the solute molecules. This is demonstrated by γ irradiation of the matrix yielding $C_2F_4^-$ radicals, which are detected by ESR. By rotating the sample tube in the field, highly resolved "perpendicular" or "parallel" features could be brought into prominence. Similar extreme orientation effects were observed by slow cooling in a tetramethylsilane (TMS) matrix, and for other solutes such as chlorotrifluoroethylene, bromotrifluoroethylene, and CF_3I.

(δ) Off-Principal-Axis Absorption; "Extra" Lines. It was shown in the previous section that, for randomly oriented radicals, the relative intensities of absorption of observed transitions is mainly determined by the orientation θ of the molecule relative to $\boldsymbol{H}$, or approximately by

$$\sin\theta \left|\frac{d\theta}{dH_r}\right|$$

This results in abrupt changes in absorption at $g_{\parallel}$ and $g_{\perp}$, i.e., when H is aligned along the principal axes, as exemplified in Figs. II-12 and II-14. Then a plot of θ versus H_r is very illuminating, since strong absorption can only possibly occur where the slope $|d\theta/dH_r|$ is its largest, and the first derivative appearing in the ESR spectrum accentuates those transitions. Such a plot is illustrated in Fig. II-15 for the four hyperfine transitions (eight including both Cu isotopes) of $Cu(NO_3)_2$. One can then understand why the four parallel lines and four perpendicular lines appear strongest in the spectrum, since the slopes are highest ($\sim\infty$) at $\theta = 0$ and $90°$. That is the usual situation; however, in this case it is also evident that there are two other values of θ (or H_r) where $|d\theta/dH_r|$ is large for two of the hf lines of a particular isotope. They occur here for $\theta \cong 80°$, $H_r = 3290$ G and $\theta \cong 63°$, $H_r = 3360$ G and provide the complications in the "perpendicular" region of the spectrum in Fig. II-15. They should more definitively be referred to as "off-principal-axis lines" in powder spectra.

These extra lines were first noticed and explained by Neiman and Kivelson (358) in the ESR spectrum of copper phthalocyanine [see also Gersmann and Swalen (179)]. They appear to often arise in polycrystalline samples when the $\boldsymbol{g}$ and $\boldsymbol{A}$ tensors are higher anisotropic. For just that reason more exaggerated cases of anomalous θ versus H_r plots occur for VO and NbO matrices; these are $^4\Sigma$ molecules, to be discussed latter in Chapter V.

θ versus H_r curves can become very complex for high-spin molecules because of fine structure effects, as opposed to hyperfine structure splittings in the above examples. The variation in the Zeeman energy level patterns with θ in molecules where the zero field splitting (zfs) parameter D is appreciable will often, because of avoided cross-

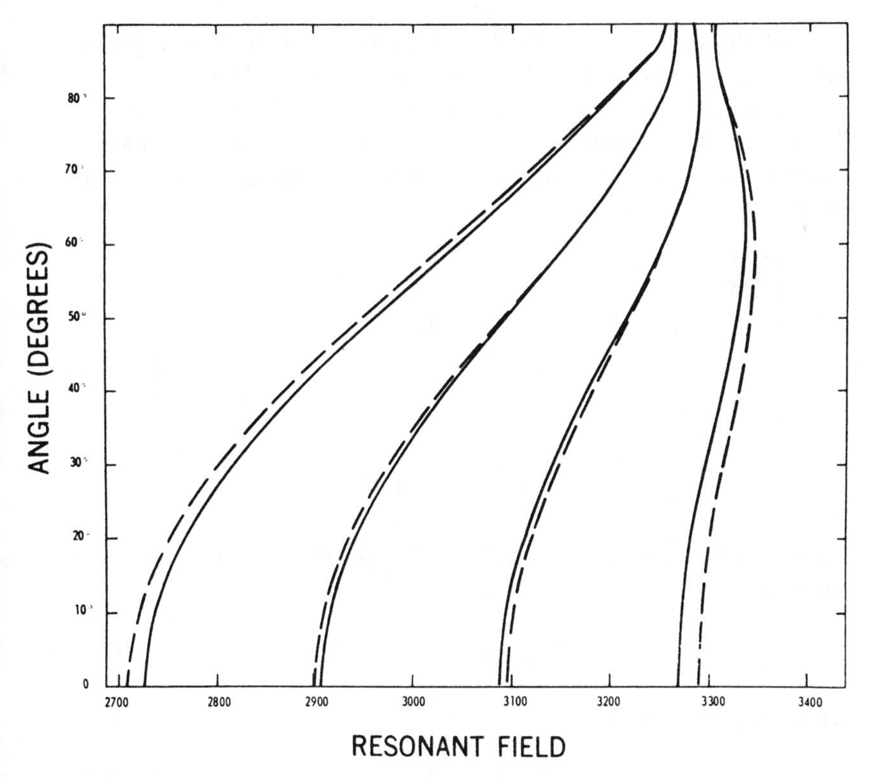

Fig. II-15. Calculated variation of the four hyperfine transitions of the copper nitrate molecule with the angle between the molecular axis and the magnetic field [from Kasai, Whipple, and Weltner (266)]. Solid lines, $^{63}Cu(NO_3)_2$; dotted lines, $^{65}Cu(NO_3)_2$.

ings, lead to extreme curvature in the pattern of a particular transition. This then can have very important effects upon the ESR spectra of high-S molecules, as will be shown in later chapters.

4. NONAXIAL DOUBLET MOLECULES

If a radical possesses no axis of trigonal or higher symmetry then its $\boldsymbol{g}$ tensor and hyperfine tensors $\boldsymbol{A}_i$ will each have three distinct principal components instead of two. More generally, in the spin Hamiltonian

$$\mathcal{H} = \beta \boldsymbol{H} \cdot \boldsymbol{g} \cdot \boldsymbol{S} + \sum_i \boldsymbol{I}_i \cdot \boldsymbol{A}_i \cdot \boldsymbol{S} - \beta \sum_i \boldsymbol{H} \cdot \boldsymbol{g}_i^{(I)} \cdot \boldsymbol{I}_i + \sum_j \boldsymbol{I}_j \cdot \boldsymbol{Q}_i \cdot \boldsymbol{I}_j \qquad \text{(II, 80)}$$

each of the tensors $\boldsymbol{g}$, $\boldsymbol{A}_i$, $\boldsymbol{g}_i^{(I)}$, and $\boldsymbol{Q}_j$ will be of orthorhombic symmetry; i.e., no two principal components will be the same.

Each hyperfine tensor will still involve isotropic contributions from s electron interactions with the nuclear moment, so that in analogy with (II, 36), when referred to its principal axes, we may write

$$\boldsymbol{A} = \begin{vmatrix} A_{xx} & 0 & 0 \\ 0 & A_{yy} & 0 \\ 0 & 0 & A_{zz} \end{vmatrix} = \begin{vmatrix} A_{\text{iso}} & 0 & 0 \\ 0 & A_{\text{iso}} & 0 \\ 0 & 0 & A_{\text{iso}} \end{vmatrix} + \begin{vmatrix} T_{xx} & 0 & 0 \\ 0 & T_{yy} & 0 \\ 0 & 0 & T_{zz} \end{vmatrix}$$

or

$$\boldsymbol{A} = A_{\text{iso}} \mathbf{1} + \boldsymbol{T} \qquad \text{(II, 81)}$$

Here $\boldsymbol{T}$ is the usual symbolism used for the anisotropic traceless tensor where

$$T_{xx} + T_{yy} + T_{zz} = 0 \qquad \text{(II, 82)}$$

which also implies that

$$A_{\text{iso}} = \tfrac{1}{3}\,(A_{xx} + A_{yy} + A_{zz}) \qquad \text{(II, 83)}$$

In general the principal axis systems for the $\boldsymbol{g}$ and $\boldsymbol{A}$ tensor may not coincide (see the footnote on page 42), but often the symmetry of the molecule will dictate that they do. The latter is true, for example, for the $\boldsymbol{g}$ and $\boldsymbol{A}(^{14}N)$ tensors of the bent $^{14}NO_2$ molecule (C_{2v} symmetry), but, on the other hand, only the direction perpendicular to the plane of the bent ^1HOO radical is a shared principal axis of its $\boldsymbol{g}$ and $\boldsymbol{A}_i$ tensors. Thus, in general, for low symmetry the anisotropic hf tensor and therefore the $\boldsymbol{A}$ tensor will have nonzero off-diagonal elements (T_{xy} and A_{xy}, etc.) when referred to the principal axes of the $\boldsymbol{g}$ tensor. This will be discussed in more detail in section (b) below.

(a) g and A Having the Same Principal Axes; $^{14}NO_2$, $^{13}CO_2^-$

If the magnetic field has direction cosines $l_x = \sin\theta\cos\phi$, $l_y = \sin\theta\sin\phi$, and $l_z = \cos\theta$ with respect to axes that are simultaneously principal axes for both g and A_i, then by a procedure analogous to that followed in sections II-1d, e one can derive an expression similar to (II, 64) but for orthorhombic $\boldsymbol{g}$ and $\boldsymbol{A}$ tensors [McClung (329)]. Defining

$$\begin{aligned} g &= [g_\alpha{}^2 \sin^2\theta + g_z{}^2 \cos^2\theta]^{1/2} \\ g_\alpha &= [g_x{}^2 \cos^2\phi + g_y{}^2 \sin^2\phi]^{1/2} \\ A &= [g^2 A_\alpha{}^2 \sin^2\theta + g_z{}^2 A_z{}^2 \cos^2\theta]^{1/2}/g \\ A_\alpha &= [g_x{}^2 A_x{}^2 \cos^2\phi + g_y{}^2 A_y{}^2 \sin^2\phi]^{1/2}/g \end{aligned} \qquad \text{(II, 84)}$$

second-order perturbation theory yields for the resonant magnetic fields for $\Delta m_S = 1 (m_S \rightarrow m_S + 1)$, $\Delta m_I = 0$ transitions:

$$\begin{aligned} H = H_0 &- Am_I - \frac{g_x g_y g_z A_x A_y A_z}{2g^3 H_0 A}(2m_S + 1) m_I \\ &- \left\{ \frac{g_z{}^2 (g^4 A_\alpha{}^2 - g_\alpha{}^2 g_z{}^2 A_z{}^2)^2}{g_\alpha{}^2 g^8 A^2} \sin^2\theta \cos^2\theta \right. \\ &+ \left. \frac{g_x{}^2 g_y{}^2 (g_x{}^2 A_x{}^2 - g_y{}^2 A_y{}^2)^2}{g_\alpha{}^2 g^6 A^2} \sin^2\theta \sin^2\phi \cos^2\phi \right\} \frac{m_I{}^2}{2H_0} \\ &- \left\{ \frac{g_z{}^2 g^4 A_\alpha{}^2 A_z{}^2}{g^4 g_\alpha{}^2 A^2} + \frac{g_\alpha{}^2 g_x{}^2 g_y{}^2 A_x{}^2 A_y{}^2}{g^6 A_\alpha{}^2} \right. \\ &+ \left. \frac{g_x{}^2 g_y{}^2 g_z{}^4 A_z{}^2 (g_x{}^2 A_x{}^2 - g_y{}^2 A_y{}^2)^2}{g_\alpha{}^2 g^{10} A_\alpha{}^2 A^2} \cos^2\theta \sin^2\phi \cos^2\phi \right\} \\ &\times \frac{[I(I+1) - m_I{}^2]}{4H_0} \end{aligned} \qquad \text{(II, 85)}$$

where all coupling constants are expressed in gauss

$$[A\ (\mathrm{G}) = A\ (\mathrm{ergs})/g\beta, \quad A_\alpha\ (\mathrm{G}) = A_\alpha\ (\mathrm{ergs})/g\beta, \quad A_x\ (\mathrm{G}) = A_x\ (\mathrm{ergs})/g_x\beta, \ \text{etc.}].$$

For $^{14}NO_2$ both the $\boldsymbol{g}$ and $\boldsymbol{A}(^{14}N)$ axes are, by symmetry, as in Fig. II-16, where the molecule lies in the *bc* plane. This radical was first observed in gas-phase magnetic resonance, where a triplet hf splitting yielded $A_{iso}(^{14}N) = 132$ MHz [Castle and Beringer (104);

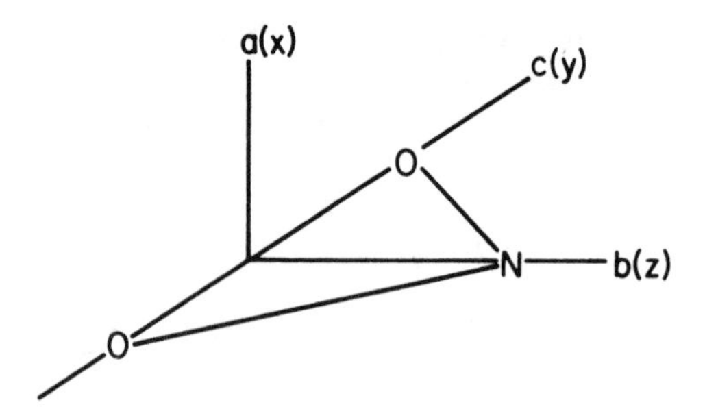

Fig. II-16. Geometry and definition of axes of the NO_2 molecule [from Luz, Reuveni, Holmberg, and Silver (319)].

see Chapter IV], and later in many crystalline solids and rare-gas matrices [see Atkins and Symons (2), pages 127 to 132]. Its orthorhombic $\boldsymbol{g}$ and $\boldsymbol{A}(^{14}N)$ tensors were first obtained by Zeldes and Livingston (483), who produced the radical in an irradiated $NaNO_2$ crystal. Their experimental data are given in columns 2 and 3 of Table II-2. These values are in essential agreement with those obtained for NO_2 trapped in neon and argon matrices [Adrian (29); Kasai, Weltner, and Whipple (264); McDowell, Nakajima, and Raghunathan (337a); Myers, Easley, and Zilles (357)]. (The gas-phase g values for NO_2 will be discussed in Chapter IV, page 231ff.) Nuclear electric quadrupole and nuclear Zeeman effects are too small to be measured for NO_2 in the solid-state spectrum. Then since A_{ii} (MHz) = $2.8025\,(g_{ii}/g_e)\,A_{ii}$ (G), from (II, 83) and

$$A_{ii} = A_{\text{iso}} + T_{ii} \qquad \text{for} \quad i = a, b, c$$

one can calculate A_{iso} and the three principal values of T_{ii}, as given in columns 4 and 5 of Table II-2. This can be done if we assume all A_{ii} to have the same sign, which is reasonable, since $|A_{\text{iso}}| \gg |T_{ii}|$. However, there still remains an ambiguity of *relative* signs; one can choose either those given for A_{iso} and T_{ii} in Table II-2 or all parameters with the opposite signs. This can be decided if the sign of just one parameter can be established. Clearly, A_{iso} is too large to be caused by spin polarization so that it must be positive (disregarding the sign of the nuclear moment, i.e., taking it as positive which is true for ^{14}N); then all signs of the T_{ii} are determined.

The usual interpretation of A_{iso} applies in that when divided by

Table II-2. ESR Data for NO_2 Formed in an Irradiated $NaNO_2$ Crystal[a]

	Experimental Data		Derived Parameters	
Axis i	g_{ii}	$\lvert A_{ii}(^{14}N)\rvert$ (MHz)	$A_{iso}(^{14}N)$ (MHz)	$T_{ii}(^{14}N)$ (MHz)
a	2.0057	138.4(5)	–	–14.8
b	2.0015	190.2(5)	153.2	+37.0
c	1.9910	130.9(5)	–	–22.3

[a] From Zeldes and Livingston (483). Note that the sign convention used here for the hf parameters is the opposite of these authors.

$A_{iso}{}^{^{14}N}$ (atom) = 1811 MHz (see Appendix B), 8.5% s character is derived for the odd electron. This is in keeping with a Walsh-type MO picture in which the odd electron occupies a $3a_1$ orbital which favors a bent molecule. The bond angle will be determined by the relative s and p character. [The reader will recall that Walsh (456) first pointed out that it is this 17th electron that causes NO_2 and CO_2^- to be bent, whereas CO_2 with 16 valence electrons is linear.]

What about the $T_{ii}(^{14}N)$ values? These are determined by the integrals

$$T_{ii} = g_e \beta_e g_I \beta_n \langle (3i^2 - r^2)/r^5 \rangle \qquad \text{where } i = x, y, z \tag{II, 86}$$

T_{zz} is exactly the same as $2A_{dip}$ defined in (II, 30) when $i = z = r \cos \alpha$ or $T_{zz} = g_e\beta_e g_I \beta_n \langle (3 \cos^2 \alpha - 1)/r^3 \rangle$. Then, as shown on page 59 for a $p\sigma$ electron, i.e., along the axis of an atomic p orbital, $\langle (3 \cos^2 \alpha - 1)/2 \rangle_{np\sigma} = \frac{2}{5}$ and therefore

$$T_{zz} = \tfrac{4}{5}\, g_e \beta_e g_I \beta_n \langle 1/r^3 \rangle_{np} \tag{II, 87}$$

Correspondingly (see page 59) at 90° to the $p\sigma$ axis

$$\langle (3 \cos^2 \alpha - 1)/2 \rangle_{np\pi} = -\tfrac{1}{5}$$

and

$$T_{xx} = T_{yy} = -\tfrac{2}{5}\, g_e \beta_e g_I \beta_n \langle 1/r^3 \rangle_{np} \tag{II, 88}$$

or

$$T_{zz} = -2T_{xx} - 2T_{yy} \tag{II, 89}$$

Note that the T_{ii} in Table II-2 approximate these conditions for a pure $p\sigma$ orbital rather well if the orbital is directed along the b molecular axis. The departure from (II, 89) is due to the fact that some spin is in a $2p$ orbital perpendicular to the plane of the molecule, i.e., along the a axis. Thus $\boldsymbol{T}(^{14}\mathrm{N})$ can be resolved into

$$\begin{vmatrix} -19.8 & & \\ & +39.5 & \\ & & -19.8 \end{vmatrix} + \begin{vmatrix} +5.0 & & \\ & -2.5 & \\ & & -2.5 \end{vmatrix}$$

where the second tensor accounts for the $2p$ electron population along the a axis [owing to the mixing in of the $2b_1$ MO in Walsh's scheme (456)]. Then with an approximate wavefunction for the odd electron given by

$$\begin{aligned} \Psi(^2A_1) &= a_{2s}{}^{\mathrm{N}}\chi_{2s}{}^{\mathrm{N}} + a_{2p_b}{}^{\mathrm{N}}\chi_{2p_b}{}^{\mathrm{N}} + a_{2p_a}{}^{\mathrm{N}}\chi_{2p_a} + a_{2p_b}{}^{\mathrm{O}}\chi_{2p_b}{}^{\mathrm{O}} \\ (a_{2s}{}^{\mathrm{N}})^2 &= 153.2/1811 = 0.085 \\ (a_{2p_b}{}^{\mathrm{N}})^2 &= 39.5/111.0 = 0.356 \\ (a_{2p_a}{}^{\mathrm{N}})^2 &= 5.0/111.0 = 0.045 \end{aligned}$$

and $(a_{2p_b}{}^{\mathrm{O}})^2$ by difference = 0.514. The nitrogen atom values of A_{iso} and $T_{zz} = \frac{4}{5}P$ were taken from Appendix B. Also, from the relative s and p character on the N atom one can derive the expected O–N–O bond angle from $\alpha = 2\cos^{-1}(\lambda^2 + 2)^{-1/2}$, where $\lambda = a_{2p}{}^{\mathrm{N}}/a_{2s}{}^{\mathrm{N}} = 2.047$, or $\alpha = 133°$, in excellent agreement with the gas-phase value of 134° [Atkins, Keen, and Symons (43); Coulson (126)].

At about the same time that the NO_2 radical was being studied, its isoelectronic counterpart, $^{13}CO_2^-$, was observed by Ovenall and Whiffen (364). The $\boldsymbol{g}$ tensors of the two radicals are quite similar, and $\boldsymbol{T}(^{13}\mathrm{C})$ has, like $\boldsymbol{T}(^{14}\mathrm{N})$, almost cylindrical symmetry about the C_2 axis. CO_2^-, however, is part of an ion-pair, and some small amount of the spin does reside on the cation to produce observable hfs [Ovenall and Whiffen (364); Atkins, Keen, and Symons (43); an educational discussion of this radical is given by Carrington and McLachlan (6), pages 138ff.].

(b) Axes of g and A_i Not Coincident; $N^{17}O_2$, $^1H^{13}CO$

Although the $\boldsymbol{g}$ tensor of the NO_2 molecule has the directions of its principal axes prescribed by symmetry and the $\boldsymbol{A}(^{14}N)$ tensor must have those same axes, also by symmetry, that is not true for $\boldsymbol{A}(^{17}O)$ if that magnetic isotope is substituted for ^{16}O in the molecule. It is required for that nucleus only that the direction perpendicular to the plane of the molecule be a principal hf axis so that only that axis would be coincident with a $\boldsymbol{g}$ tensor axis.

When $N^{17}O_2$ is formed in an ^{17}O-enriched single crystal of $NaNO_2$ by gamma irradiation, it is found that the molecule lies conveniently aligned with the crystal axes, a, b, and c as shown in Fig. II-16 [Luz, Reuveni, Holmberg, and Silver (319)]. The ESR spectrum exhibits very small ^{17}O ($I = \frac{5}{2}$) hf splittings (<5 G) if the magnetic field direction lies anywhere in the ac plane. In the bc plane the hf splitting becomes largest (~55 G) when the field is almost along b, actually at about ±7.5° to this C_{2v} axis of the molecule. The most probable disposition of the unpaired electron on the O atoms is shown in Fig. II-17. The inplane oxygen $p\sigma$ orbitals are not perpendicular to the N–O bonds but make an angle of 74.5° with them. This implies that the NO bond is "bent," since one expects the bonding oxygen orbital to be orthogonal to this $p\sigma$ orbital.

Then the principal axes of $\boldsymbol{A}(^{17}O)$ lie along a, along b' (in-plane, at an angle of 7.5° to b), and along c' (in-plane, perpendicular to b'). The hf components along these axes are 4.27, 54.90, and 1.74 G,

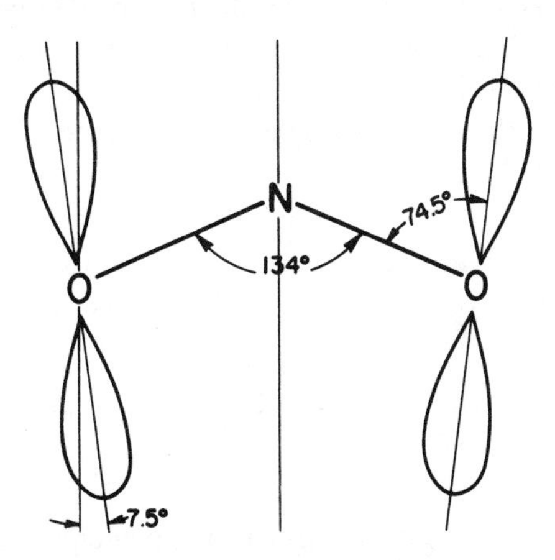

Fig. II-17. Most probable disposition of the in-plane unpaired spins on the oxygen atoms in the NO_2 molecule [from Luz, Reuveni, Holmberg, and Silver (319)].

respectively, so that the $\boldsymbol{A}(^{17}\text{O})$ tensor is essentially axially symmetric. This makes it easier to calculate the ^{17}O hyperfine spacings in the ESR spectrum, since second-order perturbation is adequate, and (II, 64) can be used.

If the samples are polycrystalline or the radical is trapped in a rare-gas matrix, it is no longer possible to measure the ESR spectrum at a definite known orientation of the magnetic field to the trapped radical. Thus, in powders one usually makes measurements of the components of the $\boldsymbol{A}_i$ tensors *relative to the principal axes of the* $\boldsymbol{g}$ *tensor* and not to the principal axes of each $\boldsymbol{A}_i$ tensor. In this reference frame the $\boldsymbol{A}_i$, if not collinear with $\boldsymbol{g}$, will have nonvanishing off-diagonal elements. It is then of concern as to exactly what is being determined when hf splittings are observed in samples containing randomly oriented radicals in the frequent case where $\boldsymbol{A}_i$ and $\boldsymbol{g}$ tensor axes are not coincident.

If x, y, and z are the principal axes of the $\boldsymbol{g}$ tensor, the spin Hamiltonian for this case would be written

$$\mathcal{H} = \beta \sum_k g_k H_k S_k + \boldsymbol{I} \cdot \boldsymbol{A} \cdot \boldsymbol{S} \qquad \text{(II, 90)}$$

where $k = x, y, z$, and quadrupole and nuclear Zeeman terms are neglected. Instead of NO_2, let us consider a radical of even lower symmetry, $^1H^{13}CO$, where for both magnetic nuclei there is only the one common axis shared by the diagonal $\boldsymbol{g}$ tensor and the hyperfine tensor. This is again the axis perpendicular to the plane of the molecule; call it x. Then for each hyperfine tensor the in-plane sets of $\boldsymbol{g}$ and $\boldsymbol{A}$ principal axes will, in general, be rotated by some angle α relative to each other. [This angle can be expected to be different for $\boldsymbol{A}(^1\text{H})$ and $\boldsymbol{A}(^{13}\text{C})$.]

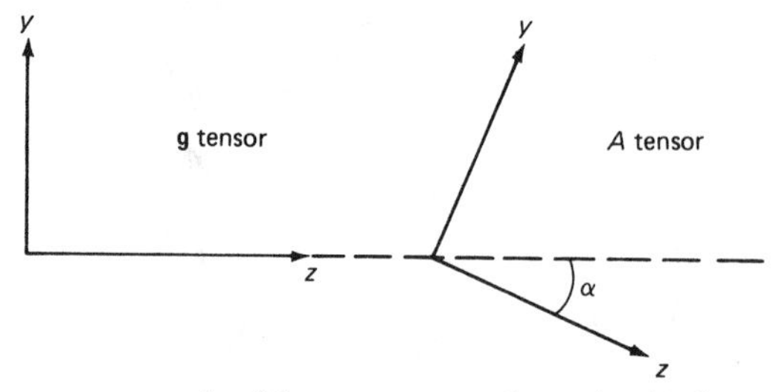

The $\boldsymbol{A}$ tensor measured with respect to the principal axes of the $\boldsymbol{g}$ tensor will be nondiagonal and have elements $A_{xx}, A_{yy}, A_{zz}, A_{yz} = A_{zy}$.

If rotated through the angle α in the molecular plane, it will be diagonalized to $\boldsymbol{A}'$ where

$$
\begin{aligned}
A_{xx}^2 &= A_{xx}'^2 \\
A_{yy}^2 + A_{yz}^2 &= A_{yy}'^2 \cos^2 \alpha + A_{zz}'^2 \sin^2 \alpha \\
A_{zz}^2 + A_{yz}^2 &= A_{yy}'^2 \sin^2 \alpha + A_{zz}'^2 \cos^2 \alpha \\
A_{yz}(A_{yy} + A_{zz}) &= (A_{yy}'^2 - A_{zz}'^2) \sin \alpha \cos \alpha \\
&= \tfrac{1}{2} \tan 2\alpha (A_{yy}^2 - A_{zz}^2)
\end{aligned}
\qquad \text{(II, 91)}
$$

Thus if the elements of the nondiagonal $\boldsymbol{A}$ tensor can be found, the principal values A_{ii}' and the angle α can be calculated. The latter are obtained directly from single crystal studies, as discussed above for $N^{17}O_2$.

Noncoincident $\boldsymbol{g}$ and $\boldsymbol{A}$ axes have been considered by several authors who have derived second-order perturbation expressions for the energy levels, from which the resonant magnetic fields may be obtained [Bleaney and O'Brien (73); Maruani, McDowell, Nakajima, and Raghunathan (327); Lin (309); Golding and Tennant (186); Sakaguchi, Arata, and Fujiwara (397); Rockenbauer and Simon (392); Iwasaki (244)]. [Recent attention has been given this subject by Pilbrow and Lowrey (374) in a review of the effects of low symmetry on the EPR of metal ions in crystals.] For $\Delta m_S = 1\,(m_S \rightarrow m_S + 1)$, $\Delta m_I = 0$ transitions and one magnetic nucleus, the resonant fields can be written

$$
\begin{aligned}
H = H_0 - \frac{K m_I}{g\beta} - \frac{1}{4g\beta h\nu} [K_1 \{I(I+1) - m_I^2\} \\
+ 2K_2 m_I^2 + 2K_3 m_I(2m_S + 1)]
\end{aligned}
\qquad \text{(II, 92)}
$$

where $H_0 = h\nu/g\beta$ and $g^2 = l_x^2 g_{xx}^2 + l_y^2 g_{yy}^2 + l_z^2 g_{zz}^2$. The l_α indicate the direction cosines of $\boldsymbol{H}$ in the principal axis system of $\boldsymbol{g}$. For $S = \frac{1}{2}$ and $I = \frac{1}{2}$

$$
H = H_0 \mp \frac{K}{2g\beta} - \frac{1}{8g\beta h\nu} [K_1 + K_2] \qquad \text{(II, 93)}
$$

One can derive for the case of one common $\mathbf{g}$ and $\mathbf{A}$ principal axis along x [Lin (309); Golding and Tennant (186)]:

$$K^2 = A_{xx}{}^2 (g_{xx}/g)^2 l_x{}^2 + A_{yy}{}^2 (g_{yy}/g)^2 l_y{}^2 + A_{zz}{}^2 (g_{zz}/g)^2 l_z{}^2$$
$$+ A_{yz}{}^2 [(g_{yy}/g)^2 l_y{}^2 + (g_{zz}/g)^2 l_z{}^2]$$
$$+ 2A_{yz}(A_{yy} + A_{zz})(g_{yy} g_{zz}/g^2)\, l_y l_z$$

$$-K_1 = \left(\frac{g_a}{B_a}\right)^2 [A_{zz} A_{yy} - A_{zy}{}^2]^2 + \frac{A_{xx}{}^2 B_a{}^2}{g_a{}^2 A^2} + \frac{l_x{}^2 g_x{}^2 A_{xx}{}^2}{g_a{}^2 g^2 A^2 B_a{}^2}$$
$$\times [(A_{yy}{}^2 - A_{zz}{}^2)\, l_y l_z g_y g_z$$
$$+ A_{yz}(A_{yy} + A_{zz})(l_z{}^2 g_z{}^2 - l_y{}^2 g_y{}^2)]^2$$

$$K_2 = \frac{1}{g_a{}^2 g^2 A^2} [l_y l_z g_y g_z (A_{yy}{}^2 - A_{zz}{}^2)$$
$$+ A_{yz}(l_z{}^2 g_z{}^2 A_{yy} - l_y{}^2 g_y{}^2 A_{zz})]^2 + \frac{l_x{}^2 g_x{}^2}{g_a{}^2 g^4 A^2}$$
$$\cdot [g^2 A_a{}^2 + l_y l_z g_y g_z A_{yz} (A_{yy} + A_{zz}) - g_a{}^2 A_{xx}{}^2]^2 \quad \text{(II, 94)}$$

where, for $i = x, y, z$,

$$g^2 = \sum_i l_i^2 g_i^2 = l_x{}^2 g_x{}^2 + g_a{}^2$$

$$g^2 A^2 = \sum_i l_i^2 g_i^2 A_{ii}^2 = g_x{}^2 l_x{}^2 A_{xx}{}^2 + g^2 A_a{}^2$$

$$B_a{}^2 = (l_y g_y A_{yy} + l_z g_z A_{yz})^2 + (l_y g_y A_{yz} + l_z g_z A_{zz})^2$$

Then, for the magnetic field along x (perpendicular to plane of molecule),

$$K = A_{xx}, \qquad K_1 = A_{yy}{}^2 + A_{zz}{}^2, \qquad K_2 = 0$$

Along y,

$$K^2 = A_{yy}{}^2 + A_{yz}{}^2$$

$$K_1 = \frac{(A_{zz} A_{yy} - A_{zy}{}^2)^2}{(A_{yy}{}^2 + A_{yz}{}^2)} + \frac{A_{xx}{}^2 (A_{yy}{}^2 + A_{yz}{}^2)}{(A_{yy}{}^2)}$$

$$K_2 = \frac{A_{yz}{}^2 A_{zz}{}^2}{A_{yy}{}^2}$$

Along z,

$$K^2 = A_{zz}^2 + A_{yz}^2$$
$$K_1 = \frac{(A_{zz}A_{yy} - A_{zy}^2)^2}{(A_{zz}^2 + A_{yz}^2)} + \frac{A_{xx}^2(A_{zz}^2 + A_{yz}^2)}{A_{zz}^2}$$
$$K_2 = \frac{A_{yz}^2 A_{yy}^2}{A_{zz}^2} \qquad \text{(II, 95)}$$

The resonant fields in (II, 93) along a particular principal axis, i, of $\boldsymbol{g}$ are given by

$$H_i = (h\nu/g_i\beta) \mp \frac{K_i}{2g_i\beta} - \frac{1}{8g_i\beta h\nu}[K_1 + K_2]_i \qquad \text{(II, 96)}$$

Then from the six ^1H hf lines observed when $\boldsymbol{H}$ is along the principal axes of $\boldsymbol{g}$, one can find, in first order,

$$g_{xx}, g_{yy}, g_{zz}, A_{xx}, (A_{yy}^2 + A_{yz}^2)^{1/2},$$

and $(A_{zz}^2 + A_{yz}^2)^{1/2}$. Second-order terms depend in a complicated way on A_{yz}, but in HCO advantage can be taken of the fact that A_{yz}^2/A_{ii}^2 is very small and all A_{ii} are of about the same size. Then one can write, for example,

$$(K^2)_y = A_{yy}^2 \left(1 + \frac{A_{yz}^2}{A_{yy}^2}\right), \quad \text{etc.}$$

and find, by expanding terms involving A_{yz}^2/A_{ii}^2,

$$(K_1 + K_2)_y \cong A_{zz}^2 \left(1 - \frac{2A_{zy}^2}{A_{yy}^2}\right)$$

If such measurements are made on two magnetic isotopes in the molecule, then the $\boldsymbol{g}$ tensor components are common to both, and in principle enough information is available to determine all of the hyperfine coupling parameters. Such was the case for the HCO radical where deuterium and ^{13}C isotopic substitutions could be made, and the ESR spectra observed for all three molecules trapped in solid CO or argon matrices [Adrian, Cochran, and Bowers (34); Cochran, Adrian, and Bowers (116)]. (Those spectra will be discussed below in section 6b.)

Subsequent to the matrix work Holmberg (226) measured the ESR spectrum of the $^1H^{13}CO$ radical formed in a single crystal of formic acid by gamma irradiation. This made possible the direct determination of the principal values of $\boldsymbol{g}$, $\boldsymbol{A}(^1\mathrm{H})$, and $\boldsymbol{A}(^{13}\mathrm{C})$ and the direction cosines of their principal axes relative to the crystal axes. As expected, the perpendicular to the plane of the molecule (called x here) is a common principal axis for all three tensors. One finds from the direction cosines that the principal axes of $\boldsymbol{A}(^1\mathrm{H})$ and $\boldsymbol{A}(^{13}\mathrm{C})$ are rotated in the plane of the molecule by $\alpha = 25.8°$ and $10.6°$, respectively, relative to the principal axes of $\boldsymbol{g}$. Then from (II, 91) the elements of these nondiagonal $\boldsymbol{A}$ tensors may be calculated and are shown in Table II-3. These data are more difficult to obtain from matrix spectra, and although the two sources agree rather well on the $\boldsymbol{g}$ components, the disagreement between the A_{ij} is hard to attribute solely to differences in environment [see page 147, and Golding and Tennant (186)].

The principal hyperfine tensors can again be resolved into isotropic and traceless dipolar tensors [Holmberg (226)]:

$$\boldsymbol{A}'(^1\mathrm{H}) = [354] + [-17, -8, +25]$$

$$\boldsymbol{A}'(^{13}\mathrm{C}) = [365] + [-24, +72, -48]$$

Table II-3. Elements of $A(^1\mathrm{H})$ and $A(^{13}\mathrm{C})$ (in MHz) of the $^1H^{13}CO$ Radical Referred to the Principal Axes of the g Tensor.
[Calculated from Table II of Holmberg (226).]

		^{1}H ($\alpha = 25.8°$)		^{13}C ($\alpha = 10.6°$)	
ij	g_{ii} [a]	A'_{ii} [a]	A_{ij} [b]	A'_{ii} [a]	A_{ij} [b]
xx	2.0037	337	337	341	341
yy	2.0023	346	352.1	438	433.9
yz	–	–	±12.5	–	±21.9
zz	1.9948	378	371.9	317	321.1

[a] Principal values.
[b] These values are not in agreement with those derived by Golding and Tennant (186).

As in $^{14}NO_2$, $\boldsymbol{T}(^{13}C)$ deviates significantly from an axially symmetric p function. It can also be resolved into two axial tensors,

$$[-24, +72, -48] = [-40, +80, -40] + [+16, -8, -8]$$

indicating that the spin occupies the out-of-plane p_x orbital to a slight extent.

According to these anisotropic elements one might then associate approximately the z axis of $\boldsymbol{A}'(^1H)$ with the C–H bond direction and the y axis of $\boldsymbol{A}'(^{13}C)$ with the axis of the σp orbital containing the unpaired electron. These sets of axes are rotated by about 15° with respect to each other according to our analysis of the single crystal data. This leads to the orientation of the magnetic tensor axes roughly as shown in Fig. II-18.

Similar considerations of noncollinearity of $\boldsymbol{g}$ and $\boldsymbol{A}$ tensor axes have been made for ^{19}F hyperfine interaction in the pyramidal CF_3 radical, as mentioned at the end of section 2, and in bent PF_2 [Wei, Current, and Gendell (468)]. In the planar σ phenyl radical the hyperfine tensor of the ortho protons is rotated in the plane by 7.5° relative to the $\boldsymbol{g}$ tensor axes [Kasai, Hedaya, and Whipple (261)].

It is clear that the study of radicals oriented in single crystals

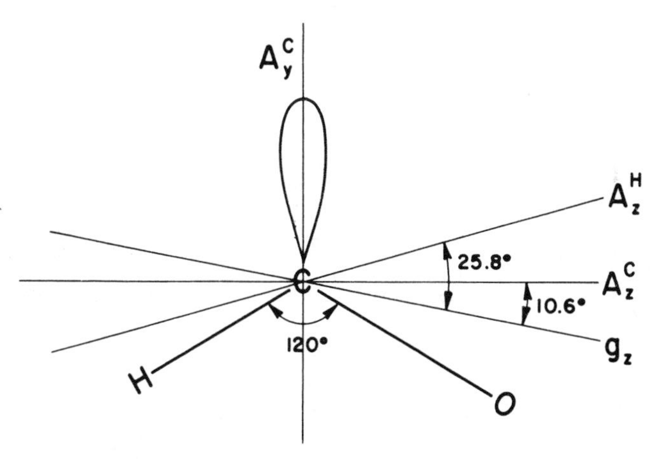

Fig. II-18. Possible relative orientations of $\boldsymbol{A}(^{13}C)$, $\boldsymbol{A}(^1H)$ and $\boldsymbol{g}$ tensor axes in the plane of the HCO radical. $A_y(^{13}C)$ has been arbitrarily taken in the direction of the σp orbital which has been assumed to lie at 120° from the CH and CO bonds. Although the angles of rotation between the three sets of axes are correct, their relative signs are unknown.

makes possible, in principle, the determination of the principal axes of the hyperfine tensor of each magnetic nucleus. This is not generally the case for randomly oriented species in matrices. Thus for low-symmetry organic radicals ESR studies of polycrystalline samples have understandably been avoided because of the common occurrence of noncollinear $\boldsymbol{g}$ and $\boldsymbol{A}$ tensors.

For a particular nucleus, let $\boldsymbol{A}'$ designate its diagonal hyperfine tensor with components along its principal axes x, y, z. Let $\boldsymbol{A}$ be the nondiagonal hf tensor resulting from rotating the axes by angle α about x. [We will assume that $\boldsymbol{A}$ is symmetric here, but this need not always be the case; see McConnell (333).] Then

$$\boldsymbol{A}' = \begin{pmatrix} A_{xx}' & 0 & 0 \\ 0 & A_{yy}' & 0 \\ 0 & 0 & A_{zz}' \end{pmatrix} \quad \text{and} \quad \boldsymbol{A} = \begin{pmatrix} A_{xx} & 0 & 0 \\ 0 & A_{yy} & A_{yz} \\ 0 & A_{yz} & A_{zz} \end{pmatrix} \tag{II, 97}$$

and

$$\boldsymbol{A}^2 = \boldsymbol{l}^{-1}\boldsymbol{A}'^2\boldsymbol{l} \tag{II, 98}$$

where

$$\boldsymbol{l} = \begin{pmatrix} 1 & 0 & 0 \\ 0 & \cos\alpha & \sin\alpha \\ 0 & -\sin\alpha & \cos\alpha \end{pmatrix} \tag{II, 99}$$

Then (II, 96) yields

$$\boldsymbol{A}^2 = \begin{pmatrix} A_{xx}'^2 & 0 & 0 \\ 0 & A_{yy}'^2\cos^2\alpha + A_{zz}'^2\sin^2\alpha & (A_{yy}'^2 - A_{zz}'^2)\cos\alpha\sin\alpha \\ 0 & (A_{yy}'^2 - A_{zz}'^2)\cos\alpha\sin\alpha & A_{yy}'^2\sin^2\alpha + A_{zz}'^2\cos^2\alpha \end{pmatrix} \tag{II, 100}$$

If the elements of the $\boldsymbol{A}^2 = [\boldsymbol{A}\cdot\boldsymbol{A}] = \boldsymbol{A}\cdot\boldsymbol{A}^+$ tensor are designated by $\boldsymbol{A}_{ij}{}^2$,

$$\boldsymbol{A}^2 = \begin{pmatrix} \boldsymbol{A}_{xx}{}^2 & \boldsymbol{A}_{xy}{}^2 & \boldsymbol{A}_{xz}{}^2 \\ \boldsymbol{A}_{xy}{}^2 & \boldsymbol{A}_{yy}{}^2 & \boldsymbol{A}_{yz}{}^2 \\ \boldsymbol{A}_{xz}{}^2 & \boldsymbol{A}_{yz}{}^2 & \boldsymbol{A}_{zz}{}^2 \end{pmatrix} \tag{II, 101}$$

Then comparison of (II, 100) and (II, 101) yields

$$\begin{aligned}
\boldsymbol{A}_{xx}{}^2 &= A_{xx}{}'^2 \\
\boldsymbol{A}_{yy}{}^2 &= A_{yy}{}'^2 \cos^2\alpha + A_{zz}{}'^2 \sin^2\alpha \\
\boldsymbol{A}_{yz}{}^2 &= (A_{yy}{}'^2 - A_{zz}{}'^2)\cos\alpha \sin\alpha \\
\boldsymbol{A}_{zz}{}^2 &= A_{yy}{}'^2 \sin^2\alpha + A_{zz}{}'^2 \cos^2\alpha
\end{aligned} \qquad \text{(II, 102)}$$

With $\boldsymbol{A}$ as in (II, 97),

$$\begin{aligned}
\boldsymbol{A}_{xx}{}^2 &= A_{xx}{}^2 \\
\boldsymbol{A}_{yy}{}^2 &= A_{yy}{}^2 + A_{yz}{}^2 \\
\boldsymbol{A}_{yz}{}^2 &= A_{yz}(A_{yy} + A_{zz}) \\
\boldsymbol{A}_{zz}{}^2 &= A_{zz}{}^2 + A_{yz}{}^2
\end{aligned} \qquad \text{(II, 103)}$$

which when combined with (II, 102) yields (II, 91).

The hyperfine parameters K, K_1, K_2, and K_3 in (II, 92) are as defined by Golding and Tennant (186); only the sign of K_1 has been changed here. In Lin's (309) symbolism they are

$$\begin{aligned}
K &= A \\
K_1 &= T_{xx}{}^2 + T_{yy}{}^2 + T_{xy}{}^2 + T_{yx}{}^2 \\
K_2 &= T_{xz}{}^2 + T_{yz}{}^2 \\
K_3 &= T_{xx}T_{yy} - T_{xy}T_{yx}
\end{aligned}$$

where

$$\boldsymbol{T} = \boldsymbol{U}^{+} \cdot \boldsymbol{A} \cdot \boldsymbol{V}$$

with $\boldsymbol{U}$ and $\boldsymbol{V}$ defined by Lin.

(c) Randomly Oriented Nonaxial Radicals.

For radicals with three distinct principal g components, the ESR absorption spectrum will have the pattern shown in Fig. II-19a if they are rigidly held but randomly oriented in a matrix [Kneubuhl (275)]. This is to be compared with Fig. II-9a for a linear molecule. The first derivative of this absorption, i.e., the observed ESR spectrum, will then appear as in Fig. II-19b. This is similar to Fig. II-9b but with two "parallel-like" lines with opposite phase.

An illustration of this pattern occurs in the spectrum of the

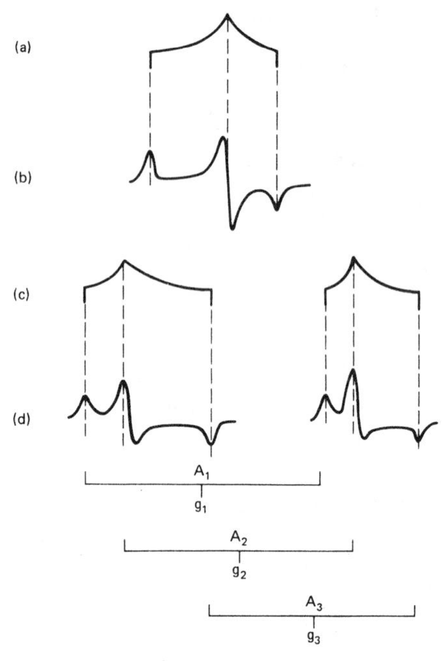

Fig. II-19. ESR spectra of randomly oriented doublet molecules with orthohombic **g** tensors. (a) Absorption spectrum, no hyperfine interaction and three principal g components; (b) ESR spectrum, first derivative of (a); (c) absorption spectrum, orthorhombic **g** and ***A*** tensors for a molecule containing one magnetic nucleus with $I = \frac{1}{2}$; (d) ESR spectrum, first derivative of (c).

$S = \frac{1}{2}$ MgOH molecule isolated in solid argon, *before* the matrix is annealed [Brom and Weltner (90)]. Two lines in that spectrum are shown in Fig. II-20; the doubling occurring within the lines is due to the small ^{1}H hfs. The spectrum provides definite evidence that MgOH is bent in unannealed sites in solid argon at 4°K (see section bb). On the other hand, if the hfs is larger relative to the **g** tensor anisotropy, then a spectrum similar to that in Fig. II-19c,d might be expected for the simple case of $S = \frac{1}{2}, I = \frac{1}{2}$ and $g_1 > g_2 > g_3$. [Other examples of typical powder patterns can be found in Atkins and Symons (2), pages 269ff.] It is clear that, depending upon the values of the **g** and $\boldsymbol{A}_i$ components and the number of magnetic nuclei in the radical, the ESR spectra in matrices can vary in their complexity and may be difficult to analyze.

If, however, one is able to discern the pattern of the transitions in the x, y, and z directions, then the components of the $\boldsymbol{g}, \boldsymbol{A}_i$, and possibly other tensors, can in principle be found from the measured

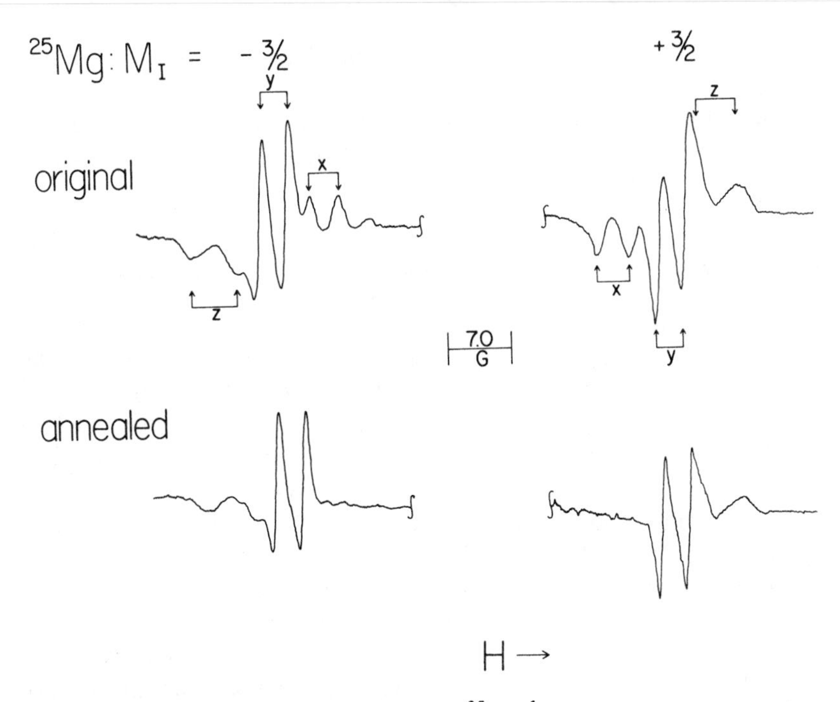

Fig. II-20. ESR lines in the spectrum of the bent $^{25}MgO^1H$ molecule in an argon matrix at 4°K [from Brom and Weltner (90)]. Two of the six ^{25}Mg ($I = \frac{5}{2}$) components are shown. Each is split into three doublets owing to ^1H hyperfine interaction before annealing.

resonant field positions if the tensor axes are collinear. This has been discussed in section 4a above. If $\boldsymbol{g}$ and $\boldsymbol{A}_i$ are not collinear, then further information, perhaps supplied by isotopic substitution, may be needed, as section 4b describes.

(α) Transition Probabilities. The considerations given in section II-1e(α) for the axial case still apply, but g_1 will have a different form if the $\boldsymbol{g}$ tensor of the radical is orthorhombic. Since the original work of Bleaney (71) there have been many derivations of g_1 for this general case [Holuj (227); Pilbrow (372); Isomoto, Watari, and Kotani (241)], and again it may be shown that g_1 is a maximum when $H_1 \perp H$, as in Fig. II-1. For this condition and when the radicals are randomly oriented, the averaged value of g_1 becomes

$$g_1^2 = [g_x^2 g_y^2 (1 - l_z^2) + g_y^2 g_z^2 (1 - l_x^2) + g_z^2 g_x^2 (1 - l_y^2)]/2g^2 \tag{II, 104}$$

where $g^2 = l_x{}^2 g_x{}^2 + l_y{}^2 g_y{}^2 + l_z{}^2 g_z{}^2$, and (l_x, l_y, l_z) are the direction cosines between $\boldsymbol{H}$ and the principal axes of the $\boldsymbol{g}$ tensor [Aasa and Vänngard (25)]. This expression reduces to (II, 62) for the case of axial symmetry when $g_x = g_y = g_\perp$ and $g_z = g_\parallel$, since $l_z{}^2 = \cos^2 \theta$ and $l_x{}^2 + l_y{}^2 + l_z{}^2 = 1$.

5. FREE AND HINDERED ROTATION OF TRAPPED RADICALS

If the constraints of the surrounding crystal or matrix cage are small, the radical may rotate in its site, and only average values of the g_i and A_i will be observed, as if the radical were tumbling in liquid solution. Then $g_{\text{av.}} = \frac{1}{3}(g_x + g_y + g_z)$, and each magnetic nucleus exhibits only its isotropic hf splitting, A_{iso}, in the spectrum. For example, CH_3 radicals trapped in solid krypton are presumably freely rotating, and four equal-intensity ^1H hyperfine lines are observed separated by $|A_{\text{iso}}| = 23$ G.

In the lowest rotational state of CH_3 the nuclear spin state of the three equivalent protons must be symmetric [McConnell (332); Morehouse, Christiansen, and Gordy (347); Jackel and Gordy (246); Jen (248)]. The H nuclei are Fermi particles so that the overall wavefunction must be antisymmetric with respect to their exchange. Since the electronic state of the radical is antisymmetric, the vibrational state is symmetric, and the lowest $J = 0$ rotational state is symmetric, the nuclear spin state must then be symmetric. The four symmetric nuclear spin functions have equal statistical weights, and their energy differences are negligible; hence the four hyperfine components have equal intensity, as shown in the ESR spectrum in Fig. II-21 (top). Therefore in this case of CH_3 trapped in solid krypton at 4°K, the ESR evidence indicates that the sites are large enough to allow essentially free rotation of CH_3, at least about its symmetry axis. However, warming up the matrix should change this intensity distribution, since in the next highest rotational state (14.5 cm^{-1} higher) only the two inner hf lines would be visible, again owing to the symmetry of the nuclear spin functions relative to that of the overall wavefunction. Then if kT approaches that energy difference, the pattern of the proton hfs will approach a 1:3:3:1 intensity distribution, as Fig. II-21 (lower) illustrated when the krypton matrix was warmed to 18°K. This pattern is also the one expected if no rotation occurs or if the motion is strongly hindered,

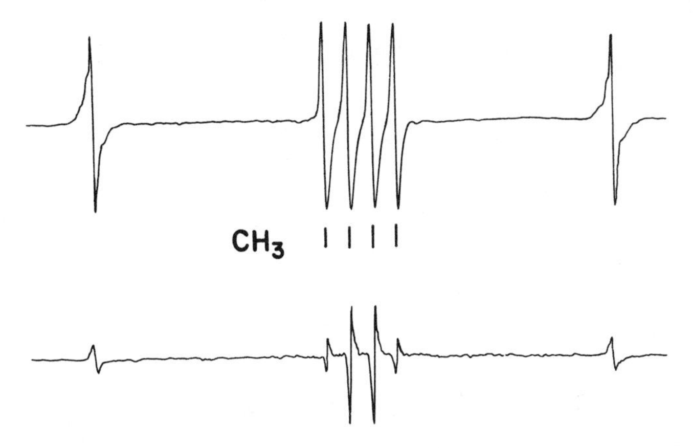

Fig. II-21. ESR spectra of CH_3 radicals obtained by γ irradiation of 1% methane in a krypton matrix [from Morehouse, Christiansen, and Gordy (347)]. (Top) At 4°K; (lower) after warming to 18°K. The two outer lines are ^1H atom signals.

which appears to be true in solid argon and solid methane at 4°K [Jen, Foner, Cochran, and Bowers (251, 248)].

Similar considerations apply to NH_2 matrix-isolated in argon at 4°K where the hyperfine splitting yields three sets of triplets, that is, nine lines, of equal intensity [Foner, Cochran, Bowers, and Jen (167)]. Here again that pattern and the nuclear statistics stipulate that the NH_2 radical is freely rotating [McConnell (332)]. This is contradicted by the spectrum of ND_2 where the intensity distribution indicates that the rotation is not free but is hindered, so that one would also expect a potential barrier to rotation to be present for NH_2 [Jen (248)]. However, it is calculated that the effect of this barrier upon the intensities in the NH_2 spectrum is so small as to be undetectable.

For a bent triatomic molecule rotation may occur preferentially about a particular axis, usually one with a low moment of inertia. This will lead to simplification of the spectrum of the rigidly trapped radical, since it averages the tensor components along the other two axes. Thus, the ESR spectrum of an immobile bent doublet radical would be expected to become that of a linear molecule as shown in Fig. II-22, if it rotated freely about one **g** tensor axis (here axis a). $^{14}NO_2$ is perhaps the most studied molecule in this regard, since apparently the barrier to rotation about an axis parallel to the O----O direction is small in many matrices. This is the axis of smallest moment of inertia, since the bond angle is 134°. Motional

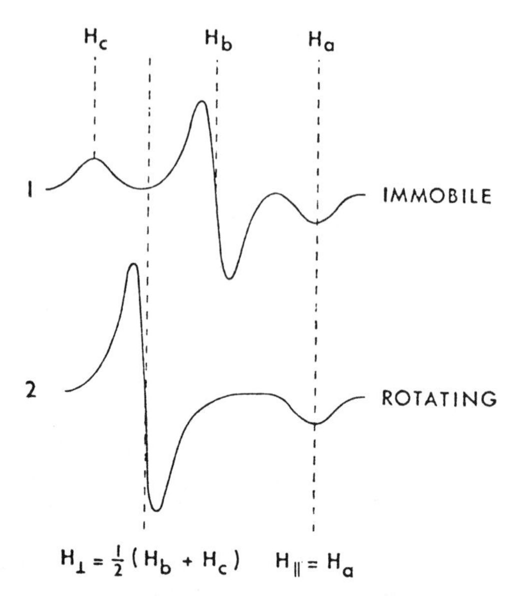

Fig. II-22. Qualitative shape of one m_I component for immobile and rotating NO_2 [from Schaafsma and Kommandeur (403)].

averaging about this axis was first proposed by Adrian (29) as an interpretation of the ESR spectrum of $^{14}NO_2$ trapped in an argon matrix [Jen, Foner, Cochran, and Bowers (251)]. The validity of that interpretation appears to be a function of the degree of annealing of the argon matrix, and spectra in solid neon indicate that both fixed oriented sites (see page 106) and rotational sites are present there [Kasai, Weltner, and Whipple (264); Myers, Easley, and Zilles (357)]. Such mixed sites also occur in a solid N_2O_4 matrix where NO_2 radicals are produced by ultraviolet photolysis [Schaafsma and Kommandeur (403)].

Although in neon at 4°K the NF_2 radical is rigidly held [Kasai and Whipple (265)], in argon the ESR spectrum appears to indicate that rotation is taking place about the axis perpendicular to the plane of the molecule, i.e., not as for trapped NO_2. This is justified for the difluoride by the fact that the bond angle is much more acute (104°) than in the dioxide [Adrian, Cochran, and Bowers (33); Farmer, Gerry, and McDowell (161)]. On the other hand, it is proposed that the similar PF_2 radical is nonrotating in an argon matrix [Wei, Current, and Gendell (468)], although this has been questioned because its g tensor has axial symmetry, suggesting again averaging about an axis of rotation [Myers, Easley, and Zilles (357)].

In the case of the CN radical in solid argon it is possible to follow the marked change in the spectrum from an anisotropic, relatively "rigid" spectrum to one approaching an isotropic pattern characteristic of a rapidly rotating molecule, as the temperature of the matrix is increased [Adrian, Cochran, and Bowers (33); Easley and Weltner (153)]. Figure II-23 shows the gradual change in the appearance of the spectrum with temperature; this behavior is completely reversible if the temperature is not taken above about 35°K. At 35°K the spectrum shows only three ^{14}N hyperfine lines of almost equal intensity. Such observations are informative and also helpful in the analysis of powder ESR spectra, since the $g_{av.}$ and A_{iso} values measured from the isotropic spectrum can serve as checks on the assignment of the tensor components. It is worth noting that the high-field line in the rapidly rotating spectrum in Fig. II-23c is broader than the other two lines. This can also provide, when interpreted by the theory of spin-relaxation [McConnell (331); Stephen and Fraenkel (422); Kivelson (274); Carrington and Longuet-Higgins (102); Freed and Fraenkel (171)], information about the sign of the isotropic hf interaction constant.

This m_I dependence of hf linewidths is illustrated strikingly in the ESR spectrum of $^{28}SiH_3$ in an argon matrix at 4°K, shown in

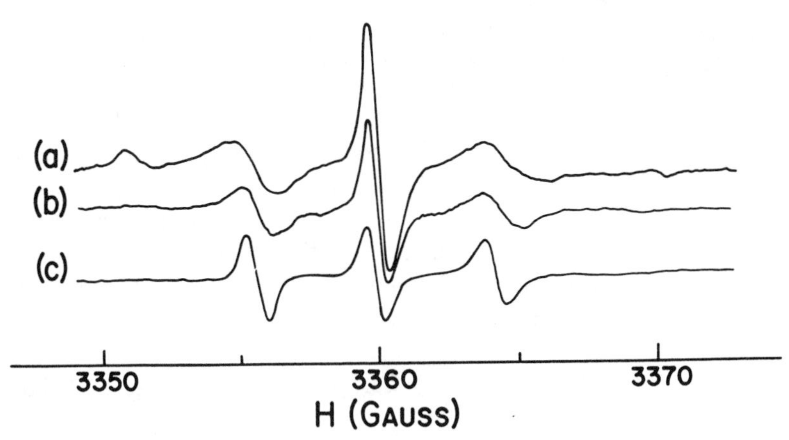

Fig. II-23. Rotation of CN in an argon matrix [from Easley and Weltner (153)]. ESR spectra of $^{12}C^{14}N$ at various temperatures: (a) 8°K, (b) 15°K, (c) 35°K.

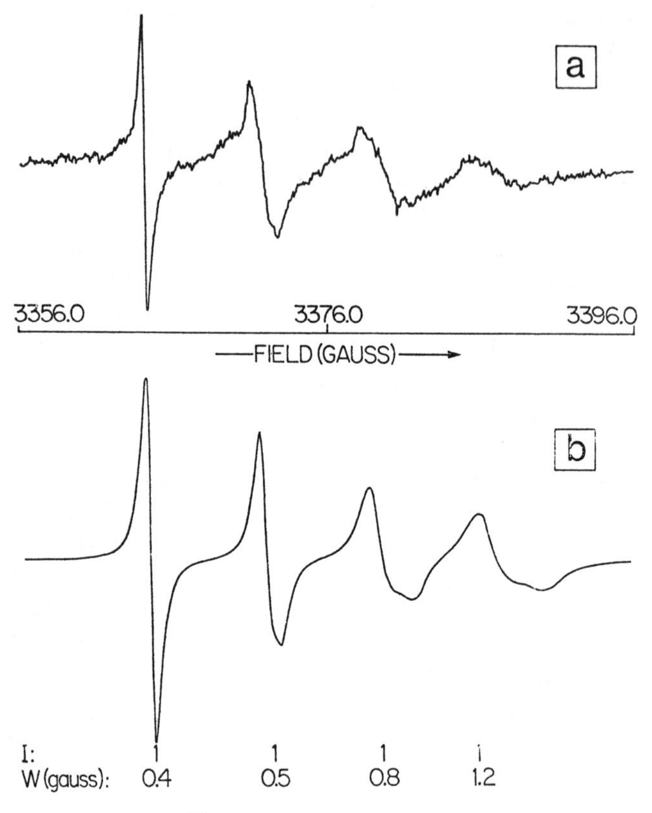

Fig. II-24. ESR spectrum of $^{28}SiH_3$ in an argon matrix at 4°K [from Raghunathan and Shimokoshi (387)]. (a) Experimental spectrum; (b) computer simulation with intensities (I) and Lorentzian linewidths (W) of the component hyperfine lines as indicated.

Fig. II-24a [Raghunathan and Shimokoshi (387)]. A simulation of that spectrum is shown in Fig. II-24b where the proton hf lines were given an intensity ratio of 1 : 1 : 1 : 1 but with a progressive increase in linewidth from 0.4 to 1.2 G in the direction of higher field. Theory indicates that the linewidths depend upon m_I in this case essentially according to a relationship

$$\text{Linewidth} = \text{Constant} + 2\beta H_z m_I^{(H)} (g_{\parallel} - g_{\perp}) t_{zz} \quad \text{(II, 105)}$$

where t_{zz} is the proton anisotropic hf constant along the axis of the molecule. McConnell (331) first pointed out that the spin relaxation for a radical in motion is affected by **g** and **A** anisotropy, and in the latter case, that it will be different for transitions with different m_I values. For SiH_3 $g_{\parallel} - g_{\perp} = -0.00116$ and $t_z = +0.87$ G, so that by (II, 105) the $m_I = +\frac{3}{2}$ hf line would be expected to be narrower than the $m_I = -\frac{3}{2}$ line. One then assigns $m_I = +\frac{3}{2}$ to the low-field line in Fig. II-24a and $m_I = -\frac{3}{2}$ to the high-field line. The $A_{\text{iso}}(\text{H})$ must be positive [see the second-order perturbation expression for the resonant fields in (II, 64)], which is in agreement with theoretical arguments [Biddles and Hudson (68); see also Higuchi, Kubota, Kumamoto, and Tokue (221)].

For a doublet molecule the upper spin state in a magnetic field has a finite lifetime because of spin-relaxation mechanisms. This is, of course, necessary in order to observe an ESR spectrum, since otherwise a spin population difference between the upper and lower states could not be maintained. Spin-relaxation rates are usually characterized by the spin-lattice or longitudinal relaxation time T_1, such that $1/T_1$ indicates the rate of radiationless transitions between the two levels, and the transverse relaxation time T_2, such that $1/T_2$ indicates the rate and the magnitude of fluctuations in the energies of these levels. Very crudely, these two rates can be added to give the total relaxation rate, which is proportional to the linewidth in the ESR spectrum [Wertz and Bolton (23), pages 192ff.; Gordy (8) pages 104ff.; Carrington and McLachlan (6) pages 176ff.].

For a radical in rapid rotation in liquid solution or in a matrix, the spin relaxation depends upon the interaction of the electron with the surrounding magnetic forces which are changing with time. These forces will induce the electron to reverse its spin and will also modulate the frequency of each transition. McConnell (331) has shown that anisotropy in the **g** tensor and in the **A** tensors combined with the rapid random motion of the radical provide the time-dependent perturbations inducing both of these effects. That theory has been extended by Stephen and Fraenkel (422), Kivelson (274), Carrington and Longuet-Higgins (102), and Freed and Fraenkel (171).

For several magnetic nuclei, i, in a molecule $1/T_1$ and $1/T_2$ are found to be proportional to (102):

$$P = \left(\beta H_z g_{\alpha\beta}' + \sum_i m_I{}^i t_{\alpha\beta}{}^i\right)\left(\beta H_z g_{\alpha\beta}' + \sum_i m_I{}^i t_{\alpha\beta}{}^i\right) \qquad \text{(II, 106)}$$

where $g_{\alpha\beta}'$ and $t_{\alpha\beta}{}^i$ are the anisotropic parts of **g** and $\boldsymbol{A}_i$, respectively. The linewidth, proportional to P and a function of the $m_I{}^i$ values, may now be written, for a single nucleus,

$$L = C + Dm_I + Em_I{}^2 \qquad \text{(II, 107)}$$

after expansion of P [de Boer and Mackor (129)]. C involves the $\boldsymbol{g}$ anisotropy, E involves the $\boldsymbol{A}$ anisotropy, and D involves both anisotropies. However, only the Dm_I term will cause an unsymmetrical variation of linewidths in the ESR spectrum, so that for our purposes

$$L = \text{Constant} + 2\beta H_z m_I g_{\alpha\beta}' t_{\alpha\beta} \qquad \text{(II, 108)}$$

or for i magnetic nuclei,

$$L = \text{Constant} + 2\beta H_z \sum_i m_I^i g_{\alpha\beta}' t_{\alpha\beta}^i \qquad \text{(II, 109)}$$

For a linear molecule with one magnetic nucleus

$$L = \text{Constant} + 2\beta H_z m_I [g_\perp (t_{xx} + t_{yy}) + g_\parallel t_{zz}]$$

or

$$L = \text{Constant} + 2\beta H_z m_I (g_\parallel - g_\perp) t_{zz} \qquad \text{(II, 110)}$$

6. MAGNETIC PROPERTIES OF SMALL DOUBLET MOLECULES IN CRYSTALS AND MATRICES

(a) Tabulated Magnetic Properties

The tables in Appendix C contain the experimental magnetic data and derived hyperfine parameters and spin density populations for diatomic, triatomic, and some tetra-atomic radicals in crystals and matrices. Many other polyatomics are considered in the books by Atkins and Symons (2) and Gordy (8), and particularly in the thorough collection by Morton and Preston (350a).

There is a variety of bonding occurring among these molecules, as well as many distinctive environmental and motional effects in the solid state. To illustrate these distinctions and the information obtainable from matrix ESR studies, representative radicals have been selected for more detailed discussion in the following section.

(b) Representative Doublet Radicals

BeH, MgH, CaH, SrH, and BaH. The simplest molecules, presumably most amenable to theoretical calculations, are the diatomic hydrides. ESR studies of the $^2\Sigma$ alkaline-earth hydrides yield detailed information about the wavefunction of the unpaired spin so that direct comparison can be made with theory, at least among the lighter members of the series. Also, trends among the members of the series and comparison with the corresponding fluorides and hydroxides, to be discussed in the following sections, are informative.

The ESR spectrum of the BeD molecule in an argon matrix at 4°K is shown in Fig. II-25a and in an expanded view in Fig. II-25b. It, rather than BeH, was chosen here for illustration because there is less overlapping of lines in the deuteride spectrum. ^{9}Be ($I = \frac{3}{2}$) atoms (100% natural abundance) were vaporized, and D (I = 1) atoms formed by dissociation of D_2 over hot tungsten. These atoms react during condensation in argon at 4°K to form the diatomic deuteride. The strong lines clearly form a quartet of triplets, and the analysis of the spectrum is relatively straightforward [Knight, Brom, and Weltner (276)]. The spectra of the other hydrides are generally even simpler because in most cases the most abundant isotope of the metal does not have a nuclear moment [Knight and Weltner (282)].

The hfs due to the hydrogen nucleus is relatively small, and spin densities derived in the usual way from (II, 43) indicate that ρ(H) is approximately 14% in BeH, rises to 19% in MgH, and then decreases down the series to 3% in BaH [(282); these values are slightly different if ionicity is taken into account—see Gordy (8), page 383]. It is apparent from these values that these molecules, particularly the heavier ones, may be fairly considered as ion-pairs, M^+H^-, where the unpaired electron is largely confined to the metal cation. If the M^+ ion were free in the gas phase the unpaired electron would be in an *ns* orbital, but one expects that the neighboring anion would severely distort that orbital in the ion-pair. As expected, this effect is even more marked in the fluorides; so we will defer further discussion of the ionic model until those molecules are considered.

Table II-4 presents an illuminating comparison of experiment and *ab initio* calculations for the two lightest hydrides. The experimental values of $|\Psi(0)|^2$ at each nucleus, obtained from A_{iso} via (II, 30), are listed in columns 2 and 4, while those calculated by Bender and Davidson (55) and by Chan and Davidson (107) are given in columns 3 and 5. For BeH, for example, a 1039 configuration-interaction (CI) wavefunction was employed. It is worth noting that such an extensive calculation is necessary, since a Hartree-Fock Self-Consistent-Field wavefunction [Cade and Huo (98)] yielded 0.2827 and 0.01646 a.u. for the two theoretical values for BeH in Table II-4.

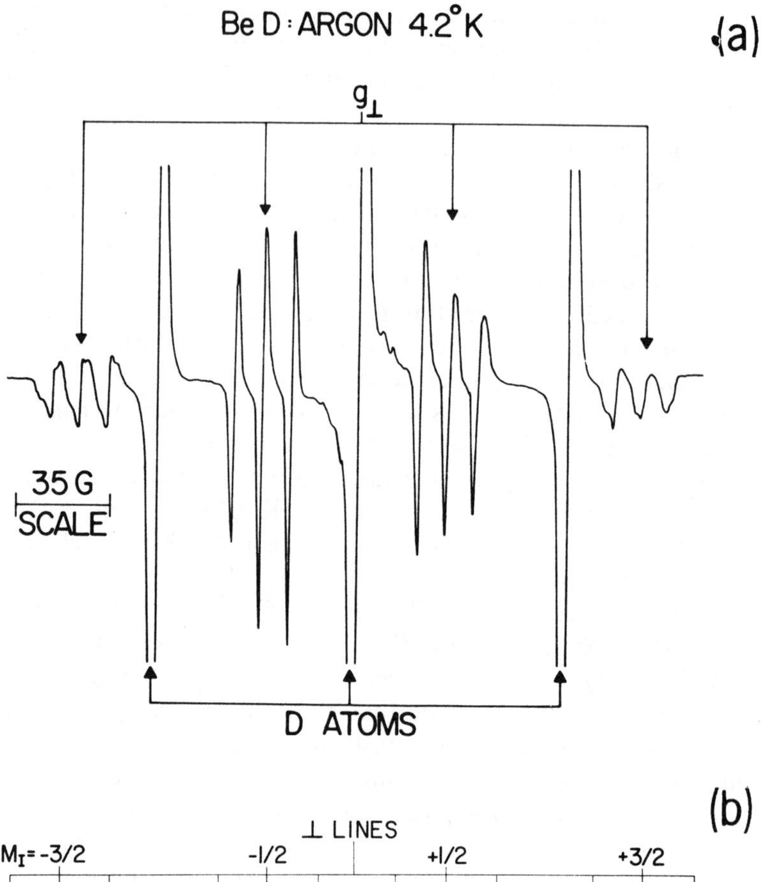

Fig. II-25. ESR spectrum of the BeD molecule in an argon matrix at 4°K [from Knight, Brom, and Weltner (276)]. View (b) is an expanded view of lines in (a). Line positions are in gauss ($\nu \simeq 9.3$ GHz).

Table II-4. Fermi Contact Term, $|\Psi(0)|^2$ in a.u.$^{-3}$,[a] at the Metal and Hydrogen Nuclei in BeH and MgH Molecules.

	Metal		Hydrogen	
	Expt.	Theory	Expt.	Theory
BeH	0.317(2)[b]	0.3223[c]	0.0434(2)[b]	0.0367[c]
MgH	0.800(8)[d]	0.7635[e]	0.0663(2)[d]	0.03789[e]

[a] 1 a.u.$^{-3}$ = 6.74873×10^{24} cm^{-3}.
[b] Knight, Brom, and Weltner (276).
[c] Bender and Davidson (55).
[d] Knight and Weltner (282).
[e] Chan and Davidson (107).

The agreement of that theory with CI calculations is even worse for MgH, where the two numbers are 0.6454 and 0.01991 a.u. [Chan and Davidson (107)]. The experimental numbers are judged to be in error relative to the gas phase by possibly a few percent (see the following fluoride study where gas-phase data have become available). Then the agreement in Table II-4 is satisfying, but it emphasizes the necessity for accurate theoretical calculations to reproduce the Fermi contact interaction.

Note that the Curl equation in (II, 26), connecting the spin-rotation constant, γ, and $\Delta g_{\perp}$, may be tested on four of these hydrides. Table II-1, page 51, shows that the agreement with observed γ values is generally within experimental error.

BeF, MgF, CaF, SrF, and BaF. The ESR spectrum of BaF has already appeared, in Fig. II-10, where it was used to illustrate a powder pattern spectrum of a $^2\Sigma$ molecule. Similar spectra of MgF trapped in a neon matrix at 4°K on a flattened sapphire rod are shown in Fig. II-26 for two orientations of the rod relative to the magnetic field. Natural abundance magnesium contains only 10% ^{25}Mg ($I = \frac{5}{2}$), and its hfs was not discernible in this spectrum, but ^{19}F ($I = \frac{1}{2}$) hfs is apparent. Note that the orientation effects indicate that the MgF molecules have their axes preferentially perpendicular to the rod surface. In argon there was essentially no orientation, as was also the case for CaF, SrF, and BaF in both matrices. BeF in an argon matrix at 12°K exhibits a rather surprising spectrum containing

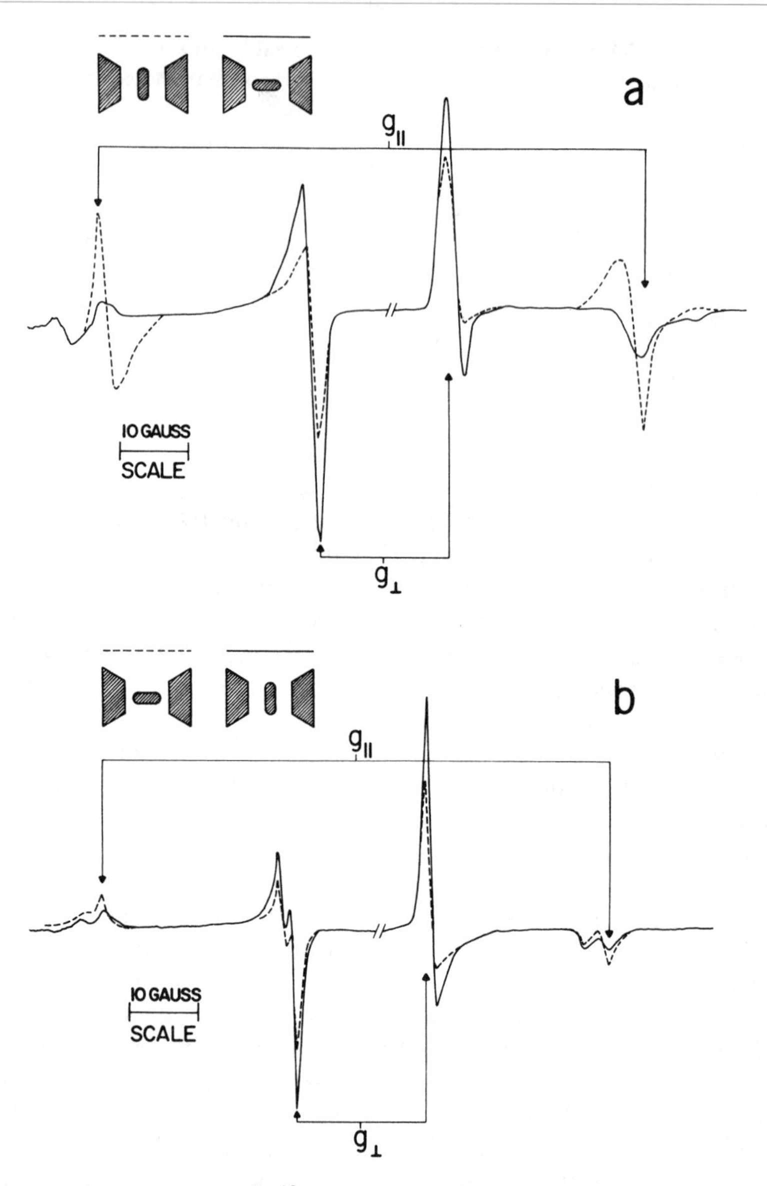

Fig. II-26. ESR spectrum of $Mg^{19}F$ in neon and argon matrices at 4°K [from Knight, Easley, Weltner, and Wilson (278)]. (a) Neon: solid line indicating the spectrum observed with the magnetic field parallel to the surface of the rod, dotted line with the magnetic field perpendicular to the surface of the rod. (b) Argon: orientations reversed, as indicated by the sketch of the rod and magnet.

Table II-5. Comparison of the ^{19}F Hyperfine Parameters (MHz) for the CaF and SrF Molecules in Neon Matrices at 4°K and in the Gas Phase.

	ESR[a]	Gas Phase[b]
CaF:		
$A_{\perp}$	106(3)	109.183[c]
$A_{\parallel}$	149(3)	149.302
SrF:		
$A_{\perp}$	95(3)	97.667[d]
$A_{\parallel}$	126(3)	127.513

[a] Knight, Easley, Weltner, and Wilson (278).
[b] $A_{\perp}$ and $A_{\parallel}$ equal b_{00} and $b_{00} + c_{00}$, respectively, in Frosch and Foley symbolism (174, 109).
[c] Bernath, Cummins, and Field (59); Childs, Goodman, and Goodman (109).
[d] Childs, Goodman, and Renhorn (110).

isotropic lines [Knight, Wise, Childers, Davidson, and Daasch (285)], and for that reason it will be considered separately and last in this series. It has been possible to observe the metal hfs in four of these five molecules so that a complete picture of the spin distribution could be obtained in those cases.

The fluorine hf data for Ca^{19}F and Sr^{19}F obtained from matrix ESR experiments are given in Table II-5 and are there compared with recent gas-phase studies utilizing intermodulated fluorescence ("Lamb-dip") spectroscopy [Bernath, Cummins, and Field (59)], and molecular-beam laser-rf double resonance [Childs, Goodman, and Goodman (109); Childs, Goodman, and Renhorn (110)]. One sees that the matrix error is a maximum of 3%. Presumably these authors or others will also observe in the future the metal hfs, such as that observed in matrices for ^{87}SrF and ^{137}BaF (278), and further confirm the accuracy of the ESR data.

If the lack of spin density on H and on F can be taken, for the sake of comparison, as a crude measure of the relative ionic character of their bonding, then Table II-6 indicates that the fluorides are more ionic, as expected. Here the spin density distribution has been simply calculated from (II, 43), but it is certainly not going to be in error enough to alter the finding that $\rho(\mathrm{F}) \ll \rho(\mathrm{H})$.

Also, if the known dissociation energies of the four heavier mole-

Table II-6. Comparison of Spin Densities and $\Delta g_\perp$ Values in Alkaline-Earth Hydrides[a, b] and Fluorides.[c, d]

	ρ(H) (%)	$-\Delta g_\perp$		ρ(F) (%)	$-\Delta g_\perp$
BeH	14	0.000	BeF	>13	0.001
MgH	19	0.002	MgF	4.5	0.001
CaH	9.5	0.006	CaF	1.2	0.002
SrH	8.6	0.016	SrF	0.90	0.005
BaH	3.2	0.028	BaF	0.31	0.007

[a] Knight, Brom, and Weltner (276).
[b] Knight and Weltner (282).
[c] Knight, Wise, Childers, Davidson, and Daasch (285).
[d] Knight, Easley, Weltner, and Wilson (278).

cules can serve as a measure of the ionicity of the bonds, then Fig. II-27 shows graphically the variation of the separate $2s\sigma$ and $2p\sigma$ character of the spin on the F atom [from (II, 34) and (II, 43); a better procedure is discussed by Gordy (8)—see section II-1c(β)]. Spin polarization may account for the small spin density, but the varying $2p\sigma$ character suggests increasing covalency in the lighter molecules (278).

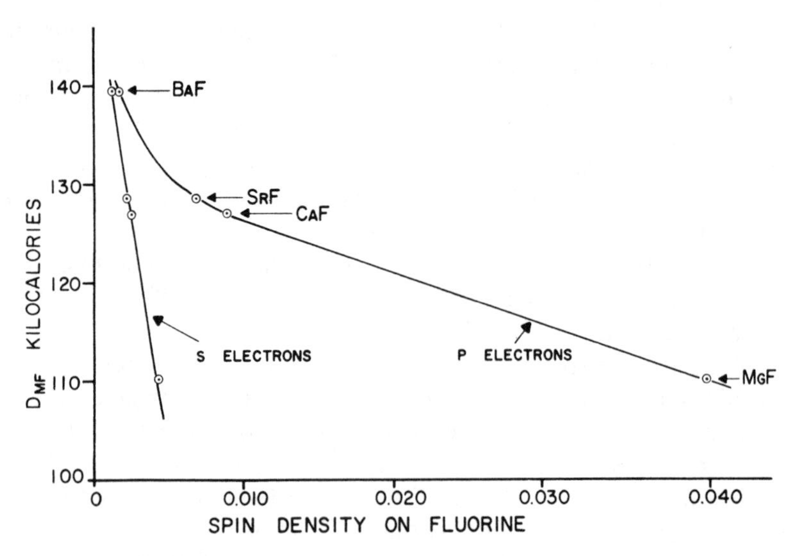

Fig. II-27. Dissociation energy (a measure of ionic bond character) of diatomic MF molecules versus the approximate spin density at the fluorine atom [from Knight, Easley, Weltner and Wilson (278)].

These then are almost ideal ion-pairs in which the nonspherical polarization of the electron cloud on the cation by the neighboring anion can be demonstrated by observation of the hyperfine interaction with the metal nucleus. This could be conveniently measured for SrF and BaF where magnetic metal isotopes could be substituted (278). The A_{iso} values for the ^2S ground states of $^{87}Sr^+$ and $^{137}Ba^+$ as measured in the gas phase are 992 and 4050 MHz, respectively, and comparison with the observed values indicates that the $5s$ and $6s$ character has dropped to 58 and 60%, respectively, in the two molecules and is confirmed by the anisotropy in $\boldsymbol{A}$ (cation) increasing slightly. [It must be remembered that for the hf coupling parameters with the heavy metal nuclei $A_{\text{dip}} \simeq A_{\text{iso}}/50$, so that large changes in the s-electron contribution can be accounted for by a small change in anisotropy.] Hence, the repulsion of the adjoining negative charge has severely disturbed the metal ions or, what is the same thing, has mixed excited np and, to a lesser extent, $(n - 1)\,d$ character into the wavefunction of the odd electron. This hydridization allows the odd electron (and probably also the inner shell electrons), to spend more time on the side of the metal ion away from the fluoride ion. A crude calculation taking F^- as a point charge yields the angular charge distribution about the metal ion shown in Fig. II-28, to be compared with a spherical distribution in the free M^+ ion.

The excited states and the spin-orbit coupling constants in both the hydrides and fluorides are determined essentially by the metal atom, so that they are comparable in the two series. Then the fact that the $\Delta g_{\perp}$ values in Table II-6 for the hydrides are two to four times larger than those for the fluorides points to a larger contribution of $p\sigma$ and $d\sigma$ character in the odd-electron molecular orbital in the ground states (and/or, less likely, to purer $p\pi$ and $d\pi$ character in the excited $^2\Pi$ states) of the hydrides. The lower electron affinity of H relative to F and the larger spin density on H in all cases means that this increased $p\sigma$, $d\sigma$ character in the hydrides is due to increased covalency.

Also, the values of $\Delta g_{\perp}$ may be used in the Curl equation (II, 26), along with the known rotational constants, to calculate the spin-rotation constants, γ, of these molecules. Table II-1, page 51 shows the excellent agreement obtained for CaF, SrF, and BaF, the only fluorides for which experimentally determined γ values are available.

As mentioned above, BeF was somewhat anomalous in exhibiting isotropic ESR lines in an argon matrix at 12°K (see Fig. II-29 and

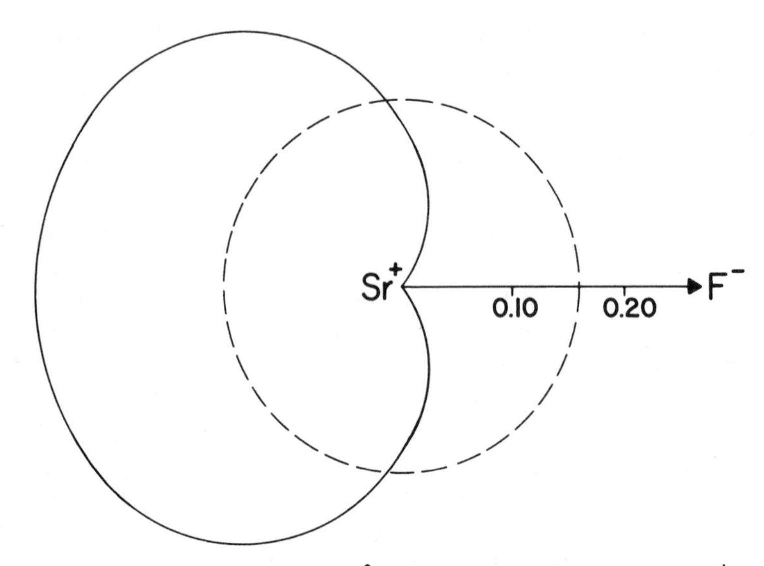

Fig. II-28. Angular charge distribution, $\Psi^2(\theta)$, of the unpaired electron on the Sr^+ ion in the SrF molecule [from Knight, Easley, Weltner, and Wilson (278)]. The electron radius is arbitrarily fixed at about the equilibrium interatomic distance $r = 3.906$ a.u. The dashed circle indicates the radius of maximum charge density of the free gas ion in the ground ^2S state.

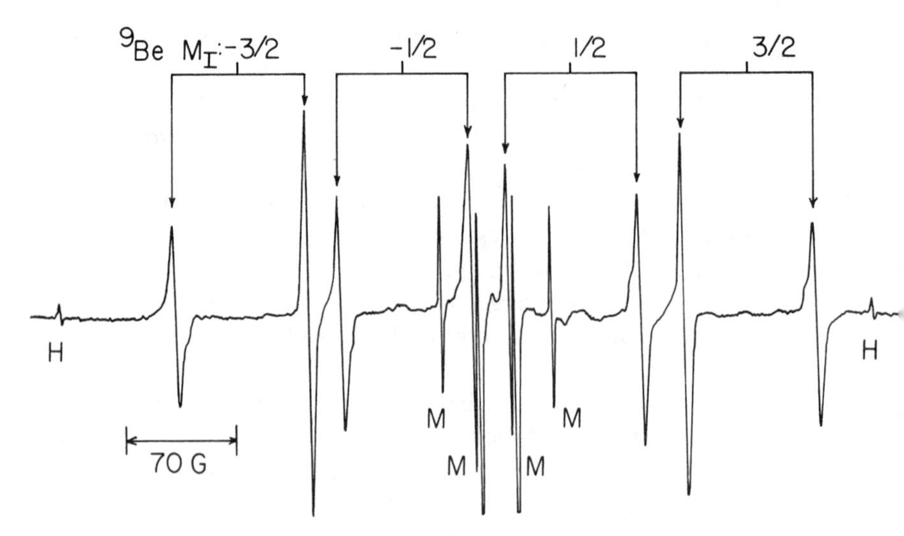

Fig. II-29. ESR spectrum of the $^9Be^{19}F$ molecule in an argon matrix at 12°K prior to annealing [from Knight, Wise, Childers, Davidson, and Daasch (285)].

compare line shapes with Figs. II-10 and II-26) [Knight, Wise, Childers, Davidson, and Daasch (285)]. The authors suggest that the molecule is either rotating in the matrix, which is unusual among these and similar diatomics, or that the hyperfine anisotropies are small. *Ab initio* calculations for BeF (and MgF) were carried out and yield essential agreement with the experimental $|\Psi(0)|^2$ values but a large diagreement with the $\langle(3\cos^2\theta - 1)/r^3\rangle$ values. Thus, a rotating molecule is indicated. This is rationalized by the authors by pointing out that the CN molecule rotates when warmed in an argon matrix (see Fig. II-23) and that the dipole moment of BeF is calculated to be smaller than that of CN.

BeOH and MgOH. The triatomic hydroxides are natural members of this series of simple alkaline-earth radicals, since OH^- is generally considered as a disguised halogen ion. They were produced and trapped in the solid rare gases by co-deposition of a beam of metal atoms and Ar or Ne containing 1% H_2O_2. The ESR spectrum of ^{25}MgOH in an argon matrix is shown in Fig. II-30 [Brom and Weltner (90)]. [^{25}Mg ($I = \frac{5}{2}$) has only 10% natural abundance so that a

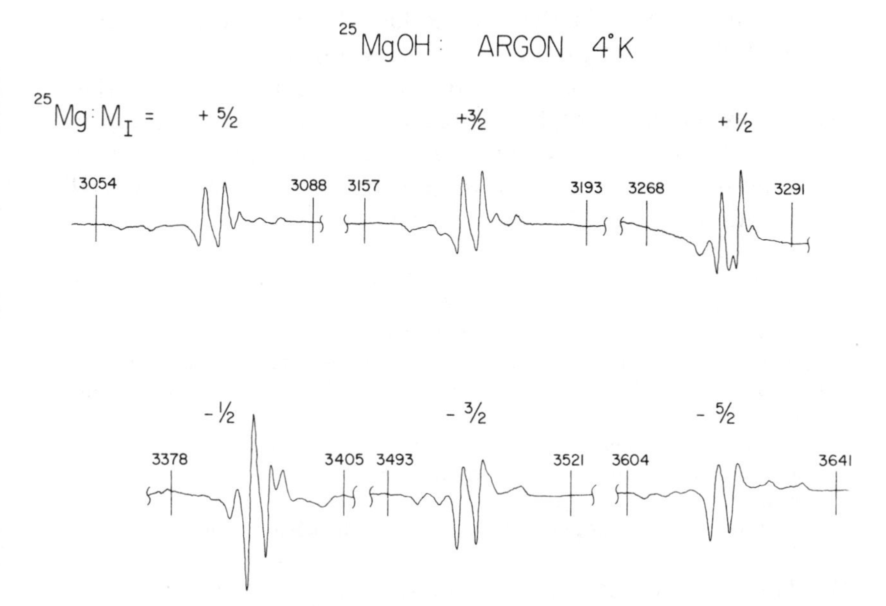

Fig. II-30. ESR spectrum of the ^{25}MgOH molecule in an argon matrix at 4°K [from Brom and Weltner (90)]. Line positions are given in gauss, $\nu \simeq 9.3$ GHz.

strong feature due to the remaining MgOH molecules also appears in the middle of the pattern in Fig. II-30, but it is not shown here.] The strong doublets are perpendicular lines split by hyperfine interaction with H ($I = \frac{1}{2}$). Close examination of these lines reveals small parallel-like lines, also doublets, on each side of the strong pair, as previously shown in Fig. II-20. As was discussed in section II-4c, such a pattern occurs for a nonlinear molecule, and in this case the $\boldsymbol{g}$, $\boldsymbol{A}(^{25}Mg)$, and $\boldsymbol{A}(^{1}H)$ components in the x, y, and z directions are quite similar. When annealed, one set of doublets apparently disappears (see Fig. II-20), indicating that the originally trapped molecule was bent in unsymmetrical sites and returns to its more stable linear form in the unstrained lattice. This is likely, since the bending force constant in such hydroxides must certainly be very low (see below). [Matrix bending effects have also been observed for several triplet molecules, such as SiCO and C_3N_2 (see Chapter III, section 6).] Although the spectra were not as ideal as for MgOH, BeOH appears also to be bent in the unannealed matrix [Brom and Weltner (91)].

As expected, the H hfs is quite small in MgOH, but it is even smaller in BeOH. Thus, roughly, the spin distribution is 0.8% and 0.3% at the protons in the hydroxides relative to about 19% and 14% in MgH and BeH, respectively. It would also be interesting to compare the spin densities on the metal atoms in the fluorides, hydroxides, and hydrides, but unfortunately the data are incomplete. However, for MgF, MgOH and MgH, one finds approximately 95, >90, and 81%, respectively, of the spin on the metal ion, which falls in line with the expected ionic character.

Ab initio calculations of the isotropic hyperfine coupling constant at each nucleus of the BeOH and MgOH molecules have been made by Hinchliffe (222). Spin properties calculated from both unrestricted Hartree-Fock (UHF) and UHF singly annihilated wavefunctions are given in Table II-7. They compare favorably with the experimental matrix data. (Neon matrix data are expected to be most gaslike, but only argon data are available for BeOH.)

The calculations indicate that the lowest-energy configurations of these molecules are linear, but that their bending force constants are very low (222). This is then in accord with the observations in the matrix ESR spectra indicating that the lattice energy in strained matrix sites was sufficient to cause bending.

CN. This molecule has been the subject of many investigations by gas-phase optical spectroscopists, perhaps because it is so ubiquitous in flames, discharges, and stellar atmospheres. For this reason it is

Table II-7. Comparison of Experimental and Theoretical Isotropic Hyperfine Coupling Parameters for the BeOH and MgOH $^2\Sigma$ Radicals.

	$(A_{iso})_i$ (G)			
	BeOH		MgOH	
Nucleus i	Calc.[a]	Expt.[b]	Calc.[a]	Expt.[c]
^{9}Be or ^{25}Mg	-105.4 -92.7	-94.2	-97.88 -87.61	-108.7 (Ne), -111.6 (Ar)
^{1}H	1.84 1.20	<2	7.52 4.19	4.18 (Ne), 3.64 (Ar)

[a] Upper value UHF, lower value after single spin annihilation [Hinchcliffe (222)].
[b] In argon matrix at 4° K (91).
[c] In neon and argon matrices at 4° K (90).

rather surprising that its hyperfine coupling constants in the ground state have only recently been determined accurately (see Chapter IV).

Its matrix ESR spectrum is interesting because it is difficult to analyze unambiguously, possibly because of motional effects. It is somewhat unusual, as discussed in section II-5, that when warmed to 35°K in solid argon CN becomes freely rotating, and when it is cooled again to 4°K an anisotropic spectrum is again obtained.

The ESR spectrum in solid argon at 4°K is shown in Fig. II-31, and the three perpendicular lines split by ^{14}N ($I = 1$) hyperfine are quite obvious. The splittings are small, as is usually the case for ^{14}N because of its small nuclear moment. The problem is, where are the corresponding parallel lines? One has clues as to $g_{\parallel}$ and $A_{\parallel}$ from the relative phases and intensities of the three perpendicular lines, and an assignment of parallel lines as in Fig. II-31 would account for those features [Easley and Weltner (153)]. However, Adrian and Bowers (32) suggest that the molecule is not rigidly trapped in argon even at 4°K, and that it is undergoing unsymmetrical motion in its trapping site, resulting in an apparent nonaxial spectrum. Their analysis is based upon the assignment of the structure on the central perpendicular line to three g components. (The small parallel-like line at 3365 G in Fig. II-31, midway between the two high-field perpendicular lines, is also present in their spectrum, but is not assigned.) As can be seen from Table II-8, the values of A_{iso}(^{14}N) and γ [from $\Delta g_{\perp}$ via (II, 26)] deduced from that analysis are in good

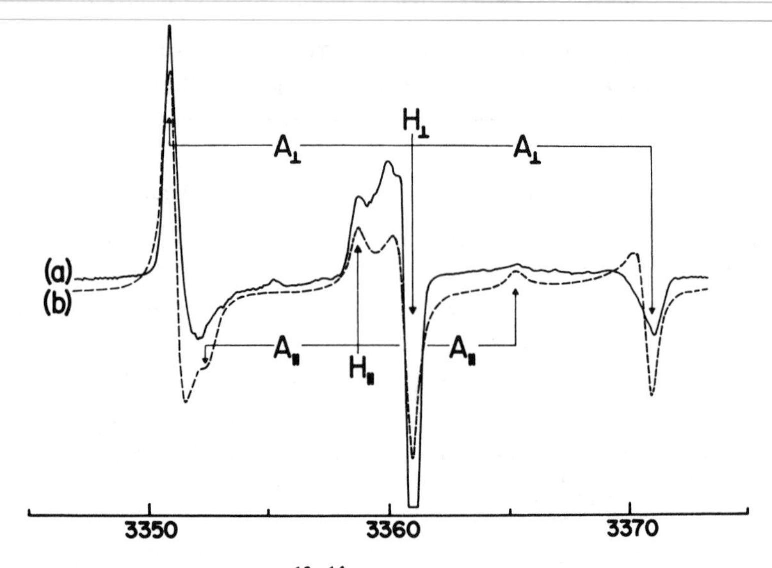

Fig. II-31. ESR spectrum of the $^{12}C^{14}N$ molecule in an argon matrix at 4°K [from Easley and Weltner (153)]. (a) (solid line) The experimental spectrum; (b) (dashed line) a computed spectrum using $g_{\parallel} = 2.0015$, $g_{\perp} = 2.0003$, $A_{\parallel}(N) = 6.5$ G, $A_{\perp}(N) = 10.0$ G, and a linewidth of 0.4 G.

Table II-8. Comparison of CN Matrix and Gas-Phase Magnetic Data.

CN Condition	*g* Tensor Components		Magnetic Nucleus	Hyperfine Tensor Components (Gauss)			Spin-Rot. Constant (MHz)
	$g_{\parallel}$	$g_{\perp}$		$A_{\parallel}$	$A_{\perp}$	A_{iso}	γ
Rigidly trapped in solid argon[a]	2.0015(5)	2.0003(5)	^{13}C	+242	+193.8	–	+228[d]
			^{14}N	(+6.5)	−10.0	(−4.5)	
In unsym. motion in solid argon[b]	(g_e)	2.00028	^{14}N	–	–	−4.43	+231[d]
Rotating in solid argon[a]	2.0008(5) 2.00095(8)[b]		^{13}C	–	–	+210.0	–
			^{14}N	–	–	−4.5	–
Gas phase[c]			^{14}N	+9.41	−12.12	−4.94	+217.5

[a]Easley and Weltner (153). See Fig. II-31.
[b]Adrian and Bowers (32).
[c]From data in Appendix C.
[d]Calculated from $\Delta g_{\perp}$ via Curl equation (II, 26).

agreement with gas-phase data. Thus, although their interpretation of the observed matrix spectrum may be questioned, comparison of the derived parameters with those in the gas justify that analysis.

Substitution of ^{13}C in the radical yields the ESR spectrum in solid argon at 4°K shown in Fig. II-32 with a calculated simulated spectrum below it (153). The observed spectrum is clearly a doublet (^{13}C hfs) of triplets (^{14}N hfs), and although the peak intensities vary, the integrated intensities appear to correctly account for the expected six equally intense perpendicular lines of opposite phase in Fig. II-32b. The observed variation in linewidth in each triplet is probably attributable to the motional effects. Although the value of $A_{\parallel}(^{13}C)$ cannot be accurately derived from the spectrum in Fig. II-32a, it can be obtained from $A_{iso}(^{13}C)$, observed when the matrix is at 35°K, and equation (II, 83).

In Table II-8 all of the hyperfine parameters measured in an argon matrix are smaller in absolute magnitude than those in the gas

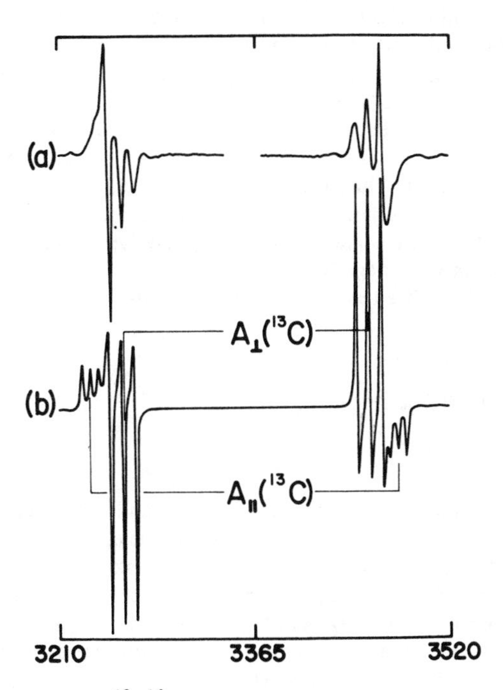

Fig. II-32. ESR spectrum of $^{13}C^{14}N$ in an argon matrix at 4°K [from Easley and Weltner (153)]. (a) Observed; (b) computed.

phase. With such small hf parameters, apparently the matrix effects are magnified (in percentage terms). Then theory (28, 31) (section I-5a) interprets these matrix effects as indicating that CN is held in a loose cage in solid argon where van der Waals attractive forces predominate, and this agrees with its tendency to rotate with only slight thermal excitation. In such a case, the g values are expected to be little affected by the matrix and are accounted for rather well by application of (II, 10) and (II, 11).

$\Delta g_\perp$ can be estimated using (II, 11) [see Stone (426)]:

$$g_\perp = g_e + 2 \sum_{n \neq p} \sum_k \frac{\langle \Psi_p | l_k{}^x | \Psi_n \rangle \langle \Psi_n | \xi_k(r_k) l_k{}^x | \Psi_p \rangle}{E_p - E_n} \qquad \text{(II, 111)}$$

where p is the MO containing the odd electron (5σ), n is any MO from which or to which excitation occurs, E_p and E_n are the energies of the MO's, and $\xi_k(r_k)$ is the spin-orbit coupling constant near atom k. The relevant excited $^2\Pi$ states are $A^2\Pi_i(1\pi^3 5\sigma^2)$ and $C^2\Pi_r(1\pi^*)$ which result from excitation from the filled 1π bonding orbital to the 5σ orbital and from the 5σ orbital to the empty $1\pi^*$ antibonding orbital, respectively. The $A^2\Pi_i \leftrightarrow X^2\Sigma$ transition is the well-known red system of CN, but the $C^2\Pi_r \leftrightarrow X^2\Sigma$ transition has not as yet been observed. At first glance, these two excited states might be expected to have opposite effects upon $\Delta g_\perp$, the $^2\Pi_i$ state producing a positive "hole" shift and the $^2\Pi_r$ state a negative "electron" shift (see section II-1b(α)). Furthermore, the $^2\Pi_i$ shift might be expected to dominate, since it is the lower-lying of the two states. However, a more detailed calculation, including overlap, shows that the sign of $\Delta g_\perp(^2\Pi_i)$ can be positive or negative.

Ψ_p and Ψ_n can be written as LCAO wavefunctions:

$$\begin{aligned}
\Psi_\sigma &= a_1 X(2s_C) + a_2 X(2p\sigma_C) - a_3 X(2s_N) - a_4 X(2p\sigma_N) \\
\Psi_\pi &= b_1 X(\Pi_C) + b_2 X(\Pi_N) \\
\Psi_\pi{}^* &= c_1 X(\Pi_C) - c_2 X(\Pi_N)
\end{aligned} \qquad \text{(II, 112)}$$

where all coefficients a_1, b_1, c_1, etc., are positive and the functions orthonormalized. (These σ, π, and π^* functions would become σ_g, π_u, and π_g in a homonuclear diatomic.) The coordinate system is here taken with the z axis lying along the CN bond and $+z$ on each atom pointing in the same direction. $+x$ and $+y$ directions (along which the Π_C and Π_N orbitals lie) are also taken the same on each atom. This choice of the z axes allows the most expeditious transformation of the origin of l_k from C to N [see II, 25a)].

Substituting these wavefunctions into (II, 111) and writing $\xi_k(r_k) = \lambda_k$, the spin orbit constant for atom k, since ξ_k is only effective near atom k, we find

$$\Delta g_{\perp}(^2\Pi_i - {}^2\Sigma) = \frac{2}{[E(^2\Pi_i) - E(^2\Sigma)]}(a_2 b_1 \lambda_C - a_4 b_2 \lambda_N)$$
$$\cdot \{a_2 b_1 - a_4 b_2 - a_3 b_1 S(2s_N, 2p\sigma_C) + (a_2 b_2 - a_4 b_1)$$
$$\cdot S(2p\sigma_C, 2p\sigma_N) + a_1 b_2 S(2s_C, 2p\sigma_N)\} \qquad \text{(II, 113)}$$

where $S(2s_N, 2p\sigma_C)$, etc., are overlap integrals, and the energy difference in the denominator is now to be obtained from observed spectral data. (Note that if the molecule became homonuclear, this $\Delta g_{\perp}$ shift would vanish, as it should, since σ_g and π_u mixing is not allowed.) $\Delta g_{\perp}(^2\Pi_r - {}^2\Sigma)$ is the same except that $-[E(^2\Pi_r) - E(^2\Sigma)]$ now enters into the denominator, and $c_1, -c_2$ replace b_1, b_2. The sum of these two $\Delta g_{\perp}$'s then constitutes the total g shift in CN, since no other states are expected to contribute.

For CN, the following set of data can be used:

$$\lambda_C = 26 \text{ cm}^{-1}, \quad \lambda_N = 76 \text{ cm}^{-1}$$
$$a_1 = 0.138, \quad a_2 = 0.69, \quad a_3 = 0, \quad a_4 = 0.57,$$
$$b_1 = 0.52, \quad b_2 = 0.85, \quad c_1 = 0.87, \quad c_2 = 0.49$$

These coefficients are obtained from (II, 43) and from Cade's wavefunctions for the excited states. The $^2\Pi_i$ state is known to lie 9200 cm^{-1} above the $X\,^2\Sigma$ state, and the $^2\Pi_r$ is estimated to lie near 55,000 cm^{-1} from analogy with isoelectronic BO or CO^+. The overlap integrals can be obtained from the tables of Mulliken, Rieke, Orloff, and Orloff (355) with $R = 2.214$ a.u., $\mu_C = 1.625$, $\mu_N = 1.95$. The calculated g-shifts are:

$$\Delta g_{\perp}(^2\Pi_i) = -0.0001, \quad \Delta g_{\perp}(^2\Pi_r) = -0.0007$$

and their sum is in fair agreement with the experimental value of -0.002.

CH_3 and Its Homologues:

CH_3. The methyl radical is easily formed and is the classic π radical, since essentially all of the electron spin is confined to a carbon $2p\pi$ orbital perpendicular to the plane of the molecule. The other carbon orbitals form three equivalent sp^2 hydridized bonds in the plane at an angle of 120° to each other. Since the $p\pi$ orbital has a node in the plane of the molecule, no isotropic hyperfine interactions are to be expected, but small splittings are actually observed, yielding $|A_{\text{iso}}(^1\text{H})| \simeq 23$ G (see Fig. II-21) and $|A_{\text{iso}}(^{13}\text{C})| \simeq 42$ G. By the usual procedure of (II, 43) these correspond to apparent s spin density distributions of less than 5%, demonstrating that the electron is largely localized in the π orbital. The small hf splittings can be

accounted for as spin polarization effects, and it has been shown that there is negative spin density at the proton arising from σ-π configuration interaction. This has been extended to proton hyperfine coupling in aromatic π radicals, where the spin may be distributed over many ring carbons, by the now familiar relationship:

$$A_{\text{iso}}(^1\text{H}) = Q\rho_c \tag{II, 114}$$

where ρ_c is the unpaired π spin density at the adjacent carbon atom, and Q is a constant with a value of -22.5 G for an average CH bond [McConnell and Chesnut (335); see other references given there].

The interesting variations of the intensities of the four proton hyperfine lines in the ESR spectrum of the CH_3 radical as a function of temperature and matrix material due to rotational effects have already been discussed, in section II-5.

CF_3. The isotropic hyperfine interaction with ^{13}C in CF_3 is much larger than in CH_3 (271 G versus 42 G), indicating that the electron is an orbital having about 21% s character, from which one infers that the molecule is pyramidal [Fessenden and Schuler (163); Rogers and Kispert (393)]. The hybridization is between sp^2 (planar) and sp^3 (tetrahedral) for the bonded orbitals, and the angle between bonds, θ, is roughly given by

$$\cos\theta = -\left(\frac{1-s}{2+s}\right) \tag{II, 115}$$

where s is the fraction of s character in the unpaired electron. For CF_3 this yields $\theta = 111°$, so that the angle between a C–F bond and the plane normal to the symmetry axis is estimated to be 17.8°, approaching closely to tetrahedral sp^3 hybridization where these angles are 109.5 and 19.5°, respectively.

The fluorine hyperfine coupling in CF_3 has been observed in irradiated liquid C_2F_6 (163), in an irradiated single crystal of trifluoroacetamide (393), and in rare gas matrices (163, 33, 327). As is often the case for ^{19}F, the hfs is relatively large because of its large nuclear moment and leads to the spectrum of CF_3 shown in Fig. II-33 containing six lines, rather than the four for CH_3 in Fig. II-21. This is so because the second-order terms in (II, 66) contribute impor-

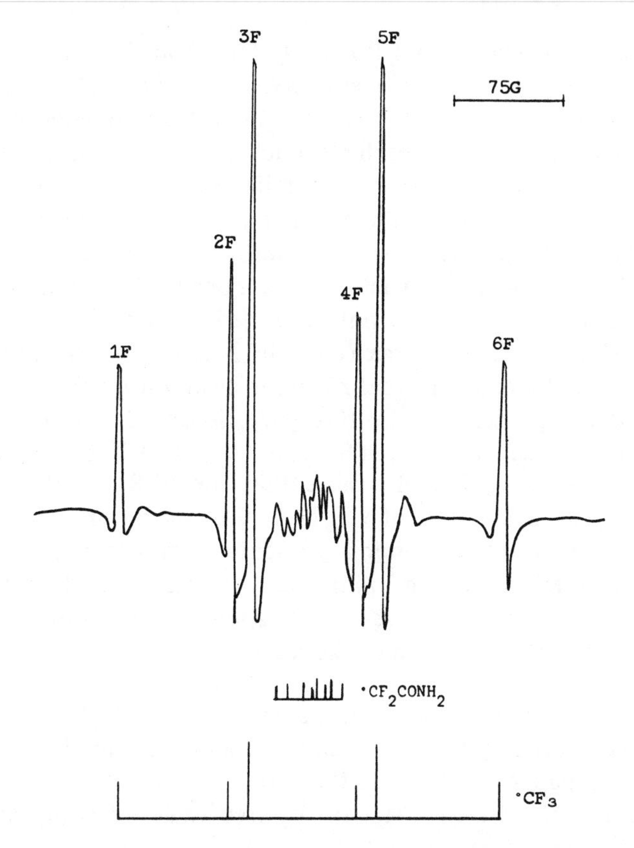

Fig. II-33. ESR spectrum of the CF_3 radical at 77°K (formed by irradiation of trifluoroacetamide single crystals) with the *c* axis parallel to ***H*** [from Rogers and Kispert (393)]. It is very similar to that for CF_3 in irradiated C_2F_6 solution, indicating that it is undergoing rapid reorientation about its threefold axis at 77°K.

tantly, and cause appreciable splitting of two inner hf lines, as Fig. II-33 shows. The relative intensities of the lines then depart from a 1:3:3:1 pattern to yield a 1:(1:2):(1:2):1 pattern, and although the peak heights are not exactly in that ratio, the varying linewidths account for the discrepancies (163).

It is also clear from the discussion in section II-4b that the principal axes of ***g*** and $\boldsymbol{A}(^{19}F)$ in the CF_3 radical are not necessarily collinear. This has been examined in some detail by Maruani, McDowell, Nakajima, and Raghunathan (327), who measured and analyzed the spectra of CF_3 obtained in rare-gas matrices [Fessenden and Schuler

(163); Adrian, Cochran, and Bowers (33)]. The single-crystal measurements of Rogers and Kispert (393) were not able to provide definitive data in this case because the radical was apparently rotating. Maruani et al. find for each ^{19}F nucleus, $A_x \cong 87$ G, $A_y \cong 80$ G, and $A_z = 263.5$ G, where z is perpendicular to the C–F bond and tilted $\alpha = 17.8°$ away from the symmetry axis of the molecule, and y is along the bond. However, if one of the principal axes of the fluorine coupling is not restricted to lie along the C–F bond as these authors assume, then A_z is not restricted to $\alpha = 17.8°$. In fact, Edlund, Shiotani, and Sohma (cited in 212a) have calculated this angle to be 26.9° and find that z leans instead *toward* the symmetry axis. This was pointed out by Hasegawa, Sogabe, and Miura (212a), who found a similar situation in SiF_3, but with $\alpha = 35.6°$.

Maruani et al. (327) also note that the ESR spectra of CF_3 (formed by uv irradiation of CF_3I) in the rare-gas matrices can be accounted for entirely by randomly oriented, nonrotating, noninverting radicals. However, the hindrance to motion must be relatively small, since a temperature rise of only a few degrees allows rotation, diffusion, and recombination of the radicals.

CH_2F, CHF_2. The partially fluorinated radicals CH_2F and CHF_2 were also observed by Fessenden and Schuler (163), and the s character of the unpaired electron on the carbon atom was shown to vary in a regular manner from CH_3 to CF_3 (see Table II-9), indicating an

Table II-9. Structures MH_3[a] and CH_xF_y[b] Radicals from ESR Spectral Data

Radical	s Character of "π" Orbital	Bond Angle, θ	Pyramidal Angle, β
CH_3	0.0	120°	90° (planar)
CH_2F	0.04	118°	82°
CHF_2	0.11	115°	76.5°
CF_3	0.21	111°	72°
SiH_3	0.14	113.5°	74°
GeH_3	0.11	115°	76.5°
SnH_3	0.07	117°	83°

[a] Jackel and Gordy (246).
[b] Fessenden and Schuler (163).

increasing nonplanarity with the substitution of fluorine atoms on CH_3.

SiH_3, GeH_3, SnH_3. For these hydrides of group IV, the hyperfine interactions with the central atoms (^{29}Si, ^{73}Ge, or $^{117,119}Sn$) again indicate an appreciable *s* character to the odd electron and therefore pyramidal molecules. Jackel and Gordy (246) then find the bond angle, θ, via (II, 115) and pyramidal angle, β, for these tetra-atomic hydrides as given in Table II-9. The large jump in *s* character from CH_3 to SiH_3, and then a gradual decrease down the series, is explained qualitatively by those authors as due to the difference in electronegativities of the central atoms relative to H, C being greater and Si, Ge, Sn being less than that of H. The rationale is that a negative or positive formal charge on the central atom will favor less or more hybridization, respectively, by stabilizing or destabilizing the lower-energy *s* orbital.

Variation in linewidths due to motional effects have been observed in the ESR spectra of SiH_3 in Ar, Kr, and N_2 matrices by Raghunathan and Shimokoshi (387). The argon spectrum was shown earlier, in Fig. II-24, and discussed in section II-5.

A clear discussion of all of the above radicals, particularly with regard to their spin density distributions, has been given by Gordy (8).

HCO. This radical was originally observed optically by Herzberg and Ramsay (218) in the gas phase and by Ewing, Thompson, and Pimentel (160) in solid carbon monoxide at 20°K. Its ESR spectrum has been observed in a single crystal of formic acid at 77°K by Holmberg (226) and was discussed earlier in section II-4 as an example of a molecule with noncollinear principal $\boldsymbol{g}$- and $\boldsymbol{A}$-tensor axes. We are concerned here with its matrix ESR spectrum and comparison of it with the other data.

Figure II-34 shows the ESR spectrum of HCO in solid CO at 11°K as obtained by Adrian, Cochran, and Bowers (34). The approximate 140 G splitting between the two patterns is due to hf interaction with the ^1H nucleus ($I = \frac{1}{2}$). This large hfs implies a large *s* character (~28%) to the electron and therefore a σ radical. However, the pattern of the high-field lines indicates that the radical has three *g* components and that there is coincidental overlap in the low-field lines, as indicated in the figure. The interpretation is then in

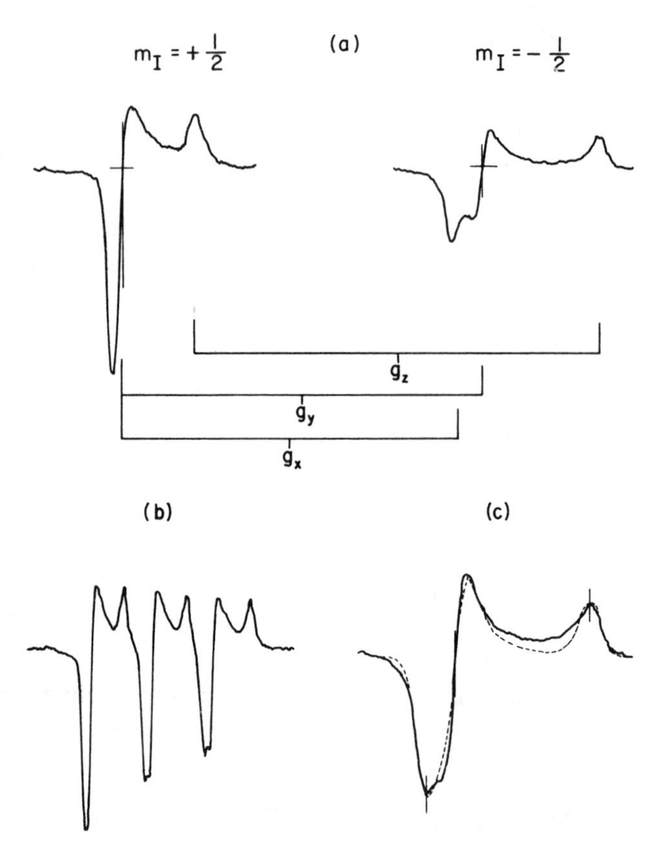

Fig. II-34. ESR spectra of HCO and DCO in CO matrices at ~10°K [from Adrian, Cochran, and Bowers (34)]. (a) HCO in CO at 11°K. (b) DCO in CO at 16°K. (c) Central line of the DCO spectrum in (b): solid curve is the observed line; dashed line is calculated.

accord with a bent radical with the odd electron in one of three in-plane $sp^2\sigma$ orbitals on the carbon atom, the other two being bonded to H and O (see Fig. II-18).

Since the principal axis (x) perpendicular to the plane of the molecule is the only one necessarily common to both $\boldsymbol{g}$ and $\boldsymbol{A}$, the $\boldsymbol{A}$ tensor will contain an off-diagonal element (A_{yz}) when referred to the principal axes of $\boldsymbol{g}$. As discussed in section II-4, it is then not possible to obtain seven parameters (three g_{ii} and four A_{ij}) from the spectrum in Fig. II-34a, and more information is required. That information is supplied by the corresponding spectrum of DCO shown in Fig. II-34b and c. Since the central line of that pattern does not have a hyperfine splitting ($m_I = 0$), it yields directly, by simulation of its line shape, the principal components of the $\boldsymbol{g}$ tensor. These, combined with the application of the theory embodied in section II-4b, then allow computation of the $\boldsymbol{A}(^1\text{H})$ tensor components, referred to the principal axes of the $\boldsymbol{g}$ tensor. $|A_{yz}(^1\text{H})|$ was found to be 16.2 MHz. Subsequently Cochran, Adrian, and Bowers (116) observed the ESR spectrum of $H^{13}CO$ trapped in an argon matrix at 4°K and derived $\boldsymbol{A}(^{13}\text{C})$ components but found $|A_{yz}(^{13}\text{C})| \simeq 0$. As discussed on page 114 comparison of the hf parameters with the single crystal data is not very good (see Table II-3), which indicates the difficulty, in this case, of obtaining accurate principal $\boldsymbol{A}$ components from matrix ESR spectra and/or that the crystal and matrix environments are having significantly different effects upon these interactions.

Na_3. This cluster molecule has been detected in sodium vapor only by mass spectrometry where its ionization potential has also been determined [Robbins, Leckenby, Willis, and Foster (390); Herrmann, Schumacher, and Wöste (216)]. Its ESR spectrum was observed while it was trapped in solid argon [Lindsay, Herschbach, and Kwiram (312, 313)]. It was best prepared by condensing sodium vapor and

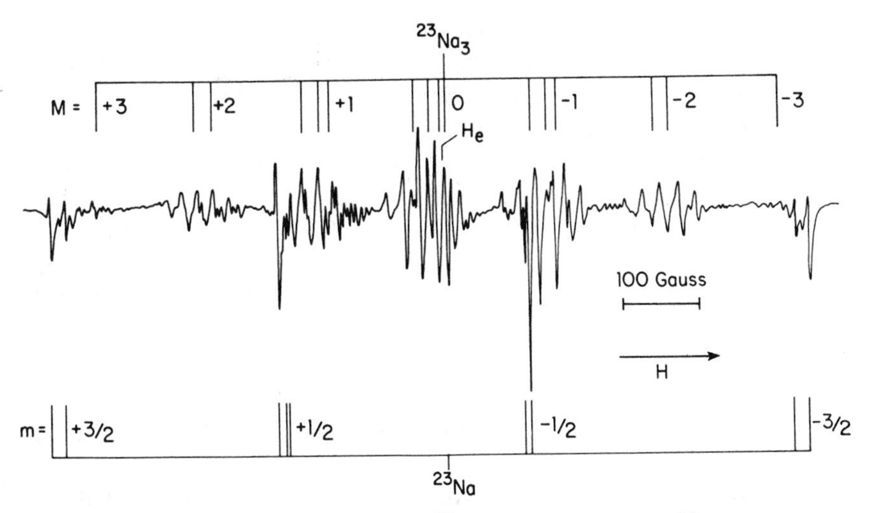

Fig. II-35. ESR spectrum of sodium trimers, $^{23}Na_3$, and sodium atoms, ^{23}Na, in an argon matrix at 4°K [from Lindsay, Herschbach, and Kwiram (312)].

argon on a sapphire plate maintained at a slightly higher temperature than when atoms were isolated. The spectrum measured in an argon matrix at 4°K is shown in Fig. II-35. Part of the spectrum arises from ^{23}Na atoms in multiple matrix sites indicated at the bottom of the figure, as discussed and illustrated in Chapter I.

The primary feature of the spectrum is the dominance of seven clusters of lines separated by about 150 G, suggesting that the observed species contains two equivalent ^{23}Na atoms, each with $I = \frac{3}{2}$. The intensity variation appears to be approximately in accord with a 1:2:3:4:3:2:1 pattern necessary for that assignment. This pattern is centered about $g = 2.00$. Then the simplest species producing such a spectrum is a trimer, Na_3, in either a linear ($^2\Sigma_u^+$) or an isoceles triangle ($^2B_2$) conformation.

Two equivalent nuclei with spin $I = \frac{3}{2}$, in first-order theory, will provide $2(I_1 + I_2) + 1 = 7$ transitions, as discussed on page 56. Then

$$H_1 = H^0 - A_1 m_{I_1} - A_2 m_{I_2}$$

where here $A_1 = A_2 = A_0$ so that

$$H_1 = H^0 - A_0(m_{I_1} + m_{I_2}) = H^0 - A_0 M \qquad \text{(II, 116)}$$

If $m_{I_1} + m_{I_2} = M$, then since $I_1 = I_2 = \frac{3}{2}$, M takes the values +3, +2, +1, 0, −1, −2, −3. This then yields seven resonance lines separated by A_0. First-order theory is sufficient if the hfs is much smaller than the electronic Zeeman energy; however, here A is large enough that second-order terms, proportional to $A^2/h\nu$, must be included. In second order these nuclear spins are coupled to form a total nuclear spin J that can take all values $I_1 + I_2 \geqslant J \geqslant |I_1 - I_2|$. In this case $J = 3, 2, 1, 0$. Each value of J is space-quantized into $2J + 1$ values of $M_J = +J, \cdots 0, \cdots -J$. Thus the total degeneracy is still $(2I_1 + 1)(2I_1 + 1)$, which is 16 here. This complete removal of degeneracy occurs in second order which adds the term to (II, 116)

$$H_2(J, M) = -\frac{1}{2} A_0^2 [J(J+1) - M^2]/H(J, M) \qquad \text{(II, 117)}$$

Including electron spin, the total wavefunction is designated by $|m_S, J, m_J\rangle$ where m_S here is equal to $\pm\frac{1}{2}$. Allowed transitions are for $\Delta m_S = \pm 1$, $\Delta m_J = 0$.

These second-order terms now cause the seven equally spaced lines to be split up into 16 lines shifted to lower field relative to the center of the first-order pattern, as indicated by the stick diagram at the top of Fig. II-35. Furthermore, each of these lines is split into a quartet due to hf interaction of the electron with the nucleus of the third, central, ^{23}Na atom given by a second hf constant A_3. This still does not account for all of the structure of each of the seven groups of lines, and it is postulated that Na_3 is trapped in two sites (designated I and II) having slightly different hf parameters. This is supported by observation of a variation in relative intensities in I and II lines in separate deposition experiments.

This analysis assumes that the anisotropy in the $\boldsymbol{g}$ and $\boldsymbol{A}$ tensors is small, which is supported by the isotropic line shapes and the absence of "parallel-like" lines in the spectrum (see, for example, Fig. II-19). This implies that there is little p electron contribution to the wavefunction, which is not unexpected for a weakly bonded Na molecule. The final "isotropic" magnetic parameters are given in Table II-10 along with the s spin densities derived in the usual approximate way via (II, 43). Thus, the ESR spectrum does not establish whether the trimer is linear or bent, but it does rule out an equilateral triangle conformation at 4°K.

Alkali Superoxide Molecules: NaO_2, KO_2, RbO_2, and CsO_2. These very ionic molecules are spontaneously formed when alkali metal atoms are vaporized and trapped with O_2 in rare-gas matrices. There is virtually complete transfer of the electron to O_2 with the formation of symmetrical triangular ion-pairs, $M^+O_2^-$. Infrared and Raman spectral studies, with unsymmetrical substitution of ^{18}O, have demonstrated that they are of C_{2v} symmetry [Andrews (41); Smardzewski

Table II-10. Magnetic Parameters of Na_3 in an Argon Matrix (312)

				3s Spin Density Distribution[a]	
Site	g_0	$\lvert A_0 \rvert$ (G)	$\lvert A_3 \rvert$ (G)	Outer Na	Middle Na
I	2.0012(12)	149.5(15)	22.6(10)	0.473(5)	0.072(3)
II	2.0012(12)	153.0(15)	20.3(9)	0.484(5)	0.064(3)

[a]Isotropic hfs of ^{23}Na atom taken as 316 G.

and Andrews (417)]. The free O_2^- radical, as discussed in section II-2a, has a $^2\Pi_i$ ground state, and therefore its ESR spectrum would not be expected to be observable isolated in a symmetrical environment in matrices because of extreme $\boldsymbol{g}$ tensor anisotropy. However, here the field produced by the neighboring cation is highly asymmetric and distorts the diatom from cylindrical symmetry, partially quenching its orbital angular momentum, and reducing its $\boldsymbol{g}$ tensor anisotropy.

The ESR spectra are then similar to those for randomly oriented linear molecules with "parallel" (g_{zz}) and "perpendicular" (g_{xx} and g_{yy}) features. The spectrum of NaO_2 in solid argon [Adrian, Cochran, and Bowers (36)] is given in Fig. II-36, and the spectra of all of the alkali-metal superoxides in krypton matrices are illustrated in Fig. II-37 [Lindsay, Herschbach, and Kwiram (310)]. The crystal fields produced by the metal ions are so effective in lifting the degeneracy that the g components are all quite close to g_e, as shown in Table II-11. The quenching is more effective in the matrix molecule than

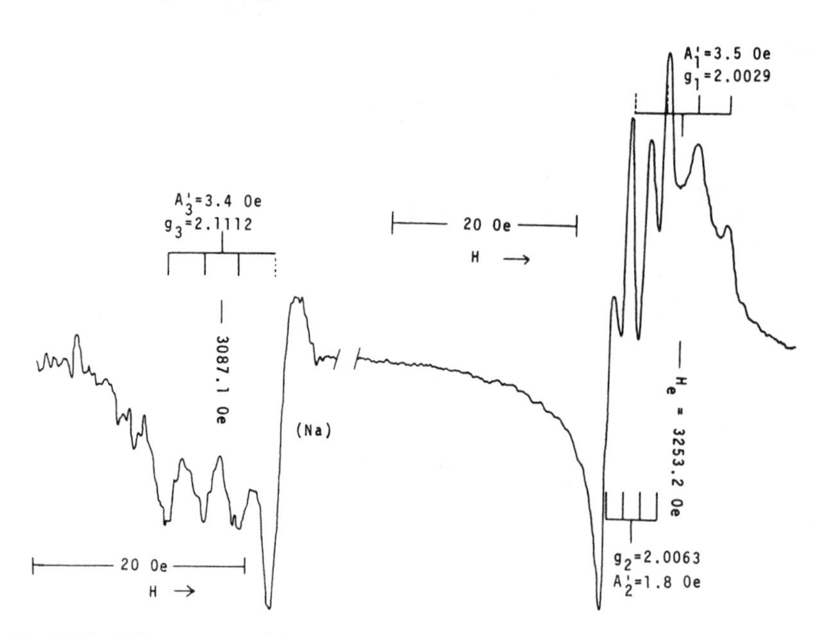

Fig. II-36. ESR spectrum of the NaO_2 molecule in an argon matrix at 4°K [from Adrian, Cochran, and Bowers (36)]. The quartet structure associated with each of the features corresponding to the three g_{ii} is attributed to ^{23}Na hfs. H_e denotes the resonant field of the free electron. Na denotes a sodium atom line.

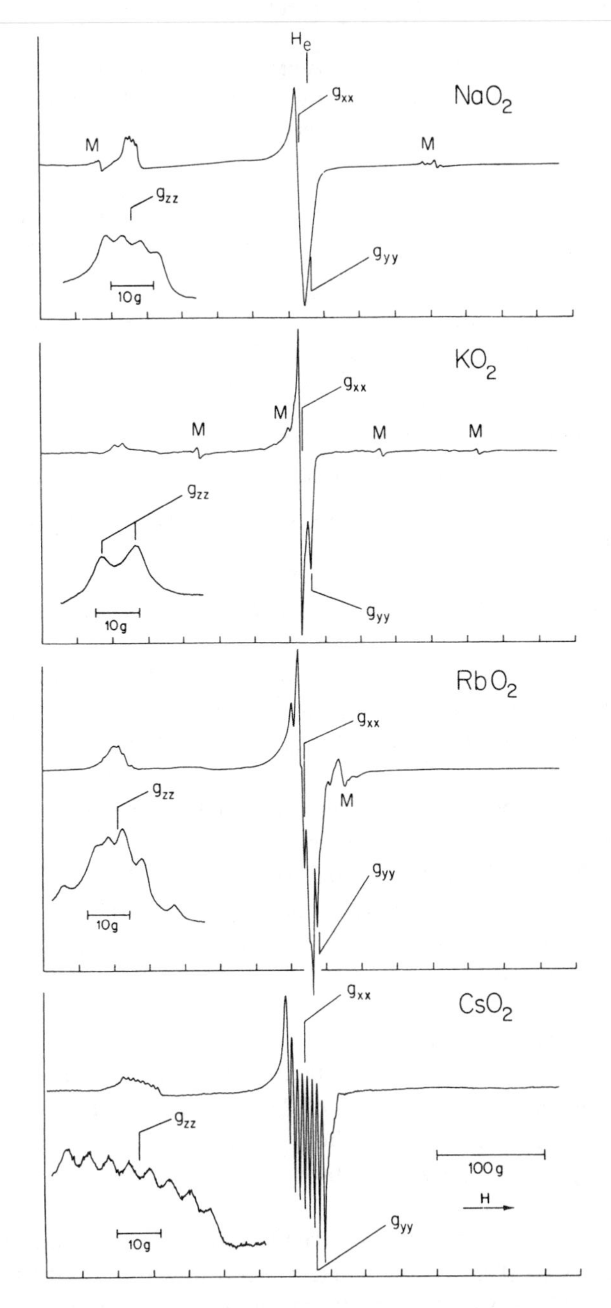

Fig. II-37. ESR spectra of alkali superoxide molecules in a krypton matrix [from Lindsay, Herschbach, and Kwiram (310)]. Lines marked "M" are due to unreacted alkali atoms. H_e = 3312.2 G for a free electron (ν = 9.2825 GHz).

Table II-11. Magnetic Parameters of Alkali Superoxides in Rare-Gas Matrices[a] [from Lindsay, Herschbach, and Kwiram (310) and Adrian, Cochran, and Bowers (36)].

	g_{xx}	g_{yy}	g_{zz}	A_{xx} (G)	A_{yy} (G)	A_{zz} (G)
$^{23}NaO_2$[b]	2.0063	2.0029	2.1112	1.8	3.5	3.4
$^{23}NaO_2$	2.0075[c]	2.0022[c]	2.1106	2.2[c]	3.8[c]	3.6
$^{39}KO_2$	2.0068	2.0007	2.1184[d]	0	0	0
$^{87}RbO_2$	2.0069[e]	1.9996[e]	2.1227	6.5[e]	8.0[e]	8.1
$^{133}CsO_2$	2.0069	2.0013	2.1069	4.84	5.3	4.93

[a] All are in Kr matrices unless otherwise indicated.
[b] In Ar matrix [Adrian et al. (36)].
[c] These numbers may be in error because of poor resolution in the pertinent spectral regions.
[d] Average of two peaks at $g = 2.1158$ and 2.1210.
[e] Because of overlap of $^{85}RbO_2$ and $^{87}RbO_2$ spectra, accurate values of these parameters may require computer synthesis.

in crystalline defects because there the O_2^- is in a more symmetrical site [Känzig and Cohen (257)].

The odd electron is not completely localized on the O_2^-, and small metal hyperfine splittings are observed. The hfs patterns for the different alkali metals are clearly distinguishable in Fig. II-37. In all cases these splittings are small and not more than 9 G, as can be seen from the experimental values listed in Table II-11.

The symmetry of the molecules is sufficiently high that the $\boldsymbol{g}$ and $\boldsymbol{A}$ tensors have the same principal axes (as in NO_2), and this is also true for the nuclear quadrupole tensor $\boldsymbol{Q}$, which should be included in the spin Hamiltonian for the ^{23}Na and ^{85,87}Rb molecules. However, the small metal hfs makes the calculation of quadrupole effects difficult here. The metal cation A_{iso} and T_{ii} values can be calculated from the experimental hfs data in exactly the way outlined on page 104ff for ^{14}N in NO_2, assuming again that all A_{ii} have the same sign. They are given in columns 2–5 of Table II-12 for the most reliable NaO_2 and CsO_2 data. Only their relative signs are established experimentally, and theory must be applied to make a choice between the two alternatives.

If the $^2\Pi_i$ O_2^- molecule is aligned along the z axis and the metal cation bonds to it along the x axis, then the π and π^* orbitals in the linear molecule are split as shown in Fig. II-38. (The reader will note the resemblance of this orbital diagram figure to Fig. II-5, used in

Table II-12. Anisotropic Metal Hyperfine Parameters (in Gauss) for NaO_2 and CsO_2; Comparison with Values Calculated from an Ionic Model
[from Lindsay, Herschbach, and Kwiram (310)].

	Experimental[a]				Calculated		
	T_{xx}	T_{yy}	T_{zz}	A_{iso}	T_{xx}	T_{yy}	T_{zz}
NaO_2 ($^2A_2$)	∓1.0	±0.6	±0.4	±3.2	+1.18	−0.62	−0.56
CsO_2 ($^2A_2$)	∓0.2	±0.3	∓0.1	±5.0	+0.31	−0.16	−0.15
CsO_2 ($^2B_1$)[b]	±0.3	∓0.2	∓0.1	±5.0	+0.46	−0.21	−0.25

[a] Assuming principal A_{ii} all have the same sign.
[b] Requires interchange of the x and y indices. See text.

discussing Cl_2^- in a V_K center.) Only the disposition of the oxygen $2p$ electrons is shown in the figure. In a simple ionic model the antibonding $\pi_x{}^*$ in-plane orbital is lowered relative to the $\pi_y{}^*$ out-of-plane orbital in the field of the cation, and the molecule of C_{2v} symmetry then has a configuration

$$\cdots a_1{}^2 a_1{}^2 b_2{}^2 b_1{}^2 a_2{}^1$$

and a $^2A_2$ ground state. The new molecular orbitals, including p and d AO's on the metal atom, are given in Fig. II-38. Using these MO's, one can calculate the g shifts as [Lindsay, Herschbach, and Kwiram (310)]:

$$\Delta g_{zz} = g_{zz} - g_e \cong \frac{2\lambda}{\delta} \alpha\beta$$

$$\Delta g_{yy} \cong -\left(\frac{\lambda^2}{\delta^2}\right) \alpha\beta + \frac{\lambda^2}{\Delta\delta} \alpha\beta\gamma$$

$$\Delta g_{xx} \cong -\left(\frac{\lambda^2}{\delta^2}\right) \alpha\beta + \frac{2\lambda}{\Delta} \alpha\gamma - \frac{\lambda^2}{\Delta\delta} \alpha\beta\gamma$$

Here λ is the spin-orbit coupling constant of oxygen, δ and Δ are the energy differences defined in Fig. II-38, and $\alpha = 2C_1{}^2$, $\beta = 2C_3{}^2$, $\gamma = 2C_6{}^2$. For the purely ionic case $\alpha = \beta = \gamma = 1$, and with the assumption made in the derivation that $\lambda << \delta << \Delta$, one sees that Δg_{zz} should be positive and Δg_{yy}, Δg_{xx} negative with Δg_{yy} more

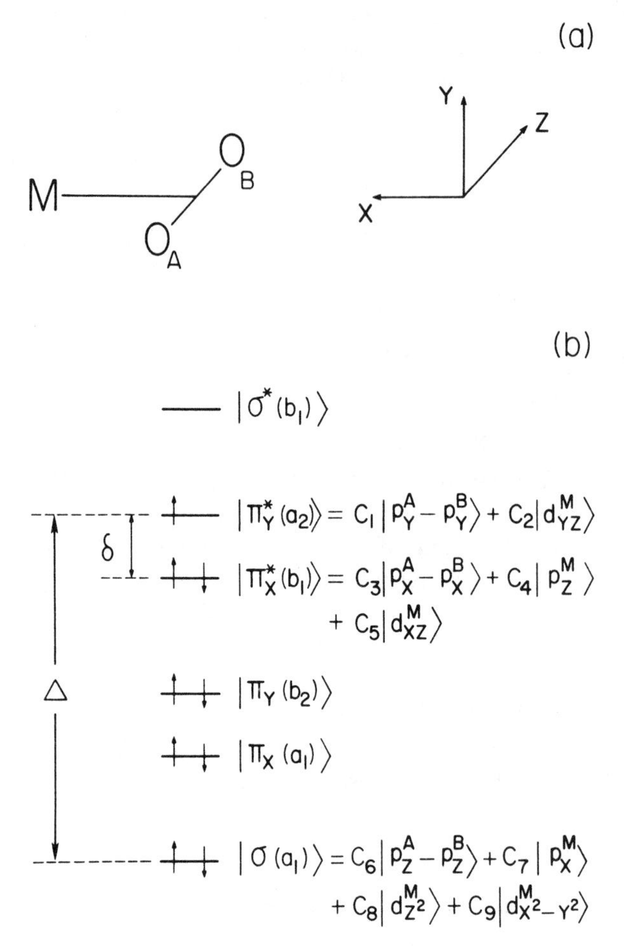

Fig. II-38. Molecular orbitals of the $M^+O_2^-$ ion-pair [from Lindsay, Herschbach, and Kwiram (310)].

negative than Δg_{xx}. This is approximately in accord with the experimental data in Table II-11.

To sort out the signs of A_{iso} and the T_{ii} of the metal ions is difficult here because of their small size. Estimates of the

$$T_{ii} = \left\langle \frac{r^2 - 3i^2}{r^5} \right\rangle$$

must be made without real knowledge of the M^+–O_2^- distance. However, this calculation has been done at various interatomic distances using the π^* orbital of O_2^- (including overlap) for NaO_2 (36) and also by a localized spin model involving distribution of the spin on O_2^- and an estimation of the distance to the nucleus (310). Both sets of workers find the lower selection of signs for $^{23}NaO_2$ to be the most likely. For CsO_2 Lindsay et al. (310) suggest from the anomalous variation in g values and from hf calculations that the $\pi_x{}^*$ and $\pi_y{}^*$ orbitals may be inverted. Thus in this case the odd electron would reside predominantly in the in-plane π^* orbital rather than the out-of-plane orbital of a purely ionic model.

It is perhaps worth noting here that Cs and Rb react with N_2O to form the $^2\Sigma$ monoxides, not as in LiO and NaO where the ground state is $^2\Pi$ [Lindsay, Herschbach, and Kwiram (311)]. In these heavier metals, this analogous "anomaly" arises because the $(n-1)\,p$ orbitals are at comparable energy to the $2p$ orbitals of O^- and can participate in the bonding.

Chapter III

$^3\Sigma$ and Other Triplet ($S = 1$) Molecules in Magnetic Fields

When a molecule contains two unpaired electrons ($S = 1$), another term must be added to the spin Hamiltonian of $S = \frac{1}{2}$ molecules, as given in (II, 15), of the form

$$\mathcal{H}_D = \boldsymbol{S} \cdot \boldsymbol{D} \cdot \boldsymbol{S} \qquad \text{(III, 1)}$$

This spin–spin interaction term leads to further splittings of the energy levels in all molecules where $S \geqslant 1$, and the additional spectral transitions due to these splittings are called *fine* structure. Splittings arising from electron–nuclear magnetic interactions are usually much smaller (hence the name *hyperfine* structure), but in some molecules the two effects may be comparable.

The interactions signified by $\mathcal{H}_D$ are present regardless of whether a magnetic field is applied, so that its effect upon the energy levels is usually referred to as zero-field splitting (zfs), and $\boldsymbol{D}$ is referred to as the zero-field tensor. [Note that hyperfine interactions also produce a splitting of the energy levels of doublet atoms and molecules at zero magnetic field (see Fig. I-3), but it is not referred to as zfs.] When $\boldsymbol{D}$ is diagonalized, it is customary to write

$$\mathcal{H}_D = -XS_x^2 - YS_y^2 - ZS_z^2 \qquad \text{(III, 2)}$$

where X, Y, Z are the eigenvalues in the three principal axis directions, with the further property that

$$X + Y + Z = 0 \qquad \text{(III, 3)}$$

Because of this traceless property of $\boldsymbol{D}$, these three principal values can be reduced to two, which have been designated as D and E.

Since $Z = -(X + Y)$, then we may write

$$X = \tfrac{1}{2}(X - Y) - \tfrac{1}{2}Z$$
$$Y = -\tfrac{1}{2}(X - Y) - \tfrac{1}{2}Z$$

and (III, 2) becomes

$$\mathcal{H}_{D,SS} = -Z[S_z^2 - \tfrac{1}{2}(S_x^2 + S_y^2)] - \tfrac{1}{2}(X - Y)(S_x^2 - S_y^2)$$

Also since

$$S_z^2 - \tfrac{1}{2}(S_x^2 + S_y^2) = \tfrac{3}{2}(S_z^2 - \tfrac{1}{3}\boldsymbol{S}^2)$$

then

$$\mathcal{H}_D = D[S_z^2 - \tfrac{1}{3}S(S + 1)] + E(S_x^2 - S_y^2) \qquad \text{(III, 4)}$$

where

$$D = -\tfrac{3}{2}Z, \qquad E = -\tfrac{1}{2}(X - Y) \qquad \text{(III, 5)}$$

Although (III, 4) is the usual form used for the fine-structure spin Hamiltonian, of course (III, 2) is completely equivalent, and one set of parameters can be transformed into the other via (III, 5) and the traceless property (III, 3). One may note in (III, 4) that the term $-\tfrac{1}{3}D\,S(S + 1)$ is a constant for a given electronic state of a molecule and therefore does not contribute to the zero-field splitting in that state, but it is retained because it puts the zero of energy at the center of gravity of the split levels.

1. ZERO-FIELD TENSOR, D

It so happens that a contribution to the spin Hamiltonian of the form of (III, 1) can arise from two distinct sources that we will label here as having zero-field tensors $\boldsymbol{D}_{SO}$ and $\boldsymbol{D}_{SS}$.

(a) D_{SO}

In the derivation of (II, 15) it was shown that spin-orbit coupling can lead to an $\boldsymbol{S} \cdot \boldsymbol{D} \cdot \boldsymbol{S}$ term and therefore indirectly to spin-spin coupling in the molecule. There it was demonstrated that

$$\boldsymbol{D}_{SO} = \xi^2 \boldsymbol{\Lambda} \qquad \text{(II, 17), (III, 6)}$$

where ξ is the spin-orbit coupling constant of the molecule, so that $\boldsymbol{D}_{SO}$ is clearly quite sensitive to the magnitude of ξ. $\boldsymbol{\Lambda}$ involves matrix elements connecting the ground and excited states via the orbital-angular-momentum operator $\boldsymbol{L}$:

$$\Lambda_{ij} = -\sum_n \frac{\langle 0|\boldsymbol{L}_i|n\rangle \langle n|\boldsymbol{L}_j|0\rangle}{E_n - E_0} \qquad \text{(II, 14), (III, 7)}$$

Spin-orbit contributions to $\boldsymbol{D}$ can be expected to be particularly important in molecules containing heavy elements where ξ is relatively large. Even in light molecules, such as O_2, such contributions can be dominant if there is a low-lying electronic state coupled strongly to the ground state via $\boldsymbol{\Lambda}$. In the calculation of $\boldsymbol{D}$ from (III, 6) all excited states n, irrespective of their multiplicities, must be considered and not just, for example, the excited triplet states if the ground state is triplet.

Note that the expression for the $\boldsymbol{g}$ tensor given in (II, 16) also contains a spin-orbit term involving ξ and Λ. It might then be mistakenly concluded that measurements of Δg_{ii} might provide, for example via (II, 10) and (II, 11), values of Λ_{ii} that could then be used in (III, 6) to calculate D_{ii}. However, it can be seen from (II, 7) that the only excited states contributing to the Λ_{ii} involved in the calculation of the $\boldsymbol{g}$ components are those having the same electronic spin as that of the ground state, whereas that is not necessarily the case in the calculation of the D_{ii} [Hameka, (205), page 26]. For example, a dominant spin-orbit-coupled excited state contributing to the zfs parameter D of the $^3\Sigma_g^-$ O_2 molecule is the low-lying $^1\Sigma_g^+$ [Kayama and Baird (270); Langhoff and Kern (305], whereas only the lowest $^3\Pi_g$ states can be expected to affect $\Delta g_\perp$ [see, for example, Easley and Weltner (153)]. If, on the other hand, the contributing excited states do have the same multiplicity as the ground state, then a relationship can be found between the zfs parameters D, E, and the $\boldsymbol{g}$ tensor components [McGarvey (339), pages 117, 118].

(b) D_{SS}

$\boldsymbol{D}_{SS}$ arises directly from spin-spin, i.e., magnetic dipole–dipole, interactions between the unpaired electrons in the molecule. [McConnell (334) has shown that closed shells do not contribute to the zero-field splitting.] Then in analogy with (I, 19)

$$\mathcal{H}_{D,SS} = g^2\beta^2 \left\{ \frac{\mathbf{S}_1 \cdot \mathbf{S}_2}{r^3} - \frac{3(\mathbf{S}_1 \cdot \mathbf{r})(\mathbf{S}_2 \cdot \mathbf{r})}{r^5} \right\} \tag{III, 8}$$

When expanded [see Wertz and Bolton (23), page 225; Carrington and McLachlan (6), p. 117],

$$\begin{aligned} \mathcal{H}_{D,SS} = \frac{1}{2}\frac{g^2\beta^3}{r^5} & [(r^2 - 3x^2) S_x^2 + (r^2 - 3y^2) S_y^2 \\ & + (r^2 - 3z^2) S_z^2 - 3xy(S_xS_y + S_yS_x) \\ & - 3xz(S_xS_z + S_zS_x) - 3yz(S_yS_z + S_zS_y)] \end{aligned} \tag{III, 9}$$

where $r^2 = x^2 + y^2 + z^2$. In tensor form this may be written as

$\mathcal{H}_{D,SS} = \mathbf{S} \cdot \mathbf{D}_{SS} \cdot \mathbf{S}$, with

$$\mathbf{D}_{SS} = \tfrac{1}{2} g^2\beta^2 \begin{bmatrix} \langle (r^2 - 3x^2)/r^5 \rangle & \langle -3xy/r^5 \rangle & \langle -3xz/r^5 \rangle \\ \langle -3xy/r^5 \rangle & \langle (r^2 - 3y^2)/r^5 \rangle & \langle -3yz/r^5 \rangle \\ \langle -3xz/r^5 \rangle & \langle -3yz/r^5 \rangle & \langle (r^2 - 3z^2)/r^5 \rangle \end{bmatrix} \tag{III, 10}$$

[Note that as defined above in (III, 2), $-X, -Y, -Z$ correspond to the eigenvalues of this matrix.] $\mathbf{D}_{SS}$ is symmetric and traceless, with each element determined by averaging over the electronic wavefunction. Thus, it is a property of each electronic state of a molecule with $S \geqslant 1$.

$\mathbf{D}$ is then the sum of $\mathbf{D}_{SO}$ and $\mathbf{D}_{SS}$, and which one of the contributions dominates depends upon the type of molecule considered. For example, in the triplet state of large unsaturated hydrocarbon molecules, such as triphenylene where $D = 0.134\ \text{cm}^{-1}$ and $E = 0$, the dominant contribution is $\mathbf{D}_{SS}$. The magnitude of the zfs among such radicals varies approximately as the inverse of the molecular volume since magnetic-dipole interaction between the two electrons, according to (III, 10), varies as $1/r^3$. On the other hand, in transition-metal complexes the spin-orbit mechanism is always the larger contributor [McGarvey (339), page 118]. Of course, analysis of ESR experiments yields only the parameters D and E, and no information about their origin in the molecule is obtained. In principle, these parameters can be theoretically derived if the electronic wavefunction of the molecule is known, and we will discuss the calculations that have been made on a few molecules in section 5 of this chapter.

2. $^3\Sigma$ MOLECULES

(a) Eigenvalues in a Magnetic Field

In a linear molecule z is the symmetry axis, x and y are equivalent directions, and $E = 0$ from (III, 5). [A molecule must have at least a threefold (trigonal) axis of symmetry for E to vanish.] Then, for $S = 1$,

$$\mathcal{H}_D = D(S_z^2 - \tfrac{2}{3}) \qquad \text{(III, 11)}$$

where components of S_z are $m_S = +1, 0, -1$. Also, from (III, 5) and the traceless property (III, 3) for a linear molecule the eigenvalues at zero magnetic field can be written down immediately as

$$X = Y = \tfrac{1}{3} D, \qquad Z = -\tfrac{2}{3} D \qquad \text{(III, 12)}$$

Neglecting the usually smaller effects of the hyperfine, nuclear Zeeman, and nuclear quadrupole terms for the moment, the spin Hamiltonian for a triplet molecule in a magnetic field,

$$\mathcal{H} = \beta \mathbf{H} \cdot \mathbf{g} \cdot \mathbf{S} + \mathbf{S} \cdot \mathbf{D} \cdot \mathbf{S} \qquad \text{(III, 13)}$$

becomes for a linear geometry,

$$\mathcal{H} = g_\parallel \beta H_z S_z + g_\perp \beta (H_x S_x + H_y S_y) + D(S_z^2 - \tfrac{2}{3}) \qquad \text{(III, 14)}$$

If we choose arbitrarily y to be perpendicular to the fixed magnetic field H [de Groot and van der Waals (131); see p. 65–this does not really limit the generality of our results], then $H_y = 0$ and

$$\mathcal{H} = g_\parallel \beta H_z S_z + g_\perp \beta H_x S_x + D(S_z^2 - \tfrac{2}{3}) \qquad \text{(III, 15)}$$

To calculate the energy levels, a natural original choice as a basis is the set of orthonormal spin functions,

$$|+1\rangle = |\alpha_1 \alpha_2\rangle$$
$$|0\rangle = \frac{1}{\sqrt{2}} |\alpha_1 \beta_2 + \beta_1 \alpha_2\rangle$$
$$|-1\rangle = |\beta_1 \beta_2\rangle \qquad \text{(III, 16)}$$

The reader will recall that for all particles with odd spin (electrons, protons, i.e., fermions) only antisymmetrical wavefunctions can occur. This requirement for antisymmetry in the coordinates of the electrons in the wavefunction requires that, in the triplet state, antisymmetric *orbital* functions be combined with the three possible symmetric *spin* functions given in (III, 16). The other possibility, occurring for two electrons in a singlet state, combines a symmetrical orbital function with the (single) antisymmetric spin function, $1/\sqrt{2}\,(\alpha_1\beta_2 - \beta_1\alpha_2)$.

The interaction of the spin operators with these functions is given by the relations

$$S_x\,|\alpha\rangle = \tfrac{1}{2}\,|\beta\rangle, \quad S_x\,|\beta\rangle = \tfrac{1}{2}\,|\alpha\rangle, \quad S_y\,|\alpha\rangle = \tfrac{1}{2}\,i\,|\beta\rangle$$
$$S_y\,|\beta\rangle = -\tfrac{1}{2}\,i\,|\alpha\rangle, \quad S_z\,|\alpha\rangle = \tfrac{1}{2}\,|\alpha\rangle, \quad S_z\,|\beta\rangle = -\tfrac{1}{2}\,|\beta\rangle$$

and $S_z{}^2\,|\alpha\rangle = \frac{1}{4}\,|\alpha\rangle$, $S_z{}^2\,|\beta\rangle = \frac{1}{4}\,|\beta\rangle$. Then for two spins, $S_x = S_{x_1} + S_{x_2}$, etc., so that $S_z{}^2 = (S_{z_1} + S_{z_2})^2$. Using these relations with the Hamiltonian (III, 15), one obtains the matrix:

$$\begin{array}{c|ccc} & |+1\rangle & |0\rangle & |-1\rangle \\ \hline |+1\rangle & \dfrac{D}{3} + G_z & \dfrac{G_x}{\sqrt{2}} & 0 \\ |0\rangle & \dfrac{G_x}{\sqrt{2}} & -\dfrac{2}{3}D & \dfrac{G_x}{\sqrt{2}} \\ |-1\rangle & 0 & \dfrac{G_x}{\sqrt{2}} & \dfrac{D}{3} - G_z \end{array}$$

(III, 17)

where $G_z = g_{\|}\beta H_z$ and $G_x = g_{\perp}\beta H_x$. Here $H_z = H\cos\theta$, $H_x = H\sin\theta$, where θ is the polar angle between $\boldsymbol{H}$ and the axis of the molecule.

For $H \parallel z$, $H_z = H$, $H_x = 0$, the matrix is diagonal so that the eigenvalues are

$$W_{+1} = \frac{D}{3} + g_{\parallel}\beta H$$

$$W_0 = -\frac{2}{3}D$$

$$W_{-1} = \frac{D}{3} - g_{\parallel}\beta H \qquad \text{(III, 18)}$$

and at zero field, the energies reduce to those given in (III, 12). The eigenvalues for $H \parallel z$ given in (III, 18) are plotted versus H in Fig. III-1a (assume D positive), and all yield, of course, straight lines, where the slopes of the $m_S = \pm 1$ levels are $\pm g_{\parallel}\beta$, respectively. The zero-field splitting (zfs) is $W_{\pm 1} - W_0 = D$, correct for any $^3\Sigma$ molecule. Thus for H along the axis of the molecule there is no mixing of the eigenfunctions (III, 16) at all magnetic fields, and m_S is always a "good" quantum number.

For $H \perp z$, $H = H_x$, $H_z = 0$, and the roots of the secular determinant are

$$W_1 = \left[-\frac{D}{3} + (D^2 + 4g_{\perp}^2\beta^2H^2)^{1/2}\right]/2$$

$$W_2 = \frac{D}{3}$$

$$W_3 = \left[-\frac{D}{3} - (D^2 + 4g_{\perp}^2\beta^2H^2)^{1/2}\right]/2 \qquad \text{(III, 19)}$$

These energies are also shown plotted versus H in Fig. III-1b. Similar to the case of $H \parallel z$, at high magnetic fields W_1 and W_3 approach $\pm g_{\perp}\beta H$, respectively, and can be correctly labeled by their $m_S = \pm 1$ values. At zero field the energies again reduce to $D/3$ and $-2D/3$, as they must. At intermediate fields it is clear that there is mixing of the $|+1\rangle$ and $|-1\rangle$ eigenvectors, leading to the opposite curvature of those energy levels.

Then at low fields, and for H at any angle to the axis of the molecule, it is preferable to use the following linear combinations, since

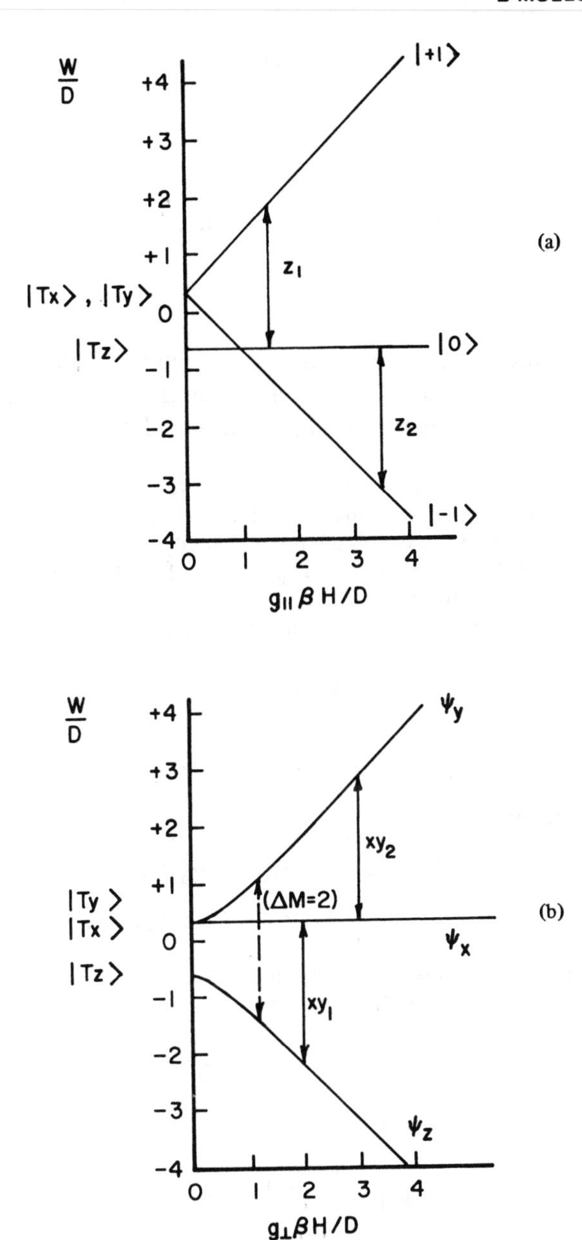

Fig. III-1. Electron spin resonance of a linear triplet molecule. (a) Magnetic field H parallel to the molecular axis, z; (b) H perpendicular to z. $\Delta m_S = \pm 1$ transitions are solid lines; $\Delta m_S = 2$ transition is dashed line (assuming the zero-field splitting parameter, D, is positive).

they diagonalize $\mathcal{H}_D$ [Hameka and Oosterhoff (206)]:

$$|T_x\rangle = -\frac{1}{\sqrt{2}}\,[\,|+1\rangle - |-1\rangle]$$

$$|T_y\rangle = \frac{i}{\sqrt{2}}\,[\,|+1\rangle + |-1\rangle]$$

$$|T_z\rangle = |0\rangle \qquad \text{(III, 20)}$$

These wavefunctions, in the limit of vanishing magnetic field, are then the eigenvectors of $\mathcal{H}_D$ with eigenvalues X, Y, Z [see (III, 2)]; i.e., their spins are quantized along the x, y, and z axes.

Note that T_x, T_y, T_z are similar to the real functions p_x, p_y, p_z for the atomic orbitals with $l = 1$. The three p orbitals are equivalent, the only distinction being the direction in which each points. Similarly to the p orbitals, $|+1\rangle = -1/\sqrt{2}\,(T_x + iT_y)$, $|-1\rangle = 1/\sqrt{2}\,(T_x - iT_y)$ and $|0\rangle = T_z$ [see, for example Carrington and McLachlan, (6), page 136].

These spin eigenfunctions have properties such that

$$S_x|T_x\rangle = S_y|T_y\rangle = S_z|T_z\rangle = 0$$

and

$$S_x|T_y\rangle = i|T_z\rangle, \quad S_x|T_z\rangle = -i|T_y\rangle, \text{ etc.}$$

by cyclic permutation, so that

$$\mathcal{H}_D|T_x\rangle = (-XS_x{}^2 - YS_y{}^2 - ZS_z{}^2)|T_x\rangle$$
$$= (0 - Y - Z)|T_x\rangle = X|T_x\rangle, \text{ etc.}$$

Using (III, 20) as a basis, again with the applied field H perpendicular to the y axis, one obtains the Hamiltonian matrix as [van der Waals and de Groot (435)]

	$\|T_x\rangle$	$\|T_y\rangle$	$\|T_z\rangle$
$\|T_x\rangle$	$\frac{D}{3}$	$-iG_z$	0
$\|T_y\rangle$	iG_z	$\frac{D}{3}$	$-iG_x$
$\|T_z\rangle$	0	iG_x	$-\frac{2}{3}D$

(III, 21)

where G_x and G_z were defined earlier. When H vanishes, this traceless matrix is diagonal, as expected.

The expansion of the secular determinant, for either basis, yields a general cubic equation whose roots are the energy levels [de Groot and van der Waals (131)]:

$$W^3 - W\left(\frac{D^2}{3} + G_x{}^2 + G_z{}^2\right) + \frac{D}{3}(G_x{}^2 - 2G_z{}^2) + \frac{2}{27}D^3 = 0 \tag{III, 22}$$

[Note that in (III, 22) the W^2 term is missing because its coefficient is the trace of the Hamiltonian matrix.] The solutions of this equation for H parallel and perpendicular to the axis of the molecule have already been given above in (III, 18) and (III, 19), respectively, but (III, 22) gives the eigenvalues for any orientation of $\boldsymbol{H}$ to the molecular axis. This equation is unwieldy and therefore not generally useful, but in principle it contains exact solutions for all eigenvalues.

If $g\beta H \gg \mathcal{H}_D$, a convenient expression for the energy levels can be found from (III, 14) using perturbation theory [Abragham and Bleaney (1), page 157]. This is done in the same way that (II, 57), which involved hyperfine interactions, was derived from (II, 49). Then for $S_z = m_S$,

$$\begin{aligned} E(m_S) = g\beta H\, m_S &+ \frac{D}{2}\,[(3g_\parallel{}^2/g^2)\cos^2\theta - 1]\left[m_S{}^2 - \frac{1}{3}S(S+1)\right] \\ &+ (g_\parallel{}^2 g_\perp{}^2/g^4)\,(D^2\cos^2\theta\sin^2\theta/2G) \\ &\cdot m_S\,[8m_S{}^2 + 1 - 4S(S+1)] \\ &+ (g_\perp{}^4/g^4)\,(D^2\sin^4\theta/8G)m_S\,[2S(S+1) - 2m_S{}^2 - 1] \end{aligned} \tag{III, 23}$$

where g is given by (II, 51), $G = g\beta H$, and $S = 1$ for the triplet case.

(b) Magnetic-Dipole Transitions

Since there are three eigenvalues, W_1, W_2, and W_3, for each value of θ (the angle between the applied field and the molecular axis), there are three possible transitions $W_1 - W_2$, $W_2 - W_3$, and $W_2 - W_3$. In high magnetic fields the magnetic moment precesses about the field axis, and each of these W_i is characterized by a value of $m_S = +1$, 0, or -1, which is a good quantum number in that limit. Then two of the possible transitions can be called $\Delta m_S = \pm 1$ and the remaining one a $\Delta m_S = \pm 2$ transition. In that limit of high field, only the usual $\Delta m_S = \pm 1$ ESR transitions are allowed; the $\Delta m_S = \pm 2$ transition is strictly forbidden.

As the applied magnetic field is lowered (for any particular value of θ), it is usual to retain this designation of the type of transition by its Δm_S value, although m_S is no longer a "good" quantum number. In this region, where the spin functions are mixed, all three transitions are allowed. There is one exception to this; when $\theta = 0°$ (H parallel to the molecular axis), there is no mixing of spin states regardless of the magnitude of H, and only the $\Delta m_S = \pm 1$ transitions are allowed. Figure III-1 indicates the allowed transitions in the two cases of $H \parallel z$ and $H \perp z$. The former have been designated as "z" transitions and the latter as "xy" transitions (since in the linear molecule the x and y directions are equivalent). Thus, in a hypothetical single crystal in which the linear triplet molecules (with a relatively low D value) were all aligned, one would observe two resonances, z_1 and z_2, in the ESR spectrum for H parallel to the molecular axis. Upon rotating the crystal in the field, the $\Delta m_S = 2$ transition would appear at low field, and the two $\Delta m_S = 1$ resonances would move toward each other with increasing θ. When $\theta = 90°$, i.e., H is perpendicular to the molecular axis, one would observe the $\Delta m_S = 2$ transition and the two resonances at xy_1 and xy_2.

For a linear molecule one can obtain the limiting resonant fields (assuming D positive) at $\theta = 0°$ and $90°$ from $W_i - W_j = h\nu$ and (III, 18) and (III, 19):

$$\Delta m_S = \pm 1 \qquad z_1 = H_\parallel = (h\nu - D)/g_\parallel \beta$$

$$xy_1 = H_\perp = [h\nu(h\nu - D)]^{1/2}/g_\perp \beta$$

$$\Delta m_S = \pm 1 \qquad z_2 = H_\parallel = (h\nu + D)/g_\parallel \beta$$

$$xy_2 = H_\perp = [h\nu(h\nu + D)]^{1/2}/g_\perp \beta$$

$$\Delta m_S = \pm 2 \qquad H_\parallel = h\nu/2g_\parallel \beta \text{ (strictly forbidden)}$$

$$H_\perp = [(h\nu)^2 - D^2]^{1/2}/2g_\perp \beta \qquad \text{(III, 24)}$$

However, as will be seen, for the $\Delta m_S = 2$ transitions, the minimum resonant field (called H_{min}) occurs at a lower field than either $H_\parallel$ or $H_\perp$.

More generally, from the resonance condition, $W_i - W_j = h\nu$, and (III, 22), the magnetic fields at which all possible transitions can occur in an ESR experiment can be calculated for specified values of $h\nu$, D, and g, and θ [de Groot and van der Waals (131)]. For an anisotropic $\boldsymbol{g}$ tensor this relationship may be shown to be

$$\cos^2\theta = \{2D^3 + 9Dg^2\beta^2H^2 \pm [4D^2 + 12g^2\beta^2H^2 - 3(h\nu)^2]^{1/2} \cdot [D^2 + 3g^2\beta^2H^2 - 3(h\nu)^2]\}/27Dg_\parallel{}^2\beta^2H^2 \quad \text{(III, 25)}$$

where $g^2 = g_\perp{}^2 \sin^2\theta + g_\parallel{}^2 \cos^2\theta$. For given values of the parameters, a cubic equation in H^2 can be obtained, from which the three roots yield the three resonant magnetic fields. However, as for (III, 22), this equation is not generally useful for that purpose, but it is informative if graphed (see section III-3).

If D is small relative to $h\nu$, then a convenient expression for the resonant magnetic fields for the allowed transitions, $|m_S\rangle \leftrightarrow |m_S - 1\rangle$, $h\nu = E(m_S) - E(m_S - 1) = g\beta H_0$, may be found from (III, 23) [Bleaney (70)]. Then with D expressed in gauss (G), i.e. $D(\text{G}) = D(\text{ergs})/g\beta$:

$$\begin{aligned} H = H_0 &- D[(3g_\parallel{}^2/g^2)\cos^2\theta - 1]\,(m_S - \tfrac{1}{2}) \\ &- (g_\parallel{}^2 g_\perp{}^2/g^4)\,(D^2\cos^2\theta\sin^2\theta/2H_0) \\ &\cdot [24m_S(m_S - 1) + 9 - 4S(S+1)] \\ &- (g_\perp{}^4/g^4)\,(D^2\sin^4\theta/8H_0) \\ &\cdot [2S(S+1) - 6m_S(m_S - 1) - 3] \end{aligned} \quad \text{(III, 26)}$$

To this fine-structure contribution are then added the hyperfine, etc., terms in (II, 57). One sees that the dominant term linear in m_S in (III, 26) contains an effective D'' which is

$$D'' = D[3(g_\parallel{}^2/g^2)\cos^2\theta - 1] \quad \text{(III, 27)}$$

This strong angular variation of the fine structure, which is also proportional to $|m_S|$, will be demonstrated quite clearly when individual triplet molecules are considered (see, for example, Fig. III-7).

An expression corresponding to (III, 26) can also be derived for the resonant fields at which $\Delta m_S = 2$ transitions occur [Abragham and Bleaney (1), page 159].

Figure III-2 illustrates the Zeeman energy level patterns for various D values (assumed positive) in the extreme cases of H parallel and perpendicular to the axis of the molecule. Also indicated are the ESR transitions (z and xy) for $\nu = 9.3$ GHz $= 0.31$ cm^{-1}. As D changes, with all other instrumental and molecular parameters held constant, the resonant magnetic fields H_r, at which these transitions occur, shift in a regular manner. Figure III-3 [similar to Fig. 9 of Wasserman, Snyder, and Yager (460) but simplified for the linear case] is a plot showing how the two parallel (z) and two perpendicular (xy) transitions depicted in Figs. III-1 and III-2 shift as the molecular D values vary. [The curve indicated as $H_{\min}$ applies to $\Delta m_S = 2$ transitions and will be discussed in more detail in section 2e.] In Fig. III-3 the transitions have been arbitrarily numbered from left to right. Then for $D < h\nu = 0.31$ cm^{-1}, z_1 and z_2 become xy_2 and xy_1, respectively, as the molecule is changed in orientation from $\theta = 0$ to $90°$ relative to H.

The change in the slope of the z_1 line in Fig. III-3 is readily explained by reference to Fig. III-1a. As D increases from zero and approaches $h\nu$, the z_1 transition ($|0\rangle \rightarrow |+1\rangle$) approaches $H = 0$, and then for $D > h\nu$ it becomes a $|0\rangle \rightarrow |-1\rangle$ transition, so that then both z_1 and z_2 occur between the same levels. For increasing D both of these transitions continue to appear but at increasingly higher values of H. On the other hand, for $H \perp z$, both the xy_1 and $\Delta m_S = 2$ transitions will not be observable for $D > h\nu$, and only xy_2 can possibly be seen in the ESR spectrum.

For example, for the triplet aromatic molecule triphenylene (threefold axis perpendicular to the plane of the molecule) $D = 0.134$ cm^{-1}, $g = 2.002$, and one expects from Fig. III-3 $\Delta m_S = \pm 1$ transitions at about $z_1 = 1900$, $z_2 = 4700$ G for $H \parallel z$ and at about $xy_1 = 2500$, $xy_2 = 4000$ G for $H \perp z$ (for $\nu = 9.3$ GHz). Of course, the exact positions can be calculated from (III, 24). On the other hand, for the $^3\Sigma$ CNN molecule, $D = 1.15$ cm^{-1}, and only one parallel (z_1) and one perpendicular (xy_2) transition can be expected to be observed

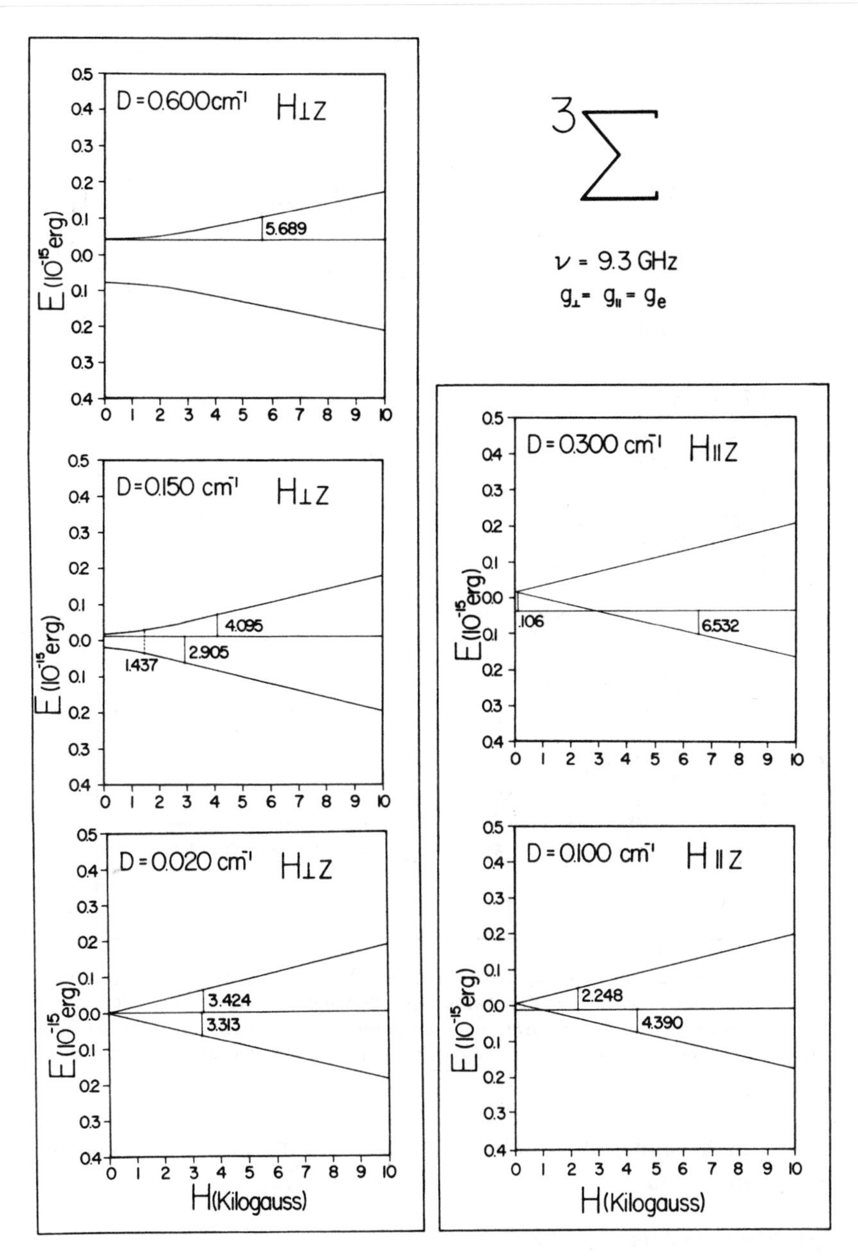

Fig. III-2. Zeeman levels and allowed $\Delta m_S = \pm 1$ transitions of an axial triplet molecule for various values of D (assumed positive). Left side, $H \perp z$; right side, $H \parallel z$, where z is the molecular axis. $\nu = 9.3$ GHz ($= 0.0616 \times 10^{-15}$ ergs $= 0.31$ cm^{-1}).

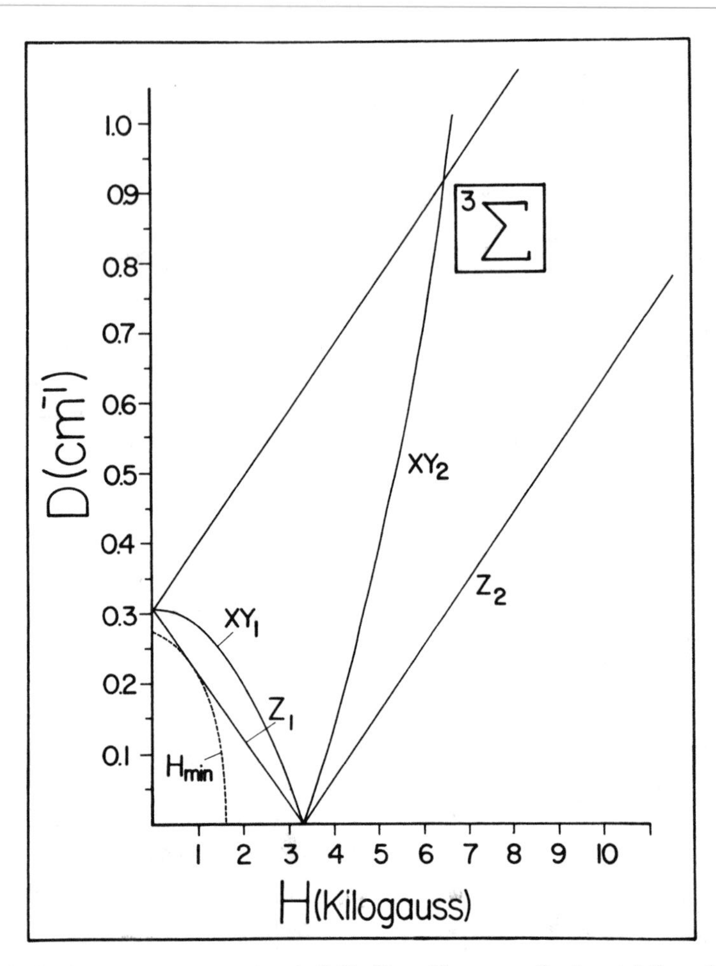

Fig. III-3. D versus resonant magnetic fields [from Wasserman, Snyder, and Yager (460)]. The intersections of a horizontal line with z (parallel) and xy (perpendicular) lines give the magnetic fields at which those $\Delta m_S = \pm 1$ transitions occur. Intersection with the dashed $H_{\min}$ curve gives the position of the $\Delta m_S = \pm 2$ transition. (It is assumed that $g_{\parallel} = g_{\perp} = g_e$ and $\nu = 9.3$ GHz.)

using an X-band ESR spectrometer having a maximum applied field of about 15 kG.

(c) Eigenfunctions

T_x, T_y, and T_z are the eigenfunctions of $\mathcal{H}_D$ and therefore of the spin Hamiltonian of (III, 14) in the absence of a magnetic field. In the presence of a field the eigenfunctions, V_i, are linear combinations of T_x, T_y, and T_z such that

$$V_i = c_{ix} T_x + c_{iy} T_y + c_{iz} T_z \tag{III, 28}$$

where $i = 1, 2, 3$ [de Groot and van der Waals (131); Kottis and Lefebvre (294)]. For $H \parallel z$ ($\theta = 0°$), with eigenvalues given by (III, 18), these are

$$\begin{aligned} V_1 &= \frac{1}{\sqrt{2}} [T_x + iT_y] = |+1\rangle \\ V_2 &= T_z = |0\rangle \\ V_3 &= \frac{i}{\sqrt{2}} [T_x - iT_y] = |-1\rangle \end{aligned} \tag{III, 29}$$

For $H \perp z$ (along x, $\theta = 90°$), with eigenvalues given by (III, 19), they are

$$\begin{aligned} V_1 &= \frac{1}{\sqrt{2}} \left[\left(1 + \frac{D}{\delta}\right)^{1/2} T_y + i\left(1 - \frac{D}{\delta}\right)^{1/2} T_z\right] \\ V_2 &= T_x \\ V_3 &= \frac{i}{\sqrt{2}} \left[\left(1 - \frac{D}{\delta}\right)^{1/2} T_y - i\left(1 + \frac{D}{\delta}\right)^{1/2} T_z\right] \end{aligned} \tag{III, 30}$$

where $\delta = \sqrt{D^2 + 4g_\perp{}^2\beta^2 H^2}$. Eigenvectors for other values of θ can be readily derived.

The eigenvectors are obtained for a particular value of θ from the requirement that the $\boldsymbol{c}_i$ satisfy the set of homogeneous equations obtained from (III, 21):

$$\begin{aligned} c_x\left(\frac{D}{3} - W\right) - ic_y G_z & & &= 0 \\ ic_x G_z + c_y\left(\frac{D}{3} - W\right) - ic_z G_x & & &= 0 \\ ic_y G_x + c_z\left(-\frac{2}{3}D - W\right) &= 0 \end{aligned} \tag{III, 31}$$

Substituting each of the three eigenvalues, corresponding to the particular value of θ, yields the three sets of c_x, c_y, c_z defining V_1, V_2, V_3 in (III, 28) [see Kottis and Lefebvre (294)].

(d) Transition Probabilities and Relative Intensities of Transitions

The relative intensities of the three possible transitions are determined by their transition probabilities P where

$$P = |\langle V_i | \boldsymbol{\mu} \cdot \boldsymbol{H}_1 | V_j \rangle|^2 \qquad \text{(III, 32)}$$

which is analogous to the expression given on page 70 for a $^2\Sigma$ molecule but involving hyperfine interaction in that case. $\boldsymbol{\mu} \cdot \boldsymbol{H}_1$ is proportional to the projection of the magnetic moment of the molecule in the direction of the weak oscillating magnetic field with a frequency in the microwave region. Since $\boldsymbol{\mu} = -\boldsymbol{g}_1 \beta \boldsymbol{S}$

$$P = g_1^2 \beta^2 |\langle V_i | \boldsymbol{S} \cdot \boldsymbol{H}_1 | V_j \rangle|^2 \qquad \text{(III, 33)}$$

or more specifically,

$$P_{ij}^k = g_1^2 \beta^2 H_1^2 \cos^2 \gamma |\langle V_i | S_k | V_j \rangle|^2 \qquad \text{(III, 34)}$$

where k can be any of the x, y, z axes of the molecule (z is axial), and γ is the angle between $\boldsymbol{H}_1$ and k. As discussed on page 70ff., g_1 is a function of the $\boldsymbol{g}$ tensor components and depends upon the direction of the oscillating field with respect to the axis of the molecule and the fixed field. For an isotropic $\boldsymbol{g}$ tensor, $g_1^2 = g^2$. Because of the $\cos \gamma$ factor, no transitions are possible for $H_1 \perp k$.

At high magnetic fields, the V_i are $|+1\rangle$, $|0\rangle$, and $|-1\rangle$ and the only nonvanishing matrix elements of $\langle V_i | \boldsymbol{S} | V_j \rangle$ are

$$\langle \pm 1 | S_x | 0 \rangle = \langle 0 | S_x | \pm 1 \rangle^* = \frac{1}{\sqrt{2}}$$

$$\langle \pm 1 | S_y | 0 \rangle = \langle 0 | S_y | \pm 1 \rangle^* = \mp \frac{i}{\sqrt{2}} \qquad \text{(III, 35)}$$

Therefore $P \neq 0$ only for $\Delta m_S = \pm 1$ transitions; $\Delta m_S = \pm 2$ transitions are strictly forbidden (see Table III-1). This is also true when $\theta = 0°$

Table III-1. Relative Transition Probabilities in Linear Triplet Molecules.

Parallel (z) transitions $H \parallel z, H_1 \parallel v$, z = molecular axis			
S_v	$\lvert\langle +1 \lvert S_v \rvert 0\rangle\rvert^2$	$\lvert\langle -1 \lvert S_v \rvert 0\rangle\rvert^2$	$\lvert\langle +1 \lvert S_v \rvert -1\rangle\rvert^2$
S_x	$\frac{1}{2}$	$\frac{1}{2}$	0
S_y	$\frac{1}{2}$	$\frac{1}{2}$	0
S_z	0	0	0
	$\Delta m_S = \pm 1$		$\Delta m_S = \pm 2$

Perpendicular (xy) transitions* $H \parallel x, H_1 \parallel v$			
S_v	$\lvert\langle z \lvert S_v \rvert x\rangle\rvert^2$	$\lvert\langle y \lvert S_v \rvert z\rangle\rvert^2$	$\lvert\langle y \lvert S_v \rvert x\rangle\rvert^2$
S_x	0	D^2/δ^2	0
S_y	$\left(1 + \frac{D}{\delta}\right)\Big/2$	0	$\left(1 - \frac{D}{\delta}\right)\Big/2$
S_z	$\left(1 - \frac{D}{\delta}\right)\Big/2$	0	$\left(1 + \frac{D}{\delta}\right)\Big/2$
		$\Delta m_S = 2$, only nonzero for $H_1 \parallel x$	

*$\delta = (D^2 + 4g^2\beta^2H^2)^{1/2}$

(H parallel to the molecular axis), regardless of the magnitude of the applied magnetic field, since m_S is always a "good" quantum number for that particular orientation.

At low and intermediate magnetic fields the V_i are linear combinations of T_x, T_y, and T_z, as in (III, 28), and $\langle V_i \lvert \mathbf{S} \rvert V_j\rangle$ vanishes except when it involves the matrix elements

$$\langle T_z \lvert S_x \rvert T_y\rangle = \langle T_x \lvert S_y \rvert T_z\rangle = \langle T_y \lvert S_z \rvert T_x\rangle = i \qquad \text{(III, 36)}$$

These, of course, are combinations of (III, 29) when H is parallel to z. When H_1 is parallel to no particular molecular axis, then all three of the matrix elements in (III, 33) will determine the transition probability. However, for the particular case of $H \perp z$, i.e., $H \parallel x$ and

H_1 along each of the three axes, relative intensities are given in Table III-1. This corresponds to consideration of the "perpendicular" (xy) transitions in Fig. III-1. One notes that only for $H_1 \parallel x$ will the $\Delta m_S = \pm 2$ transition be allowed, so that in the usual experimental arrangement with $H_1 \perp H$ (see Fig. II-1), it would again be strictly forbidden. But for all intervening angles of orientation $\Delta m_S = 2$ transitions will be allowed because the eigenfunctions will be linear combinations of all three of T_x, T_y, and T_z. [Abragham and Bleaney (1), p. 155 suggest that it would be more appropriate to call this a $\Delta m_S = 0$ transition, since it is allowed only because the same values of m_S occur in the two states involved.]

De Groot and van der Waals (131) have shown that

$$\langle V_i | \boldsymbol{S} | V_j \rangle = -i(\boldsymbol{c}_i \times \boldsymbol{c}_j^*) \qquad \text{(III, 37)}$$

where the $\boldsymbol{c}_i$ are the eigenvectors defined in (III, 28). For example, for $H \parallel x$, using the eigenfunctions in (III, 30)

$$\langle V_1 | \boldsymbol{S} | V_3 \rangle = -i(\boldsymbol{c}_1 \times \boldsymbol{c}_3^*) = -i \begin{vmatrix} \boldsymbol{i} & \boldsymbol{j} & \boldsymbol{k} \\ 0 & \frac{1}{\sqrt{2}}\left(1 + \frac{D}{\delta}\right)^{1/2} & \frac{i}{\sqrt{2}}\left(1 - \frac{D}{\delta}\right)^{1/2} \\ 0 & \frac{-i}{\sqrt{2}}\left(1 - \frac{D}{\delta}\right)^{1/2} & \frac{1}{\sqrt{2}}\left(1 + \frac{D}{\delta}\right)^{1/2} \end{vmatrix}$$

$$= \left(-i\frac{D}{\delta}\right)\boldsymbol{i}$$

or $|\langle V_1 | \boldsymbol{S} | V_3 \rangle|^2 = \langle V_1 | \boldsymbol{S} | V_3 \rangle \langle V_1 | \boldsymbol{S} | V_3 \rangle^* = D^2/\delta^2$ as in Table III-1.

The relative intensities of ESR transitions then depend upon the orientation of the molecule relative to $\boldsymbol{H}$ and $\boldsymbol{H}_1$ and to the transition-probability matrix elements for the particular transition, such as those given in Table III-1. These relative intensities are desired either for H_1 parallel ($I_\parallel$) or perpendicular ($I_\perp$) to H, depending upon the experimental arrangement. For $H_1 \perp H$ as in Fig. II-1 [Kottis and Lefebvre (294); Wasserman, Snyder, and Yager (460)],

$$\begin{aligned} I_\perp = (1 - \sin^2\theta\cos^2\phi)\, S_x{}^*S_x - \sin^2\theta\sin\phi\cos\phi\, [S_x{}^*S_y + S_y{}^*S_x] \\ - \cos\theta\sin\theta\cos\phi\, [S_x{}^*S_z + S_z{}^*S_x] + (1 - \sin^2\theta\sin^2\phi)\, S_y{}^*S_y \\ - \cos\theta\sin\theta\sin\phi\, [S_y{}^*S_z + S_z{}^*S_y] + \sin^2\theta\, S_z{}^*S_z \end{aligned} \qquad \text{(III, 38)}$$

where $S_x = \langle V_i(\theta)|S_x|V_j(\theta)\rangle$, etc. θ and ϕ fix the orientation of H relative to the molecular axis system x, y, z.

(e) $\Delta m_S = 2$ Transitions

These transitions are made important in the study of triplet molecules because they are far less dependent on the orientation of the molecule relative to the applied field than the $\Delta m_S = \pm 1$ transitions [van der Waals and de Groot (435)]. The latter transitions, as can be seen from Fig. III-1, generally occur at widely different magnetic fields as θ varies from 0° to 90°; i.e., they are very anisotropic. Then when the molecules are randomly oriented in a solid glass or matrix, the $\Delta m_S = 2$ transitions are much more localized in the ESR spectrum and may thereby be made more intense than the $\Delta m_S = 1$ transitions.

As indicated when formulas for the resonant fields for a linear triplet molecule were given in (III, 24) for $\theta = 0°$ and 90°, the lower field at which $\Delta m_S = 2$ absorption occurs is not determined by either of those angles. This may be seen by consideration of (III, 25), the general relationship connecting resonant field, H, with θ in the $S = 1$ case. Assume, for simplicity, that the **g** tensor is isotropic; then the minimum field at which resonance can occur is given by the value of H that causes the square root to vanish in the right side of (III, 25). (Any lower value of the field would lead to an imaginary term.) Then

$$H_{\text{min}} = [\tfrac{1}{4}(h\nu)^2 - \tfrac{1}{3}D^2]^{1/2}/g\beta \qquad \text{(III, 39)}$$

The molecules absorbing $h\nu$ at this minimum field lie at the angle θ' with respect to the applied field, which may be obtained from (III, 25) by substituting (III, 39):

$$\cos\theta' = \{[\tfrac{1}{9}D^2 - \tfrac{1}{4}(h\nu)^2]/[D^2 - \tfrac{3}{4}(h\nu)^2]\}^{1/2} \qquad \text{(III, 40)}$$

For example, for $h\nu = 0.31\ \text{cm}^{-1}$, and $g = 2.0023$,

$$H_{\text{min}}(\text{G}) = 1.07 \times 10^4 \left[0.0240 - \frac{D^2}{3}\right]^{1/2} \qquad \text{(III, 41)}$$

which is the curve plotted in Fig. III-3. H_{min} becomes zero when

$$D = \pm\frac{\sqrt{3}}{2}h\nu \qquad \text{(III, 42)}$$

which occurs at $|D| = 0.268\ cm^{-1}$ when $h\nu = 0.31\ cm^{-1}$. Above that value of D (for these particular values of $h\nu$ and g) $\Delta m_S = 2$ transitions can still occur, but the threshold $H = H_{min}$ no longer exists.

These transitions will be considered again in further detail for nonlinear triplet molecules, where the perhaps not obvious importance of H_{min} will be demonstrated more fully.

(f) *g* Tensor

The same expressions (II, 10, 11) apply for $\Delta g_{\parallel}$ and $\Delta g_{\perp}$ of a $^3\Sigma$ molecule as for $S = \frac{1}{2}$, except that each contains a multiplicative factor $1/S$ where S is the total spin, equal to 1 in the triplet case [Stone (426)]. However, the summation over excited $^3\Pi$ states is now also taken over two electrons instead of just one, so that the higher multiplicity has essentially no effect upon the calculation.

This also means that the Curl relationship given in (II, 26) connecting $\Delta g_{\perp}$ with the spin-rotation constant, γ, of a linear molecule is unaffected by the increase in multiplicity, at least to the approximation considered here. This, in fact, was already demonstrated in Table II-1 by the agreement between the values of γ for $^3\Sigma$ CCO derived from ESR [Smith and Weltner (421)] and gas-phase measurements [Devillers and Ramsay (139)]. Also given there was a predicted value of γ for the $^3\Sigma_g$ C_4 molecule, which will be discussed along with CCO in section 6b of this chapter.

(g) ESR Spectra of Randomly Oriented $^3\Sigma$ Molecules

The situation is analogous to that for randomly oriented $^2\Sigma$ molecules discussed in section II-3b(α) except that the number of $\Delta m_S = \pm 1$ transitions has doubled. But there is also the possibility of a $\Delta m_S = \pm 2$ transition at low fields when the zfs is not too large, which cannot occur for $S = \frac{1}{2}$.

At relatively low D values the two parallel transitions z_1 and z_2 in Fig. III-1 will appear at different magnetic fields as the orientation of the molecule changes and will eventually appear at xy_2 and xy_1, respectively, when $\theta = 90°$. Thus, there are two $\Delta m_S = \pm 1$ transitions which vary with θ as shown in the upper panel of Fig. III-4. The resulting absorption curve is then a composite of those for two $S = \frac{1}{2}$ linear molecules. For simplicity, in Fig. III-4 an isotropic $\mathbf{g}$ tensor is assumed. The actually observed ESR spectrum (for only $\Delta m_S = \pm 1$ transitions) is the derivative of the absorption curve and appears in the bottom of the figure [Wasserman, Snyder, and Yager (460)]. Notice that the high- and low-field parts of the spectrum are out of phase. One can directly obtain the zfs from this spectrum, since its overall breadth from z_1 to z_2 is $2|D|/g\beta$ or from xy_1 to xy_2 is $|D|/g\beta$.

There is some ambiguity about the exact positions of the two xy lines in Fig. III-4, since the overlapping patterns influence their shape; i.e., it is not clear that their peaks indicate the $\theta = 90°$ absorption as in the $^2\Sigma$ case. This has been

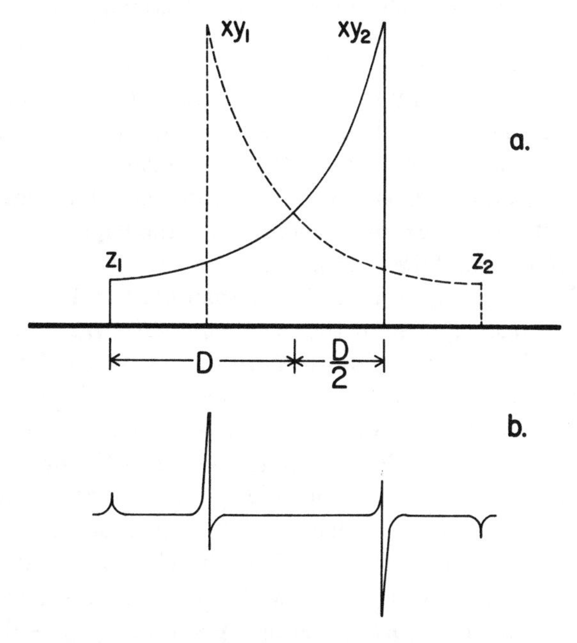

Fig. III-4. ESR absorption and spectrum of a randomly oriented axial triplet molecule [from Wasserman, Snyder, and Yager (460)]. (a) Theoretical absorption indicating variation from parallel (z_i) to perpendicular (xy_i) orientation of the magnetic field relative to the molecular axis for the two $\Delta m_S = \pm 1$ transitions (see Fig. III-1); (b) ESR spectrum (first derivative of the total absorption curve).

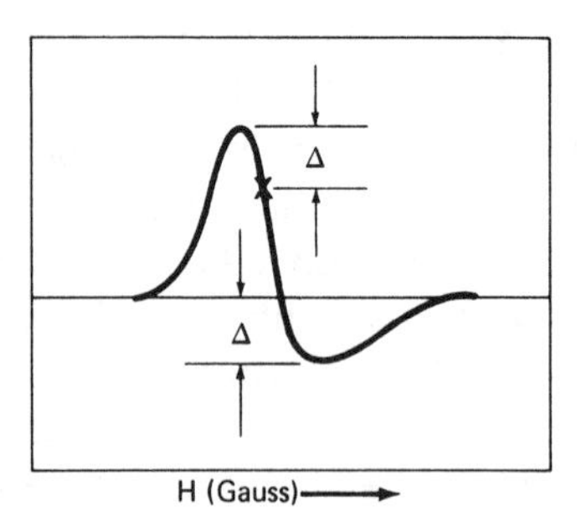

Fig. III-5. Choice of resonant magnetic field (indicated by x) for an unsymmetrical perpendicular (xy) ESR line [from Wasserman, Synder, and Yager (460)].

considered by Wasserman, Snyder, and Yager (460), who indicate that $|D|$ obtained from the xy lines will agree with that obtained from the positions of the peaks of the two z lines if the positions of the perpendicular lines are chosen as in Fig. III-5. The choice of line position is not academic, since the more intense xy lines may be the only ones observed and therefore the only source of $|D|$.

Figure III-4 shows only the $\Delta m_S = \pm 1$ transitions, but for $D < h\nu$ the $\Delta m_S = \pm 2$ transition will also appear, as discussed in section 2e, and it can be more intense than the xy transitions. Figure III-6 illustrates the complete X-band ESR spectrum, including the $\Delta m_S = 2$ transition, that one can expect to observe for the triphenylene radical where $D = 0.1341\ \text{cm}^{-1}$ [Wasserman, Snyder, and Yager (460)].

When D is about equal to or larger than $h\nu$, the $\Delta m_S = 2$ transition will no longer appear in the spectrum, since H_{min} approaches zero (see Fig. III-3). Also, the pattern of the $\Delta m_S = 1$ transitions becomes more asymmetrical. This is demonstrated in Fig. III-7 for two D values, where from top to bottom are given: the variation of the position of the two allowed transitions with the orientation of the molecule in the magnetic field, the absorbance spectrum for the randomly oriented linear molecules, and the first derivative, the X-band ESR spectrum. On the left is the case of $D = 0.14\ \text{cm}^{-1}$ (essentially the same as Fig. III-6, without the $\Delta m_S = 2$ transition) and on the right the corresponding spectrum for $D = 0.5\ \text{cm}^{-1}$ ($> h\nu = 0.31\ \text{cm}^{-1}$).

When D becomes even larger relative to $h\nu$, the "dome-shaped" portion of the θ versus H_r plot disappears. The magnetic field at which the remaining curve reaches $\theta = 90^\circ$ moves to higher fields

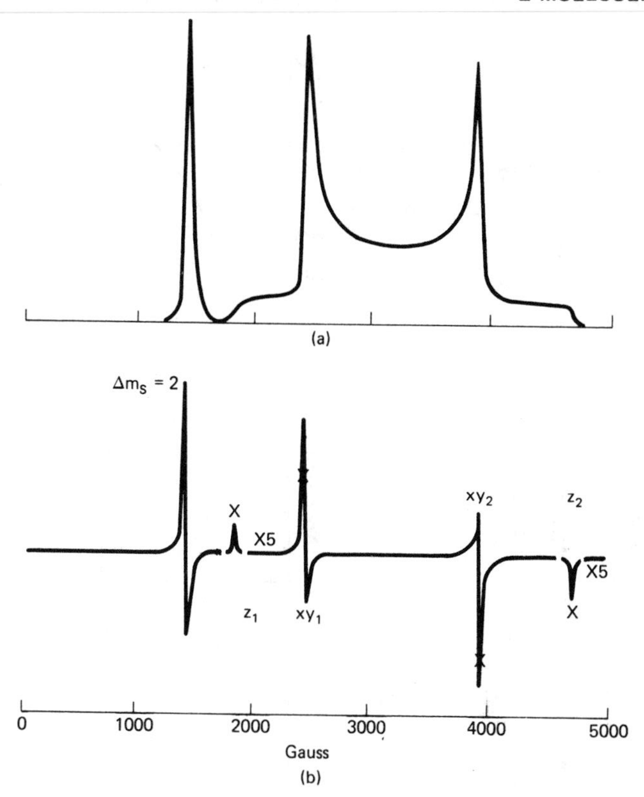

Fig. III-6. Computed absorption and ESR spectrum for the triphenylene radical [from Wasserman, Snyder, and Yager (460)]. $D = 0.1341\ \mathrm{cm}^{-1}$, $E = 0$, $g = g_e$, linewidth = 15 G. (a) Absorption; (b) ESR spectrum (first derivative of absorption).

as D increases. The result is that the spectrum is very spread out, and usually only the relatively strong perpendicular line at high fields is observable. For example, for $D = 2.2\ \mathrm{cm}^{-1}$ and $h\nu = 0.313\ \mathrm{cm}^{-1}$ (and isotropic $\boldsymbol{g}$), the strong perpendicular line will appear at 9.5 kG and a very weak, probably unobservable, parallel line at about 1.8 kG. This occurs, for instance, in the X-band ESR spectrum of the matrix-isolated $^3\Sigma$ molecule SiCO (see page 213ff).

(h) Hyperfine Interaction

The difficulties in observing hyperfine splittings in randomly oriented triplet molecules are usually caused by the large linewidths; however, such splittings are now observed routinely. Often the

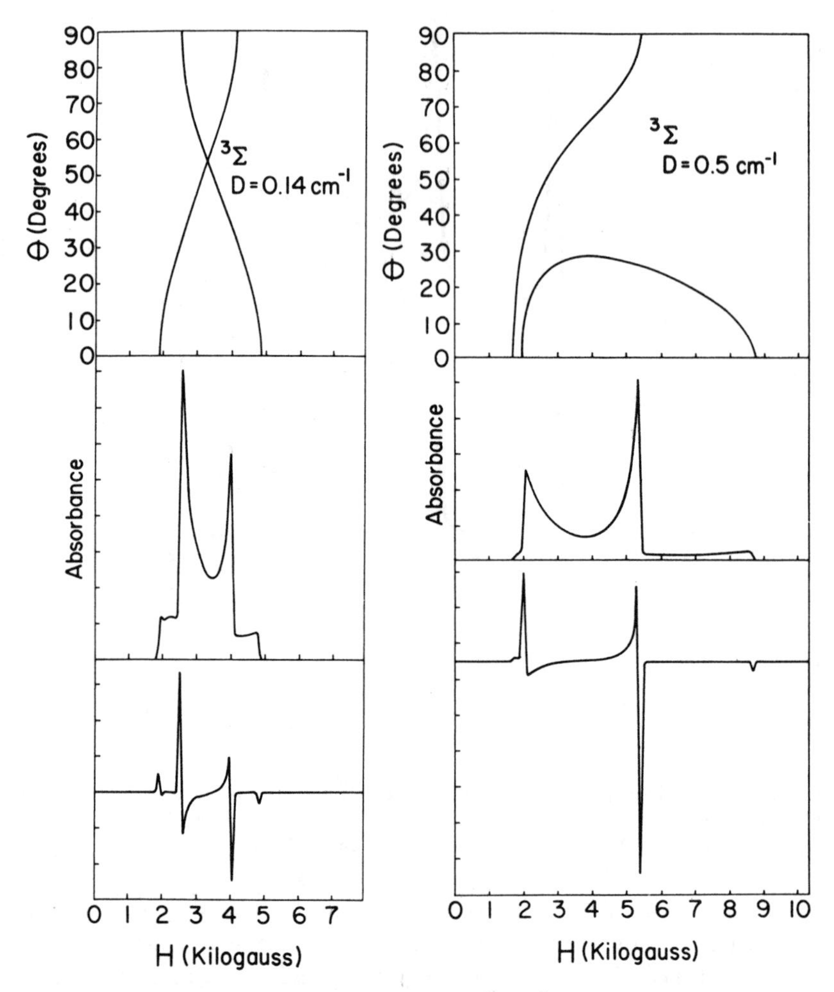

Fig. III-7. Computed ESR spectra of randomly oriented $^3\Sigma$ molecules with $D = 0.14$ and $0.5\ \text{cm}^{-1}$. Top to bottom; angular variation of $\Delta m_S = \pm 1$ transitions, absorbance spectrum, ESR spectrum. $\Delta m_S = \pm 2$ transition deleted from plot for molecule with $D = 0.14\ \text{cm}^{-1}$. (Assumed $g_{\parallel} = g_{\perp} = g_e$, $\nu = 9.3$ GHz.)

hyperfine interaction is small compared to the fine structure and electronic Zeeman terms, so that first-order perturbation is sufficient to account for the hfs. Then for one magnetic nucleus in the molecule with the hyperfine term (see section II-1c)

$$\mathcal{H}_{hf} = KI_zS_z + \frac{A_\parallel A_\perp}{K} I_xS_x + A_\perp I_yS_y$$

$$+ \frac{A_\perp{}^2 - A_\parallel{}^2}{K} \frac{g_\parallel g_\perp}{g^2} \sin\theta \cos\theta \, I_zS_z \quad \text{(III, 43)}$$

added to (III, 14), one can obtain the contribution of hf interaction from the wavefunctions given in (III, 16). In (III, 43), as before, S_x, S_y, and S_z are two-electron operators.

For $H \parallel z$, $\theta = 0$, $K = A_\parallel$, $g = g_\parallel$, and

$$(\mathcal{H}_{hf})_\parallel = A_\parallel I_zS_z + A_\perp (I_xS_x + I_yS_y) \quad \text{(III, 44)}$$

The only nonvanishing matrix elements are

$$\langle +1 | S_z | +1 \rangle = +1, \quad \langle -1 | S_z | -1 \rangle = -1, \quad \langle m_I | I_z | m_I \rangle = m_I$$

so that the hf contributions to the energy levels given in (III, 18) are

$$\begin{aligned} W_{+1} &= A_\parallel m_I \\ W_0 &= 0 \\ W_{-1} &= -A_\parallel m_I \end{aligned} \quad \text{(III, 45)}$$

For a $\Delta m_S = \pm 1$ transition, from (III, 18) and (III, 45),

$$h\nu = W_0 - W_{-1} = -\frac{2}{3}D - \frac{D}{3} + g_\parallel \beta H + A_\parallel m_I$$

so that the resonance field H_r is

$$H_r = [h\nu + D - A_\parallel m_I]/g_\parallel \beta \quad \text{(III, 46)}$$

For hyperfine interaction with a nucleus of spin $I = \frac{1}{2}$, the hf splitting is then

$$\Delta H_r = A_\parallel / g_\parallel \beta \quad \text{(III, 47)}$$

One can show that the $\Delta m_S = \pm 2$ transitions exhibit the same hfs as (III, 47).

For $H \perp z$, from (III, 19) and (III, 43),

$$W_{+1} = -\frac{D}{6} + \frac{\delta^2}{2} + A_\perp \left[1 - \frac{D^2}{\delta^2}\right]^{1/2} m_I$$

$$W_0 = \frac{D}{3}$$

$$W_{-1} = -\frac{D}{6} - \frac{\delta^2}{2} - A_\perp \left[1 - \frac{D^2}{\delta^2}\right]^{1/2} m_I \qquad \text{(III, 48)}$$

where $\delta^2 = D^2 + 4g_\perp{}^2\beta^2H^2$. Note that at zero magnetic field the hyperfine splitting *vanishes*, which is not the case in $^2\Sigma$ molecules [see also van der Waals and de Groot (436), page 109]. The expressions for the resonant fields and hfs cannot be derived explicitly for this perpendicular orientation, since the equations become too complex.

The important points to be made here are that for particular orientations of the radical in the magnetic field, the hfs is the same for $\Delta m_S = 1$ and $\Delta m_S = 2$ transitions in triplet ESR spectra, and no hfs will be detectable at zero magnetic field (Wasserman, Snyder, and Yager (460); Grivet (198); Grivet and Lhoste (199); Brandon, Gerkin, and Hutchison (83)].

A possible remaining question is how large would $A_\parallel$ and $A_\perp$ be in a triplet molecule as compared to a case where only one unpaired electron was involved; that is, are the A values twice as large as in a hypothetical $^2\Sigma$ molecule with the same effective electron–nuclear interaction? The answer is that the size of the splittings in the two cases is the same. It can be shown [Carrington and McLachlan, (6), page 123] that the splitting is proportional to the normalized spin density at the nucleus

$$\rho = \tfrac{1}{2}\{|\Psi_a|^2 + |\Psi_b|^2\} \qquad \text{(III, 49)}$$

where Ψ_a and Ψ_b are the molecular orbitals of the two electrons, so that $\rho \leqslant 1$. Another way [Hutchison (238), page 79] of looking at it is to note that although the $|-1\rangle$ energy level of a triplet molecule is split by twice the amount of either the $|+\tfrac{1}{2}\rangle$ or $|-\tfrac{1}{2}\rangle$ level of a doublet molecule, the $|0\rangle$ level in the former case is not split at all, as (III, 45) and (III, 48) indicate. Hence the observed hfs is the same.

Hyperfine splittings were first observed in photoexcited triplet naphthalene ESR spectra [Hutchison and Mangum (239); Hutchison (238), page 63; Schwoerer and Wolf (410), page 133], and since then in many other cases, even for randomly oriented molecules in solid matrices [see, for example, Grivet (198)]. As opposed to observation of $A_\perp$ when viewing the hfs of an xy line due to a $\Delta m_S = 1$ transition in a linear molecule, the hfs parameter for a $\Delta m_S = 2$ line is a mixture of $A_\parallel$ and $A_\perp$, and if $\boldsymbol{A}$ is anisotropic, the hyperfine pattern may be obscured [Grivet (198); Grivet and Lhoste (199)]. A clear example of this occurs in the ESR spectrum of the $^3\Sigma$ C_4 molecule, which is discussed in section III-6.

3. NONLINEAR TRIPLET MOLECULES

The eigenvalues in the three principal axis directions X, Y, and Z in (III, 2) are now all different, and the spin-spin interaction involves both parameters D and E:

$$\mathcal{H}_D = D[S_z{}^2 - \tfrac{1}{3}S(S+1)] + E(S_x{}^2 - S_y{}^2) \qquad \text{(III, 4)}$$

where

$$D = -\tfrac{3}{2}Z, \quad E = -\tfrac{1}{2}(X - Y) \qquad \text{(III, 5)}$$

Or, using the traceless property (III, 3),

$$X = \frac{D}{3} - E, \quad Y = \frac{D}{3} + E, \quad Z = -\frac{2}{3}D \qquad \text{(III, 50)}$$

From (III, 5) and (III, 10),

$$E = \tfrac{3}{4}g^2\beta^2\langle(y^2 - x^2)/r^5\rangle$$

as compared to

$$D = \tfrac{3}{4}g^2\beta^2\langle(3z^2 - r^2)/r^5\rangle \qquad \text{(III, 51)}$$

However, in a nonlinear molecule the z axis may be chosen as any one of the three principal axes, and the values and signs of D and E depend upon the choice. It has been suggested that this choice be made such that $-1 \leqslant (3E/D) \leqslant 0$, or $|D| \geqslant 3|E|$, which orders the levels $X > Y > Z$ for $D > 0$ and the reverse for $D < 0$ [Poole, Farach, and Jackson (381)] and results in D and E being of opposite signs. For this case, on an energy scale, for positive D, the relationship between these quantities can be roughly depicted as:

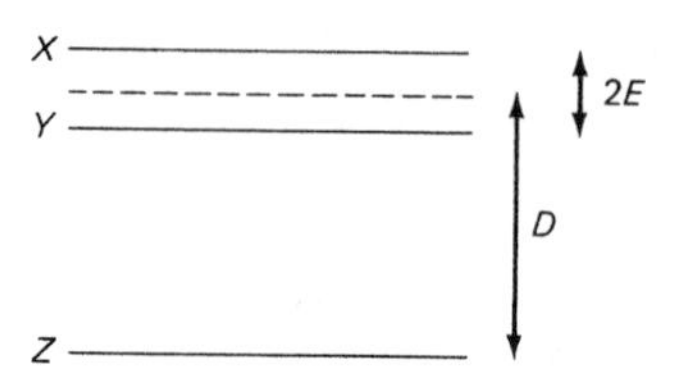

Only the sign of D (and E) relative to other parameters is usually determined in ESR experiments; line intensity variations with temperature can, in principle, specify its sign.

The classic work on naphthalene as a dilute solute in a single crystal of durene [Hutchison and Mangum (239)] was the first ESR study of an organic molecule in the triplet state. This was followed quickly by the discovery that $\Delta m_S = 2$ transitions could be readily observed for such molecules in random orientations in glasses, as discussed in the preceding section 2e for $^3\Sigma$ molecules [van der Waals and de Groot (435)]. The theory discussed there needs only the inclusion of the zero-field parameter E to adapt it to the general nonlinear molecule.

The easiest visualization of the effect of the parameter E can be seen by considering a linear triplet molecule bent slightly so that the x and y "perpendicular" axes are no longer identical. Then at zero-field the energies X and Y are no longer degenerate but are separated by $2E$, according to (III, 5). Each xy line in Figs. III-1 and III-3 now becomes two lines, one for transitions in the x-axis direction and one for transitions in the y-axis direction. The D versus H_r curves then look like those in Fig. III-8.

The spin Hamiltonian for the molecule in a magnetic field (III, 13) now is

$$\mathcal{H} = \beta \boldsymbol{H} \cdot \boldsymbol{g} \cdot \boldsymbol{S} - [XS_x{}^2 + YS_y{}^2 + ZS_z{}^2]$$

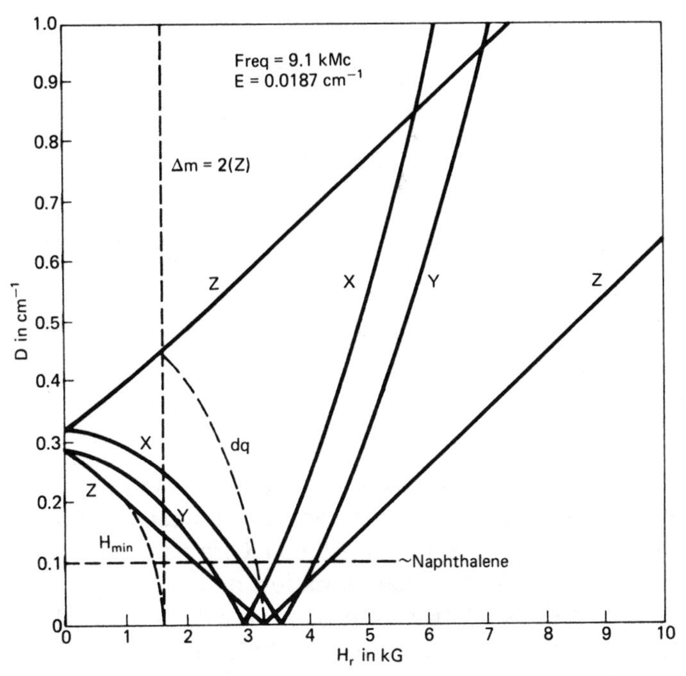

Fig. III-8. Resonant magnetic fields for varying D with $E \equiv 0.0187\ \text{cm}^{-1}$, $\nu = 9.1$ GHz [from Wasserman, Snyder, and Yager (460)]. H_{min} and double quantum (dq) [see section III-3(c)] transitions are indicated.

With T_x, T_y, and T_z, as in (III, 20), as a basis, one finds for $H \parallel x$

$$W_1 = X = \frac{D}{3} - E$$

$$W_2 = [(Y + Z) + \sqrt{(Y - Z)^2 + 4g^2\beta^2H^2}\,]/2$$

$$= \left[-\frac{D}{3} + E + \sqrt{(D + E)^2 + 4g^2\beta^2H^2}\right]\Big/2$$

$$W_3 = [(Y + Z) - \sqrt{(Y - Z)^2 + 4g^2\beta^2H^2}\,]/2$$

$$= \left[-\frac{D}{3} + E - \sqrt{(D + E)^2 + 4g^2\beta^2H^2}\right]\Big/2 \qquad \text{(III, 52)}$$

For $H \parallel z$

$$W_1 = [(X+Y) - \sqrt{(X-Y)^2 + 4g^2\beta^2H^2}\,]/2$$

$$= \frac{D}{3} - \sqrt{E^2 + g^2\beta^2H^2}$$

$$W_2 = [(X+Y) + \sqrt{(X-Y)^2 + 4g^2\beta^2H^2}\,]/2$$

$$= \frac{D}{3} + \sqrt{E^2 + g^2\beta^2H^2}$$

$$W_3 = Z = -\tfrac{2}{3}\,D \qquad \text{(III, 53)}$$

and similarly for $H \parallel y$. These levels are plotted in Fig. III-9 and for the lowest triplet state of naphthalene, and the $\Delta m_S = 1$ and $\Delta m_S = 2$ transitions are shown for $h\nu = 9.654$ GHz $= 0.3221$ cm^{-1}. This figure shows clearly the small variation of the resonant fields for $\Delta m_S = 2$ transitions relative to those for the $\Delta m_S = 1$ as the orientation of the molecule is changed relative to the magnetic field.

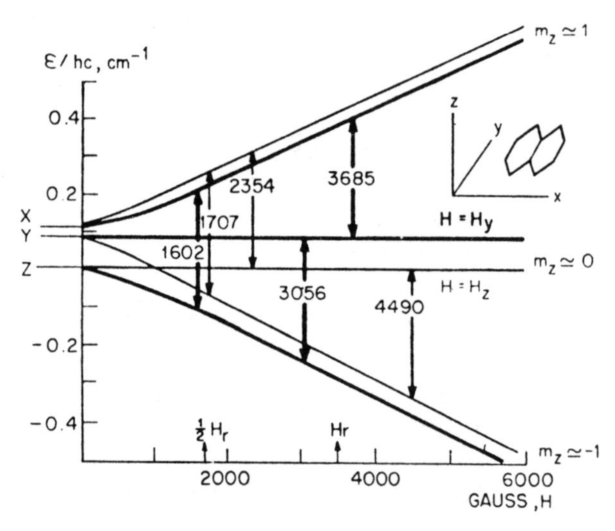

Fig. III-9. Zeeman energies of the lowest triplet state of naphthalene [from van der Waals and de Groot (435)]. Heavy lines: magnetic field (H) along y axis; thin lines: H along z axis. ($\nu = 9.654$ GHz.)

From the Hamiltonian matrix there can be derived, just as in (III, 22) for a linear molecule, a cubic equation whose roots yield the energy levels as a function of the Euler angles θ and ϕ designating the orientation of the x, y, z axes fixed in the molecule relative to the applied field.

$$W^3 - W[(g\beta H)^2 - (XY + XZ + YZ)] + (g\beta H)^2 \, [X \sin^2 \theta \cdot \cos^2 \phi + Y \sin^2 \theta \sin^2 \phi + Z \cos^2 \theta] - XYZ = 0 \quad \text{(III, 54)}$$

which may be converted to a form containing D and E instead of X, Y, Z by use of the relations (III, 50).

From (III, 54) and the resonance condition, $W_i - W_j = h\nu$, one can derive a general equation for the magnetic fields at which transitions can occur in the ESR for a triplet molecule with specified zero-field parameters X, Y, Z and specified orientation θ, ϕ with respect to the applied field [Kottis and Lefebvre (294, 295); de Groot and van der Waals (131)]:

$$X \sin^2 \theta \cos^2 \phi + Y \sin^2 \theta \sin^2 \phi + Z \cos^2 \theta$$

$$= \frac{XYZ}{(g\beta H)^2} \pm \left(\frac{1}{3}\right)^{3/2} \left[\frac{(h\nu)^2 + XY + XZ + YZ}{(g\beta H)^2} - 1\right] \times [4(g\beta H)^2 - (h\nu)^2 - 4(XY + XZ + YZ)]^{1/2} \quad \text{(III, 55)}$$

Here, for simplicity, the **g** tensor has been assumed isotropic. [That assumption was not made in deriving its counterpart for a linear molecule, given in (III, 25).] For a given molecule the equation is of the form

$$f(\theta, \phi) = F(H, h\nu)$$

and is best discussed on the basis of a graphical representation obtained by plotting $F(H, h\nu)$ versus H for a specified $h\nu$. This yields the often displayed form of Fig. III-10 for $\nu \cong 9.3$ GHz for a typical aromatic molecule, naphthalene, where $X = 0.0197$, $Y = 0.0471$, $Z = -0.0669$ cm^{-1}, and $g = 2.0030$. Thus one sees immediately that for any value of the ordinate, that is, choice of θ and ϕ, a horizontal line specifies three widely separated transitions or resonant fields

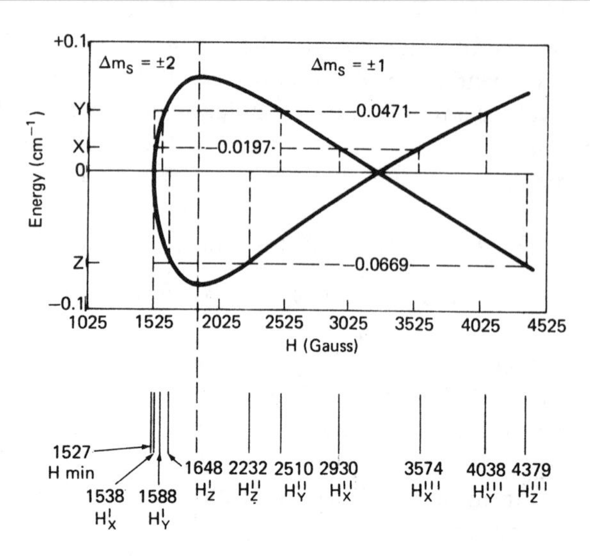

Fig. III-10. The function $F(H, h\nu)$ (in cm^{-1}) plotted versus H (in gauss) for triplet naphthalene [from Kottis and Lefebvre (294); Weissbluth (471)]. $X = 0.0197$, $Y = 0.0471$, $Z = -0.0669\ \mathrm{cm}^{-1}$, $g = 2.0030$, and $\nu = 9.279$ GHz. Resonance fields are indicated below. $H_{\min}$ is the lowest field at which resonance occurs. $H_x^{\mathrm{I}}, \cdots, H_z^{\mathrm{III}}$ are the resonance fields for canonical orientations.

for naphthalene in its triplet state. For $\theta = 90°$, $\phi = 0$, i.e., with $H \parallel x$, $f(\theta, \phi) = X$, and the three canonical resonance fields are

$$H_x^{\mathrm{I}} = 1538\ \mathrm{G}\ (\Delta m_S = 2) \quad \text{and} \quad H_x^{\mathrm{II}} = 2930\ \mathrm{G},$$

$$H_x^{\mathrm{III}} = 3574\ \mathrm{G}\ (\Delta m_S = 1).$$

For the other two canonical orientations these are

$$H_Y^{\mathrm{I}} = 1588\ \mathrm{G}\ (\Delta m_S = 2),\ H_Y^{\mathrm{II}} = 2510\ \mathrm{G},\ H_Y^{\mathrm{III}} = 4038\ \mathrm{G}\ (\Delta m_S = 1)$$

and

$$H_Z^{\mathrm{I}} = 1648\ (\Delta m_S = 2),\ H_Z^{\mathrm{II}} = 2232\ \mathrm{G},\ H_Z^{\mathrm{III}} = 4379\ \mathrm{G}\ (\Delta m_S = 1).$$

The canonical resonance fields are given by [Kottis and Lefebvre (295)]

$$H_X{}^{\mathrm{I}} = (2g\beta)^{-1}[(h\nu)^2 - (Y - Z)^2]^{1/2} = (2g\beta)^{-1}[(h\nu)^2 - (D + E)^2]^{1/2}$$

$$H_X{}^{\mathrm{II}} = (2g\beta)^{-1}[(2h\nu - 3|X|)^2 - (Y - Z)^2]^{1/2}$$

$$= \left\{\left[2h\nu - 3\left|\frac{D}{3} - E\right|\right]^2 - (D + E)^2\right\}^{1/2}$$

$$H_X{}^{\mathrm{III}} = (2g\beta)^{-1}[(2h\nu + 3|X|)^2 - (Y - Z)^2]^{1/2}$$

$$= \left\{\left[2h\nu + 3\left|\frac{D}{3} - E\right|\right]^2 - (D + E)^2\right\}^{1/2} \qquad \text{(III, 56)}$$

Those in the y and z directions are obtained by cyclic permutation of X, Y, and Z in these expressions.

Thus it is clear from the set designated I among these transitions and from the shape of Fig. III-10 at low fields that the $\Delta m_S = 2$ transitions are much more localized than the $\Delta m_S = 1$. Also, none of the canonical transitions (necessarily) occurs at the minimum field (H_{min}) possible, which lies at 1527 G in Fig. III-10. H_{min} is obtained from (III, 55) when the discriminant vanishes to yield

$$H_{\mathrm{min}} = (2g\beta)^{-1}[(h\nu)^2 + 4(XY + XZ + YZ)]^{1/2}$$

or

$$= (2g\beta)^{-1}[(h\nu)^2 - \tfrac{4}{3}(D^2 + 3E^2)]^{1/2} \qquad \text{(III, 57)}$$

This, of course, reduces to (III, 39) when $E = 0$.

Figure III-10 clearly shows the accumulation of $\Delta m_S = 2$ transitions near H_{min}, resulting in the extraordinary intensity of that transition in samples containing randomly oriented triplet molecules with relatively small zfs. This is brought out even more by Fig. III-11 [Kottis and Lefebvre (294)], where the distribution of resonance fields in the $\Delta m_S = 2$ region for naphthalene is plotted, i.e., the density of molecules having resonances between H and $H + dH$. The lower part of the figure shows how the resonances are distributed when varying the orientation and how there is an accumulation of intensity at H_{min}.

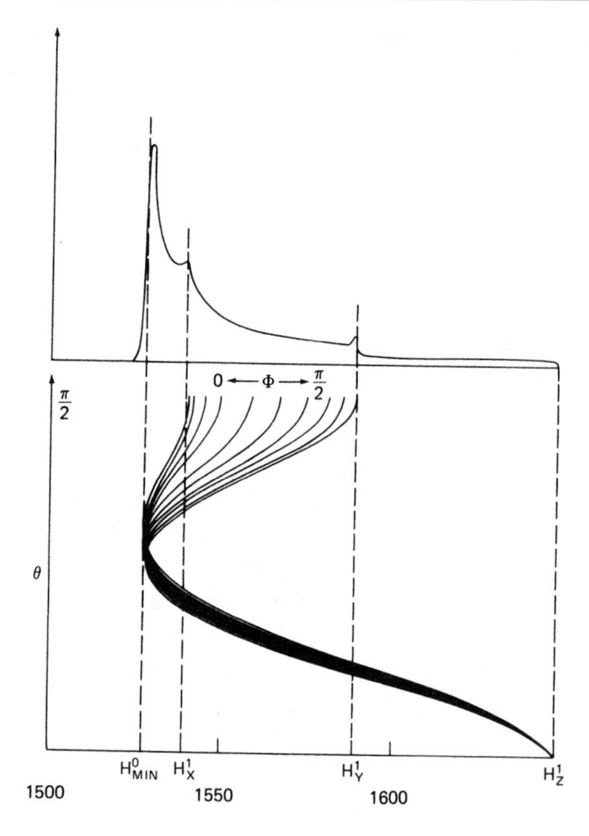

Fig. III-11. Distribution of resonance fields in the $\Delta m_S = 2$ region for triplet naphthalene [from Kottis and Lefebvre (294)]. The lower curves indicate how the resonant fields are distributed when varying the orientations ($\nu = 9.279$ GHz).

(a) Randomly Oriented Molecules

For a general triplet molecule where the zfs parameters are not too large, the ESR $\Delta m_S = 1$ absorption spectrum when summed over all orientations of molecules relative to the fixed magnetic field is given by Fig. III-12. This is similar to the idealized spectrum in Fig. III-4 (where $E = 0$) in that at each orientation zero linewidth is assumed. Finite linewidth leads to a rounding of the absorption features and to the first-derivative spectrum shown in the lower part of the figure. Thus, $E \neq 0$ leads to the appearance of two additional features besides those in Fig. III-4, arising from molecules oriented along the other "perpendicular" axis, x or y.

The $\Delta m_S = 2$ transition will appear strongly at lower field if, from (III, 57),

$$2[(D^2 + 3E^2)/3]^{1/2} < h\nu \qquad \text{(III, 57a)}$$

If D and E are so large that this is not true, then there is no threshold field $H = H_{\min}$, but there are still three $\Delta m_S = 2$ transitions for every orientation and features may appear in the low-field region at the canonical orientations. As Fig. III-3 indicates, when D (and E) becomes quite large relative to $h\nu$, some of the transitions are no longer possible, and the spectrum becomes simpler, as in the linear case.

As in the linear molecule where the weaker parallel (z) transitions

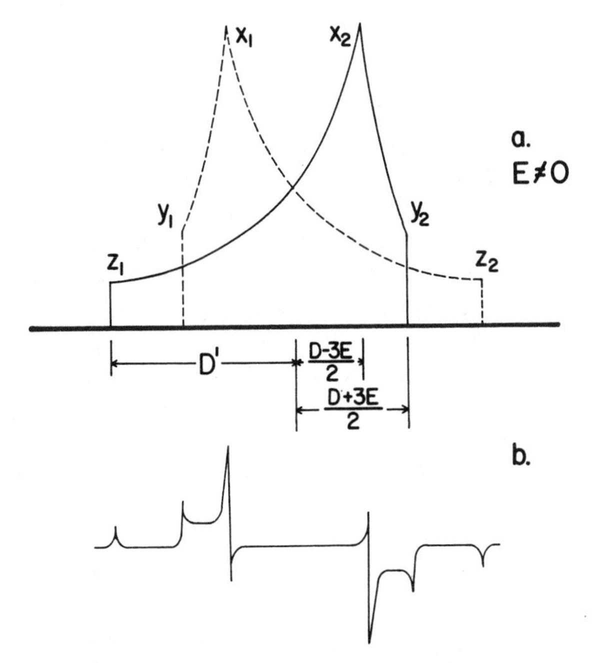

Fig. III-12. ESR absorption and spectrum of a randomly oriented nonaxial triplet molecule (D and $E \neq 0$) [from Wasserman, Snyder, and Yager (460)]. (a) Theoretical absorption for the two $\Delta m_S = \pm 1$ transitions; (b) ESR spectrum (first derivative of the total absorption curve).

were often difficult to observe in matrices, for nonlinear molecules it can also be expected that the "outer" $\Delta m_S = 1$ canonical transitions in Fig. III-12 will be the least intense. These correspond to $H \parallel z$ in napthalene.

(b) Noncollinear *g* and *D*

Just as the $\boldsymbol{g}$ and hyperfine tensors, $\boldsymbol{A}_i$, may not have the same principal axes if the symmetry of the molecule is low, similarly $\boldsymbol{g}$, $\boldsymbol{A}_i$, and $\boldsymbol{D}$ may each have a different set of principal axes in a corresponding triplet molecule. This has been considered in a few of the papers cited in section II-4b, where only noncollinear $\boldsymbol{g}$ and $\boldsymbol{A}_i$ tensors were considered [Lin (309); Golding and Tennant (186); Iwasaki (244)]. At present there appears to be no case where this available theory has been applied.

(c) Two-Quantum Transition

Two microwave quanta may be absorbed when the molecules are oriented such that each of the $|\pm 1\rangle$ levels is separated from the $|0\rangle$ level by $h\nu$. The general theory of such transitions was studied by Goeppert-Mayer (184) and applied to phosphorescent triplet molecules by de Groot and van der Waals (133). Its intensity is very dependent upon the orientation of the molecule and upon the microwave power dissipated in the cavity (133).

The resonant field (H_{dq}) at which a two-quantum transition is observed will then be located at that field where two of the three possible roots of H_r in (III, 55) coincide. This will occur when the terms preceded by the $\pm$ sign vanish. If the second term vanishes, then the condition for $H_{\min}$ in (III, 57) is obtained; if the first term vanishes, then

$$H_{dq} = (g\beta)^{-1}\left[(h\nu)^2 + XY + XZ + YZ\right]^{1/2}$$

$$H_{dq} = (g\beta)^{-1}\left[(h\nu)^2 - \tfrac{1}{3}(D^2 + 3E^2)\right]^{1/2} \qquad \text{(III, 58)}$$

This transition is also plotted in Fig. III-8.

In ESR this transition has also been observed at high microwave power levels for transition-metal ions, such as Ni^{2+}, Ti^{2+}, and Fe^{2+}, in crystals in their triplet lowest states [see Wertz and Bolton (23), pages 292 and 327]. In these cases D is relatively small, and a sharp double-quantum transition appears on top of the $|\pm 1\rangle \leftrightarrow |0\rangle$ transition broadened by crystal-field distortions from octahedral symmetry.

4. MOTIONAL AND MATRIX EFFECTS UPON THE ZFS

(a) Rotational Diffusion

Just as rotation of the molecule in a solid or liquid will lead to vanishing of the anisotropic hyperfine tensor $\boldsymbol{A}_{\text{dip}}$, the traceless $\boldsymbol{D}$ tensor will be similarly affected. Thus, with increasing motion the value of D effectively becomes smaller, and the absorption line shape in Fig. III-12 collapses, as shown in Fig. III-13 [Freed, Bruno, and Polnaszek (170)], eventually leading to one broad line in the ESR spectrum. The triphenylene molecule with $D = 0.1341\ \text{cm}^{-1}$ and $E = 0$ is the example considered in this figure in calculating the X-band ESR spectrum. Here D/R is the ratio of the zero-field-splitting parameter to the rotational diffusion constant and is then inversely proportional to the rate of rotation of the triplet molecule. Slow motion leads to a decrease in separation of the two xy lines, besides broadening of all lines. The parallel z lines then become even more difficult to observe than in the "rigid" spectrum where their intensity is already relatively low.

The theory involved in the calculation of the ESR line shapes as a function of the rate of rotational diffusion of triplet molecules is beyond the scope of our discussion here. Detailed expositions have been given by Bloembergen, Purcell, and Pound (77); Bloch (76); Korst and Khazanovitch (292); Redfield (389); Itzkowitz (243); Fixman (166); Sillescu and Kivelson (414); Saunders and Johnson (400); Norris and Weissman (362); Roberts and Lynden-Bell (391); Freed, Bruno, and Polnaszek (170). Experimental data for comparison have been obtained usually by varying the viscosity of the liquid medium as a means of altering the rotation of the triplet molecule.

(b) Torsional Oscillation; Isotope Effects

Decreasing libration, or zero-point motion, of a triplet molecule trapped in a solid when a heavier isotope is substituted in the molecule can lead to detectable increases in D. Assuming a potential

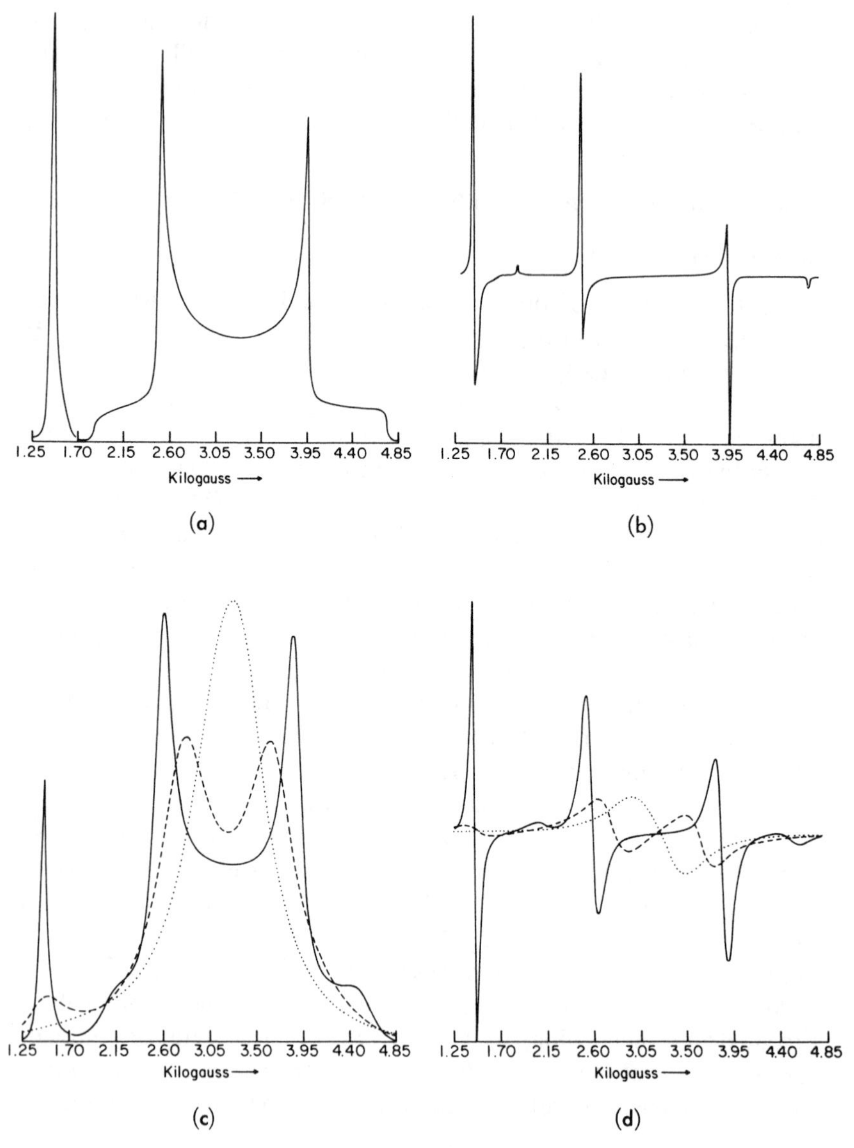

Fig. III-13. ESR line shape variation with increasing molecular motion [from Freed, Bruno, and Polnaszek (170)]. (a) and (b) are absorption and first derivative, respectively, for rigidly held but randomly oriented triphenylene (see Fig. III-6). (c) and (d) are absorption and first derivative line shapes for increasing motion in the order (———) $D/R = 200$, (----) $D/R = 20$, and (· · · ·) $D/R = 5$.

barrier restricting rotation of the form

$$V = V_0(1 - \cos 2\theta) \qquad \text{(III, 59)}$$

Meyer, O'Brien, and Van Vleck (343) were able to account for the variation of magnetic susceptibility with temperature of O_2 trapped in clathrate compounds. The theory has also been applied to O_2 in solid gases [Kon (290); Hirokawa (223)] and to the triplet molecules CH_2 [Wasserman, Yager, and Kuck (461)], CCO, and CNN [Smith and Weltner (421)] in matrices. For small torsional oscillations within the potential given in (III, 59) the effective zfs parameter D' is given by

$$D' = D_m [1 - \tfrac{3}{4} (\hbar^2/IV_0)^{1/2}] \qquad \text{(III, 60)}$$

where I is the moment of inertia of the molecule. D_m is for the rigidly held species and would be equal to D_g, the value determined from gas-phase spectra, if the radical were unperturbed by the surrounding matrix. Then for two isotopically substituted molecules the shift in D' would be given by

$$\frac{D_m - D'_1}{D_m - D'_2} = \left(\frac{I_2}{I_1}\right)^{1/2} \qquad \text{(III, 61)}$$

where D_m, to the accuracy of these measurements, is the same for both molecules. From measurements of D' for two isotopically substituted molecules with known moments of inertia, D_m may be calculated from (III, 61), and the effect of motional averaging eliminated from the zfs determination. This value for the rigidly held molecule will, in general, still differ from the gas-phase parameter, owing to matrix effects [Smith and Weltner (421)].

Just as in the expression for the anisotropic hyperfine interaction, for a linear molecule the effective zfs, D', can be expressed as

$$D' = \frac{D}{2} \langle 3 \cos^2 \theta - 1 \rangle_{\text{av.}} = D + \frac{3D}{4} [\langle \cos 2\theta \rangle_{\text{av.}} - 1] \qquad \text{(III, 62)}$$

where θ here is the angle of rotation of the molecular axis and D is the value for the rigidly held molecule. If the molecule is rapidly rotating, then

$$\frac{1}{4\pi}\int_0^{2\pi}\int_0^{\pi} (3\cos^2\theta - 1)\sin\theta\, d\theta\, d\phi = 0$$

and $D' = 0$. Meyer et al. (343) considered the case of O_2 in clathrate compounds where they assumed a potential barrier restricting rotation of the form (III, 59). For small harmonic oscillations [Sterne (423)] the energy levels are

$$E = 2(n+1)\,(V_0\hbar^2/I)^{1/2}$$

and

$$\langle 1 - \cos 2\theta\rangle_{\text{av.}} = 2\langle\theta^2\rangle_{\text{av.}} = \frac{E}{2V_0} = (n+1)\left(\frac{\hbar^2}{IV_0}\right)^{1/2}$$

At low temperature, $n = 0$ and

$$\langle\cos 2\theta\rangle_{\text{av.}} = 1 - (\hbar^2/IV_0)^{1/2}$$

Also, (III, 62) becomes

$$D' = \frac{D}{4}\,[3\langle\cos 2\theta\rangle + 1]$$

so that

$$D' = D[1 - \tfrac{3}{4}\,(\hbar^2/IV_0)^{1/2}]$$

Then

$$\frac{D' - D}{D} = -\tfrac{3}{4}\,(\hbar^2/IV_0)^{1/2}$$

or

$$V_0 = (9\hbar^2/16I)[D/(D - D')]^2$$

If D is D_m, then (III, 60) is readily derived.

(III, 61) has been applied as an approximate way to eliminate motional averaging effects when CH_2 and CD_2 are trapped in matrices [Wasserman, Yager, and Kuck (461)]. In xenon, for example, $D' =$ 0.69 and 0.759 cm^{-1} for CH_2 and CD_2, respectively [Bernheim, Bernard, Wang, Wood, and Skell (60, 61)], from which one derives $D_m = 0.93$ cm^{-1} using $(I_D/I_H)^{1/2} = 1.409$. This value is much higher than the most probable theoretical value of 0.807 cm^{-1} [Langhoff and Davidson (301); Langhoff (300)] and also experimental D_m values in other matrices, and it is suggested that it results from CH_2 being trapped in a "compressive" site in the xenon matrix [Bicknell, Graham, and Weltner (67)]. Krypton, argon, and neon matrices yield D_m values of 0.61, 0.69, and 0.76 cm^{-1}, but the least perturbing matrix appears to be octafluorocyclobutane, where $D_m = 0.775$ cm^{-1} [Wasserman, Hutton, Kuck, and Yager (458)].

In a nitrogen matrix, the single resonance of $^{18}O_2$ is shifted 37 G up-field relative to that of $^{16}O_2$ [Kon (290)]. This corresponds to a shift of the apparent D' value from 3.562 cm^{-1} for $^{16}O_2$ to 3.587 cm^{-1} for $^{18}O_2$. V_0 is calculated to be 152 cm^{-1}, and in such a potential well the first vibrational state is 42 cm^{-1} above the zero-point level. This is well above the ~6°K ≅ 4 cm^{-1} at which the measurements were made, justifying the harmonic oscillator approximation. The gas phase value is 3.96 cm^{-1}, so that there is a sizable matrix effect upon D.

Understandably, the shifts in D' when ^{13}C, ^{18}O, or ^{15}N are substituted in the linear triplet CCO or CNN radicals are small, only of the order of 0.001 cm^{-1}, corresponding to about 2-G shifts; however, there is no doubt about their occurrence [Smith and Weltner (421)]. D_m for CCO was found to be 0.7120 and 0.7395 cm^{-1} in solid argon and neon, respectively, compared to 0.772 cm^{-1} in the gas [Devillers and Ramsay (139)]. Six isotopically substituted molecules were observed in each matrix. Matrix effects upon the zfs parameter are discussed in considerable detail by Smith and Weltner (421).

(c) Matrix Effects upon *D*

Except for the case of the methylene radical trapped in solid xenon the (positive) D values for the light, small molecules discussed above that have been measured in matrices are found to be lower than those obtained from gas-phase spectra. A detailed consideration of the possible electronic interactions of an impurity triplet molecule with rare-gas matrices indicates that the dominant effect is enhancement of spin-orbit mixing of triplet and singlet molecular states [Smith and Weltner (421)]. Simple theory suggests that this effect upon D should vary as ζ^2, where ζ is the spin-orbit coupling constant of the matrix gas. Opposing, but apparently smaller, effects tending to increase D of the trapped radical are (1) exclusion forces leading to increased spin-spin interaction by effectively compressing the molecular volume (as suggested above for CH_2 in xenon), and (2) direct spin-orbit effects due to the electrons of the molecule entering the field of the electrons and nuclei of the matrix atom perturbers.

5. THEORETICAL CALCULATIONS OF ZFS PARAMETERS

In the absence of spin-spin and spin-orbit effects, the three wavefunctions for the triplet state having different m_S values, ${}^3\Psi_0{}^1$, ${}^3\Psi_0{}^0$, ${}^3\Psi_0{}^{-1}$, all have the same energy, ${}^3E_0$, in zero magnetic field. However, in the real world where these relatively weak effects are present and may be considered as small perturbations, the individual ${}^3\Psi_0{}^i$ will have the properties of excited electronic states mixed into them by these interactions, so that they may be written approximately as the ${}^3\Phi_0{}^i$:

$$ {}^3\Phi_0{}^i = {}^3\Psi_0{}^i + \sum_n{}' \sum_j b_{n,j}{}^i \, {}^3\Psi_n{}^i + \sum_k c_k{}^i \, {}^1\Psi_k \qquad \text{(III, 63)} $$

The triplet state will then no longer be degenerate at zero magnetic field but will exhibit zero-field splitting. Here i, j are m_S values, n is an index of excited triplet states, and k an index of excited singlet states. $b_{n,j}{}^i$ and $c_k{}^i$ are coefficients determined by these interactions. Using these wavefunctions one can show that the zfs parameter D can be written as

$$ D = D_{SS} + D_{SO} $$

where D_{SS} involves terms of the form [Hameka (205), page 14]

$$ \langle {}^3\Psi_0{}^i | \mathcal{H}_{SS} | {}^3\Psi_0{}^i \rangle \qquad \text{(III, 64)} $$

and D_{SO} terms of the form

$$ -\sum_n \sum_j ({}^3E_n - {}^3E_0)^{-1} \langle {}^3\Psi_0{}^i | \mathcal{H}_{SO} | {}^3\Psi_n{}^j \rangle \cdot \langle {}^3\Psi_n{}^j | \mathcal{H}_{SO} | {}^3\Psi_0{}^i \rangle $$
$$ - \sum_k ({}^1E_k - {}^3E_0)^{-1} \langle {}^3\Psi_0{}^i | \mathcal{H}_{SO} | {}^1\Psi_k \rangle \langle {}^1\Psi_k | \mathcal{H}_{SO} | {}^3\Psi_0{}^i \rangle \qquad \text{(III, 65)} $$

The interesting point about these two contributions, which was misleading for a time, is that the spin-spin contribution involves first-order perturbation terms, while the spin-orbit involves second-order terms and therefore might be considered negligible, at least in light molecules. However, it is often found, even in molecules with relatively small spin-orbit coupling constants, that $D_{SO} > D_{SS}$ [Kayama and Baird (270)]. Aromatic triplet hydrocarbons, such as naphthalene, are not in that class, since the dominant contributor is always D_{SS} [Gouterman and Moffitt (194); Hameka (204)]. For them the spin-orbit matrix elements are of the order of 1 cm^{-1} and the energy differences $E_n - E_0$ about 10,000 cm^{-1}; thus $D_{SO} \cong 10^{-4}$ cm^{-1}, whereas $D \cong 0.1$ cm^{-1}.

The three triplet molecules O_2, NH, and CH_2 are of particular interest, since they were the first small molecules for which *ab initio* calculations of zfs were made, and they have since been investigated extensively by theoretical chemists. Table III-2 gives the experimental D and E values for these molecules, and some of their analogues, compared with the presently accepted calculated contributions from spin-spin and spin-orbit interactions. Generally D_{SS} is mainly dependent upon the distribution of spin, whereas D_{SO} not only has that dependence but also depends very much upon the spin-orbit coupling constants of the atoms and upon the lowest-lying singlet states.

(a) Molecules Containing No Metal Atoms

An excellent recent review of fine-structure theory given by Langhoff and Kern (305) has more or less formed the basis for the following discussion of the calculations made on particular triplet molecules.

(α) O_2 and Analogues SO, S_2 The experimental D value for O_2 [Tinkham and Strandberg (429); Amano and Hirota (38)] in its ${}^3\Sigma_g^-$ ground state could not be accounted for on the basis of just spin-spin interaction [Kayama (269); Pritchard, Bender, and Kern (384)], and Kayama and Baird (270) showed that D_{SO} was indeed the larger contributor. This work has been followed by a series of increasingly accurate calculations of D_{SO}, resulting in the good agreement with experiments given in Table III-2 [Hall (203); Langhoff (299)]. The dominant contribution to D_{SO} arises from coupling to the low-lying

(1.64 eV) $^1\Sigma_g^+$ excited state (2.4 cm^{-1}) followed by much smaller contributions from $^3\Pi_g$ (0.16 cm^{-1}), $^1\Pi_g$ (−0.12 cm^{-1}), and $^5\Pi_g$ (−0.034 cm^{-1}) excited states.

In SO and S_2 the $^1\Sigma^+$ state lies even lower than in O_2, and when it is coupled to the ground $^3\Sigma$ state by the larger spin-orbit interaction of the heavier sulfur atom, a dramatic increase in D_{SO}, and therefore in D, occurs [Wayne and Colburn (465)].

(β) NH and Analogues PH, NX, PF The experimental $D = 1.84$ cm^{-1} of NH [Dixon (144); Wayne and Radford (467)], but as Table III-2 shows, contrary to the O_2 results, D_{SS} contributes about 85% of that value [Lounsbury (316); McIver and Hameka (340); Wayne and Radford (467)]. This can be attributed largely to localization of the spin on one center, the N atom, as compared to O_2, tending to increase the spin-spin interaction. The spin-orbit interaction is decreased because $\langle ^3\Sigma^- | \mathcal{H}_{SO} | ^1\Sigma^+ \rangle$ is correspondingly lowered by such localization; furthermore, the $^1\Sigma$ contribution is lowered because it lies at considerably higher energy (~21,000 cm^{-1}) than in O_2 [Wayne and Colbourn (465)].

In NF, PH, PF, and NCl the spins are still largely localized on N or P. Then D_{SS} drops in the phosphorous molecules because of the expanded electron cloud, whereas D_{SO} increases as the molecules become heavier [Wayne and Colbourn (465)].

(γ) CH_2 As indicated earlier, CH_2 has been shown to be a bent molecule by observation of its ESR spectrum in solid matrices [Bernheim, Bernard, Wang, Wood, and Skell (60, 61); Wasserman, Hutton, Kuck, and Yager (458)]. Gas-phase [Herzberg and Johns (217)] and matrix data then indicate an angle of about 135°. Harrison (211) has recently reviewed the structural evidence on the basis of experiment and the extensive calculations made on methylene over the last 17 years. The zfs parameters are not well determined but are now judged to be $D = 0.79(2)$ and $E = 0.05(2)$ cm^{-1} [Harrison (211); Bicknell, Graham, and Weltner (67)].

The unpaired electrons are essentially on one center here, as in the isoelectronic NH molecule, so that it is not surprising to find that spin-spin interaction accounts for 97% of the zfs (see Table III-2). Again, the increasingly accurate *ab initio* calculations of D_{SS} [Harrison (211a); Langhoff and Davidson (301)] and of D_{SO} [Langhoff

Table III-2. Comparison of Calculated and Observed Zero-Field Splitting Parameters (D and E in cm^{-1}) for some Triplet Molecules

Molecule	Expt'l D E	Calculated D_{SS} E_{SS}	D_{SO} E_{SO}	D E
$O_2(X^3\Sigma_g^-)$[a]	3.965	1.453	2.402	3.855
$SO(X^3\Sigma^-)$[b]	10.54	0.556	10.142	10.698
$S_2(X^3\Sigma_g^-)$[c]	23.54	0.264	20.330	23.54
$NH(X^3\Sigma^-)$[d]	1.8394	1.623	–	1.870
$NF(X^3\Sigma^-)$[e]	2.44	1.970	0.592	2.562
$NCl(X^3\Sigma^-)$[f]	3.56	1.618	1.278	2.896
$PH(X^3\Sigma^-)$[g]	4.42	0.332	3.492	3.824
$PF(X^3\Sigma^-)$[h]	5.92	0.484	4.926	5.410
$CH_2(X^3B_1)$[i]	0.79	0.784	0.023	0.807
	0.05	–	–	–
$H_2CO(^3A_2)$[j]	0.42	0.539	−0.224	0.304
	0.04	0.031	0.004	0.041
$C_6H_6(^3B_{1u})$[k]	0.1580	0.1676	0.0001	–
	−0.0064	–	–	–
Naphthalene[l]	0.1003	0.1008	0.0001	–
$(^3B_{2u})$	−0.0137	−0.0138	–	–

[a]Cook, Zegarski, Breckenridge, and Miller (123); Amano and Hirota (38); Langhoff (299).
[b]Watson (462); Davies, Wayne, and Stone (128); Wayne (464); Wayne and Colburn (465).
[c]Barrow, Du Parq, and Ricks (50); Wayne, Davies, and Thrush (466); Wayne (464); Wayne and Colburn (465).
[d]Dixon (144); Palmiere and Sink (367); Wayne and Radford (467); Lounsbury (316); McIver and Hameka (340).
[e]Jones (252); Wayne and Colburn (465).
[f]Colin and Jones (122); Wayne and Colburn (465).
[g]Rostas, Cossart, and Bastien (396); Palmiere and Sink (367); Wayne and Colburn (465).
[h]Colin (121); Wayne and Colburn (465).
[i]Bernheim, Bernard, Wang, Wood, and Skell (60, 61); Wasserman, Hutton, Kuck, and Yager (458); Bicknell, Graham, and Weltner (67); Harrison (211); Langhoff and Davidson (301); Langhoff, Elbert, and Davidson (304); Langhoff (300).
[j]Raynes (388); Peyerimhoff, Buenker, Kammer, and Hsu (370); Langhoff, Elbert, and Davidson (304); Langhoff and Davidson (302).
[k]de Groot, Hesselmann, and van der Waals (130); de Groot and van der Waals (132); Godfrey, Kern, and Karplus (183); Langhoff, Davidson and Kern (303). D value calculated is for a hexagonal form with 1.427-Å bond lengths. See text and Langhoff and Kern (305).
[l]Hutchison and Mangum (239); Hornig and Hyde (228); Boorstein and Gouterman (79); Gouterman and Moffitt (194); Gouterman (193); van der Waals and ter Maten (437); Godfrey, Kern, and Karplus (183).

(300)] have provided agreement with experiment within the experimental error. The dominant contributors to D_{SO} come from two low-lying $^1A_1$ states.

The hyperfine splittings detected when the molecule was enriched with the ^{13}C isotope (see Fig. III-14) may be interpreted to give an estimate of the bond angle [Bernheim, Bernard, Wang, Wood, and Skell (61); Wasserman, Kuck, Hutton, Anderson, and Yager (459)]. $A_{iso}(^{13}C)$ determines the s character of the unpaired electrons, thereby providing a measure of the s and p hybridization and yielding an angle of about 137°.

(δ) Naphthalene, Benzene and Other Aromatic Molecules. The experimental work of Hutchison and Mangum (239) and of van der Waals and de Groot (436) on photoexcited triplet aromatic molecules generated theoretical interest in the 1960s in calculating their fine-structure constants. The spin-orbit terms in such π electron molecules have been shown to be of the order of 10^{-4} cm^{-1} and therefore negligible [McClure (330); Mizushima and Koide (346); Boorstein and Gouterman (79)]. The first calculations on naphthalene by Gouterman and Moffitt (194) and Gouterman (193) showed that the zfs parameters could be attributed to spin-spin interactions. Many refinements have since been added to provide better agreement between theory and experiment for naphthalene and other aromatic molecules (see Table III-2). McGlynn, Azumi, and Kinoshita [see their Chapter 10 (16)] have admirably reviewed those calculations and compared them with experiment. A complete list of D and E values, with references to theory, for the numerous triplet aromatic molecules investigated through 1972 has been given by Denison (137).

Very extensive calculations have been made for benzene in its $^3B_{1u}$ state, assuming hexagonal geometry, beginning with those of Hameka (204) and Pitzer and Hameka (377). Apparently there is still need for inclusion of further refinements [Langhoff and Kern (305)]. However, benzene is unusual, since if it has D_{6h} symmetry in its photoexcited $^3\mathrm{B}_{1u}$ state, one expects $E = 0$, whereas the zfs parameters were found to be $D = 0.1580$ and $E = -0.0064$ for C_6H_6 in a C_6D_6 host crystal [de Groot, Hesselmann, and van der Waals (130)]. ENDOR of such crystals [Ponte Goncalves and Hutchison (379)] provided hyperfine splittings that further confirmed that

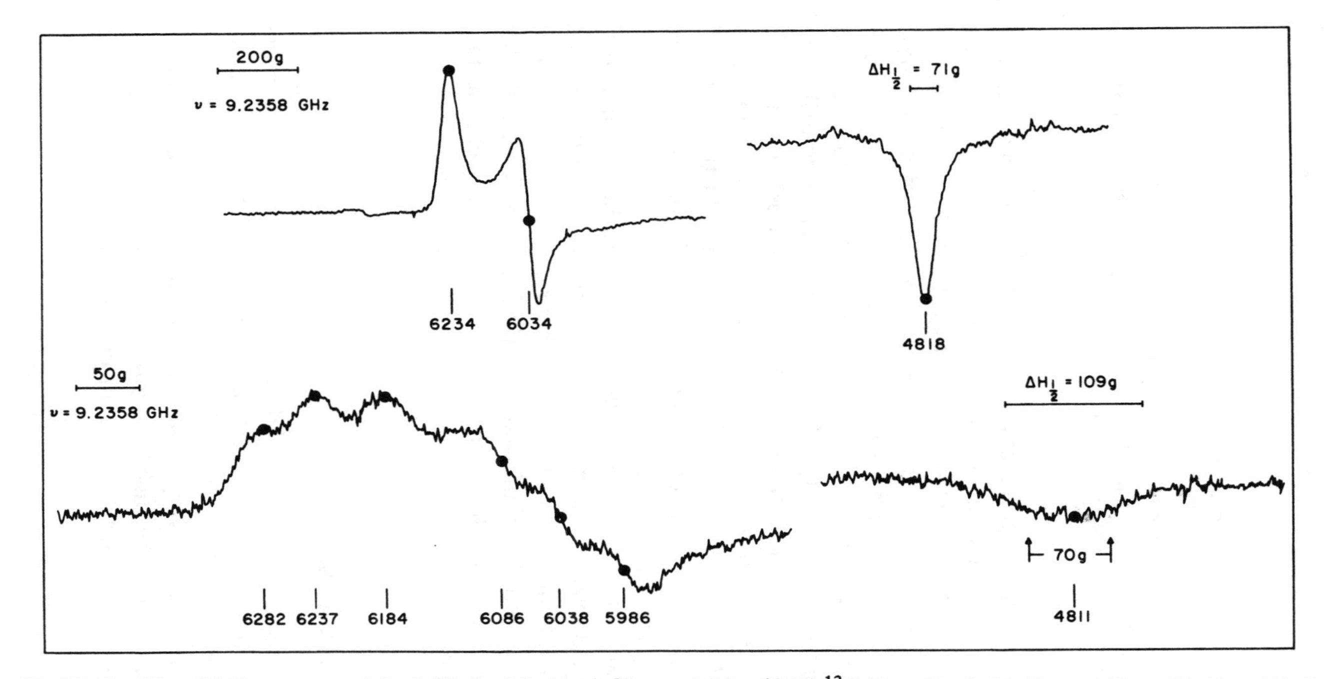

Fig. III-14. X-band ESR spectrum of (top) CD_2 and (bottom) CD_2 containing 62.1% ^{13}C [from Bernheim, Bernard, Wang, Wood, and Skell (61)]. Left are y_2 and x_2 resonance lines; right is z_1 resonance line (positions in gauss).

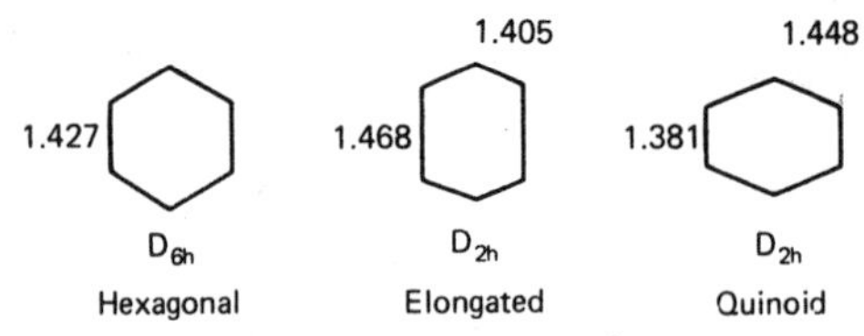

Fig. III-15. Possible conformations of benzene in its $^3B_{1u}$ state [adapted from de Groot and van der Waals (132)]. Bond lengths are in Å.

there is less than a trigonal axis of symmetry, on the average, in that triplet state. Calculations of the zfs parameter have been made for various geometries with both D_{6h} and D_{2h} symmetries (see Fig. III-15). The relative ordering and spacing of the zero-field levels obtained for either of these D_{2h} forms were not in agreement with experiment, and the conclusion was that the average conformation experimentally observed is essentially hexagonal with a slight contribution of other forms or even a small distortion by the host crystal (see section II-6). ESR spectra were also measured for triplet mesitylene, $(CH_3)_3C_6H_3$, oriented in single crystals, again with the finding that $E \neq 0$, in spite of its expected threefold axis of symmetry [de Groot, Hesselmann, and van der Waals (130)].

The important inference to be drawn from these studies is that the zfs can be very sensitive to the geometry of the molecule and might be used more generally, with theory, to establish molecular conformation [see Langhoff and Kern (305)]. Just the simple detection of a nonzero E value is significant, not only for benzene and its analogues, but perhaps even more so for triatomic molecules, such as CH_2 and TiF_2, since a bent form is then established.

(b) Transition-Metal Molecules

Here the situation is completely the opposite of that in aromatic molecules, since the large spin-orbit coupling constants of the metal atoms cause $D_{SO} >> D_{SS}$ [Griffith (10), pages 330, 354]. These molecules are usually very ionic and the incomplete d shell on the metal ion may then be highly localized on that ion. The spin-orbit coupling can then be used in its centrosymmetric form

$$\zeta \boldsymbol{L} \cdot \boldsymbol{S}$$

or for two electrons,

$$\zeta[\boldsymbol{L}_1 \cdot \boldsymbol{S}_1 + \boldsymbol{L}_2 \cdot \boldsymbol{S}_2]$$

since the spins are largely on the metal ion and not distributed over the molecule, as in O_2. As in a crystal, the symmetry of the attached ionic ligands (octahedral, tetrahedral, trigonal, linear, etc.) may split the energy levels of the metal ion and produce low-lying electronic states. Then these states can mix with the ground state via spin-orbit coupling and produce a large value of D_{SO}.

Very many calculations, usually only approximate, have been for transition-metal ions in crystals, principally with the ion in an octahedral or slightly distorted octahedral site [Abragham and Bleaney (1); Griffith (10)]. A case in point is the V^{3+} ion where the two unpaired electrons in an octahedral environment will occupy the triply degenerate t_2 orbitals d_{xy}, d_{xz}, and d_{yz}. These may be split by tetragonal distortion along the z axis such that the $d_{xz}d_{yz}$ triplet ($^3A_2$) is lowest with a higher-lying ^3E state involving the d_{xy} orbital. Spin-orbit coupling mixes the A_2 state with the E state and also mixes triplet and singlet states. A clear basic discussion of the calculation of D_{SO} for this case is given by Carrington and McLachlan [(6) pages, 159–162)].

It would be helpful if the wealth of data and theory for transition-metal ions in crystals could be used to deduce information about the zfs of transition-metal molecules. For example, the V^{3+} ion is often observed in crystals in a trigonally distorted octahedral symmetry so that comparison of its properties with the planar VF_3 molecule having a $^3A_2'$ ground state [Yates and Pitzer (482)] might be considered. However, it is clear that the crystal field effects in the two cases are still vastly different, and the zero-field splitting, essentially of the cation, in either case is very sensitive to such distortions. This will be discussed further when dealing with high-spin molecules in Chapter V, particularly when the metal ion contains a half-filled *d* or *f* shell.

It appears that crystal-field theory is useful in the interpretation of the properties of such ionic molecles as $NiCl_2$, which has a $^3\Pi_g$ ground state [Hougen, Leroi, and James (231); Gruen (200)]. With covalency properly parameterized, it can provide a semiempirical model that can account for the electronic transitions observed in the optical spectrum [Gruen (200)]. On the other hand, it is not useful in explaining why the related triplet molecule TiF_2 (^3B) has a C_{2v} bent structure [Hastie, Hauge, and Margrave (213); DeVore and Weltner (142)].

A simple well-characterized transition-metal molecule known to have a triplet ground state is TiO ($^3\Delta$). TiO is a very thoroughly investigated molecule experimentally and to a lesser extent theoretically, largely because of its prominence in some stellar atmospheres [Huber and Herzberg (12)]. The application of crystal-field theory to TiO, and similar diatomics, was suggested by Berg and Sinanoglu (58), but it is generally too simple a model, and more sophisticated theory is needed to reproduce the observed electronic properties [Carlson and Moser (99); Carlson and Nesbet (100); Scott and Richards (411)]. The calculation of the zero-field splittings is presumably the next stage in the evolution of the theory of these small transition-metal molecules.

6. MAGNETIC PROPERTIES OF TRIPLET MOLECULES

(a) Tabulated Magnetic Properties

Appendix C contains the tabulated magnetic properties of small triplet molecules. Other sources of such data are the book by McGlynn, Azumi, and Konoshita (16), and the tabulation of Denison (137).

(b) Other Representative Triplet Radicals

Propargylene and Its Homologues (Alternant Methylenes). A series of triplet alternant methylenes, i.e., $H{-}\dot{\underset{\bullet}{C}}{-}(C{\equiv}C)_n{-}R$, and the related radical $H{-}\dot{\underset{\bullet}{C}}{-}C{\equiv}N$, were studied by Bernheim, Kempf, Gramas, and Skell (62) and Bernheim, Kempf, and Reichenbecher (63). The radicals were prepared by photolysis of the corresponding diazo precursors in polychlorotrifluoroethylene matrices at 77°K. Figure III-16 shows the z and xy line observed in the ESR spectra of randomly oriented methylpentyldiynylene ($H{-}\dot{\underset{\bullet}{C}}{-}C{\equiv}C{-}C{\equiv}C{-}CH_3$) molecules at X-band (9.5 GHz) and P-band (15.7 GHz) frequencies. Table III-3 gives the zero-field-splitting parameters D and E for the homologues of propargylene and also includes the diphenylmethylene radical. One notes in Fig. III-16 that the positions of the xy lines were

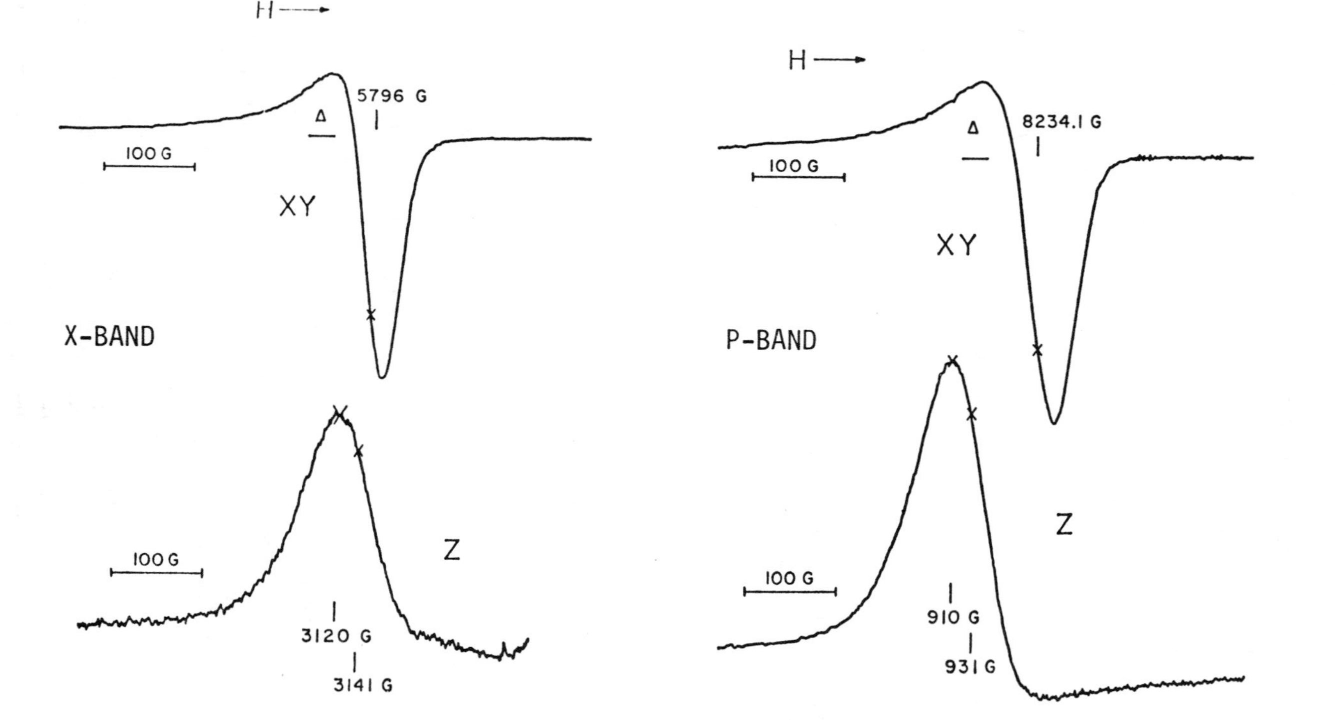

Fig. III-16. ESR spectra of methylpentyldiynylene randomly oriented in a polychlorotrifluoroethylene matrix at 77°K [from Bernheim, Kempf, and Reichenbecher (63)]. Left is at X-band (H_0 = 3391.0 G), right at P-band (H_0 = 5596.4 G) frequencies. Crosses indicate the selections of the xy (top) and z (bottom) resonance field positions.

Table III-3. Zero-Field Splitting Parameters and g-Factors for some Triplet Alternant Methylenes [from Bernheim, Kempf, and Reichenbecher (63)].

Compound[a]	D/hc (cm^{-1})	E/hc (cm^{-1})	$g_{xx} = g_{yy}$	g_{zz}	
propargylene H-↑C↑-C≡C-H	0.630	0 ± 0.001	2.005	2.356 2.082[b]	P-band
methylpropargylene H-↑C↑-C≡C-CH_3	0.623	0 ± 0.002	2.002	2.024 1.957[b]	P-band
methylpentadiynylene H-↑C↑-C≡C-C≡C-CH_3	0.609	0 ± 0.0007	2.002	2.015 1.969[b]	P-band
–	–	–	–	2.003 1.990[b]	X-band
–	–	–	–	1.998[c]	
t-butylpentadiynylene H-↑C↑-C≡C-C≡C-$C(CH_3)_3$	0.597	0 ± 0.001	1.995	1.95	P-band
phenylpentadiynylene H-↑C↑-C≡C-C≡C-C_6H_5	0.547	0 ± 0.001	1.995	1.94	P-band
cyanomethylene H-↑C↑-C≡N	0.849	0 ± 0.003	2.013	2.189 2.073	P-band
			g_{xx}	g_{yy}	g_{zz}
phenylpropargylene[d] H-↑C↑-C≡C-C_6H_5	±0.555	∓0.0042	2.014	2.019	2.04
diphenylmethylene C_6H_5-↑C↑-C_6H_5	±0.408[e] ±0.40505[f] 0.4055[g]	∓0.0188 ∓0.01918 0.0194	2.012 2.00451	1.993 2.00432	2.005 2.00251

[a] Except where noted the compounds were prepared in a polychlorotrifluoroethylene matrix.
[b] Derived from values of H_z derived in the same manner as H_{xy}, i.e., by subtracting the smaller deflection from the larger as in Fig. III-5.
[c] Derived from both X-band and P-band spectra independently of the method of picking H_z.
[d] All g-factors were determined from X-band spectra.
[e] Benzophenone solvent. All g-factors are consistent with both X-band and P-band data.
[f] Brandon, Closs, Davoust, Hutchison, Kohler, and Silbey (82); single crystal benzophenone host.
[g] Wasserman, Snyder, and Yager (460); benzophenone solvent; free electron g-factors assumed.

chosen by using Wasserman et al.'s criterion, as in Fig. III-5. The positions of z lines are usually taken at their peaks, and that leads to the most consistent results; however, the authors also chose those line positions by the perpendicular-line criterion because the z lines exhibited a small deflection on the high-field side. The latter choice was found to cause a small shift in the measured D from 0.6081 to 0.6101 cm^{-1} for methylpentyldiynylene. The zfs parameters in the table were determined from the more precisely measured xy lines and then resulted in the g values listed in column 5.

The two spins are disposed in π_x and π_y orbitals delocalized along the lengths of the linear molecules. Remembering that D for CH_2 is 0.79 cm^{-1}, it is clear that increasing delocalization when acetylenic groups or unsaturated R groups are added leads to decreasing spin-spin parameters. It is interesting that the D values for isoelectronic propargylene and cyanomethylene are so different, an indication that the –C≡N group differs from the –C≡C–H group considerably in the effect upon the electron distribution and interaction in the two molecules.

C_4. This molecule has not been observed spectroscopically except in rare-gas matrices, but theoretical work [Pitzer and Clementi (376); Clementi (112); Clementi and Clementi (113); Strickler and Pitzer (427); Hoffmann (225)] has predicted it to be linear with a $^3\Sigma_g^-$ ground state. It has been prepared from the carbon species produced in the vaporization of graphite and by the photolysis of diacetylene (C_4H_2) and trapped in solid neon and argon at 4°K [Graham, Dismuke, and Weltner (196)]. The resulting ESR spectra of $^{12}C_4$ and $^{13}C_4$ in argon are shown in Fig. III-17. No parallel (z) lines were observed, but the two out-of-phase xy lines and the $\Delta m_S = 2$ transition are quite strong. The small ^{13}C hfs on the xy lines leads only to broadening of the $\Delta m_S = 2$ line, since it is derived largely from molecules at intermediate angles of orientation; i.e., it is neither a parallel nor a perpendicular line. Table III-4 gives the line positions and calculated values of $g_\perp$ and D for the observed molecules. The small hfs and its 1:2:1 intensity pattern indicate that the unpaired spins have very little $2s$ character and therefore are in $2p\pi$ orbitals, largely interacting with one pair of equivalent carbon nuclei. (If the spins were spread equally over all nuclei in $^{13}C_4$, then a 1:4:6:4:1 hfs intensity pattern would be expected.) Also the observed $D = 0.23$ cm^{-1} appears to be smaller than one would expect for two $p\pi$ electrons on adjacent carbon atoms (as, for example, occurs in O_2 where $D \simeq 4$ cm^{-1}) so that it is inferred that the unpaired spins in C_4 are localized mainly on the two end carbon atoms.

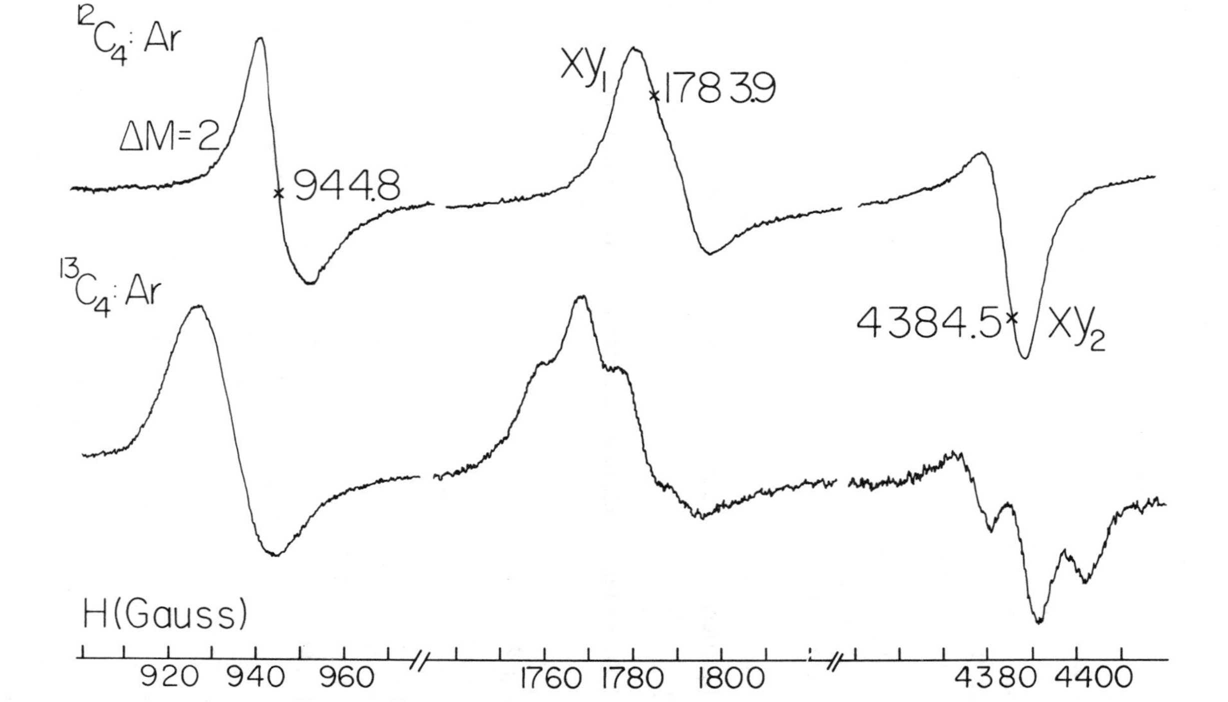

Fig. III-17. ESR spectra of triplet $^{12}C_4$ and $^{13}C_4$ molecules in argon matrices at 4° K (ν = 9.389 GHz) [from Graham, Dismuke, and Weltner (196); reprinted courtesy of The Astrophysical Journal, published by the University of Chicago Press; © 1976 The American Astronomical Society].

Table III-4.‡ Observed Line Positions and Calculated Values of $g_\perp$ and D for $^{12}C_4$ and $^{13}C_4$ in Argon, and $^{12}C_4$ in Neon at 4°K

Parameter	$^{12}C_4$:Ar	$^{13}C_4$:Ar	$^{12}C_4$:Ne
xy_1 (gauss)	1783.9(5)*	1772.0(5)	1634(1)
xy_2 (gauss)	4384.5(5)	4388.6(5)	4435(1)
$\Delta M = 2$ (gauss)	944.8(3)	936.6(3)	812(5)
D (cm^{-1})	0.2242(3)	0.2253(3)	0.238(3)
$g_\perp$	2.0041(5)	2.0040(5)	2.0046(9)
ν (MHz)	9389(1)	9387(1)	9377(1)

‡From Graham, Dismuke, and Weltner (196).
*Parentheses indicate errors in the last figure given, i.e., 1783.9(5) means 1783.9 ± 0.5.

The measured $\Delta g_\perp = +0.0018$ implies that there is a relatively low-lying $^3\Pi_g$ state, and from Curl's relationship (II, 26) also provides an estimate of the spin-rotation constant γ''. A summary of the information about C_4 obtained from solid-state ESR and optical spectroscopy is given in Table III-5.

Table III-5.‡‡. Molecular Constants* Derived for C_4 from Solid-State Data

State	Parameter	Value
$B\,^3\Pi_u$ or $B\,^3\Sigma_u^-$	T_0	19564(50)†
	ν_1'	2100(10)
$A\,^3\Pi_g$	T_0	<6000
$X\,^3\Sigma_g^-$	ν_3''	2170(5)
	λ''	0.128(3)‡
	γ''	−0.0006(2)
	$\lvert A_\perp(^{13}C)\rvert$	29.6(5) MHz
	$g_\perp$	2.0041(5)

‡‡From Graham, Dismuke, and Weltner (196).
*All constants are given in cm^{-1} except $A_\perp$ and $g_\perp$.
†Quantities in brackets indicate uncertainties in predicted gas-phase values.
‡The spin coupling parameter λ is related to the zero-field-splitting by $\lambda = D/2$.

CNN, NCN, CCO. These are isoelectronic $^3\Sigma$ molecules having D values in the 1 cm^{-1} range. CNN and NCN were first observed via ESR by Wasserman, Barash, and Yager (457) in octafluorocyclobutane matrices at 4°K where they were generated by photolysis of organic precursors. CNN and CCO were also prepared by the reaction of carbon atoms with N_2 or CO during condensation in a nitrogen or rare-gas matrix at 4°K [Smith and Weltner (421)]. In the latter work several C and ^{15}N isotopically substituted species were observed. In the case of NCN and CCO, accurate D values have been obtained from gas-phase optical spectroscopy [Herzberg and Travis (219); Devillers and Ramsay (139)].

The ESR spectrum obtained when ^{12}C vapor was co-deposited with a stream of Ne/^{12}CO (=200) is shown in Fig. III-18a. The line shapes and relative phases are those expected for a linear triplet molecule with $D > h\nu$. The split nature of the xy line at 6142 G has been attributed to a second site which disappears at higher CO dilutions (421). $^{12}C^{13}CO$ and $^{13}C^{12}CO$ can be prepared in a similar manner and their spectra are shown in Fig. III-18b. The results of the analysis of these spectra in solid neon and in other matrices are given in Table III-6. Assuming $g_\parallel = g_e$ throughout, one sees that there is a monotonic decrease in D from the gas-phase to the heavier matrices. The same is observed for CNN and NCN in the matrices in which they were observed (457). A comparison of the matrix effects upon D for the three molecules is shown in Table III-7; the sources of such effects were considered at some length by Smith and Weltner (421) [see section III-4c]. The spins were determined to be

Table III-6.[†] **Magnetic Parameters of $^{12}C_\alpha{}^{12}C_\beta{}^{16}O$ in the Gas-Phase and Various Rare-Gas matrices**

	Gas	Neon	Argon	Krypton	Xenon
D (cm^{-1})	0.772^a	0.7392(4)	0.7118(10)	0.573(10)	0.454(10)
$g_\parallel$	–	$(2.0023)^b$	$(2.0023)^b$	$(2.0023)^b$	$(2.0023)^b$
$g_\perp$	c	2.0029(4)	2.0028(10)	$(2.0023)^b$	$(2.0023)^b$
$A_\perp(C_\alpha)$ MHz^d	–	57(3)	59(3)	–	–
$A_\parallel(C_\alpha)$ MHz		17(3)	–	–	–
$A_\perp(C_\beta)$ MHz	–	–26(3)	–27(3)	–	–
$A_\parallel(C_\beta)$ MHz	–	–32(3)	–	–	–

[†] From Smith and Weltner (421).
[a] Devillers and Ramsay (139).
[b] Assumed value.
[c] The gas phase value of $\gamma = -0.0007(3)$ gives via (II, 26) a value of $g_\perp = 2.0032(3)$.
[d] Signs of A values have not been established experimentally.

Table III-7.[†] **Summary of zfs and Electronic Data for CCO, CNN, and NCN.**

Molecule	D_{gas}	D_{neon}	$D_{organic}$	D_{argon}	Lowest Observed Transition	cm^{-1}	Spin Density[a] X	Y	Z	
CCO	0.772[b]	0.7395	–	0.7120	$^3\Pi \leftarrow {}^3\Sigma$	11 650[b]	0.0219	−0.0115	0.0228	$2S$
							0.4470	−0.0123	0.5653	$2P_x$
							−0.0147	−0.0270	0.0086	$2Pz$
CNN	1.17[c]	1.159	1.153[d]	1.146	–	24 000[e]	0.0132	−0.0235	0.0296	$2S$
							0.3212	−0.1485	0.8273	$2P_x$
							0.0009	−0.0324	0.0123	$2P_z$
NCN	1.566[f]	–	1.544[d]	–	$^3\Pi \leftarrow {}^3\Sigma$	30 384[f]	0.0296	−0.0505	0.0296	$2S$
							0.6566	−0.3131	0.6566	$2P_x$
							0.0159	−0.0406	0.0154	$2P_z$

[†] From Smith and Weltner (421).
[a] Calculated from INDO at minimum energy. Bond lengths: CCO, r_{CC} = 1.28, r_{CO} = 1.26 Å; CNN, r_{CN} = 1.23, r_{NN} = 1.26 Å; NCN, 1.26 Å.
[b] Devillers and Ramsay (139).
[c] Extrapolated from neon and argon values. If the potential V_0 is assumed the same for CCO and CNN in neon matrices, a value of 1.21 cm^{-1} is derived for D_{gas} for CNN.
[d] Wasserman, Barash, and Yager (457).
[e] Milligan and Jacox (344).
[f] Herzberg and Travis (219).

predominantly in $2p\pi$ orbitals (as in O_2) in the three molecules and to be largely on the two outer atoms.

SiCO and SiNN. These silicon counterparts of the above molecules are examples of triplet molecules where D is relatively large, almost at the limit of X-band detection, owing to the larger spin-orbit coupling of Si and the greater localization of the spins on that atom. Also, because of low bending force constants, they are found to be bent in some matrix environments, as detected in their ESR spectra [Lembke, Ferrante, and Weltner (307)]. Figure III-19 shows the X-band ESR spectra of various isotopically substituted isomers of SiCO in argon matrices at 4°K. (^{29}Si is a magnetic nucleus with $I = \frac{1}{2}$.) The weaker line at higher field in those spectra, exhibiting the same hfs pattern as the major perpendicular line at 9489 G, is attributed to molecules induced to be slightly bent by the surrounding matrix. This occurs more strongly for SiN_2 in a nitrogen matrix but comparably in solid argon. [Wasserman, Barash, and Yager (457) had similarly interpreted the ESR spectrum of C_3N_2 in a hexafluorobenzene matrix.] In solid neon, assuming $g_\parallel = g_\perp = g_e$, $D = 2.28$ and 2.33 cm^{-1} for SiN_2 and SiCO, respectively. (Corresponding E values for the bent molecules

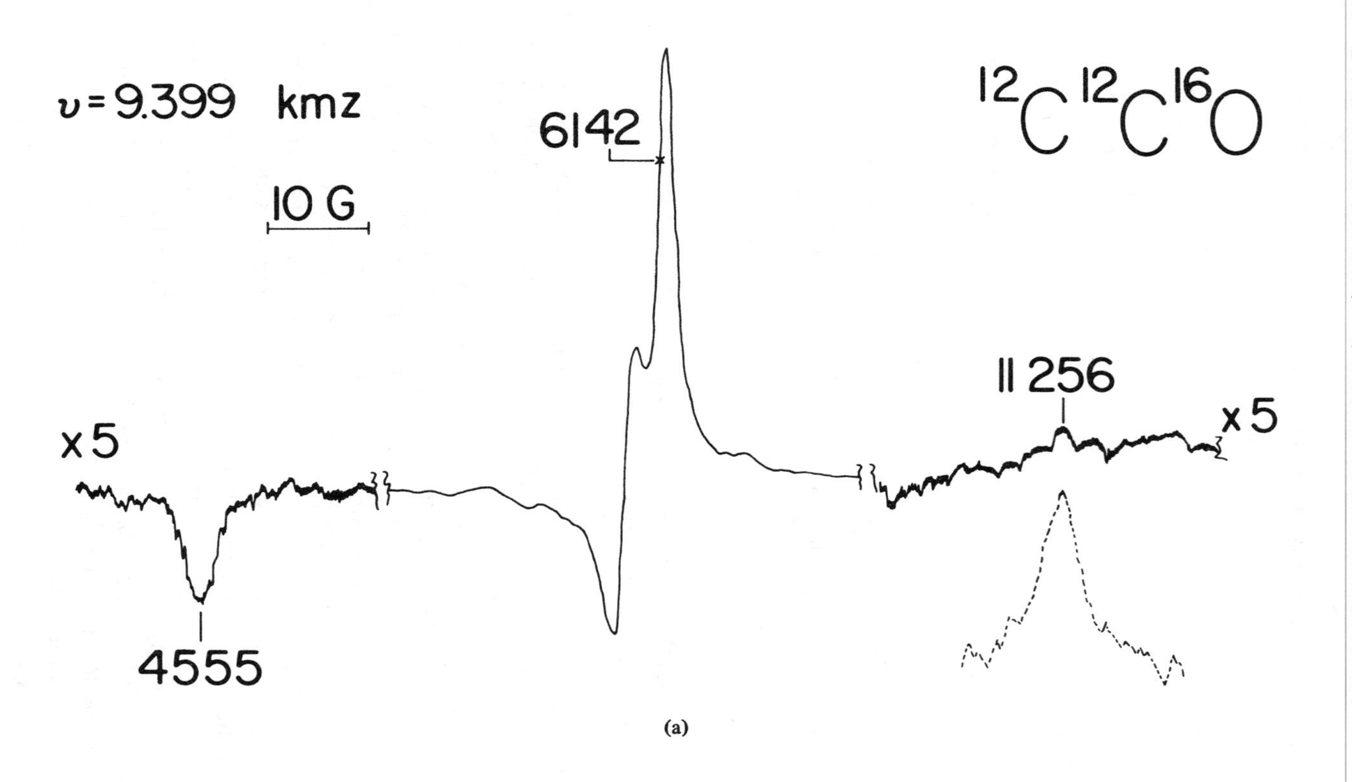

(a)

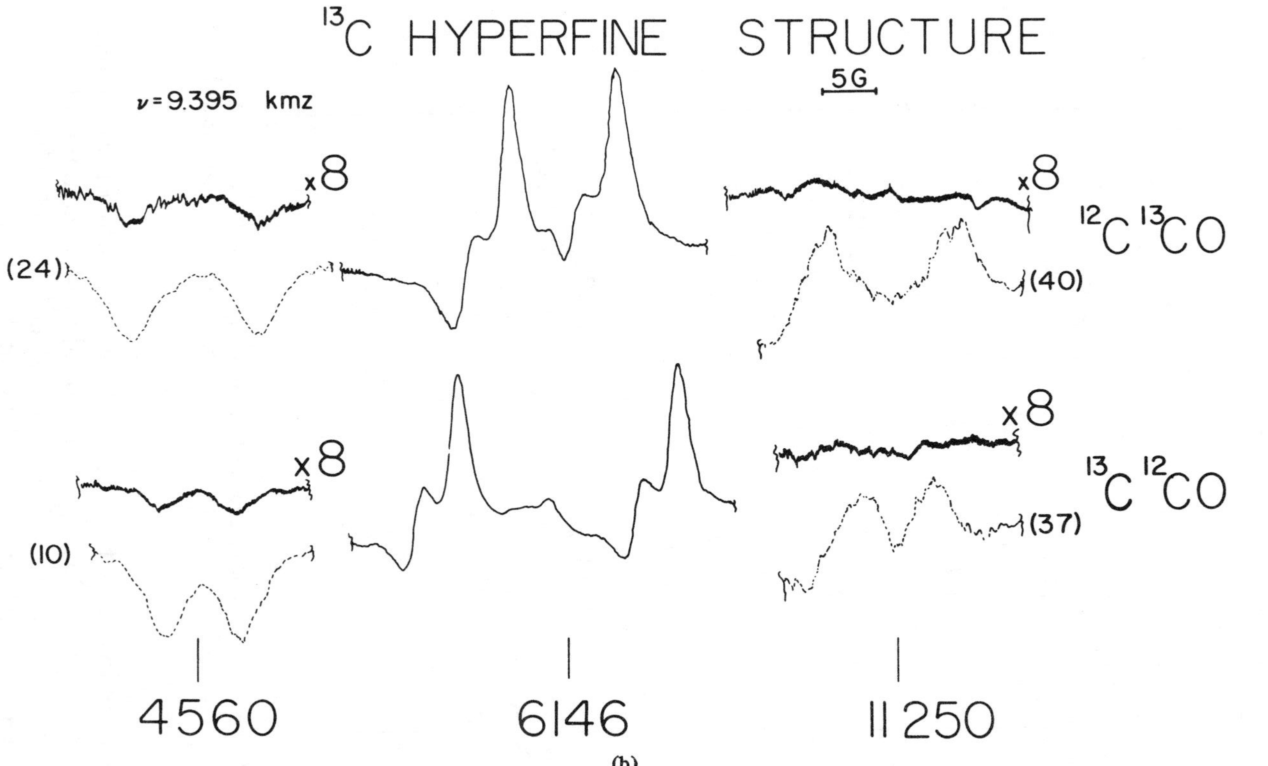

(b)

Fig. III-18. ESR spectra of $^{12}C^{12}CO$, $^{12}C^{13}CO$, and $^{13}C^{12}CO$ molecules in neon matrices at 4°K [from Smith and Weltner (421)]. Line positions are in gauss. (a) $^{12}C^{12}CO$ molecule, showing xy line and two z lines; (b) hyperfine splittings in these three lines with ^{13}C substitution.

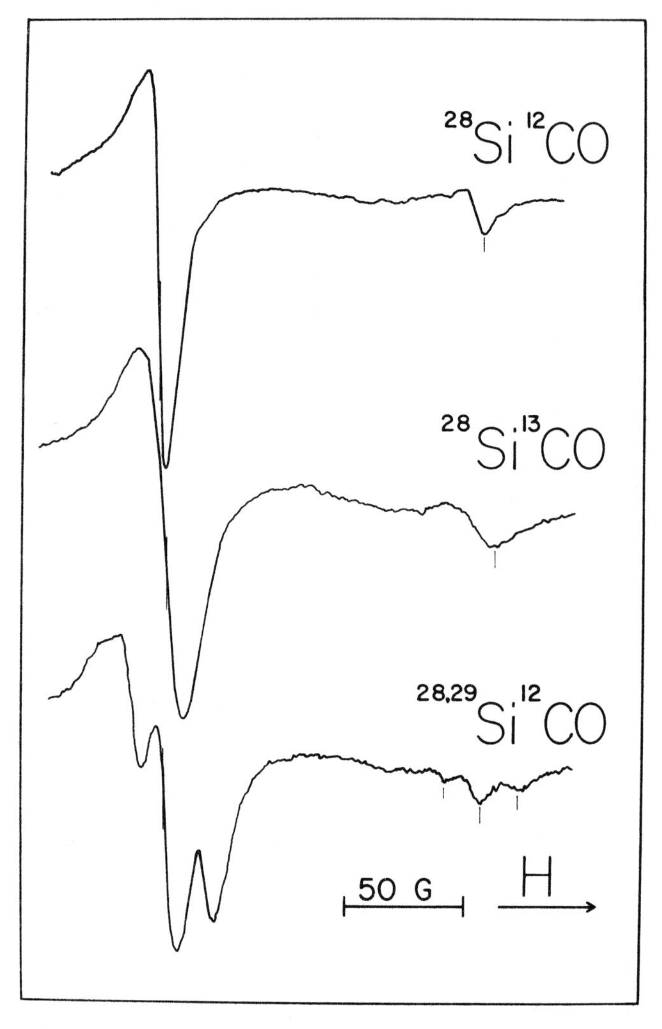

Fig. III-19. ESR spectra of SiCO molecules in argon matrices at 4°K [reprinted with permission from Lembke, Ferrante, and Weltner (307); copyright 1977, American Chemical Society].

were found to be about 1 to 2 $\times$ 10^{-3} cm^{-1}.) The hyperfine splittings obtained for ^{14}N, ^{15}N, ^{13}C, and ^{29}Si substitution confirmed CNDO calculated results, which indicated that in both molecules the electron spins are largely in the $p\pi$ orbitals of Si.

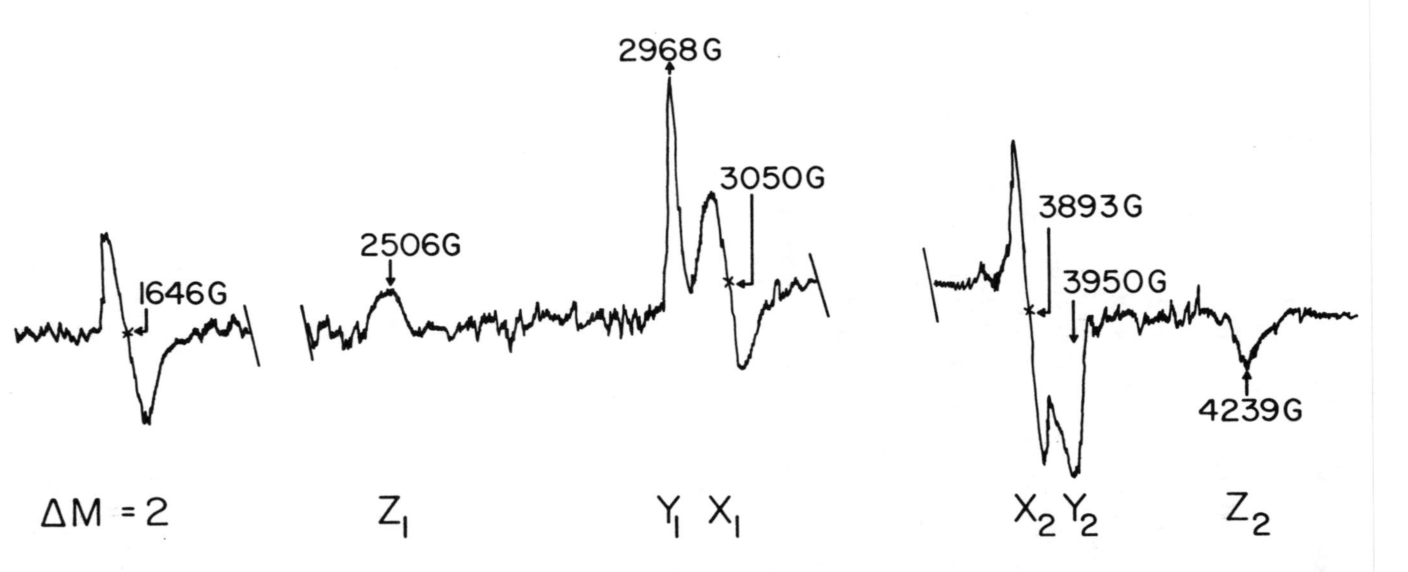

Fig. III-20. ESR spectrum of the TiF_2 molecule in a neon matrix at 4°K [reprinted with permission from DeVore and Weltner (142); copyright 1977, American Chemical Society].

TiF_2 (A Nonlinear Triatomic). From infrared studies this was found to be a bent molecule ($\theta = 130 \pm 5°$) [Hastie, Hauge, and Margrave (213)], and since the electronic configuration of titanium is $4s^2 3d^2$, one expects the ground state to be either singlet or triplet. Its ESR spectrum in solid neon at 4°K, shown in Fig. III-20, demonstrates that it is a triplet molecule with a relatively low zfs [DeVore and Weltner (142)]. It also confirms that it is nonlinear, since the more intense lines in the spectrum are split into x and y components (with z taken as an axis parallel to the F---F direction).

The absence of Ti hfs in the spectrum is not surprising, since the natural abundance of magnetic isotopes is low; however, the lack of ^{19}F hfs implies that the unpaired spins must be highly localized on Ti. The most likely orbitals are the nonbonding $d_{x^2-y^2}$ and d_{xy}, perpendicular to the plane of the molecule, and placing one electron in each gives a $^3B_1$ ground state. The $|D|$ value is only 0.078 cm^{-1} ($|E| = 0.002$ cm^{-1}), which is somewhat surprising considering the proximity of the spins and the spin-orbit coupling of the transition metal.

Chapter IV

Magnetic Resonance of Gas-Phase Molecules

When a radical, or any molecule with a sufficiently large magnetic moment, can be produced in the gas phase, electron paramagnetic resonance and, increasingly, laser magnetic resonance (LMR) may be applicable and can provide very detailed and accurate spectroscopic information. [Electron paramagnetic resonance (EPR) rather than ESR is the appropriate term here, since the magnetic properties may be derived from electron orbital, rather than spin, angular momentum, as in a $^1\Delta$ state.] For a linear molecule this information can include, in the notation usually used in this field: the rotational constant (B), fine structure constant (A), Λ doubling (ν_Λ), g values, zero-field spin-splitting (λ), magnetic hyperfine tensor components (a, b, c, d), spin-rotation constant (γ), quadrupole coupling constant (eqQ), and dipole moment (μ) in particular rotational, vibrational, and electronic states. Also, as the data become more extensive and accurate, higher-order parameters, such as centrifugal distortion corrections and the separate orbital and spin g factors, can be expected to be more routinely determined.

1. EXPERIMENTAL TECHNIQUES

Experimentally, magnetic resonance techniques may be considered as alternative methods in the general field of microwave spectroscopy [Gordy, Smith, and Trambarulo (9); Townes and Schawlow (21)]. "Pure" microwave spectroscopy, where rotational transitions are observed, molecular beam deflection, and optical double resonance techniques provide similar information of high accuracy. A general review of all of these techniques applied to free radicals has been given by Carrington (5), with electron paramagnetic resonance specifically considered by Carrington, Levy, and Miller (101).

(a) Electron Paramagnetic Resonance (EPR)

EPR in the gas phase is usually observed by passing the radical, produced directly or indirectly by gas discharge, through a high-sensitivity (high-Q) cavity in a variable magnetic field. Perhaps surprisingly, electric dipole transitions, rather than magnetic dipole transitions, have usually been observed, in spite of the implicit assumption of the latter when referring to magnetic resonance. Since electric-dipole transitions are intrinsically of the order of 1000 times more intense than magnetic dipole transitions, this is an important factor in gas-phase studies where the concentration of radicals may be low. Such $\Delta M_J = \pm 1$ transitions can be observed only for molecules having permanent electric dipole moments, but Stark modulation can then also be used instead of Zeeman modulation. This electric-field modulation has, in fact, proved to be very successful, not only in terms of sensitivity of the spectrometer but in allowing discrimination against species not possessing a dipole moment [Brown (92), page 266]. However, in the resonance experiment the molecule must also interact with the magnetic field, essentially requiring that it have an electronic magnetic moment (spin or orbital) and also one that is "tied" to the molecular orientation. $^2\Sigma$ molecules, where the electron spin angular momentum is only weakly coupled to the molecular orientation through spin-rotation interaction, are not favorable cases, and the less intense magnetic-dipole transitions must be detected. This is not true of triplet molecules where spin-spin interaction aids in coupling the electron spins to the molecular axis. Thus, most gas-phase EPR has been done on $^2\Pi$, and then $^3\Sigma$ and $^1\Delta$, molecules. [It is interesting to note that this is just the opposite of matrix-isolation ESR, where all Σ radicals are readily observed and Π are not (see page 78).]

For nonlinear doublet molecules such as HCO, the opposing effects of increasing spin decoupling versus gain in spectral tunability in increasing external magnetic fields have been discussed and quantified by Boland, Brown, Carrington, and Nelson (78a). In an increasing field the spin Zeeman interaction competes with the spin-rotation interaction and progressively decouples the spin from the molecular framework. This leads to decreasing electric dipole intensity, since it is associated with the molecular rotation. However, even though weakened by such decoupling, it is expected that the highly tunable ($\Delta m_S = 1$) transitions will be the ones most often observed, since laboratory magnetic fields can be varied over a broad range to observe such resonances. At low fields where the spin is coupled there is usually much less experimental latitude, since it is essentially required that the spectrometer frequency be (perhaps fortuitously) close to the molecular transition frequency because of the limited tuning range.

The bottleneck in the observation of more molecules in the gas phase has been the inability of the experimentalist to produce the desired radical in the spectrometer cavity [Brown (92), page 265]. It is often difficult to find a discharge-flow system that can produce a particular species in high-enough concentration. In molecular beam spectroscopy, ovens (e.g., producing LiO) and electron bombardment (e.g., exciting H_2) are used, but similar devices such as cavities at elevated temperatures (up to ~1000°K) and photolytic excitation have apparently not been successful in EPR. However, it is interesting and encouraging that the spectrum of the CH radical formed in an oxy-acetylene diffusion flame has been observed by laser magnetic resonance [Evenson, Radford, and Moran (158); Hougen, Mucha, Jennings, and Evenson (232)].

(b) Laser Magnetic Resonance (LMR)

The rapidly expanding field of LMR spectroscopy promises to provide better molecular data for many radicals, including many previously unobserved via magnetic resonance. An excellent review of the development and progress of the field up to 1980 is given by Evenson, Saykally, Jennings, Curl, and Brown (159). By using a far-infrared laser for excitation, an increase in sensitivity by a factor of more than a hundred has been attained, due to the increase in absorption coefficient with frequency.

Although there are no tunable lasers in the far infrared, near coincidence to a rotational transition of a radical can usually be achieved by taking advantage of the many optically pumped laser lines produced in a variety of gases. LMR is best explained by referring to the CO_2-transversely-pumped far-infrared spectrometer

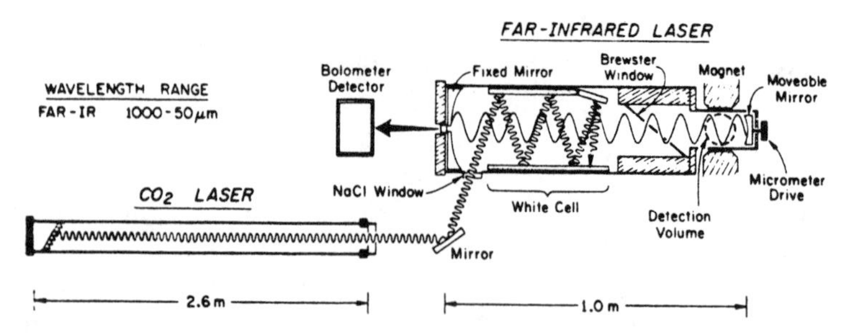

Fig. IV-1. Optically pumped laser magnetic resonance spectrometer [from Hougen, Mucha, Jennings, and Evenson (232)].

shown in Fig. IV-1, which has been used recently to measure the magnetic resonance spectrum of the CH ($^2\Pi_r$) radical [(Hougen, Mucha, Jennings, and Evenson (232)]. A portion of that spectrum is reproduced in Fig. IV-2.

The laser oscillates between the two mirrors, with tuning to maximum gain by adjusting the movable one. The Brewster window separates the intracavity sample cell on the right from the lasing medium on the left. CH_3OH, $^{13}CH_3OH$, CD_3OH, H_2O, CH_2CF_2, DCOOD, or CH_2CHCl vapor in the multiple-reflective White cell is pumped by the CO_2 laser to provide various frequencies between 17 and 86 cm^{-1}. The CH rotational transitions observed when excited by these lines are indicated in Table IV-1. Here the CH radicals are produced in the detection region by reaction of F atoms with methane while flowing through the cavity located between the poles of a scanning electromagnet, on which modulation coils are wound. A small fraction of the radiation passes out through a hole in the fixed mirror to a liquid-helium-cooled bolometer. By rotating the beam splitter about the far-IR laser axis, transitions for either $\Delta M_J = 0$ or $\Delta M_J = \pm 1$ can be selected, as Fig. IV-2 demonstrates.

Atoms have also been studied by LMR. For example ^{12}C and ^{13}C atoms were also produced in this same CH_4 + F flame, and transitions in the ^3P state were observed, yielding precise frequencies for the zero-field fine-structure transitions in both isotopes [Saykally and Evenson (402)].

If the iron core magnet in Fig. IV-1 is replaced by a solenoid magnet concentric with the laser axis and the flame is replaced by the positive column of a glow discharge running along the magnetic field, then not only molecules and atoms but, more significantly,

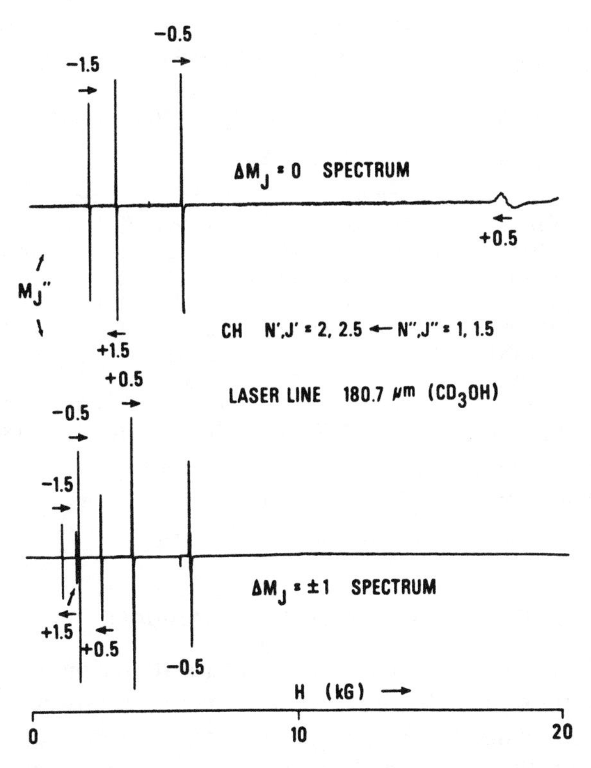

Fig. IV-2. LMR survey spectrum of the CH molecule from 0 to 20 kG, recorded with the 180.7 μm line of CD_3OH [from Hougen, Mucha, Jennings, and Evenson (232)]. The upper and lower traces correspond to parallel and perpendicular polarization, respectively, of the electric field vector of the laser radiation with respect to the external magnetic field. Lines marked above and below with a horizontal arrow were shown experimentally to shift in the indicated direction when the laser is "pulled" to slightly higher frequency. [This "laser pulling technique" is used to distinguish lines differing in M; see Evenson et al. (159)]. The lower-state quantum number M_J is indicated for each line.

molecular ions can be detected. The observation of HBr^+ in this way adds to a growing list of molecular ions observed in high resolution in the gas phase [see Saykally and Evenson (401) and references given there].

LMR is being applied successfully to the investigation of an increasing number of molecules; among those already observed are O_2, SO, NH, NH_2, PH, PH_2, HO_2, HCO, CH_2, and CCH in their ground electronic states and a few such as O_2, PH, CO, and HO_2 in excited states.

Table IV-1. Rotational Transitions of CH Observed in LMR Studies
[from Hougen, Mucha, Jennings, and Evenson (232)]

Laser Characteristics				CH Rotational Transitions		
Pump[a]	Gain Medium	λ [μm]	ν [cm^{-1}]	$(N', J') \leftarrow (N'', J'')$	$(g_J')^b$	$(g_J'')^b$
10 R(16)	$^{13}CH_3OH$	115.8	86.33845	(3, 5/2) - (2, 3/2)	+0.152	+0.062
Discharge	H_2O	118.6	84.32340	(3, 7/2) - (2, 5/2)	-0.341	-0.495
9 P(36)	CH_3OH	118.8	84.15092	(3, 7/2) - (2, 5/2)	-0.341	-0.495
9 P(22)	$^{13}CH_3OH$	149.3	66.99168	(2, 3/2) - (1, 1/2)	+0.062	+0.001
10 R(34)	CD_3OH	180.7	55.32794	(2, 5/2) - (1, 3/2)	-0.495	-0.863
10 R(16)	$^{13}CH_3OH$	203.6	49.10728	(2, 3/2) - (1, 3/2)	+0.062	-0.863
10 P(14)	CH_2CF_2	554.4	18.03865	(1, 3/2) - (1, 1/2)	-0.863	+0.001
10 P(20)	DCOOD	561.3	17.81598	(1, 3/2) - (1, 1/2)	-0.863	+0.001
10 P(16)	CH_2CHCl	567.9	17.60731	(1, 3/2) - (1, 1/2)	-0.863	+0.001

[a] The gain medium was pumped by electric discharge or by a CO_2 laser line of the 9 μm or 10 μm band.
[b] Magnetic g-values, valid for a $^2\Pi$ state with $A = 2B$ when $H \simeq 0$.

2. EXPERIMENTAL DATA FOR GASEOUS RADICALS

Molecules investigated in the gas phase by magnetic resonance, but often additionally by other forms of microwave spectroscopy, are listed in Appendix D. The parameters chosen for tabulation there are generally those relevant to the magnetic properties of the molecules. (The uncertainties in the numbers listed are not given, and one should consult the original references cited for a reliable judgment of their accuracy if that is a concern.) If undetermined, the values of B_e and A have been taken from Huber and Herzberg (12). Hyperfine parameters d and $h_{3/2}, h_{1/2}$ [to be defined in section (a) below] may sometimes be given in addition to, or in lieu of, Frosch and Foley's a, b, c defined in (II, 29). a_F is used to symbolize the isotropic Fermi constant contribution, and in bent triatomic molecules the T_{ii} designate the anisotropic components in the three molecule-fixed axes (as in Chapter III). g values have been determined accurately in the gas phase for only a few molecules and therefore were not tabulated in Appendix D. However, the accuracy obtained in those few cases makes it possible to calculate the separate electronic, orbital, etc., contributions to the g values (the Zeeman parameters), and they are discussed in some detail below in section (b).

(a) Hyperfine Parameters

Most of the molecules studied by gas-phase EPR are in $^2\Pi$ electronic states. In such states $^2\Pi_{1/2}$ substates are nominally nonmagnetic (see page 78), but rotational mixing of the $^2\Pi_{3/2}$ and $^2\Pi_{1/2}$ substates may cause this not to be true; for example, magnetic resonance of the OH radical in both the $^2\Pi_{3/2}$ and $^2\Pi_{1/2}$ substates has been observed [Radford (385, 386)]. [Another way of expressing this is to say that the rotation induces a magnetic moment by spin uncoupling, i.e., departure from the Hund case (a) limit.] This is not the case for the heavier molecules SH, SeH, and TeH where the rotational distortion constant A/B is small enough to leave the $^2\Pi_{1/2}$ state not susceptible to magnetic resonance studies. From measurements in various J rotational levels in these two substates, the hyperfine parameters a, b, c, and d can be derived. d was not introduced earlier, when the other Frosch and Foley parameters were given in (II, 29). It is

$$d = 3g_e\beta_e g_I\beta_n \langle \sin^2 \alpha/2r^3 \rangle \qquad \text{(IV, 1)}$$

where the average is taken over the distribution of unpaired electrons. This parameter can be said to arise because of the difference in the I, S dipolar interaction in the two components of a Λ doublet, which results from lack of cylindrical symmetry in the electron wavefunction caused by admixture of other remote electronic states.

If it is assumed that the same electrons are responsible for both orbital and spin contributions to the hfs, then from the definitions of a, c, and d, the relationship

$$c = 3(a - d) \qquad \text{(IV, 2)}$$

would also hold [Townes and Schawlow (21), page 245]. However, whenever measurements of all four parameters have been made for $^2\Pi$ states, this relationship has been found to be unreliable, indicating that configuration interaction must necessarily be considered [Brown (92)]. (This is analogous to the necessity for definition of P_L and P_S in (I, 28) as discussed on page 16.) In the case of a molecule such as OH this amounts to saying that the ground-state electron configuration,

$$(1\sigma)^2 (2\sigma)^2 (3\sigma)^2 (\pi^+)^2 \pi^-$$

is substantially mixed with other excited configurations involving higher, normally vacant, σ orbitals. These unpaired σ electrons contribute to the spin part of the hf interaction (b, c, d) but not to the orbital part (a), thereby invalidating (IV, 2).

When insufficient measurements are available to completely establish all of the Frosch and Foley hf parameters (a, b, c, d), one or both of the axial components

$$h_{3/2} = a + \tfrac{1}{2}(b + c) \quad [^2\Pi_{3/2} \text{ state}]$$

and

$$h_{1/2} = a - \tfrac{1}{2}(b + c) \quad [^2\Pi_{1/2} \text{ state}] \tag{IV, 3}$$

may still be obtained from the spectral analysis. The determination of b may be difficult because it depends upon A/B, which may not be known with sufficient accuracy.

Other hyperfine parameters defined by Radford (385) are A_1 and A_2, which are functions of J:

$$A_1 = [\pm 4J(J+1)X]^{-1} \{2a(\pm X + 2 - \lambda) + b[\pm X + 4 - 2\lambda - 4(J + \tfrac{3}{2})(J - \tfrac{1}{2})] + c(\pm X + 4 - 2\lambda)\}$$

$$A_2 = d[\pm 4J(J+1)X]^{-1} (\pm X - 2 + \lambda)(J + \tfrac{1}{2})$$

where $X = +[4(J + \frac{1}{2})^2 + \lambda(\lambda - 4)]^{1/2}$ and $\lambda = A/B_P$. Although, in principle, by measuring A_1 for three different J values, the hf parameters a, b, and c can be determined, this cannot be done accurately without also knowing d [Radford (386)]. If the data were insufficient, then values of A_1 in particular J states were often tabulated for $^2\Pi$ molecules [Carrington, Levy, and Miller (101); Brown (92)].

(b) *g* Values; the Curl Equation

An effective $\mathbf{g}$ tensor and the concept of a spin Hamiltonian (see page 40ff) are not useful in the gas phase. One is concerned instead with accounting individually for the $\mathbf{g}$ parameters involved in the interaction of the external magnetic field with the electron-spin, the electron-orbital, and the nuclear-rotational magnetic moments. There appear to be only four radicals, O_2 and SO in their $^3\Sigma$ ground states, OH in its $^2\Pi$ ground state, and H_2 in its $^3\Pi_u$ excited states, for which there are sufficiently accurate data to determine these parameters. Since $L = 0$ for O_2 and SO, the theory is a little simpler, so we will consider them first and then OH as an example for $L = 1$.

The Zeeman Hamiltonian for these molecules can be written [Tinkham and Strandberg (429); Kawaguchi, Yamada, and Hirota (268)]:

$$\mathcal{H}_Z = \beta[g_s{}^e \boldsymbol{S} \cdot \boldsymbol{H} + g_l{}^e (S_x H_x + S_y H_y) - g_r \boldsymbol{N} \cdot \boldsymbol{H}] \qquad \text{(IV, 4)}$$

[The signs before the g factors vary among the different authors. Those in this chapter are chosen to be in accord with the latest work on OH by Brown, Kaise, Kerr, and Milton (93).] Here $\boldsymbol{N}$ is the end-over-end rotational angular momentum. The three terms then account for the three interactions named above. (IV, 4) can also be written:

$$\mathcal{H}_Z = \beta[(g_s{}^e + g_l{}^e) \boldsymbol{S} \cdot \boldsymbol{H} - g_l{}^e S_z H_z - g_r \boldsymbol{N} \cdot \boldsymbol{H}] \qquad \text{(IV, 5)}$$

or [Evenson and Mizushima (157)],

$$\mathcal{H}_Z = \beta[g_\perp \boldsymbol{S} \cdot \boldsymbol{H} + (g_z - g_\perp) S_z H_z - g_r \boldsymbol{N} \cdot \boldsymbol{H}] \qquad \text{(IV, 6)}$$

so that $g_s{}^e + g_l{}^e = g_\perp$ and $-g_l{}^e = g_z - g_\perp$, where $g_z = g_\parallel \cong g_s$. (Here we follow the usual symbolism in the gas phase in letting g_s rather than g_e designate the free electron spin g value 2.002319.) To a first approximation the three g parameters in (IV, 5) are $g_s{}^e \cong g_s$, $g_l{}^e \cong 0$ for a Σ state, and $g_r \cong 0$, since nuclear rotation produces such a small magnetic moment. However, the experimentally observed small divergences from these values are significant. Christensen and Veseth (111) have carefully considered all of the experimental data (references in Appendix D) and have obtained the "fitted" g-factors for the two molecules given in Table IV-2. We will consider each of these parameters in turn.

Table IV-2. Fitted g-Factors for the $^3\Sigma$, $v = 0$ Ground States of O_2 and SO [a]

Molecule	g_s^e [b]	g_l^e	g_r^e [c]
O_2	2.002089(16)	$2.77(2) \times 10^{-3}$	$3.9(1) \times 10^{-4}$
SO	2.002143(29)	$3.37(7) \times 10^{-3}$	$3.8(4) \times 10^{-4}$

[a] From Christensen and Veseth (111).
[b] Includes small corrections $\sim 10^{-5}$ to the fitted values for third-order spin-orbit terms.
[c] Sign is opposite of that of Christensen and Veseth (111).

g_s^e is lowered from the free electron value by a small amount ($\sim 2 \times 10^{-4}$), which can be accounted for by relativistic corrections when the electrons occupy molecular orbitals (111).

g_l^e differs from zero in the molecules because of the mixing in of orbital angular momentum by the spin-orbit coupling. The dominant term determining g_l^e is essentially the same second-order perturbation term producing the g shift $g_\perp - g_e$ (see page 43). The Curl equation then applies in the form

$$\gamma \cong -2Bg_l^e \qquad \text{(II, 26) (IV, 7)}$$

A test of this relationship for these two molecules is given in Table IV-3 where "fitted" experimental values of $-\gamma/2B$ are compared with corresponding "fitted" values of g_l^e. Column 4 of that table includes a third-order correction to g_l^e, and column 5 gives a theoretical estimate. All agree reasonably well.

Table IV-3. Comparison of g_l^e and γ for O_2 and SO

Molecule	$-\dfrac{\gamma(\text{fitted})}{2B}$	g_l^e (fitted)	$g_l^{e(2)}$ [a]	$g_l^{e(2)}$ (th.) [b]
O_2	2.93×10^{-3}	$2.77(2) \times 10^{-3}$	2.96×10^{-3}	2.6×10^{-3}
SO	3.91×10^{-3}	$3.77(7) \times 10^{-3}$	4.34×10^{-3}	5.1×10^{-3}

[a] Accounting for third-order contributions to g_l^e (fitted), yielding improved agreement with column 2.
[b] Theoretical estimates based on calculation of spin-orbit matrix elements and pure precession approximation.

The rotational magnetic moment is made up of contributions from the nuclei (g_r^n) and from the electrons rotating with the nuclei (g_r^e). g_r^n is usually overshadowed slightly by the opposing effect of the electron motion, as is the case for each of these molecules. For these homonuclear diatomics g_r^n is calculable, to sufficient accuracy (see small print section below), from $m/2M$ where m is the mass of the electron and M the mass of the proton, so that in each case it is equal to 2.74×10^{-4}. (Thus g_r^n would be just $\frac{1}{2}$ if the magnetic moment were expressed in terms of the nuclear magneton rather than the Bohr magneton.) The rotational $g_r = g_r^n - g_r^e$ then results in a negative g_r because of the larger values of g_r^e in Table IV-2. [Generally, except for protonated radicals such as NH_2, one can expect that $g_r^e > g_r^n$ because the electrons are more extended in space than the nuclei (see below)].

It has also been shown [Eshbach and Strandberg (156)], to the accuracy of second-order perturbation theory, that

$$(g_r^e)_{gg} = \frac{2}{I_g} \sum_n{}' \frac{|\langle 0|L_g|n\rangle|^2}{E_n - E_0} \qquad \text{(IV, 8)}$$

Then from (II, 11) and (II, 26) one expects that g_r^e will be related to the spin rotation constant γ and $\Delta g_\perp = g_\perp - g_e$ in a diatomic molecule approximately by

$$g_r^e \cong \frac{2B|\Delta g_\perp|}{\zeta} \cong \frac{|\gamma|}{\zeta} \qquad \text{(IV, 9)}$$

Table IV-4 tests this relationship for O_2 and SO. The agreement is excellent for O_2 but not as good for the heavier SO molecule.

Eshbach and Strandberg (156) were the first to be concerned with the magnitude of rotational g factors and to consider a rigid charge approximation. g_r^n is obtained simply from a classical calculation after recognizing that the molecular angular momentum is $\boldsymbol{N}\hbar$ [Townes and Schawlow (21), page 291]. The fraction of this momentum carried by the protons in a nucleus at distance r and rotating in a spherical orbit about the center of mass would be ZMr^2/eI where Z is the nuclear charge provided by the protons, each of mass M. I is the moment of inertia of the nucleus (plus electrons) about the center of mass. Then in a magnetic field $E = -\boldsymbol{\mu}_L \cdot \boldsymbol{H} = \beta_n \boldsymbol{L} \cdot \boldsymbol{H}$ where $\boldsymbol{L} = -(ZMr^2/eI)\ \boldsymbol{N}$ and $E = -\beta_n(ZMr^2/eI)\ \boldsymbol{N} \cdot \boldsymbol{H}$ for each rotating nucleus. From this then

Table IV-4. Comparison of the Experimental Values of g_r^e and those Calculated from (IV, 9)

Molecule	γ[a] (cm^{-1})	ζ(molec.) (cm^{-1})	g_r^e(calc.)	g_r^e(exptl.)[b]
O_2	-0.00843	21[c]	4.0×10^{-4}	$3.9(1) \times 10^{-4}$
SO	-0.00562	80[d]	7.0×10^{-4}	$3.8(4) \times 10^{-4}$

[a] Experimental, from Appendix D.
[b] From Christensen and Veseth (111) with changed sign.
[c] Tinkham and Strandberg (429).
[d] Carrington, Levy, and MIller (101).

$g_r^{n'} = ZMr^2/eI$ if the magnetic moment is expressed in nuclear magnetons or $g_r^n = Zmr^2/eI$ if expressed in Bohr magnetons (m is the mass of the electron), for each nucleus in the molecule. For a diatomic molecule, XY, one can show that

$$g_r^n = \frac{m}{M_x + M_y} \frac{Z_x M_y^2 + Z_y M_x^2}{eM_x \cdot M_y} \tag{IV, 10}$$

In general,

$$(g_r^n)_{gg} = \frac{m}{eI_g} \sum_k Z_k (r_k^2 - g_k^2) \tag{IV, 11}$$

where $(r_k^2 - g_k^2)^{1/2}$ is the distance of the kth nucleus from the principal axis g, and I_g is the moment of inertia of the molecule about principal axis g.

It is more convenient to use $g_r^{n'}$, since the numbers are simpler. For protonated compounds $g_r^{n'}$ will be of the order of 1.0, and as molecules containing heavier atoms are considered it will approach 0.5 because the ratio of charge to mass is about $\frac{1}{2}$. This is the case for both O_2 and SO.

g_r^e can be calculated approximately by assuming that the electrons are fixed rigidly in their bonds and then treating them in the same manner as nuclei [Eshbach and Strandberg (156); Weltner (473)]. Then

$$(g_r^e)_{gg} I_g = m \sum_i (r_i^2 - g_i^2)$$

where the sum is over all electrons at a distance $(r_i^2 - g_i^2)^{1/2}$ from the principal rotational axis g. The inner shell electrons are simply subtracted from the nuclear charge which they surround ["slip" effect; see Townes and Schawlow (21), page 213]; the valence electrons are put at the midpoints of bonds. The electron contribution then largely cancels the nuclear rotational effect and, in fact, usually surpasses it in magnitude so that $g_r = g_r^n - g_r^e$ is then negative.

For $^2\Pi$ OH, where the orbital angular momentum is not zero, the effective Zeeman Hamiltonian may be written [Brown, Kaise, Kerr, and Milton (93)]:

$$\mathcal{H}_Z = \beta[g_S{}^e \boldsymbol{S} \cdot \boldsymbol{H} + (g_L{}' + g_r) \boldsymbol{L} \cdot \boldsymbol{H} - g_r \boldsymbol{N} \cdot \boldsymbol{H} + g_l(S_x H_x + S_y H_y) - g_r{}^{e'}(e^{-2i\phi} N_+ H_+ + e^{2i\phi} N_- H_-) + g_l{}'(e^{-2i\phi} S_+ H_+ + e^{2i\phi} S_- H_-)] \qquad \text{(IV, 12)}$$

where only the nuclear Zeeman term has been omitted and all signs are taken as positive to conform to our earlier notation. Comparison with (IV, 4–6) shows that several terms have been added in the departure from a Σ state. $g_L{}'$ is the electron orbital g-factor which can be expected to differ only slightly from unity owing to small relativistic, diamagnetic, and nonadiabatic effects. N now includes the orbital contribution, i.e., $\boldsymbol{N} = \boldsymbol{L} + \boldsymbol{R}$, where R indicates molecular rotation. $g_l{}'$ and $g_r{}^{e'}$ enter because of Λ-type doubling in Π states and are analogues of $g_l{}^e$ and $g_r{}^e$ [Carrington and Lucas (103)]. ϕ is the electron orbital azimuthal angle and $N_\pm = (N_x \pm iN_y)$.

As with the other Zeeman parameters, initial estimates of the Λ-doubling parameters can be made from

$$g_l{}' \simeq p/2B \quad \text{and} \quad g_r{}^{e'} \simeq -q/B \qquad \text{(IV, 13)}$$

where p and q are the corresponding parameters introduced by Mulliken and Christy (354).

Table IV-5 gives the experimental values of the six Zeeman parameters for OH in its $X^2\Pi$ state [Brown, Kaise, Kerr, and Milton (93)] as compared with estimates from the Curl relationship (IV, 7) and from (IV, 13).

In nonlinear molecules the Curl equation has the more general form

$$g_S{}^{ii} = g_S - \epsilon_{ii}/2B_{ii} \qquad \text{(IV, 14)}$$

where the ϵ_{ii} and B_{ii} are the principal values of the spin-rotation and rotational constants respectively. The $g_S{}^{ii}$ have been measured in the gas phase for HO_2 [see Barnes, Brown, Carrington, Pinkstone, Sears, and Thistlethwait (45)] and for NO_2 [see Brown, Steimle,

Table IV-5. Zeeman Parameters for OH in the $X^2\Pi$ State [from Brown, Kaise, Kerr, and Milton (93) and Evenson et al. (159); See also Veseth (454)]

Parameter	Experimental[a]	Estimated[b]	Calculated[c]
g_L'	1.00107(15)	–	1.00093
g_S	2.00152(36)	–	2.00206
$10^3 g_l$	4.00(56)	3.20	4.29
$10^3 g_r^n$	–	–	0.528[d]
$10^3 g_r^e$	1.161	–	–
$10^3 g_r$	–0.633(19)	–	–0.55
$10^3 g_l'$	6.386(30)	6.347	8.58
$10^3 g_r^{e'}$	2.0446(23)	2.087	2.19

[a]Values from Brown, Kaise, Kerr, and Milton (93). The numbers in parentheses represent one standard deviation, in units of the last quoted decimal place.
[b]Values calculated from (IV, 7) and (IV, 13).
[c]Values calculated assuming that the $X^2\Pi$ and $A^2\Sigma^+$ states are in pure precession. The relativistic correction was taken as 1.30×10^{-4}.
[d]From (IV, 10).

Coles, and Curl (97)], and it can be seen from Table IV-6 that there is good agreement with the corresponding g components derived from (IV, 14). g values for NO_2 from solid-state studies, from both crystal and matrix sources, also agree satisfactorily with the gas-phase data. For HO_2, the agreement is less satisfactory, which might be attributable to a disparity between the principal g and rotational tensor axes; however, it is known that the O–O bond is essentially axial in both cases (45). The solid-state data given in the table were obtained in an ESR study of a single crystal of a H_2O_2/urea addition compound where environmental effects may be present [Ichikawa, Iwasaki, and Kuwata (240)]. Similar effects may also be present in a H_2O_2/H_2O matrix at 77°K where it was found that $g_a = 2.0353(5)$, $g_b = 2.0086(5)$, and $g_c = 2.0042(5)$ [Wyard, Smith, and Adrian (481)]. Unfortunately, in the more gaslike environment of an argon matrix at 4°K it appears that the molecule is rotating about the O–O bond so that only an average of g_b and g_c is obtained: $g_\parallel = g_a = 2.0393$ and $g_\perp = (g_b + g_c)/2 = 2.0044$ [Adrian, Cochran, and Bowers (35)]. The latter is derived as 2.0047 from the gas-phase values so that one can say, not unexpectedly, that the best

Table IV-6. Comparison of Experimental g Values for HO_2 and NO_2 with those Calculated from Curl's Equation

	Experimental		Curl's
	Gas-Phase	Solid-State	Equation
$HO_2(X^2A'')^a$			
g_S^{aa}	2.04204(19)	2.0495	2.04293
g_S^{bb}	2.00790(24)	2.0081	2.00863
g_S^{cc}	2.00152(24)	2.0018	2.00218
$NO_2(X^2A_1)^b$			
g_S^{aa}	1.9906	1.9910	1.9910
g_S^{bb}	2.0028	2.0015	2.0020
g_S^{cc}	2.0061	2.0057	2.0062

[a] For HO_2: gas-phase data (g and ϵ values) taken from Barnes et al. (45); solid-state data from Ichikawa et al. (240) (see text).
[b] For NO_2: gas-phase g values derived from Brown et al. (97), ϵ values from Brown and Sears (95); solid-state data from Zeldes and Livingston (483) for NO_2 in single-crystal $NaNO_2$ [essentially the same as those found by Adrian (29) in an argon matrix and those of Myers, Easley, and Zilles (357) in a neon matrix].

solid-state values are probably obtained in the argon matrix where the hydrogen-bonding interactions are minimized.

In the same manner the approximate relationships given by (IV, 8) and (IV, 9) between the electron rotational $(g_r^e)_{gg}$ values and the corresponding spin-rotational constants can also be tested for HO_2 (45). Since the electron has only small spin density on H, the authors use the spin-orbit coupling constant of the O atom, $\zeta = 151\ \text{cm}^{-1}$, and derive the values given in column 3 of Table IV-7.

Table IV-7. Comparison of Experimental $(g_r^e)_{gg}$ values for HO_2 with those Calculated Using (IV, 9)[a]

	$(g_r^e)_{gg}$(expt'l)	$\lvert\epsilon_{ii}\rvert/\zeta$
$10^3(g_r^e)_{aa}$	9.857(41)	10.9
$10^4[(g_r^e)_{bb}+(g_r^e)_{cc}]$	1.85(41)	0.94

[a] From Table IV of Barnes et al. (45). There the sign of g_r^e is the opposite of here, since there $g_r = g_r^n + g_r^e$.

The agreement is better than expected. g_r^{ii} values have also been measured for the NH_2 radical, and the comparison with calculated values is similar to that in Table IV-6 [Brown and Sears (96)].

(c) Spin-Rotation Interaction in High Spin Molecules

The usefulness of the Curl equation (IV, 14) has been demonstrated by its application to doublet and triplet diatomics and triatomics (Tables II-1, IV-3, and IV-6). For a molecule with $S \geqslant \frac{3}{2}$, Hougen (230) has shown that more than one spin-rotation constant is strictly required. For $^4\Sigma$ states, Martin and Merer (325) and, more thoroughly, Brown and Milton (94) have discussed the higher-order spin dependence of the spin-rotation interaction. Their conclusion was that the use of two γ's is not needed except when measurements of the highest precision are made or where another nearby electronic state interacts strongly by spin-orbit coupling. It appears that the $^4\Sigma$ VO molecule is such a case, as was first pointed out by Barrow (47) and confirmed by Cheung, Hansen, and Merer (108). An additional parameter γ_S was needed to account for its laser-induced fluorescence spectrum, even though its value is only -1.0×10^{-5} and -23.1×10^{-5} cm^{-1} in the $X^4\Sigma^-$ and $C^4\Sigma^-$ states. γ in these states was found to be +0.022516(66) and -0.018444(69) cm^{-1}. The ground-state value was shown earlier in Table II-1 to be in accord with $\Delta g_\perp$ obtained from Kasai's (258) argon matrix ESR spectrum when the Curl equation was used.

Then it seems that in most cases the Curl relationship will also be useful for high spin molecules. In Table II-1, the measured value of $\Delta g_\perp$ has been used to predict the spin-rotation constant for the heavier $^4\Sigma$ molecule NbO, and comparison with a future gas-phase measurement will be interesting.

Chapter V
High Spin Molecules

We have arbitrarily chosen to designate all molecules with more than two unpaired spins ($S \geqslant \frac{3}{2}$) as high spin. In this category fall mostly molecules containing transition-metal or rare-earth-metal atoms, since, as will be seen below, there are only a few examples of organic molecules with these higher multiplicities. Correspondingly, there appear to be no published theoretical calculations of the zero-field splittings in molecules in this class, although this situation may change rapidly, since experimental data are now available for diatomic molecules that are amenable to accurate computations. As mentioned in section III-5b, the presence of metal atoms leads to the domination of the spin-orbit contribution to the zfs, and probably the spin-spin term may even be neglected. For example, for the quartet Cr^{3+} ion in distorted octahedral symmetry in molecular complexes the dipole-dipole term contributes only about 5% of the spin-orbit contribution [McGarvey (338); Lohr and Lipscomb (314)]. Also, in such complexes the zfs parameters are sensitive to symmetry changes that affect the extent of coupling of excited quartet states with the ground state.

Half-filled shells, such as occur in $Mn^{2+}(3d^5)$ and $Gd^{3+}(4f^7)$, because of their spherical symmetry and the traceless property of the $\boldsymbol{D}$ tensor, are expected to exhibit small zfs [similar to hfs; see section I-5a(β)], and that is generally the case in crystals [Abragham and Bleaney (1), pages 440, 335]. However, for the lower symmetry and perhaps increasing covalency present in many molecules D can have relatively large values: in the series of molecules MnH_2, MnF_2, MnO, and GdF_3, D varies from 0.26 to >1 cm^{-1} [Van Zee, Brown, Zeringue, and Weltner (447)].

In molecules containing more than one metal atom there is the possibility of exchange coupling of unpaired electrons either through direct overlap or indirectly through the attached ligands. This then can lead to a ground state that has a higher or lower total spin than that of the individual metal atoms. However, even if the lowest state is diamagnetic, triplet and higher spin states may be accessible as the temperature is raised. Such exchange-coupled systems will also be briefly considered here in section 6.

1. $^4\Sigma$ AND OTHER QUARTET ($S = \frac{3}{2}$) MOLECULES IN MAGNETIC FIELDS

The three unpaired electrons in an $S = \frac{3}{2}$ molecule, although producing increasing fine structure in its spectrum, do not lead to any necessity for changes in the basic theory discussed for triplet molecules. The spin Hamiltonian in (III, 13) is still applicable:

$$\mathcal{H} = \beta \boldsymbol{H} \cdot \boldsymbol{g} \cdot \boldsymbol{S} + D[S_z^2 - \tfrac{1}{3} S(S+1)] + E(S_x^2 - S_y^2) \quad \text{(V, 1)}$$

with the same meanings for D and E as given in (III, 5). E is zero for a linear radical or one with a trigonal axis of symmetry. The spin basis functions now become

$$\begin{aligned}
|+\tfrac{3}{2}\rangle &= |\alpha_1 \alpha_2 \alpha_3\rangle \\
|+\tfrac{1}{2}\rangle &= (1/\sqrt{3})|\alpha_1 \alpha_2 \beta_3 + \alpha_1 \beta_2 \alpha_3 + \beta_1 \alpha_2 \alpha_3\rangle \\
|-\tfrac{1}{2}\rangle &= (1/\sqrt{3})|\alpha_1 \beta_2 \beta_3 + \beta_1 \alpha_2 \beta_3 + \beta_1 \beta_2 \alpha_3\rangle \\
|-\tfrac{3}{2}\rangle &= |\beta_1 \beta_2 \beta_3\rangle \qquad \text{(V, 2)}
\end{aligned}$$

and $S_z^2 = (S_{z_1} + S_{z_2} + S_{z_3})^2$, etc. Then the 4×4 eigenvalue matrix can be constructed from (V, 1) and can, in fact, be solved exactly for the eigenvalues with the static magnetic field along any principal axis [Vinokurov, Zaripov, Stepanov, Pol'skii, Chirkin, and Shekun (455); Hou, Summit, and Tucker (229); Pilbrow (373)].

A. The Axial Case, $^4\Sigma$ Molecules

(a) Eigenvalues and Magnetic-Dipole Transitions. For a linear $^4\Sigma$ molecule one derives the matrix, similar to (III, 17):

$$\begin{array}{c|cccc} & |+\frac{3}{2}\rangle & |+\frac{1}{2}\rangle & |-\frac{1}{2}\rangle & |-\frac{3}{2}\rangle \\ \hline |+\frac{3}{2}\rangle & D+\frac{3}{2}G_z & \sqrt{\frac{3}{2}}\,G_x & 0 & 0 \\ |+\frac{1}{2} & \sqrt{\frac{3}{2}}\,G_x & -D+\frac{1}{2}G_z & G_x & 0 \\ |-\frac{1}{2}\rangle & 0 & G_x & -D-\frac{1}{2}G_z & \sqrt{\frac{3}{2}}\,G_x \\ |-\frac{3}{2}\rangle & 0 & 0 & \sqrt{\frac{3}{2}}\,G_x & D-\frac{3}{2}G_z \end{array} \quad (V, 3)$$

where $G_z = g_{\parallel} \beta H_z$ and $G_x = g_{\perp} \beta H_x$. For H parallel to the molecular axis, z, the eigenvalues are

$$W_{\pm 3/2} = D \pm \tfrac{3}{2}\, g_{\parallel} \beta H$$
$$W_{\pm 1/2} = -D \pm \tfrac{1}{2}\, g_{\parallel} \beta H \quad (V, 4)$$

and the zero-field splitting is then $2D$. Representative Zeeman plots of these energy levels for two choices of D (assumed positive) are shown in the right-hand side of Fig. V-1. For $h\nu = 9.3$ GHz (X-band), the magnetic field positions of the $\Delta m_S = \pm 1$ transitions are also shown.

For $H \perp z$:

$$W_{\pm 3/2} = \pm \tfrac{1}{2}\, g_{\perp} \beta H + [(D \pm \tfrac{1}{2}\, g_{\perp} \beta H)^2 + \tfrac{3}{4}\,(g_{\perp} \beta H)^2]^{1/2}$$
$$W_{\pm 1/2} = \pm \tfrac{1}{2}\, g_{\perp} \beta H - [(D \pm \tfrac{1}{2}\, g_{\perp} \beta H)^2 + \tfrac{3}{4}\,(g_{\perp} \beta H)^2]^{1/2} \quad (V, 5)$$

Accurate plots of the energy levels for $H \perp z$ for several values of D are shown on the left-hand side of Fig. V-1. $\Delta m_S = \pm 1$ transitions are also indicated between these xy levels. For any angle θ at *high* fields, m_S is a "good" quantum number such that increasing values of $m_S = -\frac{3}{2}, -\frac{1}{2}, +\frac{1}{2}, +\frac{3}{2}$ designate levels of increasing energy in Fig. V-1.

For low $D \ll g\beta H$, these "perpendicular" energy levels may be expanded to yield [Seidel, Schwoerer, and Schmid (412); Brickmann and Kothe (85)]:

$$W_{\pm 3/2} = \pm \tfrac{3}{2}\, g_{\perp} \beta H - \frac{D}{2} \mp \frac{3D^2}{8 g_{\perp} \beta H} + \cdots$$

$$W_{\pm 1/2} = \pm \tfrac{1}{2}\, g_{\perp} \beta H + \frac{D}{2} \mp \frac{3D^2}{8 g_{\perp} \beta H} + \cdots \quad (V, 6)$$

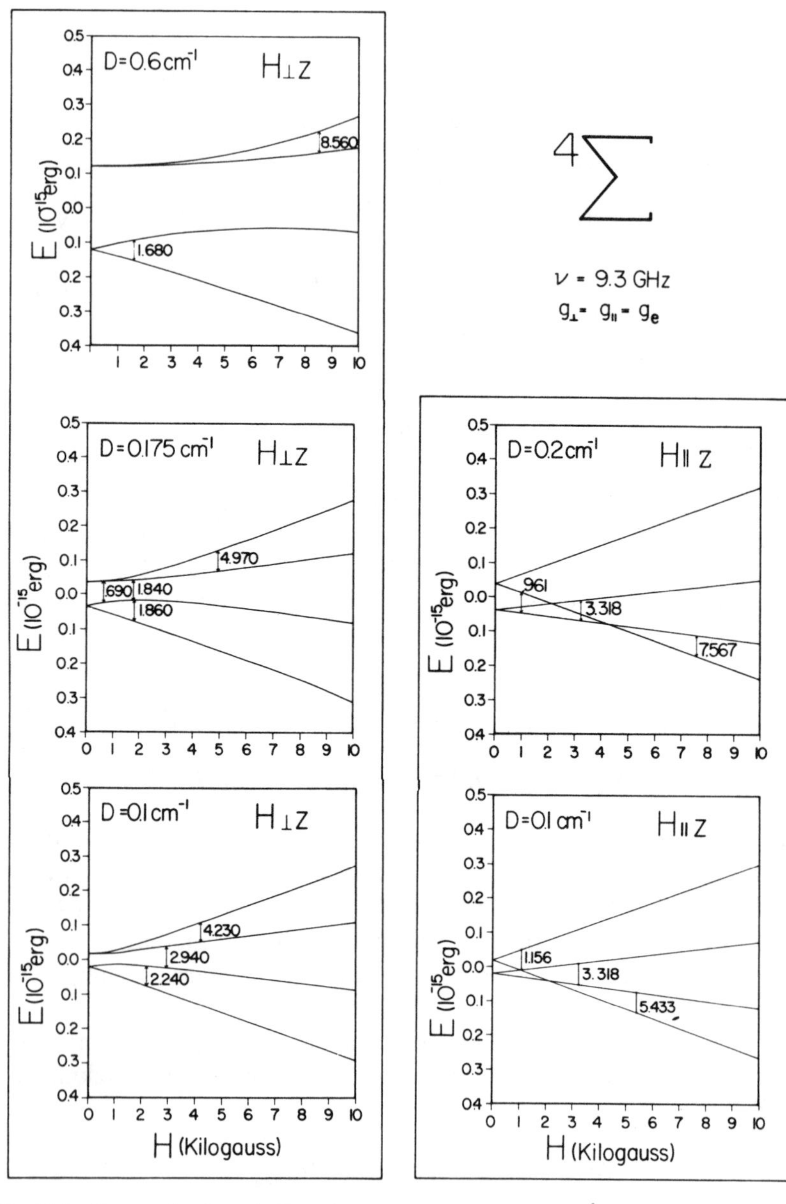

Fig. V-1. Zeeman levels and allowed $\Delta m_S = \pm 1$ transitions of a $^4\Sigma$ molecule for various values of the zero-field splitting parameter D [reprinted with permission from Van Zee, Brown, Zeringue, and Weltner (447); copyright 1980, American Chemical Society]. Left side, $H \perp z$, right side $H \parallel z$, where z is the molecular axis. $\nu = 9.3$ GHz ($= 0.0616 \times 10^{-5}$ erg $= 0.31$ cm^{-1}), $g_\perp = g_\parallel = g_e$. The energy level spacing at zero magnetic field is $2D$.

For the large D limit ($D \gg g\beta H$), Singer (415) has obtained for $H \perp z$:

$$W_{\pm 3/2} = D + \frac{3}{8D}(g_\perp \beta H)^2 \pm \frac{3}{16}\frac{(g_\perp \beta H)^3}{D^2} + \cdots$$

$$W_{\pm 1/2} = -D \pm g_\perp \beta H - \frac{3}{8D}(g_\perp \beta H)^2 \mp \frac{3}{16}\frac{(g_\perp \beta H)^3}{D^2} + \cdots$$

(V, 7)

These equations (V, 7) then apply at low magnetic fields, and at increasingly higher magnetic fields for larger D, for $H \perp z$. (This can be verified visually by observing the behavior of the energy levels in Fig. V-1 near $H = 0$, particularly for $m_S = \pm\frac{3}{2}$.)

Then if the parallel (z) and perpendicular (xy) transitions for linear molecules with D values varying from 0 to 1.0 cm^{-1} are calculated, the graph in Fig. V-2 may be drawn. H here gives the resonant fields, for $\nu = 9.3$ GHz, at which an axial quartet molecule with a given (positive) D value will have $\Delta m_S = \pm$ transitions. The numbering of the parallel and perpendicular transitions in the figure is arbitrarily made from left to right at low D values. (It is then an analogous plot to that given in Fig. III-3 for triplet molecules.) Even for a radical with a very large zero-field splitting, it can be seen that the Kramers' doublet transitions $m_S = +\frac{1}{2} \leftrightarrow -\frac{1}{2}$ (xy_1 and z_2) should be observable.

If D is negative, the energy level schemes (and the m_S designations at high fields) are inverted in Fig. V-2, but of course the transitions occur at the same fields. However, a low Boltzmann factor could prevent observation of even the xy_1 and z_2 transitions for a molecule with a large negative D. Thus, at 4°K, $D \cong -3.5$ cm^{-1} would cause the intensities of these transitions to be lowered by a factor of about 10, relative to those observed if D were positive.

(b) Effective *g* Values. When $|D| \gg g\beta H$ (or in the low-field limit) where the Kramers' doublet separation is much larger than the Zeeman splittings, one can refer to *effective* g values (g^e) for each doublet such that

$$g^e = \frac{h\nu}{\beta H} \tag{V, 8}$$

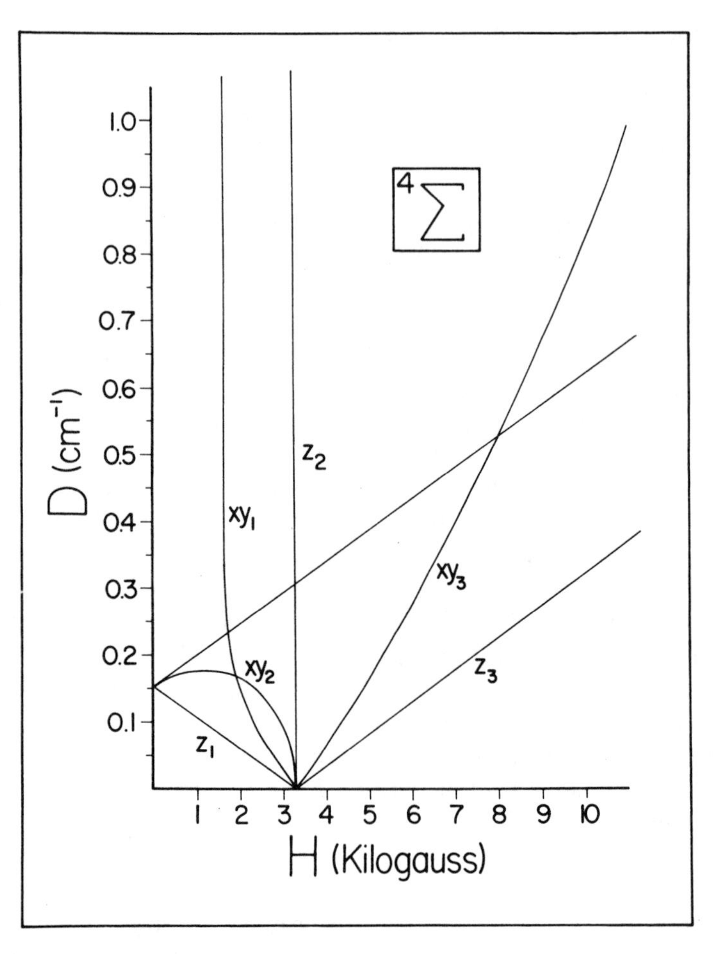

Fig. V-2. Resonant magnetic field positions at which $\Delta m_S = \pm 1$ transitions occur for various values of the zero-field splitting parameter D for a $^4\Sigma$ molecule [reprinted with permission from Van Zee, Brown, Zeringue, and Weltner (447); copyright 1980, American Chemical Society]. It is assumed that $g_\perp = g_\parallel = g_e$ and $\nu = 9.3$ GHz.

Thus for $|\pm\frac{1}{2}\rangle$

$$g^e \beta H = W_{+1/2} - W_{-1/2} \tag{V, 9}$$

and from (V, 5) and (V, 2)

$$g_\perp{}^e \cong 2g_\perp [1 - \tfrac{3}{16} (g_\perp \beta H/D)^2] \tag{V, 10}$$

and $g_\parallel{}^e \cong g_\parallel$ [Geusic, Peter, and Schulz–DeBois (180); Ludwig and Woodbury (317); Kirkpatrick, Muller, and Rubins (272); Pilbrow (373)]. The perpendicular transition xy_1 in Fig. V-2 at high D Then has $g^e \cong 4$ and the parallel $g^e \cong 2$. Correspondingly, for the $|\pm\frac{3}{2}\rangle$ doublet

$$g^e \cong 2g[\tfrac{3}{16}(g_\perp \beta H/D)^2]$$
$$g_\parallel{}^e = 3g_\parallel \qquad \text{(V, 11)}$$

If D were negative, this doublet would lie lowest in energy [as is the case for Cr^{3+} in emerald (180)].

(c) Randomly Oriented Molecules. From Fig. V-2 one sees that at very low D values the parallel (z) and perpendicular (xy) transitions of a linear molecule would be spaced approximately as in Fig. V-3 (top), i.e., z_1 and z_3 would lie at lower and higher fields than xy_1 and xy_3, respectively [Brickmann and Kothe (85)]. If $g_\parallel$ and $g_\perp$ were each in the neighborhood of 2.0, then also the z_2 and xy_2 transitions would lie close to each other. (In Fig. V-3 $g_\parallel$ has been arbitrarily chosen slightly less than $g_\perp$.) Then in analogy with the triplet spectrum shown in Fig. III-4 the ESR spectrum of such quartet molecules, rigidly held but randomly oriented with respect to the magnetic field, would be expected to appear as in Fig. V-3 (bottom). Here the central $m_S = -\frac{1}{2} \leftrightarrow +\frac{1}{2}$ transition, which is almost independent of orientation, will be much stronger than the others, since they will be spread out over a larger range of magnetic field. The exact shape of the curves in Fig. V-3 can, of course, be calculated by finding the eigenvalues and eigenvectors of (V, 3), and $\Delta m_S = \pm 1$ transitions for values of θ other than just $\theta = 0$ and $90°$.

Brickmann and Kothe (85) have also derived the positions of the "half-field" $\Delta m_S = \pm 2$ transitions and their line shapes for randomly oriented linear molecules. These transitions would occur in Fig. V-3 at

$$H(xy) = \frac{H_0}{2} \pm \frac{D}{2}$$

$$H(z) = \frac{H_0}{2} \pm D$$

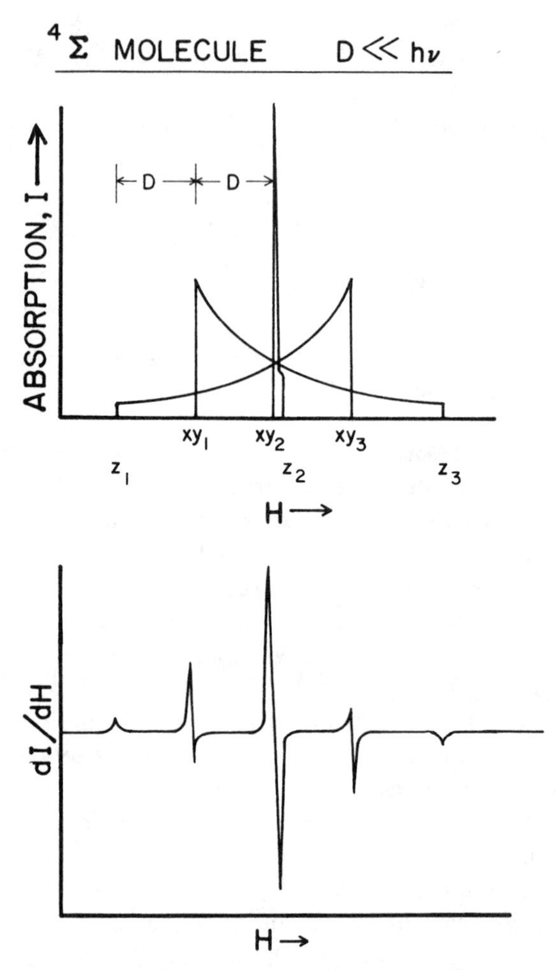

Fig. V-3. ESR absorption and spectrum of randomly oriented $^4\Sigma$ molecules for $D << h\nu$ [reprinted with permission from Van Zee, Brown, Zeringue, and Weltner (447); copyright 1980, American Chemical Society]. (Top) Theoretical absorption indicating variation from parallel (z_i) to perpendicular (xy_i) orientation of the magnetic field relative to the molecular axis for the three $\Delta m_S = \pm 1$ transitions (see Fig. V-1); (bottom) ESR spectrum (first derivative of the total absorption curve).

where H_0 is the position of the central strong line. Note that such transitions are between Kramers' doublets, $m_S = -\frac{3}{2} \leftrightarrow +\frac{1}{2}$ and $m_S = -\frac{1}{2} \leftrightarrow +\frac{3}{2}$, and can be expected to be weaker by a factor of about $(D/g\beta H)^2$ than the $\Delta m_S = \pm 1$ transitions. Hence they do not have the relative intensity that occurs in triplet molecules since there is not the "piling up" of intensity at a particular field (H_{min}) as in the case of the $m_S = +1 \leftrightarrow -1$ transition. $\Delta m_S = \pm 3$ transitions have also been considered and observed, but their intensities are only $(D/g\beta H)^4$ relative to the $\Delta m_S = \pm 1$ [see Weissman and Kothe (472)].

For D values greater than $h\nu$ only two xy transitions can be observed in the X-band spectral region (see Fig. V-2), and the intensity of the xy line at $g^e \cong 4$ will dominate. For $D > \sim +1$ cm^{-1}, the only observable line can be expected to be this Kramers' doublet $m_S = -\frac{1}{2} \leftrightarrow +\frac{1}{2}$ line. Its position for large zero-field splittings is very insensitive to D; hence the zfs could be determined only approximately from its observed position, even if $g_\perp$ were accurately known. The parallel lines are often too weak to observe (see Fig. V-3), even z_2, since it lies near $g = 2$ where other radicals often have spectral lines.

Departure from low D values leads to a corresponding departure from the symmetrical absorbance and ESR spectrum depicted in Fig. V-3, as previously illustrated in Figs. III-4 and III-7 for the $^3\Sigma$ case. Keeping other parameters fixed and increasing D to 0.08 cm^{-1} yields the computed results in Fig. V-4 for a linear $^4\Sigma$ molecule. Because of the direct dependence of the absorption intensity on $d\theta/dH$ (see p. 90) for randomly oriented molecules one sees that the middle line (xy_2, z_2 line in Fig. V-3) has an additional strong absorption branch besides the parallel ($\theta = 0°$) and perpendicular ($\theta = 90°$) features. $d\theta/dH$ also becomes infinite at $\theta \simeq 45°$, causing this "extra" line to appear at about 3.6 kG. Figure V-5 shows the same graphs for $D = 0.40$ cm^{-1} with all other parameters remaining the same. Here the dominant xy_1 line and the other xy_3 line show up clearly (see Fig. V-2), but there is an off-principal-axis transition at about 6.4 kG. It is then apparent that computer simulations will generally be necessary to identify all of the observed lines in such spectra.

(d) Hyperfine Interation. The hyperfine coupling $\boldsymbol{I} \cdot \boldsymbol{A} \cdot \boldsymbol{S}$, for a linear molecule adds the usual terms

$$A_\parallel I_z S_z + A_\perp (I_x S_x + I_y S_y) \tag{V, 12}$$

to the spin Hamiltonian in (V, 1). If D and A are both small relative to $g\beta H$, then to first order in D and second order in A [Low (15)],

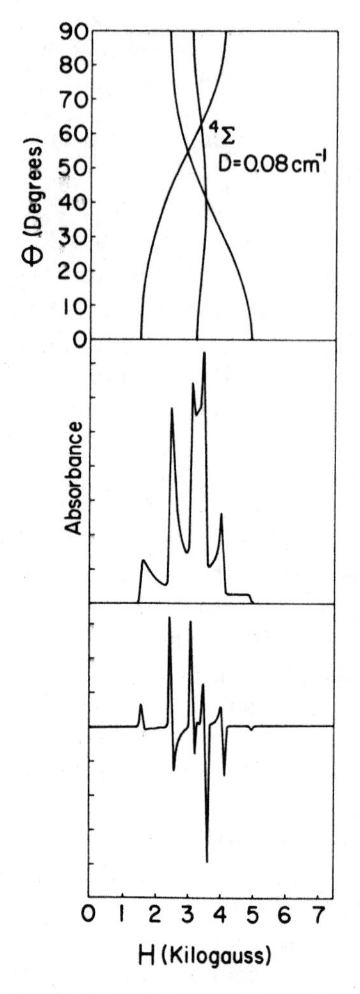

Fig. V-4. Computed variation of allowed ($\Delta m_S = \pm 1$) transitions with orientation (top), absorption spectrum (middle), and ESR spectrum (bottom) of randomly oriented $^4\Sigma$ molecules with $D = +0.08$ cm^{-1}. ($g_\perp = g_\parallel = g_e$, $\nu = 9.3$ GHz.)

for $H_0 \parallel z$:

$$H = H_0 - 2D\left(m_S - \frac{1}{2}\right) - A_\parallel m_I - \frac{(A_\perp{}^2)}{2H_0} \cdot [I(I+1) - m_I{}^2 - m_I(2m_S - 1)] \qquad \text{(V, 13)}$$

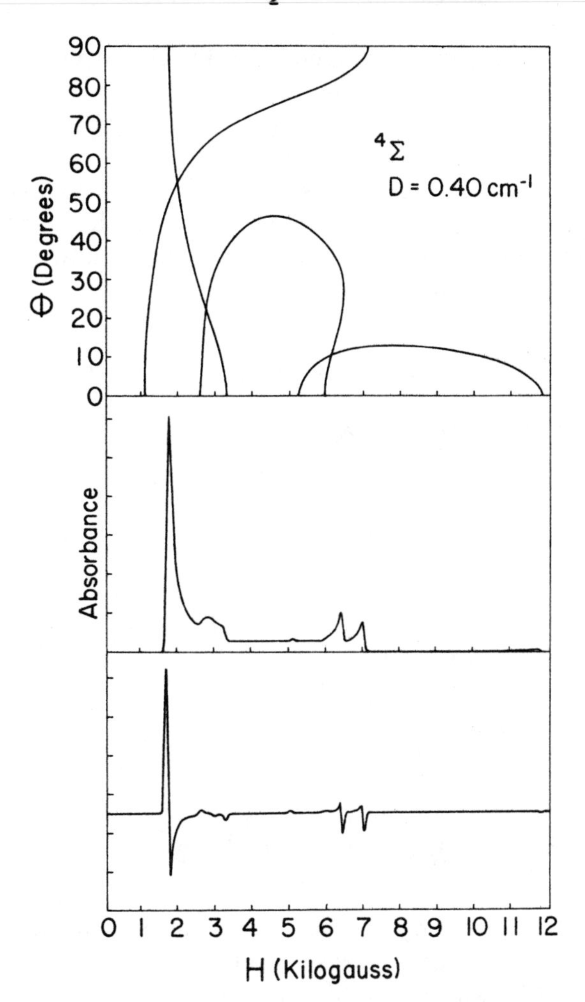

Fig. V-5. Computed variation of allowed ($\Delta m_S = \pm 1$) transitions with orientation (top), absorption spectrum (middle), and ESR spectrum (bottom) of randomly oriented $^4\Sigma$ molecules with $D = +0.40\ \text{cm}^{-1}$. ($g_\perp = g_\parallel = g_e$, $\nu = 9.3$ GHz.)

and for $H_0 \perp z$:

$$H = H_0 + D\left(m_S - \frac{1}{2}\right) - A_\perp m_I - \frac{(A_\parallel{}^2 + A_\perp{}^2)}{4H_0}\,[I(I+1) - m_I{}^2]$$

$$- \frac{(A_\perp A_\parallel)}{2H_0}\,[m_I(2m_S - 1)] \qquad \text{(V, 14)}$$

Thus, in this limit, for each orientation the fine structure produces $2S$ equidistant lines separated by $2D$ (parallel) and D (perpendicular) as in Fig. V-3. Each of these lines is then split by hyperfine interaction, $\sim A_\parallel$ for parallel and $\sim A_\perp$ for perpendicular.

In the special case where D is positive and so large that only the transition between $m_S = \pm\frac{1}{2}$ levels need be considered, an effective spin Hamiltonian (with $S' = \frac{1}{2}$) can be written [Ludwig and Woodbury (317); Kasai (258)]:

$$\mathcal{H}' = \beta[g_\parallel^e S_z' H_z + g_\perp^e (S_x' H_x + S_y' H_y)] + A_\parallel' S_z' I_{iz} + A_\perp'(S_x' I_x + S_y' I_y) \quad \text{(V, 15)}$$

where

$$g_\parallel^e = g_\parallel$$

$$g_\perp^e = g_\perp \left(S + \frac{1}{2}\right)\left[1 + \frac{\left(S - \frac{1}{2}\right)\left(S + \frac{3}{2}\right)}{(2S+1)^2}\left(\frac{h\nu}{2D}\right)^2\right] \quad \text{(V, 16)}$$

$$A_\perp' = \left(S + \frac{1}{2}\right) A_\perp$$

$$A_\parallel' = A_\parallel - \left(\frac{A_\perp^2}{4D}\right)\left(S - \frac{1}{2}\right)\left(S + \frac{3}{2}\right) \quad \text{(V, 17)}$$

and S, $g_\parallel$, $g_\perp$, $A_\parallel$, and $A_\perp$ are the "true" values. Ludwig and Woodbury (317) also include nuclear quadrupole and nuclear Zeeman terms in (V, 15), which have been neglected here. The relationships (V, 15–17) are given here for the general case where $2S$ is odd, since they are also applicable to molecules where $S = \frac{5}{2}$ and $\frac{7}{2}$, which will be discussed in later sections of this chapter.

The specific application to the $S = \frac{3}{2}$ VO molecule [Kasai (258)] is considered below. In that case Kasai assumed D was large enough that the terms involving D in (V, 16) and (V, 17) were negligible, so that $g_\perp^e = 2g_\perp$ and $A_\parallel' = A_\parallel$. The spin Hamiltonian in (V, 15) can then be transformed by rotation to diagonalize the Zeeman term [Bleaney (70)] and yields:

$$
\begin{aligned}
\mathcal{H} = g\beta H S_z &+ A I_z S_x + [(4A_\perp{}^2 - A_\parallel{}^2)/A](2g_\parallel g_\perp/g^2) \sin\theta \cos\theta \, I_z S_x \\
&+ (\tfrac{1}{2} A_\perp)[(A_\parallel + A)/A](I_+S_- + I_-S_+) \\
&+ (\tfrac{1}{2} A_\perp)[(A_\parallel - A)/A](I_+S_+ + I_-S_-) \qquad \text{(V, 18)}
\end{aligned}
$$

where

$$
\begin{aligned}
g^2 &= g_\parallel{}^2 \cos^2\theta + 4g_\perp{}^2 \sin^2\theta, \\
A^2 &= (A_\parallel{}^2 g_\parallel{}^2/g^2)\cos^2\theta + (16A_\perp{}^2 g_\perp{}^2/g^2)\sin^2\theta
\end{aligned}
$$

θ is the angle between the z axis and the applied magnetic field. This equation can be solved analytically for $\theta = 0$ but not for other values of θ. Kasai (258) has used a continued-fraction method for its solution for $\theta = 90°$, since the hyperfine splittings are rather large compared to the magnetic fields at which the fine structure appears.

Having obtained $A_\parallel$ and $A_\perp$ from the analyzed spectra by the above methods, one can then calculate A_{iso} and A_{dip} according to (II, 34) and infer the s and p, d character of the wavefunctions of the three unpaired electrons. For example, for VO (see below) the three electrons are essentially in a $4s\sigma^1 3d\delta^2$ configuration, and crudely it can be assumed that all s character is due to the one electron in the σ orbital and all anisotropy arises from the two electrons in the δ orbitals. Then [Kasai (258)]:

$$
\begin{aligned}
A_{\text{iso}} &\simeq \tfrac{1}{3}\,(A_{\text{iso}}, \text{atomic})_{4s} \\
A_{\text{dip}} &\simeq \tfrac{2}{3}\,(A_{\text{dip}}, \text{atomic})_{3d\delta}
\end{aligned}
$$

where the atomic values are perhaps best approximated as those for the V^{2+} ion. [Thus, for three *equivalent* electrons this partitioning is not needed, as for example for the three interacting F centers in a crystal, as mentioned below.]

B. Individual Molecules

VO and NbO. Kasai (258) made the first investigation of a $^4\Sigma$ molecule with the study of the VO molecule trapped in solid argon at 4°K. The molecule has a very high D value, now established to be +4.11

cm^{-1} [Veseth (453); Cheung, Hansen, and Merer (108)], so that its X-band spectrum involves only a powder pattern extending from xy_1 (see Fig. V-2) at about 1500 G to z_2 at about 3300 G; that is, only the transition within the lowest Kramers' doublet ($m_S = +\frac{1}{2} \leftrightarrow -\frac{1}{2}$)

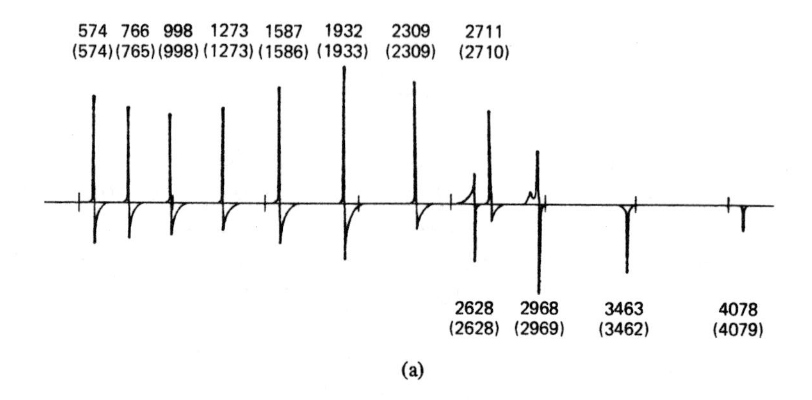

(a)

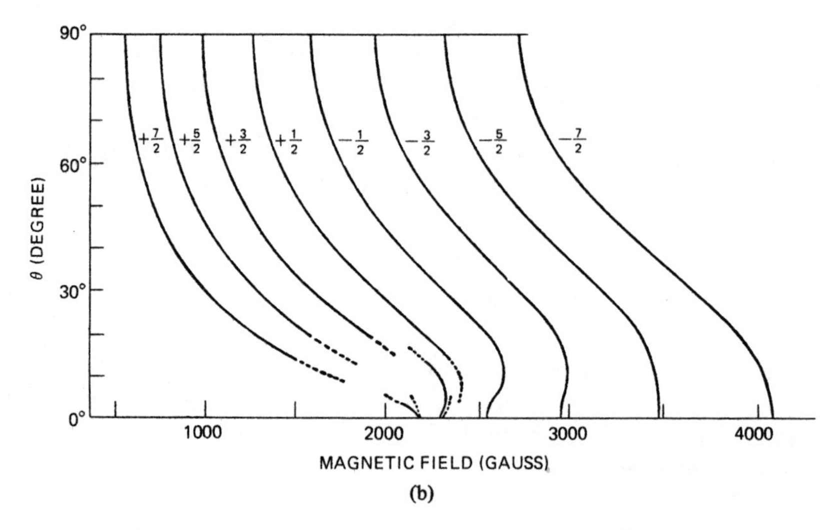

(b)

Fig. V-6. ESR spectrum of the VO molecule in an argon matrix at 4°K [from Kasai (258)]. (a) Observed and calculated (in parentheses) positions of the ⊥, the ∥, and the off-principal-axis transitions. (b) Resonance field versus θ; each line is identified with a particular m_I value. (c) ESR spectrum in the 0–3000 G range with H parallel to the plane of the sapphire rod; the arrows indicate the eight ⊥ signals. (d) ESR spectrum in 2000–5000 G range with H perpendicular to the plane of the sapphire rod (gain × 2); open arrows indicate "extra" lines. (ν = 9.4326 GHz.)

could be observed. However, ^{51}V is ~100% natural abundance with a nuclear spin of $\frac{7}{2}$ so that this single line pattern is split into eight hyperfine components. The observed spectrum is shown in Fig. V-6, c and d, and its interpretation is given in Fig. V-6, a and b. Two

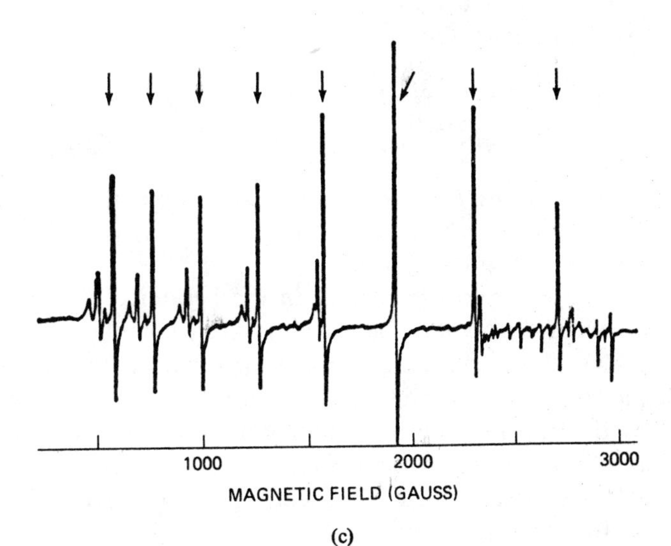

(c)

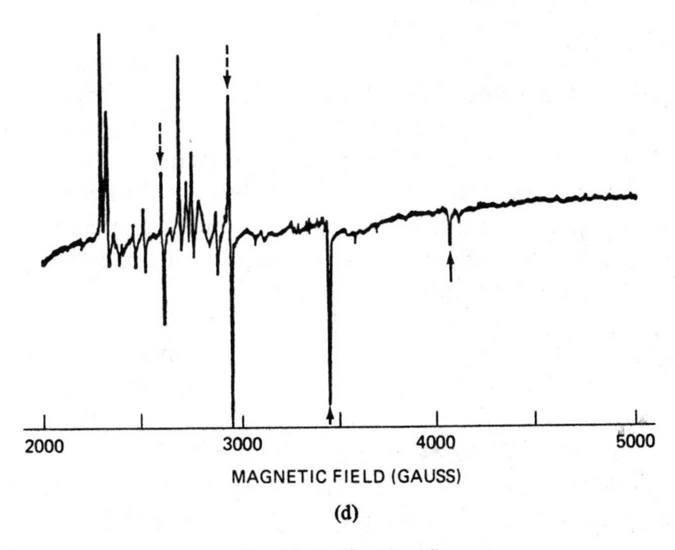

(d)

Fig. V-6. *(Continued)*

parallel hf components are observed along with all of the eight strong perpendicular hf components. Also appearing are two hyperfine "extra" lines due to inflections in the $m_I = -\frac{1}{2}$ and $\frac{3}{2}$ curves near $\theta \simeq 15°$ as shown in Fig. V-6b. These off-principal-axis transitions appear as intensely as the two parallel features. Applying (V, 15), Kasai derived the parameters:

$$g_{\parallel} = 2.0023(10), \qquad g_{\perp} = 1.9804(10)$$

$$A_{\parallel}(^{51}\mathrm{V}) = 714.7(5), \quad A_{\perp}(^{51}\mathrm{V}) = 837.1(5)\ \mathrm{MHz}$$

and $D \gg g\beta H$. [This is a very good assumption, since from the known $D = 4.11\ \mathrm{cm}^{-1}$ the second term in (V, 16) is only 0.0011 cm^{-1}.] The effective $g^e(\pm\frac{1}{2})$ is 3.963.

Niobium is just below vanadium in the Periodic Table, and the spectrum of NbO is found to be a slightly more complicated version of that of VO. Although undetermined, one expects the zero-field-splitting parameter, D, for NbO to be even larger than that for VO, because of the larger spin-orbit coupling of Nb. The molecule was trapped by Brom, Durham, and Weltner (87) in neon and argon matrices at 4°K and yielded the spectrum shown in Fig. V-7a in neon. The four major lines are indicated there and can be correlated with the derived θ versus H plot given in Fig. V-7b. It is evident that the ^{93}Nb (100% natural abundance, $I = \frac{9}{2}$) hyperfine splitting is too large to allow observation of all of the $\Delta m_I = 0$ transitions when $\nu \simeq 9.3$ GHz, but some weaker forbidden $\Delta m_I = \pm 1$ transitions are also observed. Matrix site effects, $\Delta m_I = \pm 2$ transitions, and evidence of preferential molecular orientation, lead to other complications in the spectra. However, the four lines indicated in Fig. V-7 could be fit satisfactorily, considering that nuclear quadrupole and Zeeman effects were neglected. A comparison of the derived parameters for VO and NbO is given in Table V-1.

VF_2 and MnO_2. These isoelectronic molecules have $S = \frac{3}{2}$, are symmetrical, and are probably linear. The observed spectrum of each consists of just one fine-structure line at $g^e \simeq 4$, indicating that D is greater than or equal to $h\nu$ (see Fig. V-2) [DeVore, Van Zee, and Weltner (141)]. No other lines xy_3 or z_2 were detected, perhaps because of overlap of z_2 with impurity lines at $g \simeq 2$, weakness of the xy_3 line at higher field, or possibly a high enough D value to place xy_3 above 12,000 G.

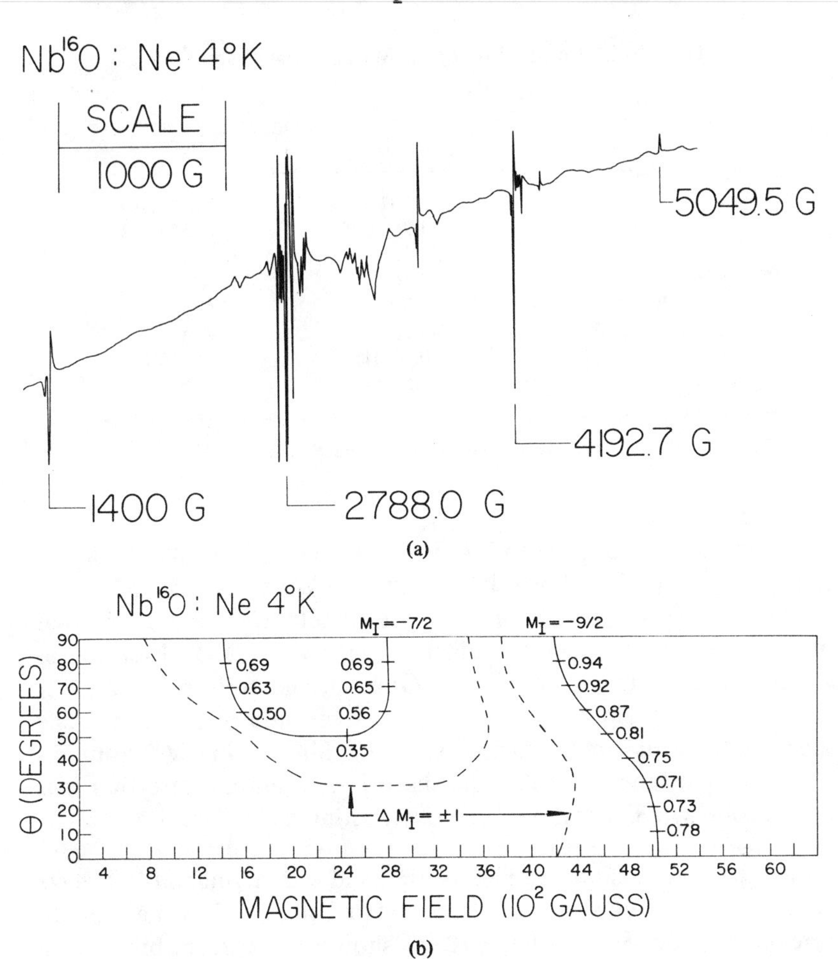

Fig. V-7. ESR spectrum of the ^{93}NbO molecule in a neon matrix at 4°K [from Brom, Durham, and Weltner (87)]. (a) Observed spectrum; $\nu = 9.4050$ GHz. (b) Calculated θ versus H (resonant magnetic field) variation from magnetic parameters in Table V-1. Numbers on $\Delta m_S = 0$ transitions are relative transition probabilities.

The observed spectrum for $^{51}V^{19}F_2$ trapped in solid argon at 4°K is shown in Fig. V-8; note that the hyperfine pattern of an octet of triplets identifies it as such. The triplet pattern is shown in an expanded view lower in the figure and appears to exhibit a clear 1:2:1 intensity pattern due to two equivalent ^{19}F nuclei. The lack of splitting (into x and y) or distortion of these triplets indicates that the molecule is probably linear or nearly so. $g_\perp{}^e$ for this line is found

Table V-1. Comparison of the Magnetic Parameters[a] of VO and NbO

	VO	NbO
$g_{\parallel}$	2.0023(10)	2.0023(10)
$g_{\perp}$	1.9804(10)	1.9577(10)
$A_{\parallel}$ (MHz)	714.7(5)	1587(1)
$A_{\perp}$ (MHz)	837.1(5)	1647(1)
A_{iso} (MHz)	796.3(5)	1627(1)
A_{dip} (MHz)	−40.8(3)	−24(1)
$\rho(0)$ (a.u.)	0.541(1)	1.489(5)
$\langle(3\cos^2\theta - 1)/r^3\rangle$ (a.u.)	−0.582(1)	−0.368(5)

[a] All hyperfine splitting parameters are for interaction with the metal atom nucleus.

to be 3.922 as compared to 3.963 for VO, which suggests, according to Fig. V-2, that $|D|$ is lower in VF_2 than in VO but ≥ 0.5 cm^{-1}.

MnO_2 yields a corresponding spectrum to that of VF_2 in that only the lowest Kramers' doublet transition (xy_1) is observed for $\nu \simeq 9.3$ GHz [Ferrante, Wilkerson, Graham, and Weltner (162)]. It is split into six lines by ^{55}Mn ($I = \frac{5}{2}$) hyperfine interaction. The spectrum appears similar to that given for MnO in Fig. V-29 but lies at a higher magnetic field. The sharp lines again indicate that it is most probably a linear molecule. $|D|$ is estimated to be >1.4 cm^{-1}.

Organic Radicals. Schmauss, Baumgärtel, and Zimmermann (404) have synthesized the hydrocarbon radical shown below, and Kothe, Ohmes, Brickmann, and Zimmermann (293) showed by susceptibility mea-

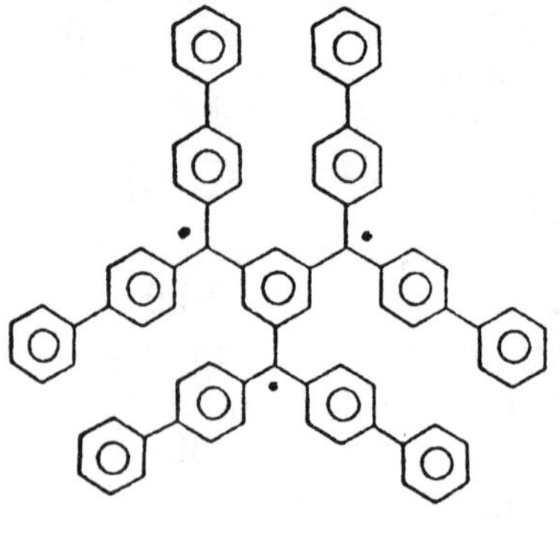

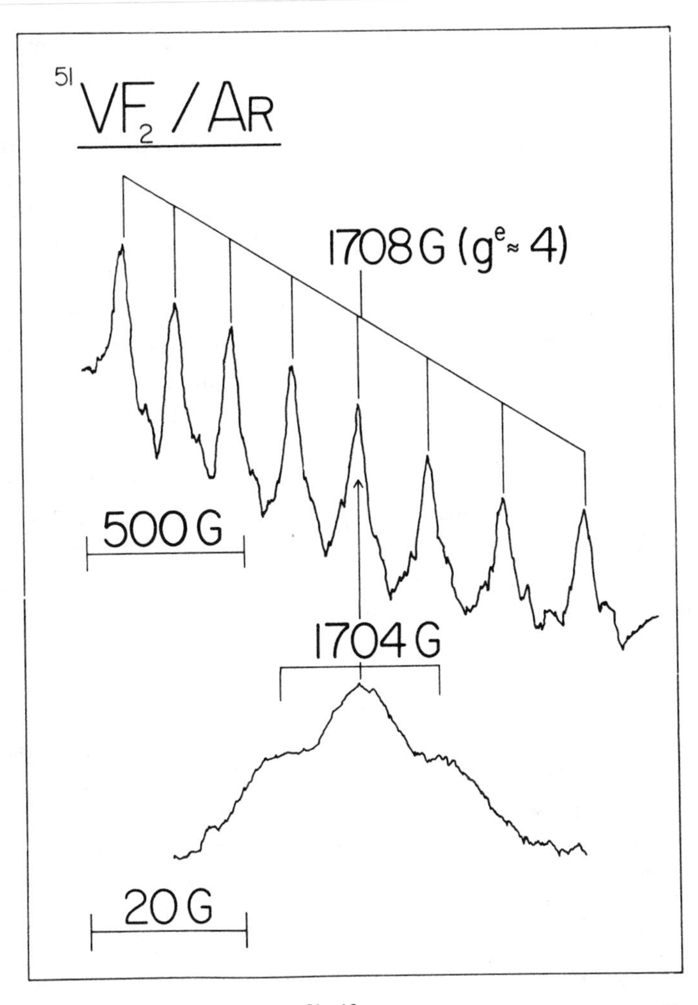

Fig. V-8. X-band ESR spectrum of the $^{51}V^{19}F_2$ molecule in an argon matrix at 4°K [from DeVore, Van Zee, and Weltner (141). (This figure was originally presented at the Fall 1977 Meeting of The Electrochemical Society, Inc. held in Atlanta, Georgia.)].

surements that it had a quartet ground state. ESR studies by Brickmann and Kothe (85) of the radical in a frozen toluene glass yielded a spectrum similar to that depicted in Fig. V-3 with the intense central line at 3274 G and a weak $\Delta m_S = 2$ transition at 1637 G. $|D|$ was found to be 44.1 G = 0.0041 cm^{-1} with an isotropic $\boldsymbol{g}$ tensor.

Impurity Pairs in Silicon Crystals. Ludwig, Woodbury, and Ham (318) made an interesting study of transition-metal donor plus acceptor pairs in silicon at room temperature. Cr, Mn, and Fe are typical donors and B, Al, Zn, Cu, Pt acceptors, and not only $S = \frac{3}{2}$ pairs are formed but also $S = \frac{5}{2}$ and $\frac{1}{2}$. The transition metal ion occupies an interstitial site neighboring to a negatively charged acceptor ion in a substitutional site. Generally the axis of the pair (z axis) is the [111] crystalline direction.

Cr, Mn, and Fe ions in silicon exhibit full cubic symmetry, and pair formation introduces noncubic terms in the potential, mainly through a DS_z^2 term, which may be dominant. In the latter case, the Hamiltonian is then that given in (V, 15) but with nuclear quadrupole and Zeeman terms included, if necessary (317).

For example, for $(MnPt)^0$ (an uncharged pair) $S = \frac{3}{2}$ and $D >> h\nu$ so that only the resonance between $m_S = \pm\frac{1}{2}$ states is observed. The ESR spectrum with the magnetic field parallel to the pair axis is shown in Fig. V-9. The sextet due to ^{55}Mn ($I = \frac{5}{2}$) hyperfine interaction is further split by interaction with the 34% natural abundance of ^{195}Pt ($I = \frac{1}{2}$). Other $S = \frac{3}{2}$ pairs observed in this way are $(^{57}Fe^{11}B)^0$, $(^{53}Cr^{197}Au)^0$, and $(^{55}Mn^{197}Au)^+$ (317).

In the silicon matrix the two atoms are considered to be in adjacent substitutional and interstitial sites and to form an ion pair by virtue of their electronegativity difference. The crystal-field effects of the silicon determine their lowest states and coupling between the two impurity atoms may then determine the ground state of the pair (317). The simplest case is $(MnB)^+$ containing Mn^{2+} in an interstitial site of tetrahedral symmetry for which $S = J = \frac{5}{2}$, $L = 0$, according to the well-accepted theory of Ludwig and Woodbury (317).

The ground states of the corresponding neutral diatomic molecules trapped in rare-gas matrices are expected to be different from those formed in silicon. This is exemplified by the CrAu molecule, which in solid argon has an $S = \frac{5}{2}$, $^6\Sigma$ ground state [Baumann, Van Zee, and Weltner (51a)], whereas in silicon this neutral pair has $S = \frac{3}{2}$. However, these two species appear to have comparable values of D, since the molecule has been shown to have $|D| \geqslant 2$ cm^{-1}, while the "impurity pair" in silicon has $D = +6.7$ cm^{-1}.

Similar $S = \frac{1}{2}$ metal diatomics such as AgM, where M = Mg, Ca, Zn, Hg, etc., were observed by Kasai and McLeod (263) in rare-gas matrices.

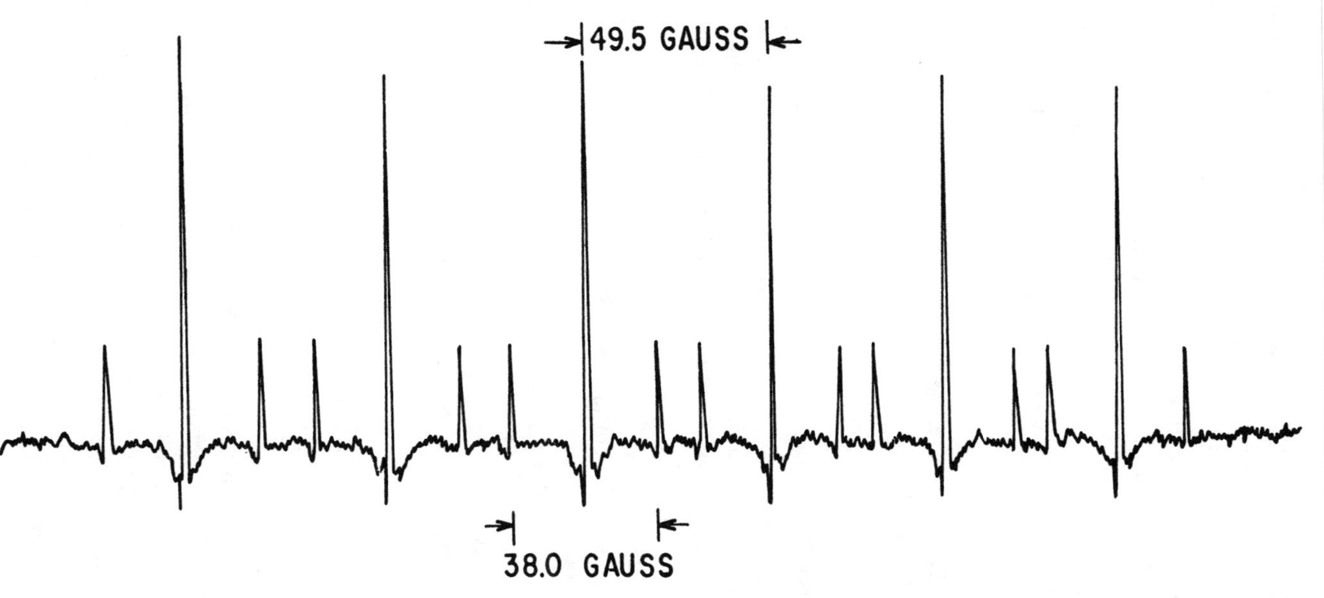

Fig. V-9. X-band ESR spectrum of a $^{55}Mn^{195}Pt$ impurity pair in crystalline silicon, with H parallel to the pair axis [from Ludwig and Woodbury (317)].

R-Center in KCl Crystals. A classic axial $S = \frac{3}{2}$ case occurs in the so-called R-center in single KCl crystals [Seidel, Schwoerer, and Schmid (412); van Doorn and Haven (438); Pick (371)]. It is an association of three F centers forming an equilateral triangle in the (111) plane of the crystal. Thus, its ESR spectrum can be analyzed according to the spin Hamiltonian in (V, 1) with $E \equiv 0$ and with the z axis perpendicular to the [111] direction (of course, there are four such axes in the crystal). These authors then find that g is isotropic and equal to 1.996 and $D = +168.5$ G $= +0.0157$ cm^{-1}, a small value.

2. $^5\Sigma$ AND OTHER QUINTET ($S = 2$) MOLECULES IN MAGNETIC FIELDS

(a) Theory:

(a) General Spin Hamiltonian. For $S \geqslant 2$, theory dictates that the zero-field-splitting parameters are not limited to just D and E, as in the spin Hamiltonian (III, 5) and (V, 1). [Abragham and Bleaney (1), p. 140; Hutchings (236)]. Other terms involving higher powers of S_z, S_x, and S_y may be necessary, depending upon the symmetry of the potential energy function. The Hamiltonian may then be written [Stevens (424); Hutchings (236)]:

$$\mathcal{H} = \beta \boldsymbol{H} \cdot \boldsymbol{g} \cdot \boldsymbol{S} + \sum_{m,n} B_n{}^m O_n{}^m \qquad \text{(V, 19)}$$

where $m \leqslant n$ and the first two coefficients are, as before: $B_2{}^0 \equiv D/3$ and $B_2{}^2 \equiv E$, and

$$\begin{aligned} O_2{}^0 &= 3S_z{}^2 - S(S+1) \\ O_2{}^2 &= \tfrac{1}{2}\,[S_+{}^2 + S_-{}^2] \end{aligned} \qquad \text{(V, 20)}$$

Thus the $O_n{}^m$ are spin operators, and the number of them to be included in (V, 19) is dependent upon the symmetry and the value of S. The lower the symmetry the more terms that should be included; however, this is not to say that they will all be necessary (or detectable) for an adequate description of the potential function. Usually the coefficients decrease rapidly in magnitude as n increases.

In general $O_n{}^m$ will be zero for $n > 2S$ [Hutchings (236)] so that for $S = 2$ only terms up to $B_4{}^m$ need be included. Which of these may be nonvanishing is determined by the symmetry [Griffith (10);

Prather (383)]; thus, for $S = 2$, the relevant coefficients for some point groups are:

$D_{\infty h}, C_{\infty v}$	$B_2{}^0, B_4{}^0$
D_2, D_{2h}, C_{2v}	$B_2{}^0, B_2{}^2, B_4{}^0, B_4{}^2, B_4{}^4$
C_{3h}, D_{3h}	$B_2{}^0, B_4{}^0$
C_3, C_{3v}	$B_2{}^0, B_4{}^0, B_4{}^3$
O_h, T_d	$B_4{}^0, B_4{}^4$

where

$$O_4{}^0 = [35S_z{}^4 - 30S(S+1)S_z{}^2 + 25S_z{}^2 - 6S(S+1) + 3S^2(S+1)^2]$$

$$O_4{}^2 = \tfrac{1}{4}\,[(7S_z{}^2 - S(S+1) - 5)\,(S_+^2 + S_-^2) + (S_+^2 + S_-^2)\,(7S_z{}^2 - S(S+1) - 5)]$$

etc. [see Hutchings (236)].

Other symbols are also used for the coefficients, a prevalent one being $b_n{}^m$:

$$b_2{}^0 = 3B_2{}^0 = D \qquad b_2{}^2 = 3B_2{}^2 = 3E$$
$$b_4{}^0 = 60B_4{}^0 = \tfrac{1}{3}F \qquad b_4{}^m = 60B_4{}^m$$
$$b_6{}^m = 1260B_6{}^m$$

(b) Eigenvalues and Magnetic-Dipole Transitions. With the spin basis functions:

$$|+2\rangle = |\alpha_1\alpha_2\alpha_3\alpha_4\rangle$$

$$|+1\rangle = \frac{1}{2}\,|\beta_1\alpha_2\alpha_3\alpha_4 + \alpha_1\beta_2\alpha_3\alpha_4 + \alpha_1\alpha_2\beta_3\alpha_4 + \alpha_1\alpha_2\alpha_3\beta_4\rangle$$

$$|-0\rangle = \frac{1}{\sqrt{6}}\,|\beta_1\beta_2\alpha_3\alpha_4 + \beta_1\alpha_2\beta_3\alpha_4 + \beta_1\alpha_2\alpha_3\beta_4 + \alpha_1\beta_2\beta_3\alpha_4 + \alpha_1\beta_2\alpha_3\beta_4 + \alpha_1\alpha_2\beta_3\beta_4\rangle$$

$$|-1\rangle = \frac{1}{2}\,|\beta_1\beta_2\beta_3\alpha_4 + \beta_1\beta_2\alpha_3\beta_4 + \beta_1\alpha_2\beta_3\beta_4 + \alpha_1\beta_2\beta_3\beta_4\rangle$$

$$|-2\rangle = |\beta_1\beta_2\beta_3\beta_4\rangle \qquad \text{(V, 21)}$$

the matrix of (V, 19) with only D and E included is [Máckowiak and Kurzyński (322); Schwoerer, Huber, and Hartl (409)]:

	$\lvert -2\rangle$	$\lvert -1\rangle$	$\lvert 0\rangle$	$\lvert +1\rangle$	$\lvert +2\rangle$
$\lvert -2\rangle$	$2D - 2G_z$	$G_x + iG_y$	$\sqrt{6}\,E$	0	0
$\lvert -1\rangle$	$G_x - iG_y$	$-D - G_z$	$\sqrt{\frac{6}{2}}\,(G_x + iG_y)$	$3E$	0
$\lvert 0\rangle$	$\sqrt{6}\,E$	$\sqrt{\frac{6}{2}}\,(G_x - iG_y)$	$-2D$	$\sqrt{\frac{6}{2}}\,(G_x + iG_y)$	$\sqrt{6}\,E$
$\lvert +1\rangle$	0	$3E$	$\sqrt{\frac{6}{2}}\,(G_x - iG_y)$	$-D + G_z$	$G_x + iG_y$
$\lvert +2\rangle$	0	0	$\sqrt{6}\,E$	$G_x - iG_y$	$2D + 2G_z$

(V, 22)

Here, as earlier, $G_x = g_x \beta H_x$, $G_y = g_y \beta H_y$, and $G_z = g_z \beta H_z$, where x, y, z are the principal axes of $\mathbf{g}$. Baranowski, Cukierda, Jezowska-Trzebiatowska, and Kozlowski (46) rearrange this matrix and obtain solutions for its eigenvalues as the roots of a quadratic and a cubic equation. These are then used to directly deduce the resonant magnetic fields at which transitions can occur between the five levels for various values of D, assuming $E \neq 0$. At zero field the energy levels are given by

$$\begin{aligned} W_5 &= 2D(1+\alpha)^{1/2} \\ W_4 &= 2D \\ W_3 &= -D + 3E \\ W_2 &= -D - 3E \\ W_1 &= -2D(1+\alpha)^{1/2}, \quad \text{where } \alpha = 3E^2/D^2 \end{aligned} \qquad \text{(V, 23)}$$

As in the triplet case, in the limit of low magnetic fields the eigenfunctions are the linear combinations of the above spin functions that will diagonalize the matrix [see (III, 20)]:

$$|T_5\rangle = \frac{1}{2}(1+\beta)^{1/2}[\,|+2\rangle + |-2\rangle] + \frac{1}{\sqrt{2}}(1-\beta)^{1/2}|0\rangle$$

$$|T_4\rangle = \frac{1}{\sqrt{2}}[\,|+2\rangle - |-2\rangle]$$

$$|T_3\rangle = \frac{1}{\sqrt{2}}[\,|+1\rangle + |-1\rangle]$$

$$|T_2\rangle = \frac{1}{\sqrt{2}} [\,|+1\rangle - |-1\rangle]$$

$$|T_1\rangle = \frac{1}{2}(1 - \beta)^{1/2} [\,|+2\rangle + |-2\rangle] - \frac{1}{\sqrt{2}}(1 + \beta)^{1/2}|0\rangle \qquad \text{(V, 24)}$$

where $\beta = 1/[1 + (3E^2/D^2)]^{1/2}$ [Schwoerer et al. (409)].

For linear quintet molecules where $E \equiv 0$ (and assuming the $B_4{}^m$ terms are negligible) the energy levels at zero field from (V, 22) become $+2D$, $-D$, and $-2D$. The zero-field splittings are then $W_{\pm 1} - W_0 = |D|$ and $W_{\pm 2} - W_{\pm 1} = |3D|$. Thus for positive D and $H \parallel z$ ($\theta = 0°$), where z is the axis of the molecule, the Zeeman patterns would be as shown on the right side of Fig. V-10 [Dowsing (149)]. When the molecule is turned in the magnetic field, the levels are mixed, as discussed in section III-1b for the triplet case, and in the extreme case when $H \perp z$ ($\theta = 90°$), typical Zeeman patterns are shown on the left side of that figure. The magnetic fields at which $\Delta m_S = \pm 1$ transitions occur for an X-band frequency of 9.3 GHz are also indicated for both parallel (z) and perpendicular (xy) orientations. Then Fig. V-11, analogous to Figs. III-3 and V-2, gives a plot of these resonant field positions for linear molecules with D values ranging up to 1.0 cm^{-1} [Dowsing (149); Van Zee et al. (447); Baranowski et al. (46)]. One notes that as for $S = 1$ in Fig. III-3 the xy_2 transition becomes dominant for molecules with large D values (see the upper left-hand frame in Fig. V-10). Even more so than in the triplet case, this transition should be observable via X-band even if D has a value much greater than 1 cm^{-1}. (It is, however, also necessary that the temperature be high enough to populate the lower level of the transition, since it will be of the order of D above the lowest spin level.)

For very low D values ($D \ll g\beta H$), or high magnetic fields, Máckowiak and Kurzyński (322) obtain from first-order perturbation theory:

$$W_{\pm 2} = \pm 7g\beta H + u$$

$$W_{\pm 1} = \pm g\beta H - \tfrac{1}{2}u$$

$$W_0 = -u \qquad \text{(V, 25)}$$

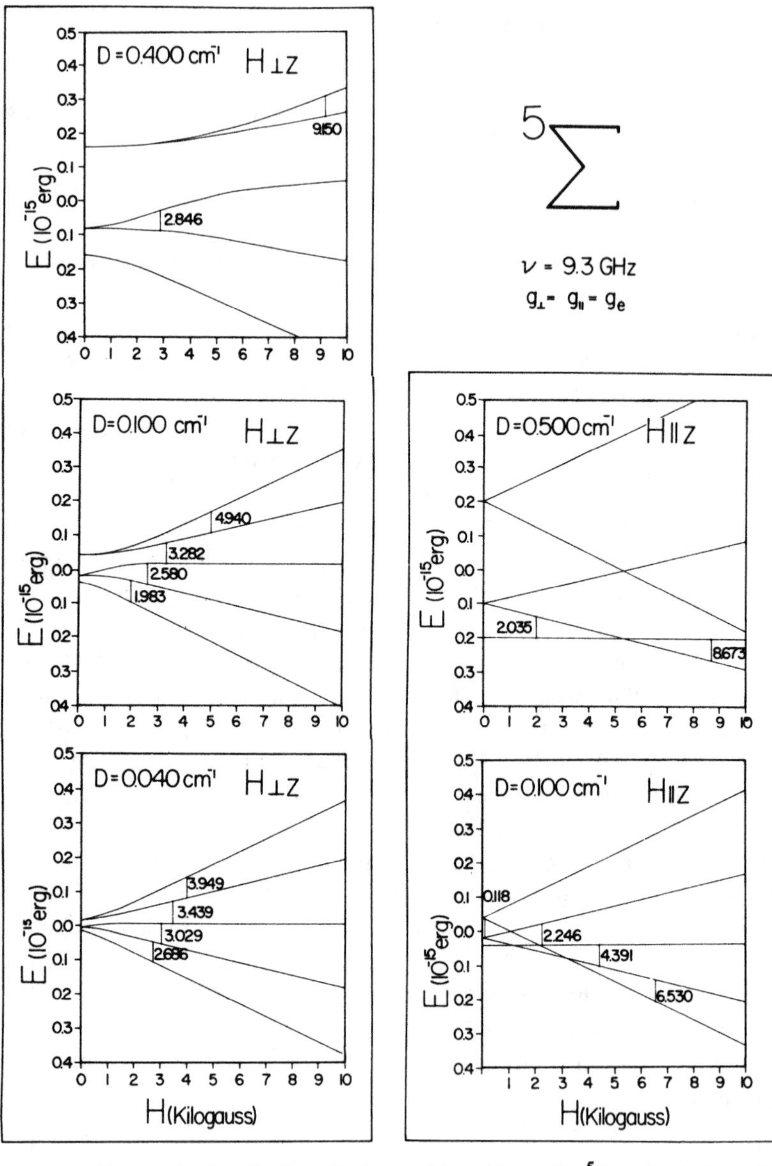

Fig. V-10. Zeeman levels and allowed $\Delta m_S = \pm 1$ transitions of a $^5\Sigma$ molecule for various values of the zero-field-splitting parameter D. Left side, $H \perp z$; right side, $H \parallel z$, where z is the molecular axis. $\nu = 9.3$ GHz (0.0616×10^{-15} erg = 0.31 cm^{-1}), $g_\perp = g_\parallel = g_e$, and D is assumed positive. The energy level spacings at zero magnetic field are D and $3D$.

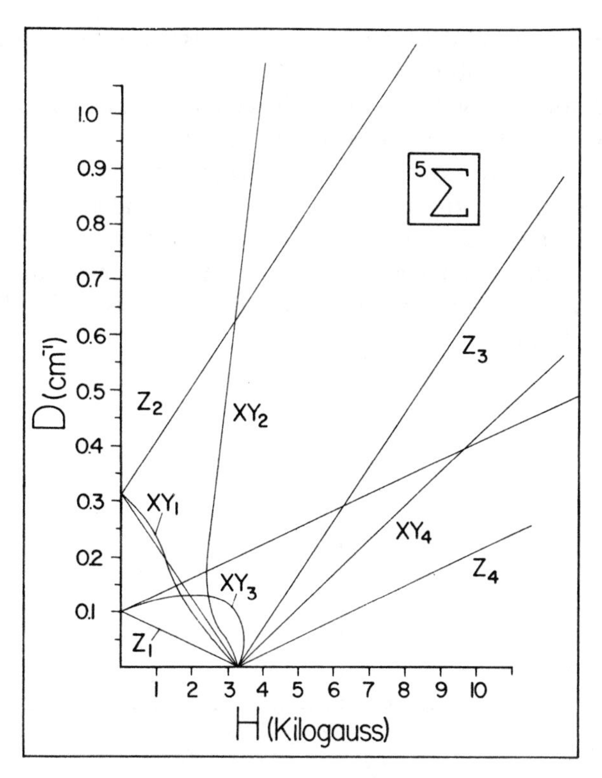

Fig. V-11. Resonant magnetic field positions at which $\Delta m_S = \pm 1$ transitions occur for various values of the zero-field-splitting parameter D for a $^5\Sigma$ molecule [from Van Zee, DeVore, and Weltner (448)]. It is assumed that $g_\perp = g_\parallel = g_e$ and $\nu = 9.3$ GHz.

with $u = D(3 \cos^2 \theta - 1) + 3E \cos 2\phi \sin^2 \theta$, where θ and ϕ describe the orientation of the molecule or crystal in the magnetic field, and g is assumed isotropic. In this approximation the four $\Delta m_S = \pm 1$ transitions are:

$$H_r\begin{pmatrix} 1 \longrightarrow 0 \\ 0 \longrightarrow 1 \end{pmatrix} = \frac{h\nu}{g\beta} \pm \frac{u}{2g\beta}$$

$$H_r\begin{pmatrix} 2 \longrightarrow 1 \\ -1 \longrightarrow -2 \end{pmatrix} = \frac{h\nu}{g\beta} \pm \frac{3u}{2g\beta} \qquad \text{(V, 26)}$$

(c) Randomly Oriented Molecules. If D is very small, then application of (V, 26) yields the absorption pattern and derived ESR spectrum in Fig. V-12 for randomly oriented linear molecules with $S = 2$, analogous to Fig. III-4 for the triplet case. Thus, this is the expected spectrum for quintet organic radicals where the zero-field splittings arise predominantly from spin-spin interaction and are usually small. If a small E term also arises, because the molecule lacks at least a trigonal axis of symmetry, a splitting of the xy perpendicular lines would occur, and additional lines would appear, as in Fig. III-12 for a triplet molecule.

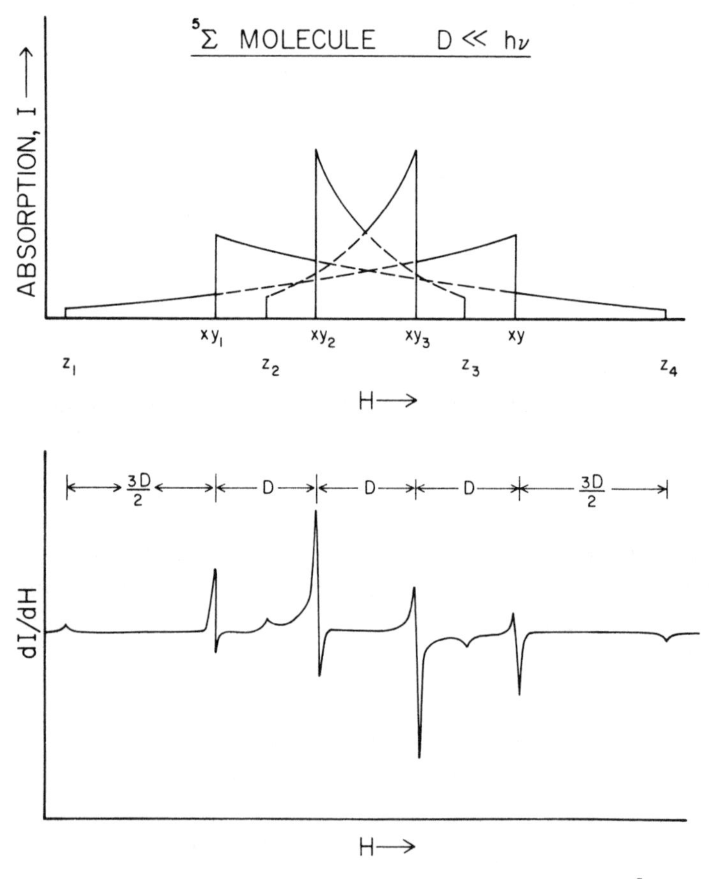

Fig. V-12. Absorption (upper) and ESR (lower) spectra of randomly oriented $^5\Sigma$ molecules with $D << g\beta H$. Perpendicular (xy_i) lines dominate.

As D becomes larger, the spectrum will become asymmetric relative to g_e. Fig. V-13 gives a computed spectrum for linear quintet molecules with $D = 0.09$ cm^{-1}. The variation of the $\Delta m_S = \pm 1$ transitions with θ now also has a bulge on the $|0\rangle \leftrightarrow |+1\rangle$ line at $\theta \simeq 60°$ which produces an "extra" line in the calculated ESR spectrum shown in the lowest panel. Baranowski et al. (46) have computed

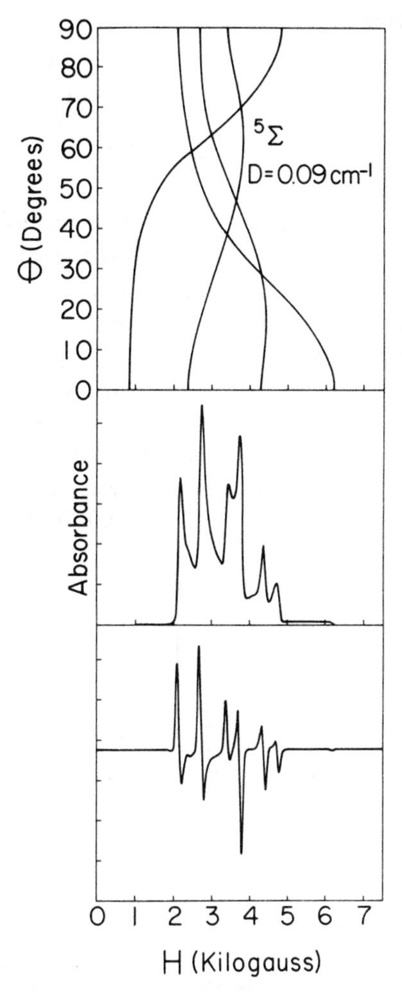

Fig. V-13. Computed variation of allowed ($\Delta m_S = \pm 1$) transitions with orientation (top), absorption spectrum (middle), and ESR spectrum (bottom) of randomly oriented $^5\Sigma$ molecules with $D = +0.09$ cm^{-1}. ($g_\parallel = g_\perp = g_e$, $\nu = 9.3$ GHz.)

and plotted the ESR absorbance and angular dependence for several values of D and E for the case of $S = 2$, and all also show these off-principal-axis transitions.

B. Individual Molecules

Organic Radicals. The only quintet radical known for many years was *m*-phenylene-bis-phenyl-methylene:

First observed by Itoh (242) in 1967, it was formed by photolysis of 1,3-bis-(α-diazobenyl)-benzene oriented in a single crystal of benzophenone. The magnetic field was applied perpendicular to the [001] and [110] axes, and the main signals were the four lines indicated by A, A', B, B' in Fig. V-14. The primed and unprimed pairs arise from two magnetically nonequivalent sites in the crystal. These quartets correspond to the four xy lines shown in Fig. V-14, and, as indicated there, the separation of A (or A') is just three times as large as that of the B (or B') pair. Itoh's complete analysis of the spectrum yielded $g = 2.0023$, $D = 0.07131\ \text{cm}^{-1}$, and $E = 0.01902\ \text{cm}^{-1}$. In an earlier

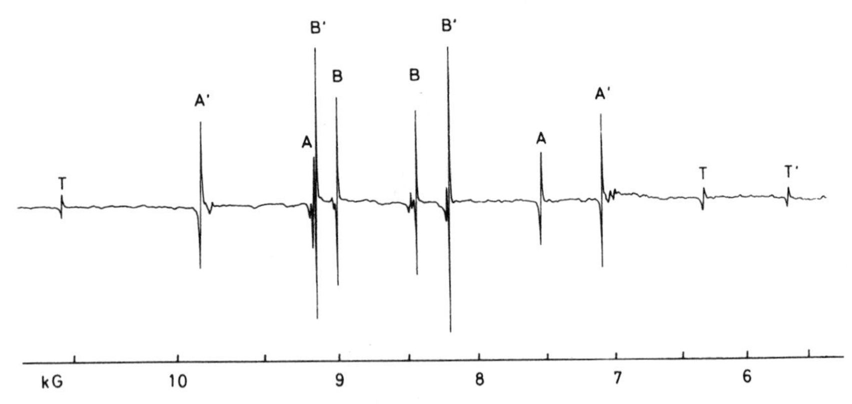

Fig. V-14. ESR spectrum of *m*-phenylene-bis-phenyl-methylene at 77°K [see Itoh (242); this is a more recent spectrum kindly provided by Professor K. Itoh]. A and B refer to lines of this radical; primed and unprimed pairs are separate sites. T and T' refer to another, triplet, molecule. ($\nu \cong 24$ GHz.)

theoretical calculation, Higuchi (220) had obtained $D = 0.0501$ and $E = 0.0099\ \text{cm}^{-1}$, in satisfactory agreement with the observed values in view of the approximate nature of the calculation.

Huber and Schwoerer (233) have observed the triplet and quintet states of the dicarbenes formed when diacetylene single crystals are photopolymerized. The quintet states are considered to arise from structures such as

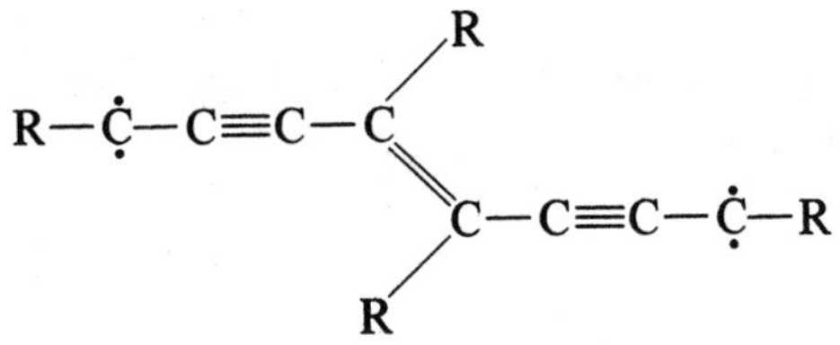

```
                        R
                       /
R—Ċ—C≡C—C
    ·            \\
                  C—C≡C—Ċ—R
                 /        ·
                R
```

Analysis [(233); Schwoerer et al. (409)], utilizing a spin Hamiltonian including $B_4{}^0$ and $B_4{}^2$, yielded the results

$$D = 0.0944\ \text{cm}^{-1}, \quad E = -0.0024\ \text{cm}^{-1}$$

$$B_4{}^0 = 0.00012\ \text{cm}^{-1}, \quad B_4{}^2 = 0.00000\ \text{cm}^{-1}$$

$$g_x = g_y = g_z = 2.0023$$

where D was assumed positive. $\Delta m_S = 1$, 2, and 3 transitions were observed.

Crystals and Inorganic Molecules. Relatively few observations of $S = 2$ systems in inorganic crystals have been made, owing in part to lack of Kramers' degeneracy for even spins and to large zero-field splittings. Cr^{2+}, Mn^{3+}, and Fe^{2+} in various lattices have been discussed by Abragham and Bleaney (1) and more recently reviewed by Máckowiak (321).

The corresponding molecules CrO, CrH_2, and FeF_2 probably have $S = 2$ but have not been observed in the ESR in matrices. It is probable that CrO has a $^5\Pi$ ground state and FeF_2 is linear and also orbitally degenerate. This would make the $\boldsymbol{g}$ tensor very anisotropic in each case and prevent the observation of their ESR spectra as randomly oriented molecules [Van Zee et al. (447)].

Recently, Sc_2 has been shown to have a $^5\Sigma$ ground state with $|D| = 0.11\ \text{cm}^{-1}$ [Knight, Van Zee, and Weltner, see Table C-8].

3. $^6\Sigma$ AND OTHER SEXTET ($S = \frac{5}{2}$) MOLECULES IN MAGNETIC FIELDS

This multiplicity occurs among inorganic molecules and crystals containing particularly Cr^+, Mn^{2+}, and Fe^{3+} ions. Its ubiquity in ESR spectroscopy signifies that perhaps it deserves more detailed consideration, but the extensive studies in crystals will be discussed only briefly here [see Abragham and Bleaney (1); Wertz and Bolton (23)].

The spin Hamiltonian for $S = \frac{5}{2}$, as indicated when discussing quintet species, will be of the form of (V, 19):

$$\mathcal{H} = \beta \boldsymbol{H} \cdot \boldsymbol{g} \cdot \boldsymbol{S} + \sum_{m,n} B_n{}^m O_n{}^m \qquad \text{(V, 19)}$$

where the spin operators characterizing the crystal field expansion can, in principle, include terms up to $n = 2S = 5$. However, other restrictions (invariance under time reversal [Abragham and Bleaney (1), p. 140]) prescribe that odd values of n are excluded. Therefore, as for $S = 2$, only $O_2{}^m$ and $O_4{}^m$ terms will occur, where m is dictated by symmetry, as indicated on page 257.

A. The Axial Case, $^6\Sigma$ Molecules

For axial symmetry where $D = 3B_2{}^0$ and $E \equiv 0$, and neglecting $O_4{}^m$ fine-structure terms, the spin Hamiltonian is

$$\mathcal{H} = g_\parallel \beta H_z S_z + g_\perp \beta (H_x S_x + H_y S_y) + D(S_z{}^2 - \tfrac{35}{12}) + A_\parallel I_z S_z + A_\perp I_x S_x + I_y S_y \qquad \text{(V, 27)}$$

Hyperfine interaction with one nucleus is included, but nuclear quadrupole and Zeeman effects are also neglected. This Hamiltonian applies satisfactorily to the ^{55}MnO molecule (where $I_{\text{Mn}} = \frac{5}{2}$) and to $^{55}Mn^{2+}$ ions in an axial crystal field in ionic crystals. Using the basis functions $|m_S, m_I\rangle$, one can then construct the eigenvalue matrix in analogous manner to that used in obtaining (V, 22). Or the axes may be rotated to diagonalize the Zeeman energy, i.e., the direction of H chosen as the z axis at an angle θ to the molecular axis (also the axis of $\boldsymbol{D}$). The matrix then becomes that given in Table V-2 for the

Table V-2. Elements of the Eigenvalue Matrix for $S = \frac{5}{2}$, $I = \frac{5}{2}$ for an Axial Molecule (Neglecting $O_4{}^m$ Terms) [Courtesy of Dr. R. J. Van Zee].

$$G = g\beta H \qquad D'' = \frac{D}{4}\left(\frac{g_\perp}{g}\right)^2 \sin^2\theta$$

$$g^2 = g_\parallel{}^2 \cos^2\theta + g_\perp{}^2 \sin^2\theta \qquad X_+ = \frac{A_\parallel A_\perp}{4K} + \frac{A_\perp}{4}$$

$$K^2 = g_\parallel{}^2 A_\parallel{}^2 \cos^2\theta + g_\perp{}^2 A_\perp{}^2 \sin^2\theta \qquad X_- = \frac{A_\parallel A_\perp}{4K} - \frac{A_\perp}{4}$$

$$D' = \frac{D}{2}\left[3\left(\frac{g_\parallel}{g}\right)^2 \cos^2\theta - 1\right]$$

Diagonal Matrix Elements ($i = j$, i = 1 to 36)

i	$\lvert M_S, M_I\rangle$	Matrix Element	i	$\lvert M_S, M_I\rangle$	Matrix Element
1 36	$\lvert\pm\frac{5}{2}, \pm\frac{5}{2}\rangle$	$\pm\frac{5}{2}G + \frac{10}{3}D' + \frac{25}{4}K$	10 27	$\lvert\mp\frac{1}{2}, \pm\frac{5}{2}\rangle$	$\mp\frac{1}{2}G - \frac{8}{3}D' - \frac{5}{4}K$
2 35	$\lvert\pm\frac{5}{2}, \pm\frac{3}{2}\rangle$	$\pm\frac{5}{2}G + \frac{10}{3}D' + \frac{15}{4}K$	11 26	$\lvert\pm\frac{5}{2}, \mp\frac{3}{2}\rangle$	$\pm\frac{5}{2}G + \frac{10}{3}D' - \frac{15}{4}K$
3 34	$\lvert\pm\frac{3}{2}, \pm\frac{5}{2}\rangle$	$\pm\frac{3}{2}G - \frac{2}{3}D' + \frac{15}{4}K$	12 25	$\lvert\pm\frac{3}{2}, \mp\frac{1}{2}\rangle$	$\pm\frac{3}{2}G - \frac{2}{3}D' - \frac{3}{4}K$
4 33	$\lvert\pm\frac{5}{2}, \pm\frac{1}{2}\rangle$	$\pm\frac{5}{2}G + \frac{10}{3}D' + \frac{5}{4}K$	13 24	$\lvert\pm\frac{1}{2}, \pm\frac{1}{2}\rangle$	$\pm\frac{1}{2}G - \frac{8}{3}D' + \frac{1}{4}K$
5 32	$\lvert\pm\frac{3}{2}, \pm\frac{3}{2}\rangle$	$\pm\frac{3}{2}G - \frac{2}{3}D' + \frac{9}{4}K$	14 23	$\lvert\mp\frac{1}{2}, \pm\frac{3}{2}\rangle$	$\mp\frac{1}{2}G - \frac{8}{3}D' - \frac{3}{4}K$
6 31	$\lvert\pm\frac{1}{2}, \pm\frac{5}{2}\rangle$	$\pm\frac{1}{2}G - \frac{8}{3}D' + \frac{5}{4}K$	15 22	$\lvert\mp\frac{3}{2}, \pm\frac{5}{2}\rangle$	$\mp\frac{3}{2}G - \frac{2}{3}D' - \frac{15}{4}K$
7 30	$\lvert\pm\frac{5}{2}, \mp\frac{1}{2}\rangle$	$\pm\frac{5}{2}G + \frac{10}{3}D' - \frac{5}{4}K$	16 21	$\lvert\pm\frac{5}{2}, \mp\frac{5}{2}\rangle$	$\pm\frac{5}{2}G + \frac{10}{3}D' - \frac{25}{4}K$
8 29	$\lvert\pm\frac{3}{2}, \pm\frac{1}{2}\rangle$	$\pm\frac{3}{2}G - \frac{2}{3}D' + \frac{3}{4}K$	17 20	$\lvert\pm\frac{3}{2}, \mp\frac{3}{2}\rangle$	$\pm\frac{3}{2}G - \frac{2}{3}D' - \frac{9}{4}K$
9 28	$\lvert\pm\frac{1}{2}, \pm\frac{3}{2}\rangle$	$\pm\frac{1}{2}G - \frac{8}{3}D' + \frac{3}{4}K$	18 19	$\lvert\pm\frac{1}{2}, \mp\frac{1}{2}\rangle$	$\pm\frac{1}{2}G - \frac{8}{3}D' - \frac{1}{4}K$

Off-Diagonal Matrix Elements $(i, j) = (j, i)$

(i, j)	Matrix Element
(1, 5), (11, 22), (15, 26), (32, 36)	$5X_-$
(1, 6), (2, 9), (4, 13), (7, 18), (10, 21), (11, 23), (14, 26), (16, 17), (19, 30), (24, 33), (28, 35), (31, 36)	$2\sqrt{10}\,D''$
(2, 3), (16, 17), (20, 21), (34, 35)	$5X_+$
(2, 8), (3, 9), (7, 17), (10, 20), (17, 27), (20, 30), (28, 34), (29, 35)	$2\sqrt{10}\,X_-$

Table V-2. (Continued)

Off-Diagonal Matrix Elements $(i, j) = (j, i)$

(i, j)	Matrix Element
(3, 10), (5, 14), (6, 15), (8, 19), (9, 20), (12, 24), (13, 25), (17, 28), (18, 29), (22, 31), (23, 32), (27, 34)	$6\sqrt{2}\, D''$
(4, 5), (5, 6), (11, 12), (14, 15), (22, 23), (24, 26), (31, 32), (32, 33)	$2\sqrt{10}\, X_+$
(4, 12), (6, 14), (23, 31), (25, 33)	$3\sqrt{5}\, X_-$
(5, 13), (12, 23), (14, 25), (24, 32)	$8X_-$
(7, 8), (9, 10), (27, 28), (29, 30)	$3\sqrt{5}\, X_+$
(8, 9), (17, 18), (19, 20), (28, 29)	$8X_+$
(8, 18), (9, 19), (18, 28), (19, 29)	$6\sqrt{2}\, X_-$
(12, 13), (13, 14), (23, 24), (24, 25)	$6\sqrt{2}\, X_+$
(13, 24)	$9X_-$
(18, 19)	$9X_+$

important case of an axial sextet molecule containing a Mn atom ($I = \frac{5}{2}$). The 184 nonzero elements of the 36 X 36 eigenvalue matrix are listed; again the basis is arranged according to decreasing values of $|m_S + m_I\rangle$.

(a) Eigenvalues and Magnetic-Dipole Transitions. Neglecting the hyperfine interaction, the 6 X 6 eigenvalue matrix has the roots $+\frac{10}{3}D$, $-\frac{2}{3}D$, $-\frac{8}{3}D$ at zero magnetic field. These then correspond to the eigenfunctions $|\pm\frac{1}{2}\rangle$, $|\pm\frac{3}{2}\rangle$, $|\pm\frac{5}{2}\rangle$, respectively, for positive D, so that the zero-field splittings are $2D$ and $4D$.

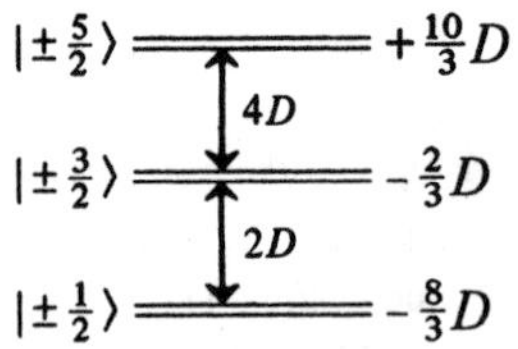

$|\pm\frac{5}{2}\rangle$ ——— $+\frac{10}{3}D$

$4D$

$|\pm\frac{3}{2}\rangle$ ——— $-\frac{2}{3}D$

$2D$

$|\pm\frac{1}{2}\rangle$ ——— $-\frac{8}{3}D$

Solution of this eigenvalue matrix with the magnetic field at an angle θ relative to the axis of the molecule yields the Zeeman levels and from them the positions of the $\Delta m_S = \pm 1$ transitions. These levels for linear molecules with various values of D (assumed positive) and $g_{\parallel} = g_{\perp} = 2.0023$ are shown in Fig. V-15 for the two extreme orientations H ($\theta = 0°$) and H ($\theta = 90°$). As in the other molecules

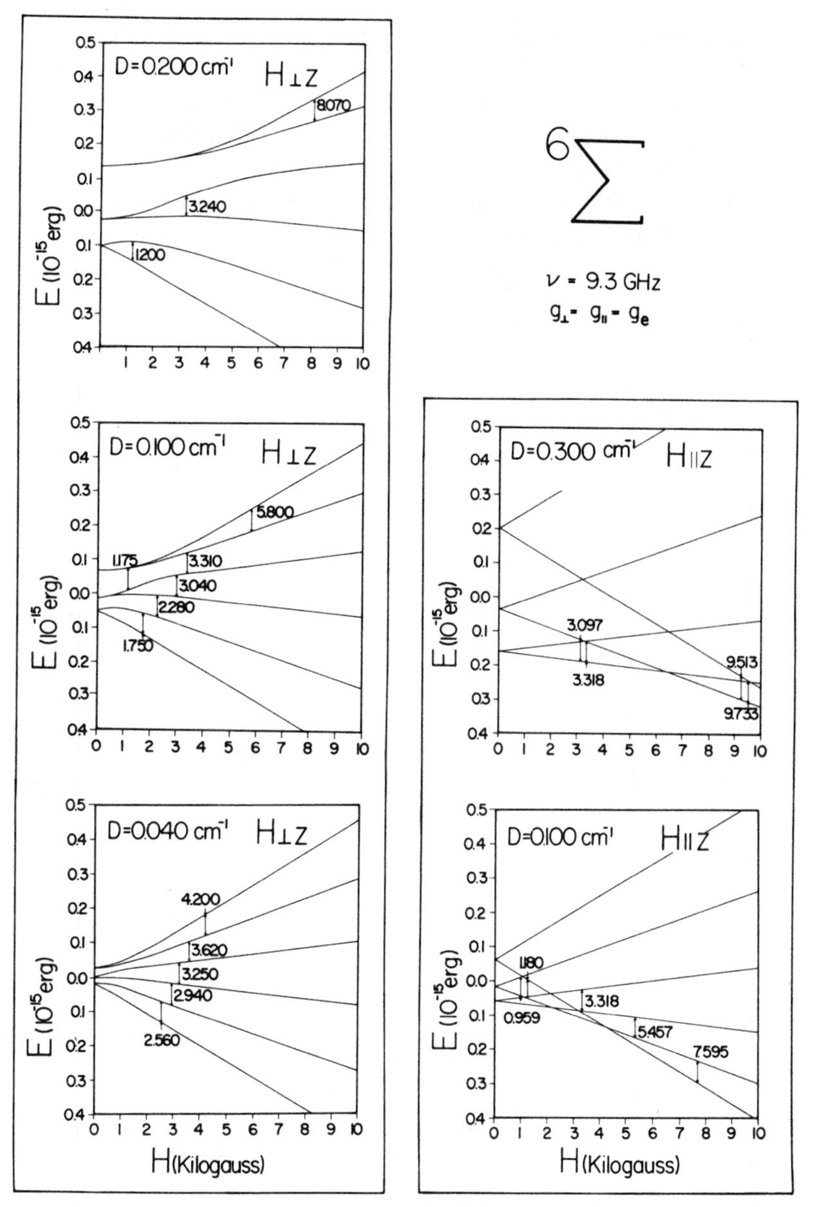

Fig. V-15. Zeeman levels and allowed $\Delta m_S = \pm 1$ transitions of a $^6\Sigma$ molecule for various values of the zero-field-splitting parameter D. Left side, $H \perp z$; right side $H \parallel z$, where z is the molecular axis. $\nu = 9.3$ GHz ($=0.0616 \times 10^{-15}$ erg $= 0.31$ cm^{-1}), $g_{\parallel} = g_{\perp} = g_e$. The energy level spacings at zero magnetic field are $2D$ and $4D$.

with an odd number of electrons, a series of Kramers' doublets occur, and the lowest such $|\pm\frac{1}{2}\rangle$ doublet should always provide an ESR $\Delta m_S = \pm 1$ signal. If D is negative, the levels are inverted in Fig. V-15, and the signals observed will depend upon the magnitude of D, the transition probabilities, and the temperature.

From plots like those in Fig. V-15 one may trace the parallel (z_i) and perpendicular (xy_i) transitions of a $^6\Sigma$ molecule as a function of D, as shown in Fig. V-16. As for $S = \frac{3}{2}$ (Fig. V-2), at very

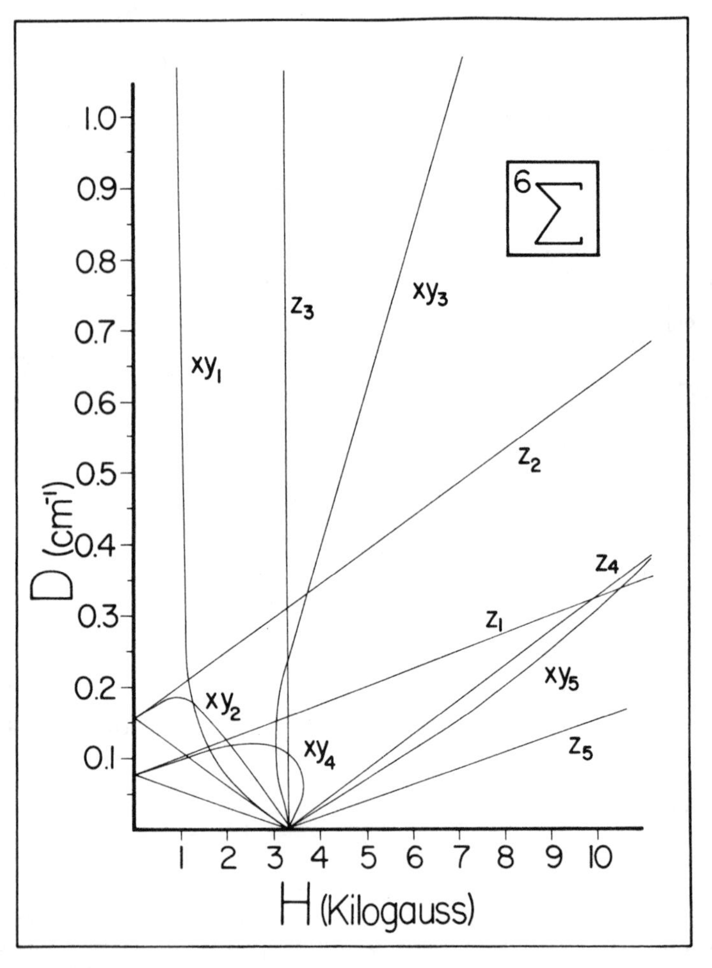

Fig. V-16. Resonant magnetic field positions at which $\Delta m_S = \pm 1$ transitions occur for various values of the zero-field-splitting parameter D for a $^6\Sigma$ molecule [from DeVore, Van Zee, and Weltner (140)]. It is assumed that $g_\perp = g_\parallel = g_e$ and $\nu = 9.3$ GHz.

large positive D values only the xy_1 transition, corresponding to the $|\pm\frac{1}{2}\rangle \leftrightarrow |-\frac{1}{2}\rangle$ transition, will be observed, although up to $D \simeq 2$ cm^{-1} the xy_3 transition may also be observable by X-band ESR.

(b) Effective *g* Values. Thus, in the limit of $D >> g\beta H$, for positive D values, only the lowest Kramers' doublet $|\pm\frac{1}{2}\rangle$ is important (see the upper left-hand panel in Fig. V-15), and the use of an effective g value, g^e, again becomes convenient. As discussed for $S = \frac{3}{2}$, g^e values for the $|\pm\frac{1}{2}\rangle$ doublet when $S = \frac{5}{2}$ are found to be [Kirkpatrick, Muller, and Rubins (272); Pilbrow (373)]:

$$g_{\parallel}{}^e \simeq g_{\parallel}$$

$$g_{\perp}{}^e \simeq 3g_{\perp}\,[1 - \tfrac{1}{2}\,(g\beta H/D)^2] \qquad \text{(V, 28)}$$

Then for large D, and $g_{\parallel} \simeq g_{\perp} \simeq 2.00$, $g_{\parallel}{}^e = 2.0$ and $g_{\perp}{}^e = 6.0$.

(c) Randomly Oriented Molecules. For small zero-field splittings and an approximately isotropic $g = 2$, the ESR spectrum is expected to be symmetrical about xy_3 and to drop off rapidly in intensity at low and high fields as indicated in Fig. V-17. xy_3 and z_3 correspond to the $m_S = +\frac{1}{2} \leftrightarrow -\frac{1}{2}$ fine-structure transition in this limit of low D, and often it will be the only transition observed. If other m_S transitions are observed, D can be directly obtained by difference, since for $g\beta H >> D$, successive xy_i lines are split by $|D|$ and z_i lines by $2|D|$ [Low (15)].

As D becomes larger, the overall intensity of the lines shifts to lower fields, but also a spectrum rapidly becomes complicated by extra lines. This is illustrated by Fig. V-18, which shows a simulated spectrum for $D = +0.10$ cm^{-1}. This has been calculated by the same procedure described for $S = \frac{3}{2}$ linear molecules. Of the five strongest lines, one at 4.2 kG is an off-principal-axis transition. As expected, for $D > h\nu$, the most intense line becomes the $|\pm\frac{1}{2}\rangle$ Kramers' doublet transition at $\sim g^e = 6$. This is shown in the simulated ESR spectrum in Fig. V-19 for $D = +0.34$ cm^{-1}, which is the value found for the linear MnF_2 molecule trapped in an argon matrix (see below). The large number of extra lines in that spectrum is noteworthy and indicates clearly the necessity for comparison of observed and computer-simulated spectra.

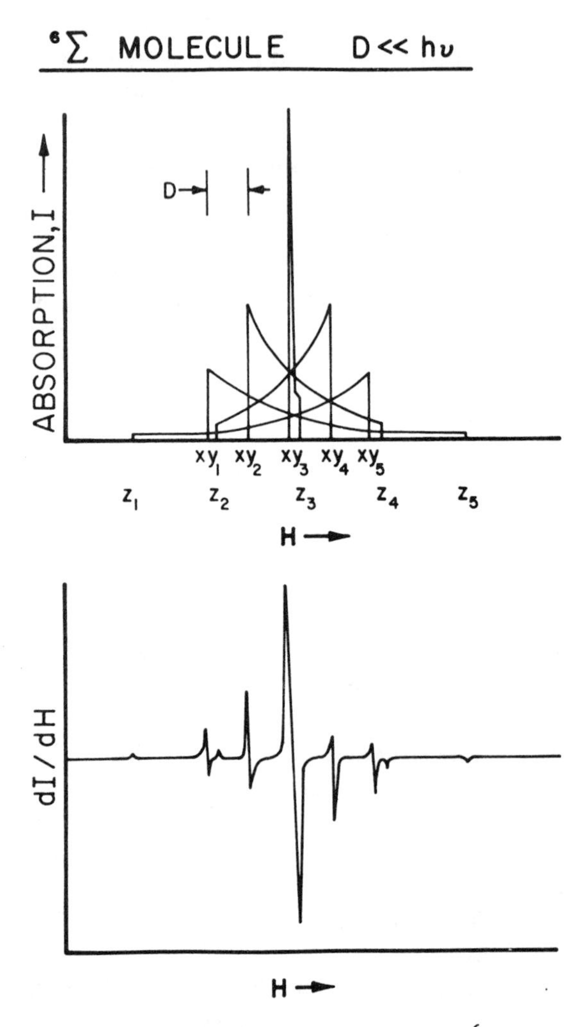

Fig. V-17. ESR absorption and spectrum of randomly oriented $^6\Sigma$ molecules for $D << h\nu$ [from Van Zee and Weltner (452)]. (Top) Theoretical absorption indicating variation from parallel (z_i) to perpendicular (xy_i) orientation of the magnetic field relative to the molecular axis for the five $\Delta m_S = \pm 1$ transitions (see Fig. V-15); (bottom) ESR spectrum (first derivative of the total absorption curve). Perpendicular (xy_i) lines dominate, particularly the $m_S = -\frac{1}{2} \leftrightarrow +\frac{1}{2}$ transition.

(d) Hyperfine Interaction. As indicated above, in glassy and polycrystalline solids one can expect only the one allowed fine-structure transition $m_S = +\frac{1}{2} \leftrightarrow -\frac{1}{2}$ to be strong in the ESR spectrum if D is small. If hyperfine interaction also occurs, for example, with an

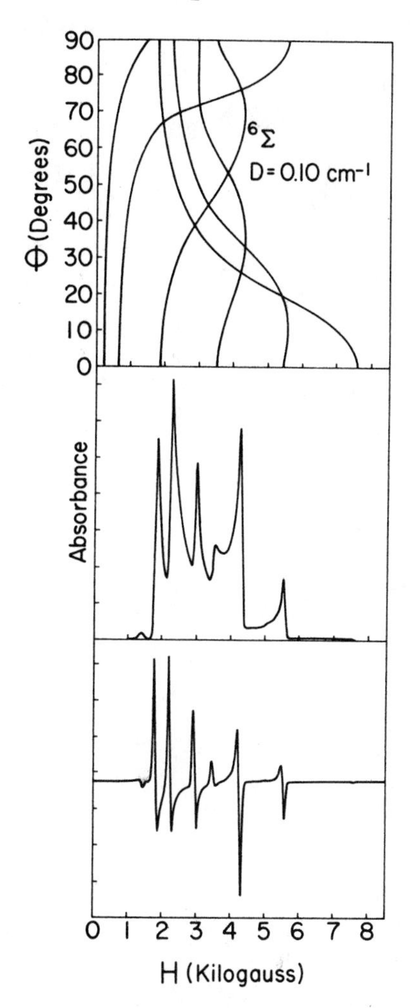

Fig. V-18. Computed variation of allowed ($\Delta m_S = \pm 1$) transitions with orientation (top), absorption spectrum (middle), and ESR spectrum (bottom) of randomly oriented $^6\Sigma$ molecules with $D = +0.10\ \text{cm}^{-1}$. ($g_{\parallel} = g_{\perp} = g_e$, $\nu = 9.3$ GHz.)

$I = \frac{5}{2}$ nucleus such as ^{55}Mn, then this transition will be split into six hf lines identified by their m_I values, $-\frac{5}{2}, -\frac{3}{2} \cdots, +\frac{5}{2}$ separated by perhaps $|A| \simeq 100$ G. However, for the condition where D (assuming it is the dominant crystal-field term) is of the same magnitude as A, i.e., $\sim 0.01\ \text{cm}^{-1}$, three exceptional features of the spectrum may be observed for random orientation in solids: (1) unequal intensities of the six hf lines, (2) positive to negative peak splittings of these

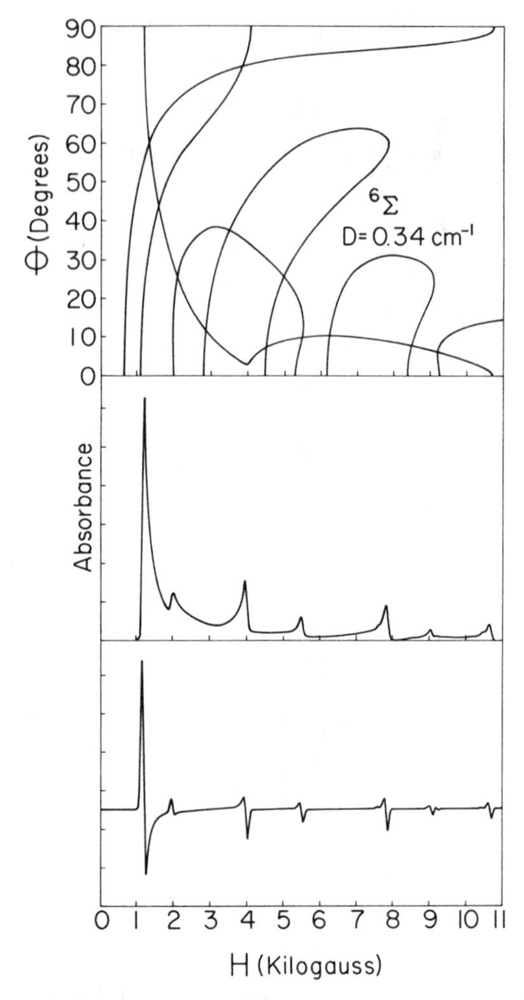

Fig. V-19. Computed variation of allowed ($\Delta m_S = \pm 1$) transitions with orientation (top), absorption spectrum (middle), and ESR spectrum (bottom) of randomly oriented $^6\Sigma$ molecules with $D = +0.34$ cm^{-1}. ($g_\parallel = g_\perp = g_e$, $\nu = 9.3$ GHz.)

strong lines, (3) weak doublets appearing between the six hf lines. Such a spectrum is shown in Fig. V-20; this example has unusually good resolution, in that not all of these features are so clearly resolved in most solids.

(1) Intensity Variation. Allen (37) first showed that if D and A were small relative to $g\beta H$ and of similar magnitude, then there

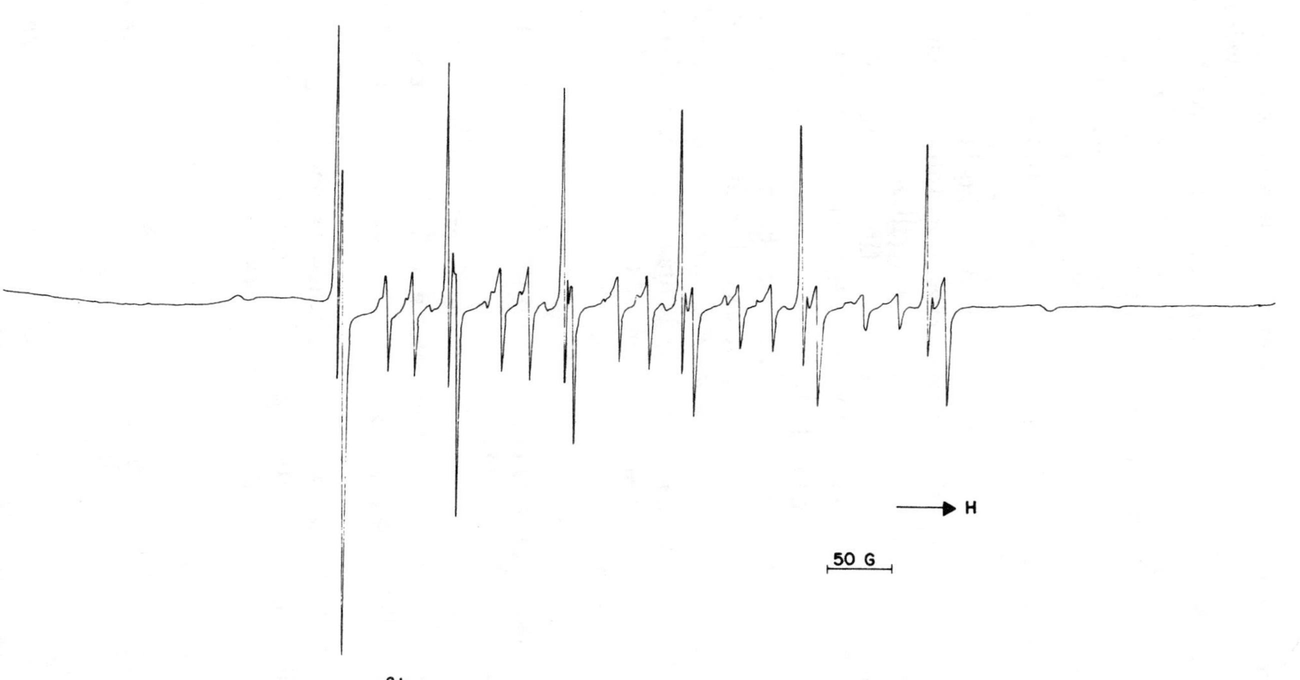

Fig. V-20. ESR spectrum of Mn^{2+} doped into polycrystalline $CaCO_3$ [from Eidels-Dubovoi and Beltrán-López (154)].

would be a variation of the intensities of the allowed $\Delta m_I = 0$ transitions with $|m_I|$. The perturbation treatment was later extended to third order by de Wijn and van Balderen (143). The effect of D is to mix the $|\pm\frac{5}{2}\, m_I\rangle$, $|\pm\frac{3}{2}\, m_I\rangle$, $|\pm\frac{3}{2}, m_I \pm 1\rangle$, etc., and thereby affect the transition probabilities for the $\Delta m_I = 0$ transitions within the $m_S = \pm\frac{1}{2}$ doublet. In fact, from the measurement of the relative intensities $I_{5/2}/I_{1/2}$, etc., one can derive a value of $|D|$. Allen's theory predicts that the intensities are independent of the sign of m_I, that is, that the hf intensity pattern should be symmetrical about its center. Obviously this is not true of the relative peak heights in Fig. V-20, but it is true if the areas of the absorption lines are measured. Quadrupole coupling was estimated to be much less significant in such samples than the mixing by D. [The latest calculations of this effect are those of Eidels-Dubovoi and Beltrán-López (154) and of Beltrán-López and Castro-Tello (54)].

(2) Peak-to-Peak Splitting. Each of the allowed transitions shows such a splitting which is due to a small angular variation of the $m_S = +\frac{1}{2} \leftrightarrow -\frac{1}{2}$ transition in powder samples [Bleaney and Rubins (74); de Wijn and van Balderen (143); Beltrán-López and Castro-Tello (54)]. The peak-to-peak separation in third-order depends upon m_I and is given by

$$\Delta H = (2D^2/H_i)\{1 + (16/H_i)\,(H_i - 8Am_I)^2/(9H_i - 64Am_I)\}$$

where H_i is the position of the hf line indicated by m_I. Thus for $D = 80$ G and $A = -94$ G, which apply in Fig. V-20, the splitting of the highest field line ($m_I = +\frac{5}{2}$) is 14.6 G, while that for the lowest field line ($m_I = -\frac{5}{2}$) is 6.7 G.

(3) Forbidden Transitions. The ten weak lines lying between the strong $\Delta m_I = 0$ transitions are due to $\Delta m_I = \pm 1$ transitions whose positions are given by [Bleaney and Rubins (74); Friedman and Low (172)]

$$H = H_0 + A\left(m_I - \frac{1}{2}\right) \pm \frac{17}{4}\,\frac{A^2}{H} + \frac{A^2}{2H}\left[\frac{33}{4} - m_I(m_I - 1)\right] \pm g_I H/g$$

where the small quadrupole effect has been neglected. This gives a series of five doublets for $I = \frac{5}{2}$ split by

$$\delta_I = \frac{17A^2}{2H} + \frac{2g_I H}{g} \qquad \text{(V, 29)}$$

Thus their positions are independent of D, but the relative intensities of forbidden to allowed lines for $m_S = \frac{1}{2}$ are given by [Shaffer, Farach, and Poole (413)]

$$(I_f/I_a) = (512/15)\left(\tfrac{35}{4} - m_I^2 + m_I\right)(D/g\beta H)^2 \qquad \text{(V, 30)}$$

For typical parameters for $^{55}Mn^{2+}$ with $g_I = 0.7514 \times 10^{-3}$, $D = 64 \pm 10$ G, $A = 83 \pm 1$ G, $H_0 = 3200$ G, the splittings are about 21 G.

In the limit of high (positive) D the fine-structure transition within the $m_S = +\frac{1}{2} \leftrightarrow -\frac{1}{2}$ Kramers' doublet is again the only one observed, but it now occurs at an effective $g^e \simeq 6$, as discussed earlier. Since $A \ll g\beta H$ is usually the case, a "normal" hyperfine pattern is then expected. Anisotropy in the A tensor will cause increasingly wider spacings between the hf lines toward higher fields, and it may then be possible to obtain both hf components for a linear molecule from this variation in hfs (see Fig. V-29 and the discussion of the MnO molecule below).

B. Nonaxial Molecules; The $g' = 4.3$ Signal

The most widely discussed, and still somewhat controversial, $S = \frac{5}{2}$ molecules are probably those associated with a so-called $g' = 4.3$ signal. (Here $g' = h\nu/\beta H_0$ is just another way of indicating the resonant field. It is not meant to be as specific as g^e when referring to a particular Kramers' doublet.) This signal has been identified as due to the Fe^{3+} ion in a large number of biologically interesting molecules such as transferrin and ferrichrome-A, and also in complexes with EDTA and in some silicate glasses. The environment of the Fe^{3+} ion in ferrichrome A is shown in Fig V-21.

This broad signal centered at about 1500 G in the X-band ESR spectra of powders and glasses was originally interpreted by use of the spin Hamiltonian (V, 27) with $D = 0$ and $E \gg g\beta H$ representing extreme rhombic symmetry [Castner, Newell, Holton, and Slichter (106)]. For this case the three Kramers' doublets in zero field are then spaced apart by $2\sqrt{7}\,E$, and to first order the middle doublet has an isotropic $g' = 30/7$. However, Wickman, Klein, and Shirley

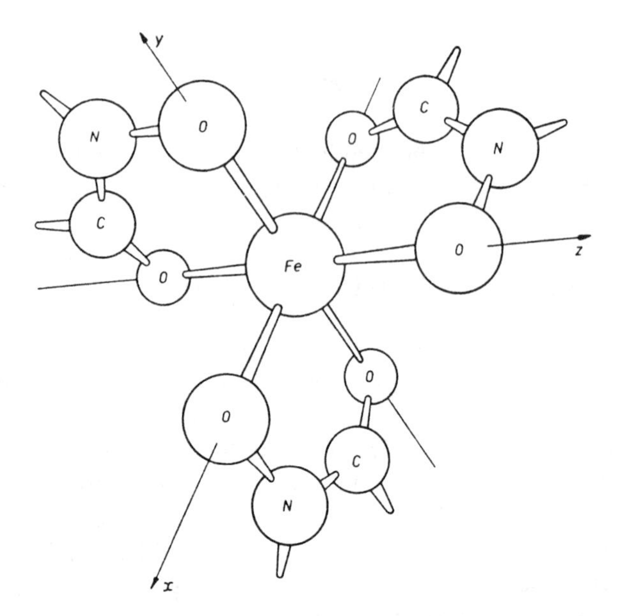

Fig. V-21. Surroundings of the iron atom in ferrichrome-A [from Ingram (13)].

(479) and Blumberg (78) showed that after rotation of axes this is equivalent to $|E/D| = \frac{1}{3}$, for both E and D large with respect to $g\beta H$. This latter interpretation is physically more reasonable, since it is expected that there will be an axial crystal-field (D) contribution along the direction of the greatest (or least) metal–ligand interaction. This $0 \leqslant |\lambda = E/D| \leqslant \frac{1}{3}$ suffices to cover the entire range of values of D and E from a linear to an orthorhombic field [see Blumberg (78)].

If the spin Hamiltonian (V, 27) containing the fine-structure parameters D and E is solved assuming the Zeeman term is relatively small, the resulting energy levels can be numbered 1 to 6 in increasing energy [Wickman, Klein, and Shirley (479); Dowsing and Gibson (150)]. Then the transitions $3 \leftrightarrow 4$ for $\lambda \simeq \frac{1}{3}$ for the x, y, and z principal directions very nearly overlap at $g' \simeq 4.3$, thus accounting for the broad strong band observed. Such a signal for Fe^{3+} in a powdered crystal of $NH_4CoY \cdot 2H_2O$ (H_4Y = ethylenediaminetetra-acetic acid, EDTA) is shown in Fig. V-22; the weak signal near $g' = 10$ is also expected from the $1 \leftrightarrow 2$ and $5 \leftrightarrow 6$ transitions.

It is understandable that these spectra can be very difficult to analyze exactly because of the possible necessity of inclusion of fourth-order fine-structure terms $O_4{}^m$ [Kedzie, Lyons, and Kestigan

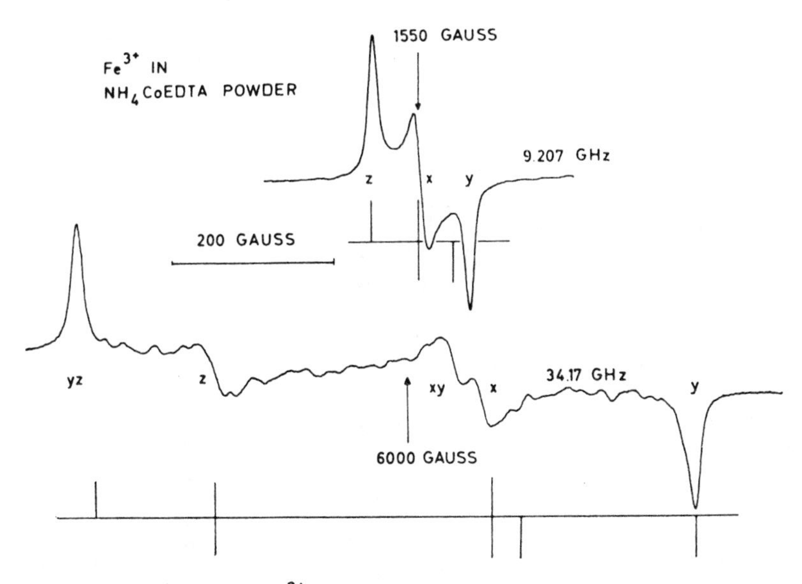

Fig. V-22. ESR spectra of Fe^{3+} in a crystalline powder of $NH_4CoEDTA$ obtained at 9.207 GHz (upper) and at 34.17 GHz (lower) [from Aasa (24)]. Below each experimental spectrum, theoretical line positions are given assuming $|D| = 0.83$ cm^{-1} and $|E/D| = 0.31$.

(271)]. In the case of powder spectra the appearance of "extra" fine-structure lines, which occur so frequently even for linear molecules, makes the analysis even more formidable. A *tour de force* calculation of the expected transitions over a broad range of the parameters D and E/D (0, 0.15, 0.25, and $\frac{1}{3}$) has been carried out by Aasa (24) [see also Sweeney, Coucouvanis, and Coffman (428)]. His detailed results for the $3 \leftrightarrow 4$ transition responsible for the line at $g' \simeq 4.2$ are shown in Fig. V-23 for $E/D = 0.32$. In this figure the numbers in parentheses are relative transition probabilities and lines designated by xz, xy, etc. indicate extra powder lines arising when H lines in the xz, xy plane, etc. (The ordinate can be converted to D in cm^{-1} by dividing those numbers into 0.3 and the abscissa into gauss by multiplying by 3330 for X-band.) Similar curves are obtained for all values of E/D near $\frac{1}{3}$. The most noteworthy observation from this figure is that the $g' \simeq 4.2$ line may consist of three to six powder lines. The theoretical fit to the EDTA case is shown in Fig. V-22.

The $g' = 4.3$ signal is rather ubiquitous among high spin d^5 Fe^{3+} compounds, and the circumstances under which it may appear are

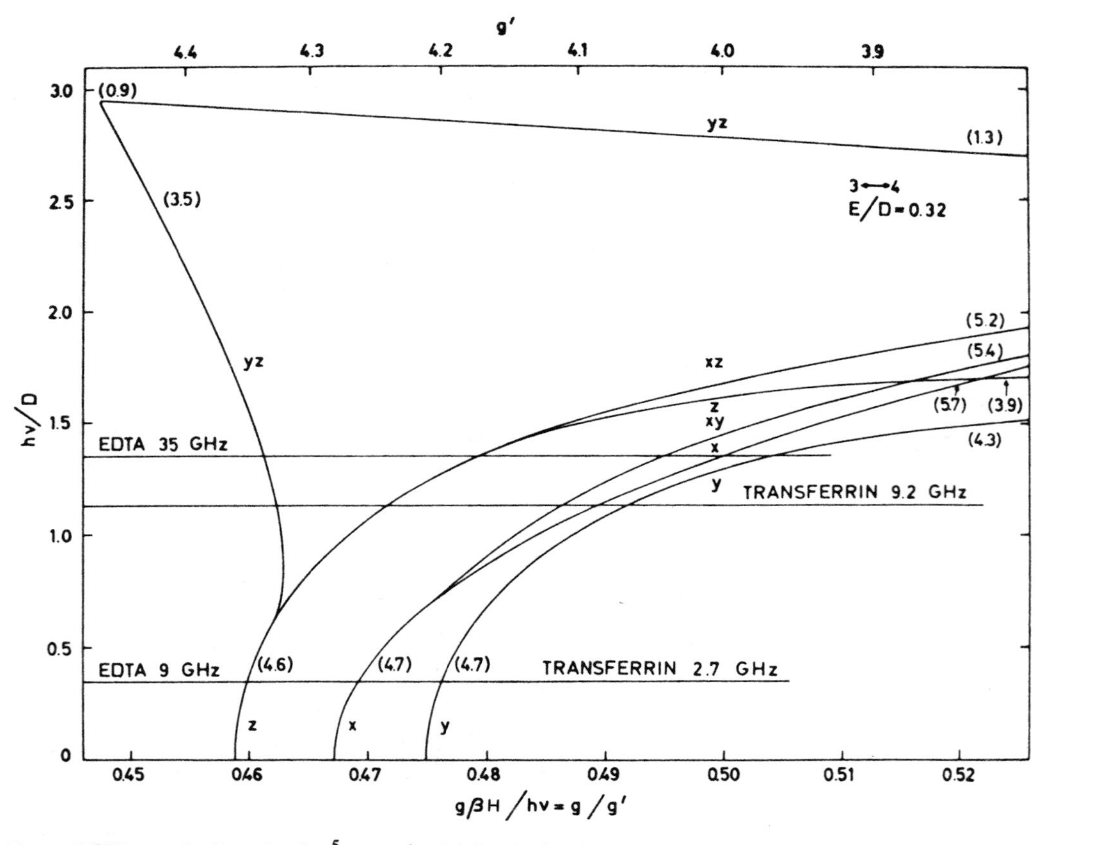

Fig. V-23. Positions of ESR powder lines for $S = \frac{5}{2}$ near $g' = 4.2$ for the $3 \leftrightarrow 4$ transition and $E/D = 0.32$ [from Aasa (24)]. The intersections of the horizontal lines and the curves give the approximate positions of the experimental powder lines for Fe^{3+}-EDTA and Fe^{3+}-transferrin.

not the same throughout [Oosterhuis (363); Golding, Singhasuwich, and Tennant (185)]. It has even been observed for Mn^{2+} in an As-Te-I glass [Nicklin, Poole, and Farach (361)].

C. Individual Molecules

Ferrihemoglobin. The Fe^{3+} ion in symmetrical surroundings yields a more conventional ESR signal that that observed in ferrichrome, etc., discussed above. It is not within the scope of this book to pursue biological molecules, but Fe^{3+} in a square planar symmetry is ideally given by the ferrihemoglobin complex indicated in Fig. V-24. When R is H_2O or F^-, the spectrum is that of a $3d^5$ ion with $D \gg g\beta H$, i.e., only the $|+\frac{1}{2}\rangle \leftrightarrow |-\frac{1}{2}\rangle$ transition is observed. This appears between $g^e = 6$ and $g^e = 2$ in X-band, as expected from Fig. V-16. However, when $R = OH^-$ or N_3^-, the axial symmetry along the normal to the plane is lost, and the complex is more nearly octahedral. The five electrons have entered the three t_{2g} orbitals lying appreciably below the e_g orbital, and the three g components are in the range of $g_z \simeq 3$, $g_y \simeq 2.2$, $g_x \simeq 1.7$ [Gibson, Ingram, and Schonland (181)].

FeF_3. This molecule was vaporized from the solid salt and trapped in an argon matrix at 4°K [DeVore, Van Zee, and Weltner (141)]. Its observed X-band ESR spectrum contained only the one strong line at 1125 G shown in Fig. V-25, indicating that $D > h\nu$ and that it is the transition within the lowest $m_S = \pm\frac{1}{2}$ Kramers' doublet. The line exhibits a small hyperfine splitting (26 G) due to the three equivalent ^{19}F nuclei ($I = \frac{1}{2}$). This small hfs implies that the molecule is very ionic and that the spins are confined largely to the d^5 configuration of the Fe^{3+} ion. Theory indicates that FeF_3 is planar and of D_{3h}

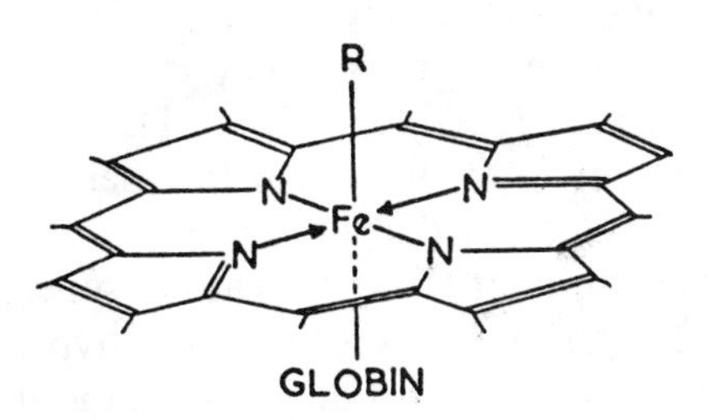

Fig. V-24. Surroundings of the iron atom in the center of the hemoglobin molecule [from Gibson, Ingram, and Schonland (181)].

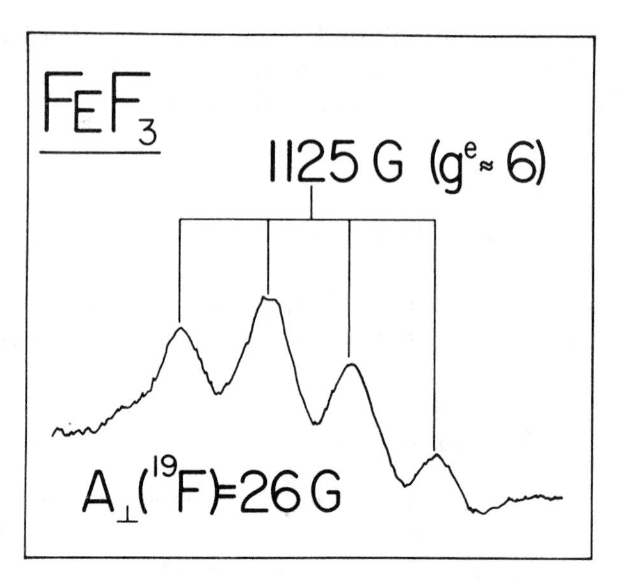

Fig. V-25. ESR spectrum of the FeF_3 molecule in an argon matrix at 4°K. The ^{19}F hyperfine splitting is 26 G. [from DeVore, Van Zee, and Weltner (141). (This figure was originally presented at the Fall 1977 Meeting of The Electrochemical Society, Inc. held in Atlanta, Georgia)].

symmetry so that it would have a $^6A_1'$ ground state [Hand, Hunt, and Schaefer (207); Yates and Pitzer (482)]. The center of the observed ESR line has an effective $g^e = 5.960$, and if (V, 28) is applied, assuming $g_\perp \leqslant g_e$, a lower limit of $|D| \geqslant 0.84$ cm^{-1} is obtained.

MnF_2, MnH_2, and MnO. Two strong out-of-phase transitions are observed in the X-band ESR spectrum of MnF_2 when it is trapped in neon and argon matrices at 4°K [DeVore, Van Zee, and Weltner (140)]. They are shown in an argon matrix in Fig. V-26 and are assigned to the perpendicular xy_1 and xy_3 transitions of a $^6\Sigma$ molecule with $D = 0.34$ cm^{-1} (see Fig. V-16). The two lines have the same overall width, but only the one at low field exhibits resolved hfs. The sextet of triplets on that line is in accord with nuclear interaction with two equivalent ^{19}F ($I = \frac{1}{2}$) atoms superimposed on the six-line ^{55}Mn ($I = \frac{5}{2}$) hfs. The lack of the same hyperfine pattern on the high-field (~4000 G) line is not understood, but it may be due to overlapping with an expected nearby parallel line. The assignment of

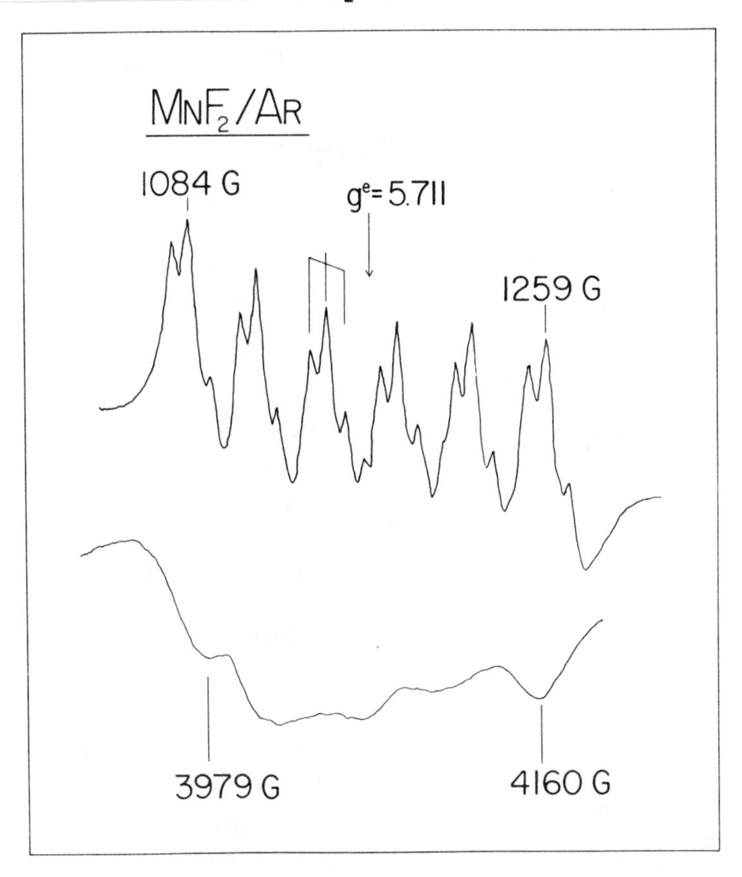

Fig. V-26. Strong perpendicular lines in the ESR spectrum of the MnF_2 molecule in an argon matrix at 4°K (ν = 9.3820 GHz) [from DeVore, Van Zee, and Weltner (140)].

the two lines and the value of D was corroborated by the observations of three relatively weak off-principal-axis transitions at 5.56, 7.92, and 9.11 kG as predicted by the θ versus H plot in Fig. V-27 [Van Zee, Brown, and Weltner (446)]. Expanded views of these three weak broad lines are also shown in that figure; their calculated positions are rather sensitive to the assumed values of $|D|$ and $g_\perp$. If $g_\parallel$ is assumed equal to g_e, then the final derived parameters in solid argon were $|D| = 0.341(5)$ cm^{-1}, $g_\perp = 1.999(5)$. Again the small ^{19}F hfs of 23 G indicates very small spin density on fluorine and a very ionic molecule.

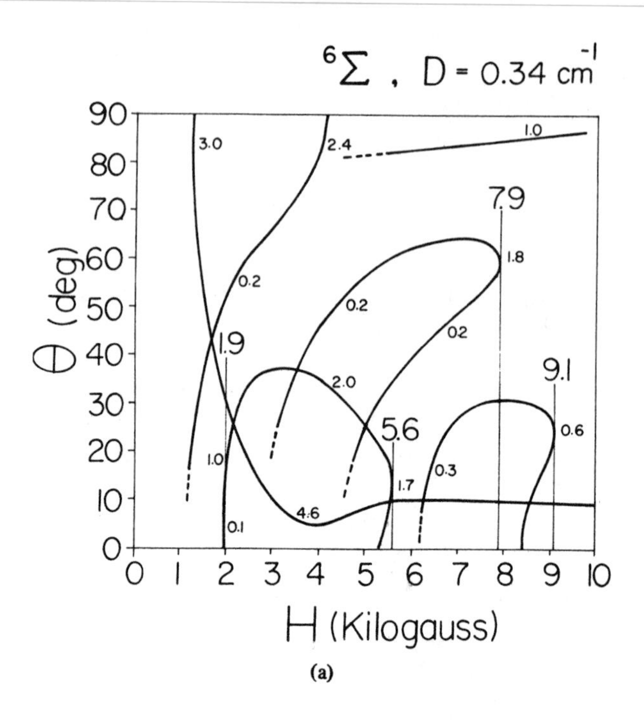

(a)

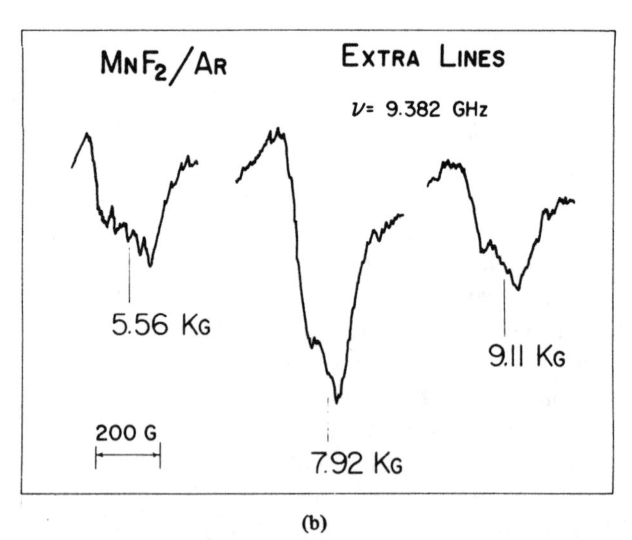

(b)

Fig. V-27. Off-principal-axis transitions in the ESR spectrum of the MnF_2 molecule in an argon matrix at 4°K [from Van Zee, Brown, and Weltner (446)]. (a) Computed θ versus H plot for a $^6\Sigma$ molecule with $D = 0.34$ cm^{-1}, $g_{\parallel} = g_{\perp} = g_e$ and $\nu = 9.382$ GHz. Small numbers indicate relative transition probabilities; large numbers indicate positions (in kilogauss) expected for "extra" lines. (b) Observed three weak "extra" lines.

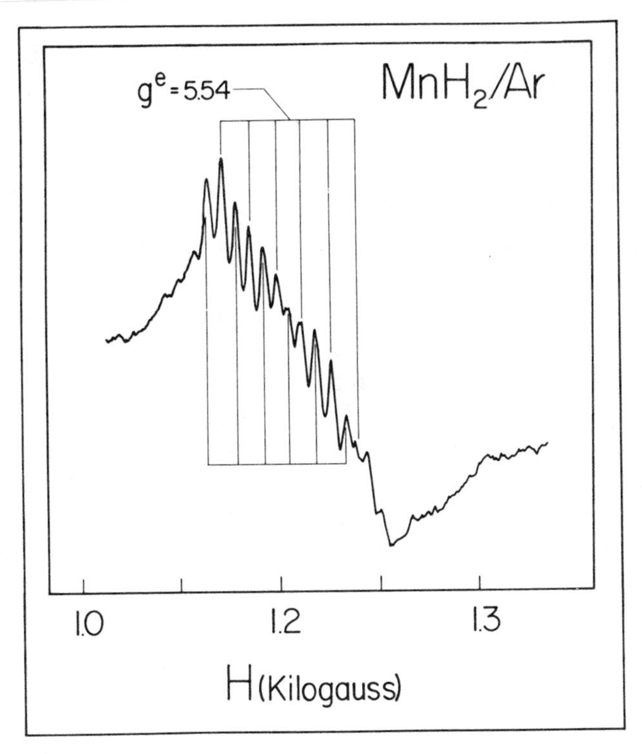

Fig. V-28. ESR spectrum of the MnH_2 molecule in an argon matrix at 4°K [from Van Zee, DeVore, Wilkerson, and Weltner (449)]. (ν = 9.385 GHz.)

Similar ESR observations were made of the MnH_2 molecule formed by the reaction of Mn and H atoms during the condensation of an argon matrix at 4°K [Van Zee, DeVore, Wilkerson, and Weltner (449)]. Here only one strong line (xy_1) was observed at 1211 G ($g^e \simeq 6$), as shown in Fig. V-28; the other strong perpendicular line is judged to lie under the very intense Mn atom signals near $g_e = 2$. Then a value of $D = 0.26$ cm^{-1} is indicated, and again several weak "extra" lines were predicted and observed [Van Zee, Brown, and Weltner (446)]. The fitting of all observed lines leads to a set of parameters in solid argon: $|D| = 0.258(5)$ cm^{-1}, $g_\perp = 2.013(5)$, with $g_\parallel$ assumed equal to g_e. The only anomalous feature of the spectrum is the appearance of 12 hf lines in Fig. V-28 rather than the 18 lines expected for the linear $^6\Sigma$ molecule containing two equivalent ^{1}H ($I = \frac{1}{2}$) atoms. The corresponding transition in the

ESR spectrum of MnD_2 also had complex hf structure. It was proposed that the molecule might be slightly bent so that a small splitting of the low-field line was producing the observed hyperfine effects (449). Infrared absorptions also favored a bent structure, but subsequent theoretical studies by Demuynck and Schaefer (136) favored a linear structure, although the potential curve for bending is very flat. It is also possible that an off-principal-axis line might occur in the neighborhood of the 1211 G line of the linear molecule, contributing to the observed line shape (446). However, even though the molecular configuration is not definitely established, the zfs parameter $|D|$ should be reasonably accurate.

Only the one strong fine-structure line at $g^e \simeq 6$ was definitely observed for the MnO molecule, as shown in Fig. V-29 [Ferrante, Wilkerson, Graham, and Weltner (162)]. The well-resolved ^{55}Mn hfs is relatively large, and its spacing exhibits a strong "second-order" shift. Contrary to the analysis of Ferrante et al., an exact treatment

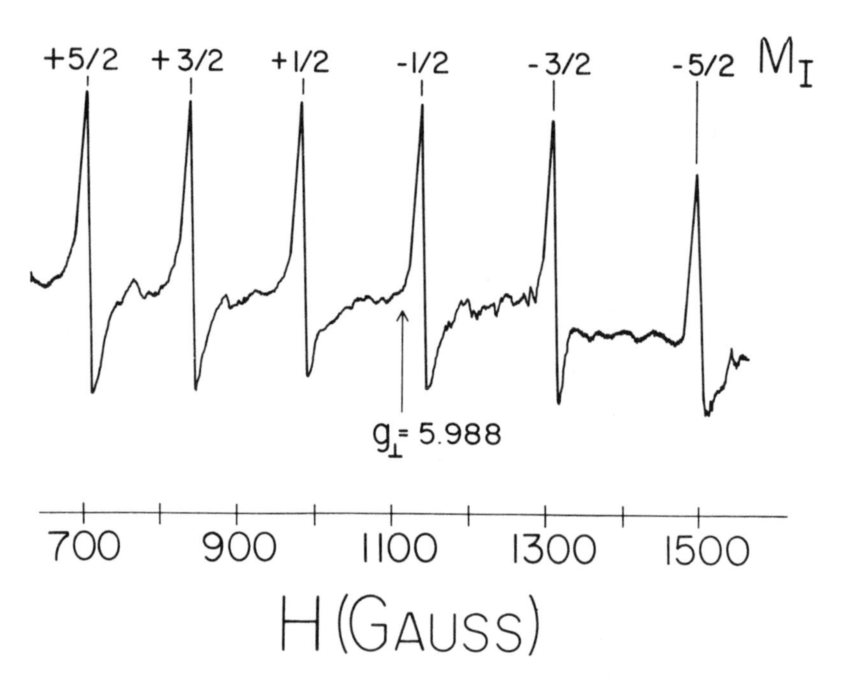

Fig. V-29. ESR spectrum of the ^{55}MnO molecule in an argon matrix at 4°K [from Ferrante, Wilkerson, Graham, and Weltner (162)]. (ν = 9.380 GHz.)

of this hfs is necessary, i.e., using the matrix given in Table V-2. Even then an unambiguous determination of the g values and D could not be made from the observation of just this one fine-structure line. For example, comparable fits to the line positions (± 2 G) were obtained, assuming $g_{\parallel} = g_e$, for $g_{\perp} = 2.000(1)$, $|D| = 1.0$ cm^{-1}, $A_{\perp} = 154.9(1)$ G, $A_{\parallel} = 170(10)$ G and for $g_{\perp} = 1.991(1)$, $|D| = 2.26$ cm^{-1}, $A_{\perp} = 154.2(1)$ G, $A_{\parallel} = 168(10)$ G [Van Zee, Baumann, and Weltner (445)]. It is reasonable to assume that both A values are positive, and one notes that their magnitudes are little affected by the variations in g and D values. These hf parameters are considerably changed from those obtained in (162) and used (assuming them accurate) by Gordon and Merer (191) in their analysis of the optical spectrum of gaseous MnO. One finds from the ESR data that $A_{\text{iso}} = 160$ G and $A_{\text{dip}} = 5$ G for ^{55}Mn. Using a simple model, we can calculate the contributions to A_{iso} and A_{dip} for the configuration $\sigma\pi^2\delta^2$ where σ is an sd_{z^2} hybrid, and the other orbitals are essentially Mn atomic d orbitals. Then $A_{\text{dip}} \simeq (1/2)\,(1/5)a_{\text{dip}}{}^{d\sigma} + (2/5)a_{\text{dip}}{}^{d\pi} + (2/5)a_{\text{dip}}{}^{d\delta}$ and $A_{\text{iso}} \simeq (1/2)\,(1/5)a_{\text{iso}}{}^{4s}$. a_{dip} and a_{iso} are atomic values obtained from the $3d^5 4s^2$ state of the atom. For atomic ^{55}Mn according to Appendix B, $a_{\text{iso}} = 1797$ G and $a_{\text{dip}} = +64$, $+32$, and -64 G for d_{z^2}, $d_{xz,yz}$, and $d_{x^2-y^2,xy}$, respectively, and one calculates $A_{\text{dip}} = -6$ G and $A_{\text{iso}} = 180$ G, in fair agreement with the observations. [See Ferrante et al. (162) for further discussion.]

However, it is not clear whether the ESR spectrum of MnO supports the zfs value of $D = 1.32$ cm^{-1} found from the emission spectrum of the gas molecule by Pinchemel and Schamps (375), or whether the true value is higher than that. This would be settled if a high-field line in the X-band spectrum were observed; however, such a line can be expected to be relatively weak. $D = 1.32$ cm^{-1} would place this xy_3 line near 8000 G.

4. $^7\Sigma$ MOLECULES ($S = 3$) IN MAGNETIC FIELDS

Extension of the $S = 2$ spin Hamiltonian in (V, 19) requires only that $O_6{}^m$ terms also be permitted when $S = 3$; however, the extent and accuracy of the experimental data to date have not warranted inclusion of either $O_4{}^m$ or $O_6{}^m$ terms in the analysis (see, however, section V-5B). Hence only the g tensor components and $D = 3B_2{}^0$ enter as parameters, unless hyperfine interaction is to be included.

For molecules containing ^{55}Mn, with $I = \frac{5}{2}$, this would lead to a 42 × 42 eigenvalue matrix, since $m_S = 0, \pm 1, \pm 2, \pm 3$; $m_I = \pm\frac{1}{2}, \pm\frac{3}{2}, \pm\frac{5}{2}$. If the hfs is not included, the calculated Zeeman levels are as given in Fig. V-30 for linear molecules with various values of D (assumed positive), $g_{\parallel} = g_{\perp} = g_e$, and $\nu = 9.3$ GHz. The zfs between successive levels is now D, $2D$, and $3D$. As earlier, two orientations of the static field with respect to the molecular axis are considered, $\theta = 0°$ (parallel) and $\theta = 90°$ (perpendicular), and the $\Delta m_S = \pm 1$ transitions indicated. From these data are then obtained the D versus H_r curves shown in Fig. V-31. The analogy with the $S = 2$ case is evident, with the addition of the possibility of observing two more transitions at each orientation. Even for molecules with D considerably greater than 1 cm^{-1}, one could expect the xy_2 transition to be observable.

A. Theory

(a) Randomly Oriented Molecules. For $D \ll h\nu$, the expected absorption pattern of six Δm_S transitions would be spread out about $g =$ 2.0 as shown in Fig. V-32. The relative intensities of the out-of-phase $1 \leftrightarrow 0$ and $0 \leftrightarrow -1$ perpendicular lines would then be expected to be much greater than the other fine-structure transitions. For the MnH molecule, for example, where $D = -0.002(1)$ cm^{-1}, transitions involving $m_S = \pm 3$ were not observable in argon matrix spectra [Van Zee, De Vore, Wilkerson, and Weltner (449)]. As D becomes greater than about 0.05 cm^{-1}, Fig. V-31 shows that the field positions of some allowed transitions will become very sensitive to D. In Fig. V-33 a computer-simulated spectrum for randomly oriented $^7\Sigma$ molecules with $D = 0.09$ cm^{-1} indicates clearly the complexity of the spectrum that occurs. Many of the relatively strong lines are due to off-principal-axis transitions, as can be seen from the shapes of the curves in the θ versus H diagram.

B. Individual Molecules

MnH, MnF. Since the Mn^+ ion has a $(3d^5 4s)\,^7S$ ground state, these ionic molecules are expected to have $^7\Sigma$ ground states. Gas-phase optical spectroscopy, particularly by Nevin and co-workers (360), have established this and have supplied the zfs parameter $D = -0.002$ cm^{-1} for MnH [see also Kovacs and Pacher (296)]. Bagus and

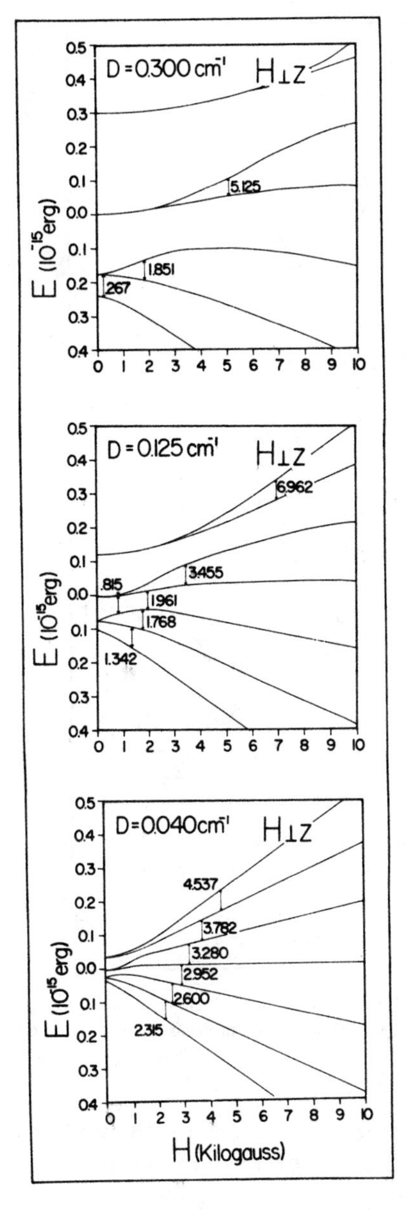

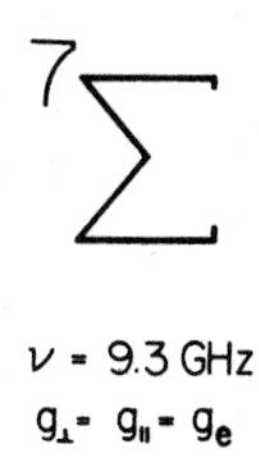

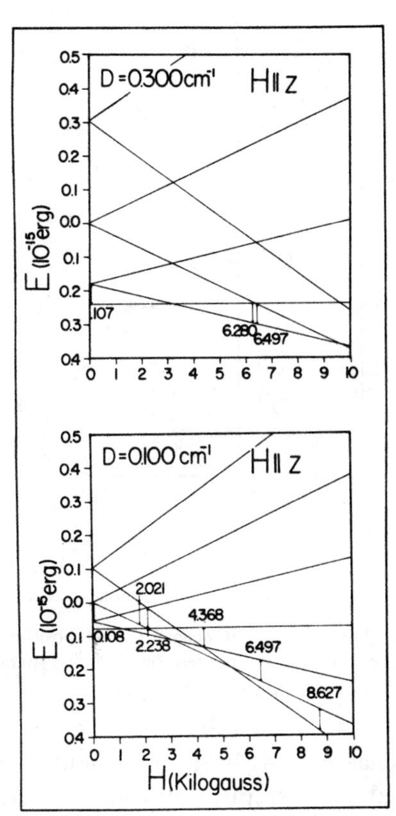

Fig. V-30. Zeeman levels and allowed $\Delta m_S = \pm 1$ transitions of a $^7\Sigma$ molecule for various values of the zero-field-splitting parameter D. Left side, $H \perp z$; right side, $H \parallel z$, where z is the molecular axis. $\nu = 9.3$ GHz ($= 0.0616 \times 10^{-15}$ erg $= 0.31$ cm^{-1}), $g_\parallel = g_\perp = g_e$. The energy level spacings at zero field are D, $2D$, and $3D$.

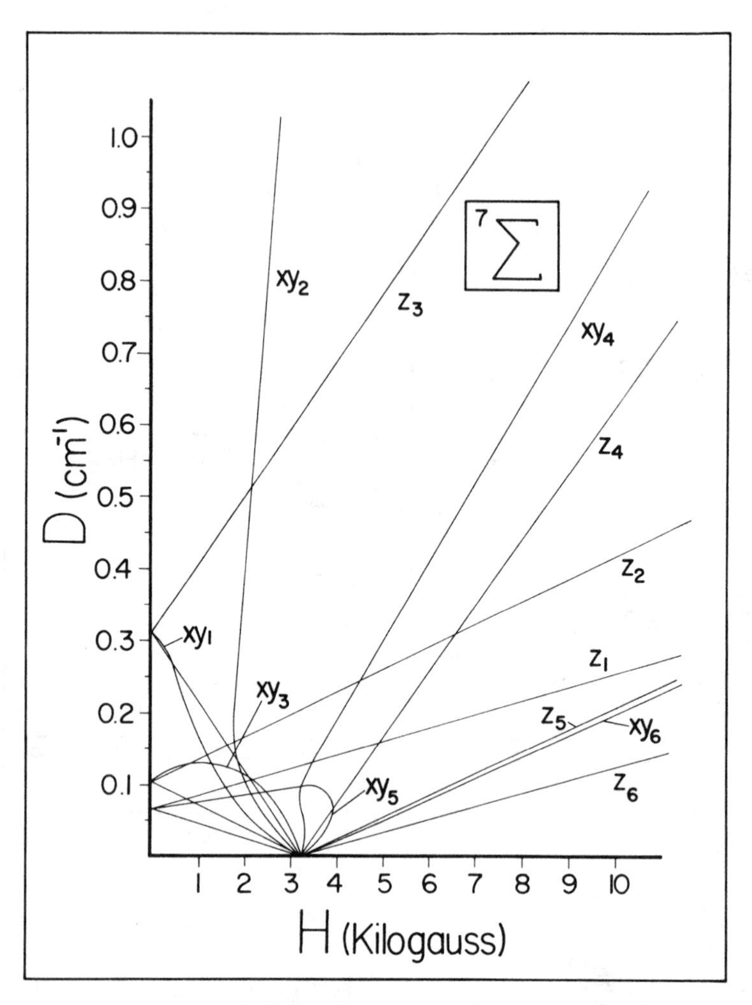

Fig. V-31. Resonant magnetic field positions at which $\Delta m_S = \pm 1$ transitions occur for various values of the zero-field-splitting parameter D for a $^7\Sigma$ molecule. It is assumed that $g_{\parallel} = g_{\perp} = g_e$ and $\nu = 9.3$ GHz.

Schaefer (44) have carried out *ab initio* calculations on the hydride and derived the electronic wavefunctions of the ground and first excited $^7\Pi$ state.

The ESR spectra of these molecules trapped in rare-gas matrices indicate that in both cases the zfs is quite small. Then, because of the strong angular dependence of the intensities (see Fig. V-32)

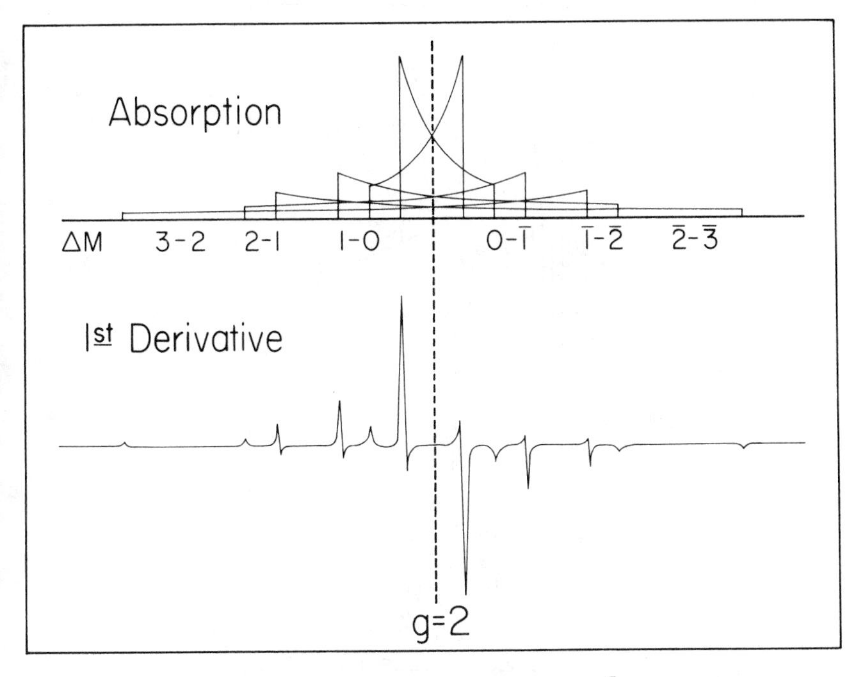

Fig. V-32. ESR absorption and spectrum of randomly oriented $^7\Sigma$ molecules for $D << h\nu$ [from DeVore, Van Zee, and Weltner (140)]. (Top) Theoretical absorption indicating variation from parallel (z_i) to perpendicular (xy_i) orientation of the magnetic field relative to the molecular axis for the six $\Delta m_S = \pm 1$ transitions (see Fig. V-30); (bottom) ESR spectrum (first derivative of the total absorption curve). Perpendicular (xy_i) lines dominate, particularly the out-of-phase $m_S = +1 \leftrightarrow 0$ and $-1 \leftrightarrow 0$ transitions.

only the $\Delta m_S = \pm 1$ transitions involving the $m_S = 0$ and ± 1 levels were observable in both molecules; however, weak $\pm 1 \leftrightarrow \pm 2$ transitions were also detected for MnH. The center of each spectrum is obscured by strong Mn atom absorptions extending over several hundred gauss at $g = 2.0$.

A portion of the observed X-band spectrum of $^{55}Mn^{19}F$ in an argon matrix at 4°K is shown in Fig. V-34; other lines were observed outside of that range of magnetic field, but they were much weaker [DeVore, Van Zee, and Weltner (140)]. The 20 G ^{19}F hfs is indicated. All of the lines are perpendicular (xy) lines, and no parallel transitions could be definitely identified. Essentially the same set of $m_S = 0 \leftrightarrow \pm 1$ transitions were observed in a neon matrix. The spectra

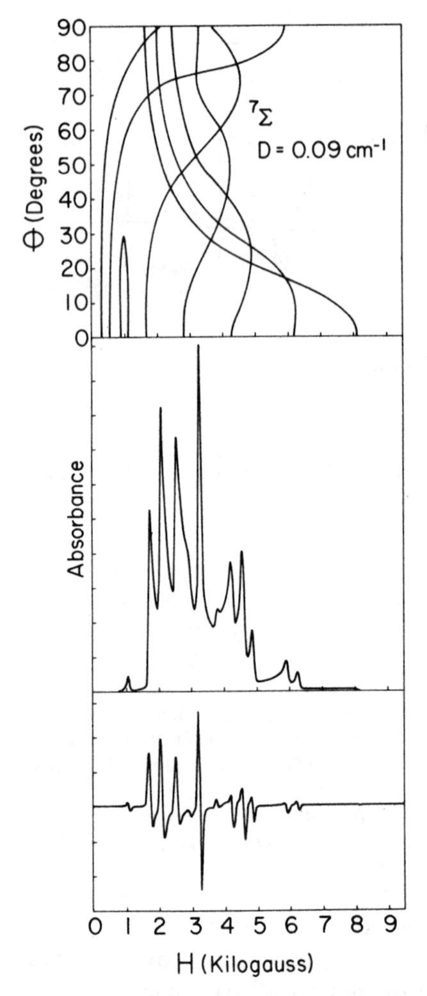

Fig. V-33. Computed variation of allowed ($\Delta m_S = \pm 1$) transitions with orientation (top), absorption spectrum (middle), and ESR spectrum (bottom) of randomly oriented $^7\Sigma$ molecules with $D = +0.09$ cm^{-1}. ($g_\parallel = g_\perp = g_e$, $\nu = 9.3$ GHz.)

of MnH and MnD in argon matrices were similar to these, but the hydrogen and deuterium hfs were too small to be resolved, indicating that $A_\perp(H) \leqslant 7$ G [Van Zee, DeVore, Wilkerson, and Weltner (449)].

The small D and A values allow the perturbation procedure given in Abragham and Bleaney (1) and Low (15) to be used in the analysis. A comparison of the resulting parameters in argon matrices is given

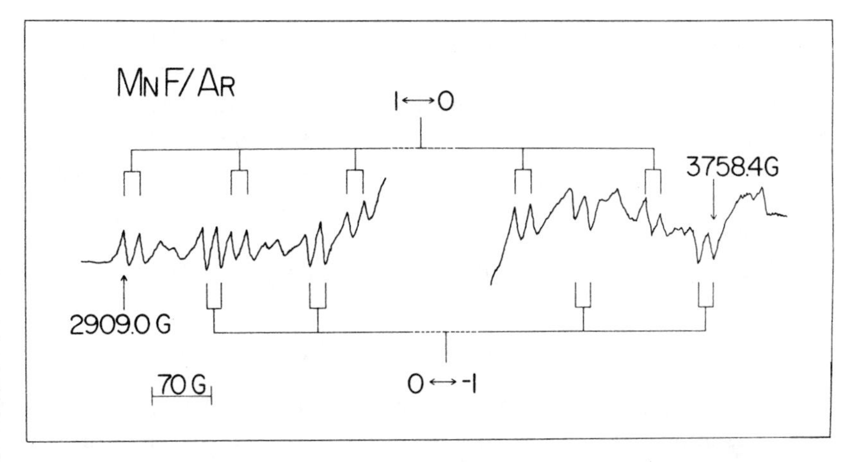

Fig. V-34. A portion of the ESR spectrum of the $^{55}Mn^{19}F$ molecule in an argon matrix at 4°K [from DeVore, Van Zee, and Weltner (140)].

in Table V-3. The spin density on H or F is quite small, amounting to only a few percent in each case. The isolated Mn^+ ion, which has a significant 4*s* contribution to its $3d^5 4s$ configuration, may be estimated to have $A_{iso}(Mn^+) = 770$ MHz [Kasai (259)]. Then the lowering of the 4*s* character in the molecular ion relative to that in

Table V-3. Magnetic Parameters of the $^7\Sigma$ MnH and MnF Molecules in Argon Matrices at 4°K

	MnH[a]	MnF[b]
$D(\mathrm{cm}^{-1})$	−0.0027	−0.0083
$g_\perp$	2.001(1)	2.001(1)
$\lvert A_\perp(\mathrm{Mn})\rvert$ (MHz)	299(2)	427(1)
$\lvert A_\parallel(\mathrm{Mn})\rvert$ (MHz)	322(6)	475(5)
$\lvert A_\perp(\mathrm{X})\rvert$ (MHz)	⩽20	56(1)
$\lvert A_\parallel(\mathrm{X})\rvert$ (MHz)		85(2)
$\lvert A_{iso}(\mathrm{Mn})\rvert$ (MHz)	306(3)	443(6)
$\lvert A_{dip}(\mathrm{Mn})\rvert$ (MHz)	8(2)	16(6)
$[A_{iso}(\mathrm{Mn})/770]$ (%)[c]	40	58

[a]From Van Zee, DeVore, Wilkerson, and Weltner (449).
[b]From DeVore, Van Zee, and Weltner (140).
[c]Comparison of the Mn^+ 4*s* character in the molecule with that in the isolated atomic ion (see text).

the atomic ion is given in the last row of Table V-3. This is to be expected for an ion pair where each ion suffers distortion (polarization) due to the charge on the other. The spherical $4s$ orbital of Mn^+ in the molecule is thereby mixed with $3d_{z^2}$ and $4p_z$ (z = molecular axis) to form a hybrid orbital such that electron density is shifted away from the anion, as has been shown to be the case for the alkaline-earth monofluorides [see section II-6b].

The calculations of Bagus and Schaefer (44) on MnH give the ground configuration of MnH (over closed shells) as

$$\cdots (6\sigma)(7\sigma)^2 (8\sigma)(3\pi)^2 (1\delta)^2$$

where 6σ, 3π, and 1δ are nearly pure $3d$ orbitals. The 7σ bonding orbital is about 16% Mn $4s$, 8% Mn $4p$, and 72% H $1s$. The orbital of concern here is the 8σ nonbonding orbital with 60% Mn $4s$ (and 2% H $1s$), which may be compared with the 40% estimated from the ESR.

Organic Radicals. Two septet ground state molecules have been observed. The ESR spectrum of the trinitrene (I) was first measured at 77°K by Wasserman, Schueller, and Yager (459a) and found to have $D = 0.0548(80)\ \text{cm}^{-1}$ and $E = 0$ in accord with the threefold axis of symmetry. Later, Takui and Itoh (428a) observed the odd alternant

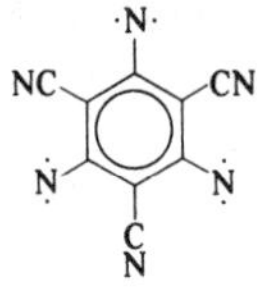

(I)

hydrocarbon benzene-1,3,5-tris-phenylmethylene (II) and established that it also has a septet ground state. Its parameters are $g = 2.0038$, $D = +0.04158$, and $E = \pm 0.01026\ \text{cm}^{-1}$. Variation of the

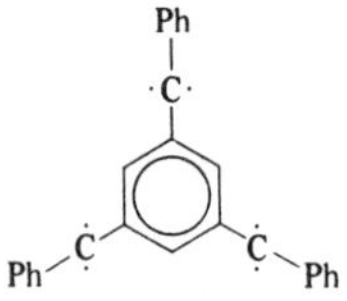

(II)

temperature determined the positive sign of D. This molecule does not have a threefold axis because the bonding at each "carbene" is not linear.

Thus, as expected, these aromatic high spin molecules, with their low spin-orbit coupling constants, have small zero-field-splitting constants, and their ESR spectra are essentially exemplified by Fig. V-32.

5. $^8\Sigma$ ($S = \frac{7}{2}$) AND $^9\Sigma$ ($S = 4$) MOLECULES

Molecules with these high multiplicities can be expected to be rare and will probably usually arise because they contain a rare-earth atom.

By the same procedures used above, one calculates the Zeeman energy level variation with magnetic field shown in Fig. V-35 for an $S = \frac{7}{2}$ $^8\Sigma$ molecule, with the positions of the $\Delta m_S = \pm 1$ transitions indicated for X-band. As in the $S = \frac{5}{2}$ case, for D positive the lowest Kramers' doublet should allow the observation of an ESR absorption regardless of the magnitude of the zfs. This is shown in Fig. V-36 as the xy_1 ($+\frac{1}{2} \leftrightarrow -\frac{1}{2}$) transition at $g^e \simeq 8$. Even for quite large D values the xy_3 transition should also be observable in the 0 to 10 kG range.

A. Axial and $^8\Sigma$ ($S = \frac{7}{2}$) Molecules

GdF_3. This pyramidal (C_{3v}) molecule is essentially a $Gd^{3+}(4f^7)$ ion with three F^- ligands around an axis of trigonal symmetry. It therefore conforms to the linear case where $E \equiv 0$, but higher terms in $O_n{}^m$ in the spin Hamiltonian could be nonvanishing. Its observed ESR spectrum in a neon matrix at 4°K is shown in the lower part of Fig. V-37, and the simulated spectrum for $g_\perp = 1.990$, $g_\parallel = 1.995$, $D = +0.433$ cm^{-1}, $S = \frac{7}{2}$, in the upper part [Baumann, Van Zee, Zeringue, and Weltner (52)]. The complete calculated θ versus H variation, absorbance, and its first derivative (same as in Fig. V-37) are shown in Fig. V-38. xy_1 and xy_3 determine the two most intense lines, and the remainder are caused by off-principal-axis absorptions. It was found here that inclusion of terms involving $B_4{}^m$, etc., coefficients was not necessary to obtain a satisfactory fit of the line positions.

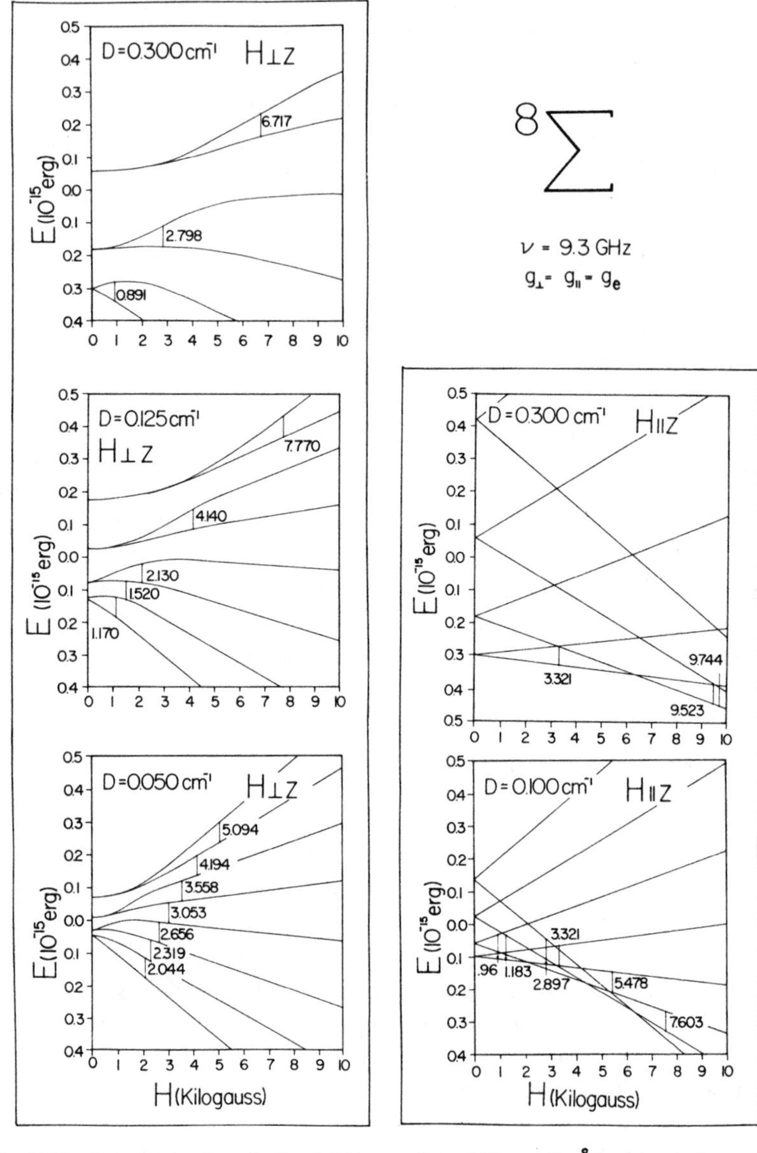

Fig. V-35. Zeeman levels and allowed $\Delta m_S = \pm 1$ transitions of a $^8\Sigma$ molecule for various values of the zero-field-splitting parameter D [from Baumann, Van Zee, Zeringue, and Weltner (52)]. Left side, $H \perp z$; right side, $H \parallel z$, where z is the molecular axis. $\nu = 9.3$ GHz ($=0.0616 \times 10^{-15}$ erg $= 0.31$ cm^{-1}), $g_{\parallel} = g_{\perp} = g_e$. The energy level spacings at zero magnetic field are $2D$, $4D$, and $6D$.

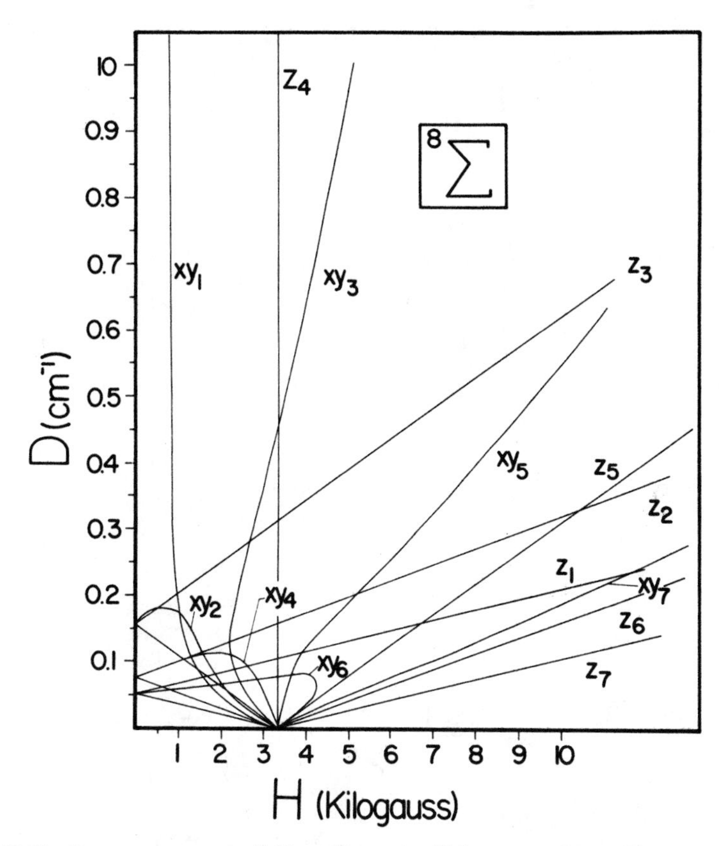

Fig. V-36. Resonant magnetic field positions at which $\Delta m_S = \pm 1$ transitions occur for various values of the zero-field-splitting parameter D for a $^8\Sigma$ molecule [from Baumann, Van Zee, Zeringue, and Weltner (52)]. It is assumed that $g_\perp = g_\parallel = g_e$ and $\nu = 9.3$ GHz.

No hfs, due to either ^{155,157}Gd or ^{19}F, was discernible on any of the lines. This is not surprising for Gd isotopes, since such splittings are small for Gd^{3+} in crystals. The natural abundance of each isotope is only about 15%, but hfs on the intense line at 871 G should have been detectable if larger than the linewidth. For ^{19}F it is also not unexpected, since the molecule is highly ionic, and other fluorides (e.g., MnF_2, TiF_2) show either small splittings or none at all.

CrCu. This molecule was prepared by the simultaneous vaporization and condensation of Cr and Cu atoms in solid krypton at 4°K. It is possible that it has a $^6\Sigma$ ground state, but a corrected analy-

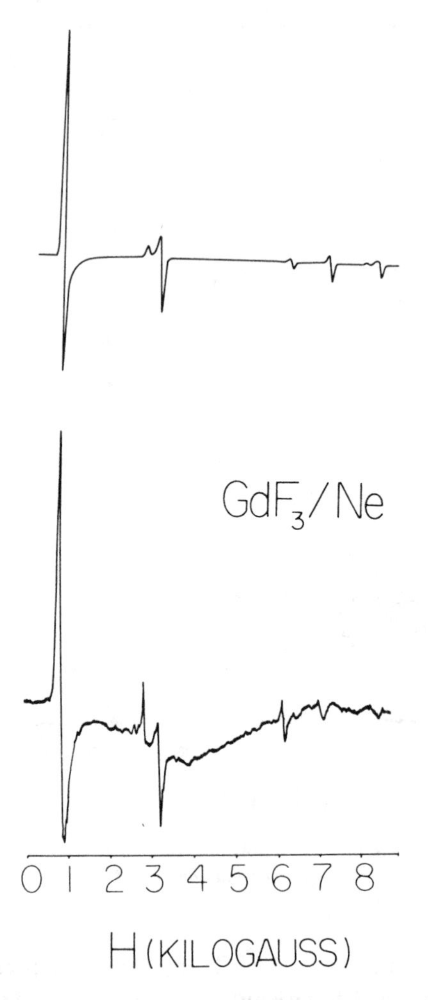

Fig. V-37. Computed (upper) and observed (lower) ESR spectrum of the GdF_3 molecule in a neon matrix at 4°K [from Baumann, Van Zee, Zeringue, and Weltner (52)]. $g_\perp = 1.990$, $g_\parallel = 1.995$, $D = +0.433\ \mathrm{cm}^{-1}$, $\nu = 9.3985$ GHz.

sis of the ^{63,65}Cu hyperfine structure indicates that it is more probably $^8\Sigma$ [Van Zee and Weltner (452)]. The observed X-band spectrum in Fig. V-39 shows clearly the ^{63}Cu and ^{65}Cu hfs with approximately the expected relative intensity ratio of 0.4 and hyperfine splitting ratio of about 0.9. The spectrum is essentially centered at $g = 2$ so that it is in accord with the observation of a $^6\Sigma$ or an $^8\Sigma$ molecule

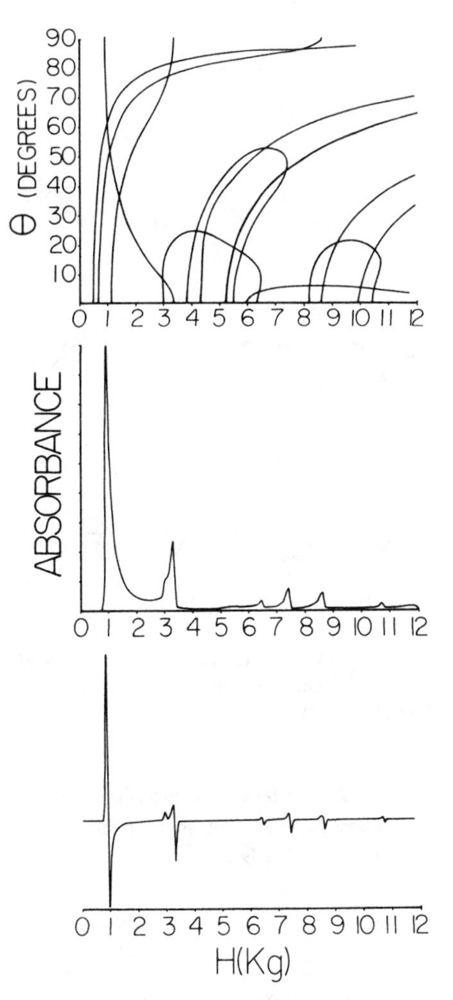

Fig. V-38. Simulated ESR spectrum of the GdF_3 molecule in a neon matrix at 4°K [from Baumann, Van Zee, Zeringue, and Weltner (52)]. (Top) θ versus resonant magnetic field variaton; (middle) absorption spectrum; (bottom) ESR spectrum. Parameters used: $g_\perp$ = 1.990, $g_\parallel$ = 1.995, D = +0.433 cm^{-1}, ν = 9.3985 GHz, and $S = \frac{7}{2}$

with a low D value. In either case the central strong xy transition (see Fig. V-17) would be dominant in intensity. The observed $A_{\text{iso}}(^{63}\text{Cu})$ = 781(25) MHz, when multiplied by $2S = 2(\frac{7}{2}) = 7$ and divided by A_{iso}(atom) = 5995 MHz [Morton and Preston (350)], yields 91% s character on Cu. This would arise if a $3d^5 4s$ Cr atom

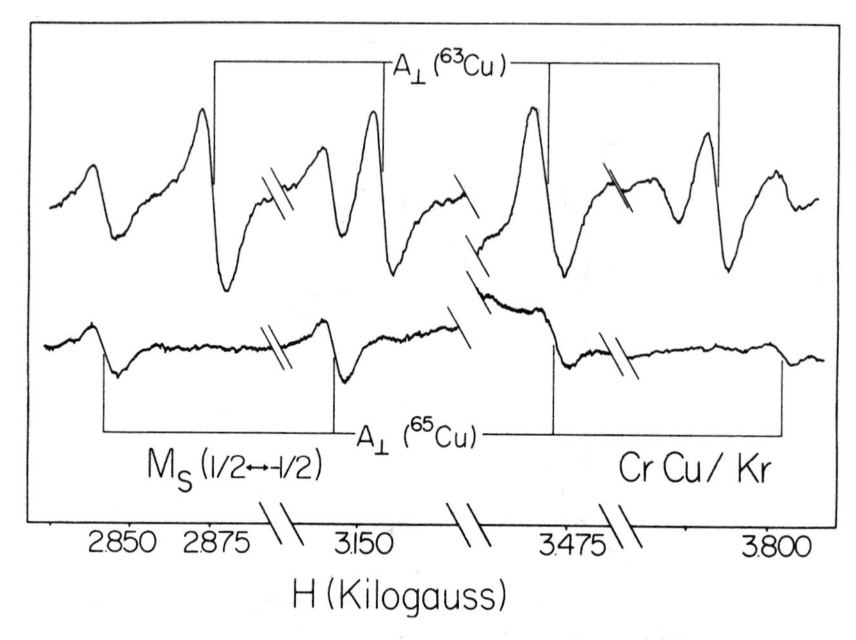

Fig. V-39. ESR spectra of the CrCu molecule in a krypton matrix at 4°K containing (top spectrum) natural abundance ^{63,65}Cu and (bottom spectrum) isotopically enriched ^{65}Cu [from Van Zee and Weltner (452)]. ($\nu \simeq 9.385$ GHz.)

Table V-4. Electronic and Magnetic Parameters for CrCu in a Krypton Matrix at 4°K

D (cm^{-1})	−0.007(5)
$g_{\parallel}$ [a]	(2.0023)
$g_{\perp}$	2.0030(10)
$\lvert A_{\parallel}(^{63}\text{Cu})\rvert$ (MHz)	631(60)
$\lvert A_{\perp}(^{63}\text{Cu})\rvert$ (MHz)	836(9)
$A_{iso}(^{63}\text{Cu})$ (MHz) [b]	781(25)
$A_{dip}(^{63}\text{Cu})$ (MHz) [b]	−75(20)
% s character (^{63}Cu) [c]	91
$\lvert\Psi(0)_{Cu}\rvert^2$ (a.u.)	0.659(20)
$\lvert eqQ(^{63}\text{Cu})\rvert$ (MHz)	310(120)

[a] Assumed value.
[b] Calculated assuming $A_{\parallel}$ and $A_{\perp}$ are positive.
[c] Atomic values for comparison taken from Appendix B. This value assumes an $^8\Sigma$, rather than a $^6\Sigma$, ground state.

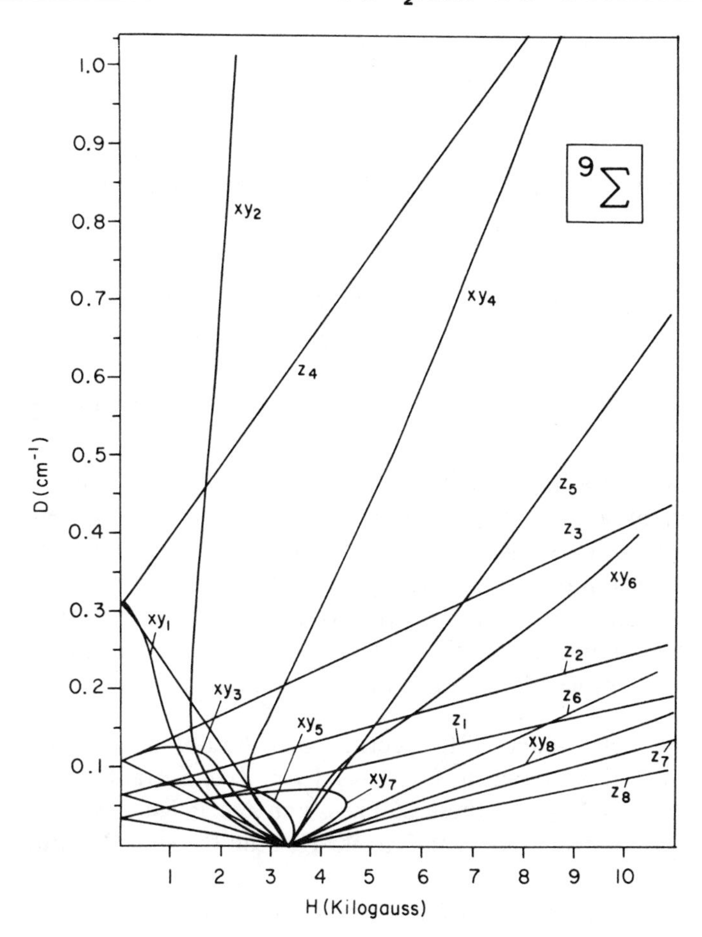

Fig. V-40. Resonant magnetic field positions at which $\Delta m_S = \pm 1$ transitions occur for various values of the zero-field-splitting parameter D for a $^9\Sigma$ molecule [from Van Zee, Ferrante, Zeringue, and Weltner (450)]. It is assumed that $g_\perp = g_\parallel = g_e$ and $\nu = 9.3$ GHz.

bonded with a $3d^9 4s^2$ Cu atom through the $4s$ electrons and left a $3d$ hole and an unpaired $4s$ electron on Cu. The final derived parameters are given in Table V-4. A large nuclear quadrupole coupling constant $eqQ(^{63}Cu)$ was found necessary to fit the observations for the natural abundance and ^{65}Cu-enriched molecules.

B. $^9\Sigma$ ($S = 4$) Molecules

A D versus H_r plot is given in Fig. V-40 for the $^9\Sigma$ case where $O_4{}^m$ and higher terms in the spin-Hamiltonian have been neglected.

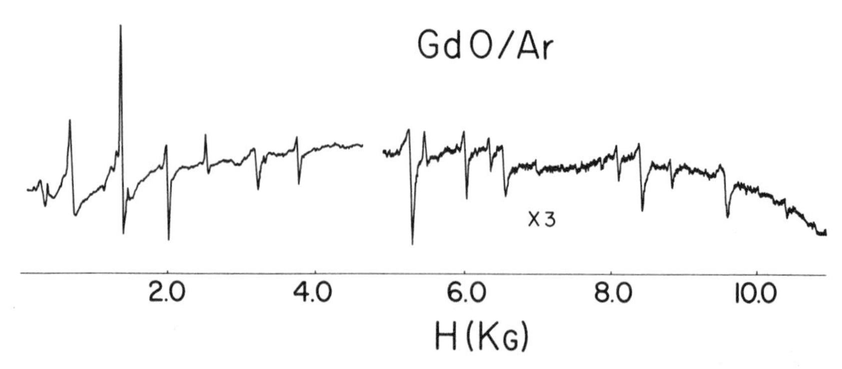

Fig. V-41. ESR spectrum of the GdO molecule in an argon matrix at 4°K [from Van Zee, Ferrante, Zeringue, and Weltner (450)]. ($\nu = 9.3790$ GHz.)

From this figure it can be seen that for small zero-field splitting the spectrum of a randomly oriented molecule will be very complex, since there will be eight allowed on-axis transitions plus perhaps a comparable number of off-principal-axis transitions. However, for $D >> h\nu$ again great simplification occurs, and only xy_2 and xy_4 would be observable. The GdO molecule appears to be the only example of a molecule with this high multiplicity.

GdO. The rich X-band ESR spectrum of GdO obtained in an argon matrix at 4°K is shown in Fig. V-41 [Van Zee, Ferrante, Zeringue, and Weltner (450)]. Of these 16 lines, Van Zee assigned 4 as xy lines and the remaining 12 as "extra" lines. He used the spin Hamiltonian:

$$\mathcal{H} = g_{\parallel}\beta H_z S_z + g_{\perp}\beta(H_x S_x + H_y S_y) + b_2{}^0(S_z{}^2 - \tfrac{20}{3}) + (b_4{}^0/60)(35S_z{}^4 - 575S_z{}^2 + 1080) \quad (V, 31)$$

where, in the earlier symbolism, $b_2{}^0 = D = 3B_2{}^0$ and $b_4{}^0 = 60B_4{}^0$. The resulting parameters are given in Table V-5.

From the lack of ^{155,157}Gd hfs, it is implied that the 6s character of the unpaired spins is small and that the Gd^{2+} ion configuration is largely $4f^7 5d\sigma$. (There is probably some covalent bonding so that reference to Gd^{2+} is only an approximation.)

6. HYPERFINE INTERACTION IN HIGH-SPIN MOLECULES

A 4s electron on the Mn atom is calculated (see Table B1) to provide a hyperfine splitting of 5036 MHz, but the $(3d^5 4s^1)$ Mn^+ ion

Table V-5. Magnetic Parameters for $X^9\Sigma$ GdO in Argon and Neon Matrices at 4°K

	Argon	Neon
$b_2{}^0(\text{cm}^{-1}) = D$	±0.2078(3)	±0.1960(5)
$b_4{}^0(\text{cm}^{-1})$	±0.00040(5)	±0.00040[a]
$g_\parallel$	1.990(9)	–
$g_\perp$	1.986(5)	1.992(5)

[a] Neon value assumed from argon.

is observed to have an hfs of 770 MHz [Kasai (259); Ferrante et al. (162)]. (The charge on the ion will make this value somewhat higher than predicted from the neutral atom.) The electron density at the nucleus has been effectively "diluted" by the five $3d$ unpaired spins so that in place of (II, 43) we would calculate the s electron contribution to the spin density from the approximate expression:

$$\rho_s = \frac{A_{\text{iso}}(\text{Mn}^+)}{A_{\text{iso}}(\text{atom})/6} \qquad \text{(V, 31a)}$$

where one-sixth of the unpaired spins have s character. Then in the MnF molecule where A_{iso} (^{55}Mn) = 454 MHz [DeVore, Van Zee, and Weltner (140)], one derives $\rho_s = 0.58$.

The wavefunction for the $^7S_3$ Mn^+ ion would be written as

$$\psi(\text{Mn}^+) = \frac{1}{\sqrt{6!}} |\chi(4s)(1)\chi(3d_1)(2)\chi(3d_2)(3)\chi(3d_3)(4)\chi(3d_4)(5)\chi(3d_5)(6)|$$

where the right-hand side is a Slater determinant accounting for the permutation of the six electrons among the six orbitals. This will also be multiplied by seven spin factors, but these will be degenerate for this spherical ion in the absence of a magnetic field. Since only the $4s$ electron contributes to $|\psi(0)|^2$ and the permutation of the five equivalent $3d$ electrons yields a factor of $\sqrt{5!}$, then per electron

$$|\psi(0)|^2(\text{Mn}^+) = \tfrac{1}{6}|\psi(0)|_{4s}{}^2$$

In the $^4\Sigma$ VO molecule, as previously discussed in section V-1, the three unpaired electrons can be considered to be approximately in the configuration $4s\sigma^1(3d\delta)^2$ on the vanadium atom, or one-third of the electrons are $4s$ [Kasai (258)]. Thus the same reasoning applies as in calculating the spin density in MnF. Correspondingly, A_{dip} for VO would be expected to be given approximately by $\frac{2}{3} A_{dip}(3d\delta)$ (atom) $= \frac{2}{3} [-\frac{2}{7} P]$ where $P = g_e\beta_e g_I\beta_n \langle 1/r^3 \rangle_{3d}$ is obtained from Table B1 for the vanadium atom. A similar rationalization of the ^{55}Mn hfs observed for MnO was discussed on page 360.

The above discussion may seem to be in contradiction to the statement made earlier that in triplet molecules one does not expect the hyperfine splitting to be twice that for a molecule with only one unpaired electron [see section III-2h]. However, this will only be true if the addition of an unpaired equivalent spin is also accompanied by addition of an equivalent nucleus. Thus, in the $^3\Sigma_g^{+}$ $^{17}O_2$ molecule the equivalent spins in the π_x and π_y molecular orbitals are interacting with the two equivalent oxygen nuclei.

In cases where electron exchange coupling occurs between relatively distant atoms, one observes a halving of the hfs of the two atoms when the exchange coupling is rapid relative to the hyperfine interaction. Thus, as will be seen in the next section, the hfs of ^{55}Mn in the van der Waals molecule Mn_2 is one-half that in an isolated Mn atom and is simply an indication that the spin density in the molecule is distributed over two atoms. For a loose cluster of n atoms, one then expects the observed hfs to be A (atom)/n.

7. ELECTRON EXCHANGE COUPLING

Spins on neighboring metal atoms or ions may exchange either through direct overlap of their orbitals or via overlap with intervening atoms. The latter is called superexchange [Goodenough (7); Anderson (40)]. More than two atoms may undergo exchange, and they may not be the same type of atom, but here we will restrict the discussion to just "dimers" containing the same two kinds of exchanging metal atoms. Included in exchange interactions are contributions from magnetic dipole-dipole coupling between unpaired electrons on the neighboring metal atoms, which can, in fact, be the dominant contributor when the interatomic distance is large.

Such dimer systems can arise in at least three ways: (1) through the natural stability of the dimer in the pure crystal (e.g., in copper acetate crystals); (2) by dilution of a magnetic salt with an isomorphous diamagnetic one such that there are isolated clusters of two magnetic neighbors (e.g., Mn^{2+} in MgO); (3) as dimers formed from atoms during trapping in an inert solid matrix (e.g., Mn_2 in solid argon). In category (1) there are now an exceedingly large number of dimers that have been observed and their magnetic properties established either through measurements of the magnetic susceptibility or by ESR of crystals [Kokoszka and Duerst (288); Smith and Pilbrow (420); Hodgson (224); Doedens (146); Sinn (416); Hatfield (214)]. Category (2) has been reviewed by Owen and Harris (366) and includes more than 40 examples. Only Mn_2 in the solid rare gases has been observed in category (3) [VanZee, Baumann, and Weltner (444)].

Exchange interaction is generally represented by the *isotropic* or Heisenberg-type coupling:

$$\mathcal{H}_{ex} = -J\mathbf{S}_1 \cdot \mathbf{S}_2 \qquad \text{(V, 32)}$$

(Alternative forms of this equation are sometimes used containing a positive sign and/or $2J$ instead of J.) Here negative J means a singlet (antiferromagnetic) lowest state, and a positive J indicates a high spin state (ferromagnetic) as having the lowest energy. Most dimeric complexes are antiferromagnetic.

The form of (V, 32) is deceiving in implying that isotropic exchange is a direct consequence of the spin angular momentum operators of the two electrons. It is actually an electrostatic interaction but determined by the antisymmetry requirement of the Pauli exclusion principle. Thus, for two electrons on separate atoms the separation of the singlet and triplet states effectively defines the exchange coupling, since these states must have symmetric and antisymmetric orbital wave functions, respectively, which differ in energy through their electrostatic coulomb and exchange integrals. This may perhaps be made clearer by considering the case of two spins where $S_1 = S_2 = \frac{1}{2}$:

$$(\mathbf{S}_1 + \mathbf{S}_2)^2 = \mathbf{S}_1{}^2 + \mathbf{S}_2{}^2 + 2\mathbf{S}_1 \cdot \mathbf{S}_2$$

or

$$\mathbf{S}_1 \cdot \mathbf{S}_2 = \tfrac{1}{2}\,[S(S+1) - S_1(S_1+1) - S_2(S_2+1)]$$

Then for $S = 1$,

$$-J\mathbf{S}_1 \cdot \mathbf{S}_2 = \frac{-J}{4}$$

and for $S = 0$,

$$-J\mathbf{S}_1 \cdot \mathbf{S}_2 = +\frac{3J}{4}$$

$$J = -(E_{\text{triplet}} - E_{\text{singlet}}) \qquad \text{(V, 33)}$$

A more general form of (V, 32) would include a biquadratic term

$$+j(\mathbf{S}_1 \cdot \mathbf{S}_2)^2$$

But it is small and often neglected.

Isotropic exchange tends to align neighboring spins either parallel or antiparallel to each other. However, in most cases the spins are also coupled indirectly through spin-orbit interaction and are also affected by the surrounding ligands or lattice, which leads to what is usually termed *anisotropic exchange* [Van Vleck (443)]. This will now be shown to arise directly from (V, 32) in the more general case where J is not a constant but a tensor, while also considering the effects of a magnetic field.

(a) Anisotropic Exchange

If (V, 32) is taken to have the form [Abragham and Bleaney (1), page 502ff.].

$$\mathcal{H}_{\text{ex}} = J_{xx} S_{1x} S_{2x} + J_{yy} S_{1y} S_{2y} + J_{zz} S_{1z} S_{2z} \qquad \text{(V, 34)}$$

where the two interacting atoms have $S_1 = S_2 = \frac{1}{2}$ and a magnetic field is along the z axis (assumed the same as the g_z axis), then

$$\mathcal{H}_{\text{ex}} = g_z \beta H_z (S_{1z} + S_{2z}) + J_{xx} S_{1x} S_{2x} + J_{yy} S_{1y} S_{2y} + J_{zz} S_{1z} S_{2z}$$

If it is then noted that

$$J_{xx} S_{1x} S_{2x} = \frac{1}{2} J_{xx} [(S_{1x} + S_{2x})^2 - S_{1x}{}^2 - S_{2x}{}^2] = \frac{1}{2} J_{xx} S_x{}^2 - \frac{1}{4} J_{xx}$$

since $S_{1x}{}^2 = S_{2x}{}^2 = \frac{1}{4}$,

$$\mathcal{H}_{\text{ex}} = g_z \beta H_z S_z + \frac{1}{2} (J_{xx} S_x{}^2 + J_{yy} S_y{}^2 + J_{zz} S_z{}^2) - \frac{1}{4} (J_{xx} + J_{yy} + J_{zz})$$

Now separating the interaction into an isotropic part J and an anisotropic (traceless) part J' such that

$$J_{xx} = J_{xx}' + J, \quad \text{etc}$$

where

$$J = \frac{1}{3} (J_{xx} + J_{yy} + J_{zz})$$

and

$$J_{xx}' + J_{yy}' + J_{zz}' = 0 \tag{V, 35}$$

then

$$\begin{aligned} \mathcal{H}_{\text{ex}} = g_z \beta H_z S_z &+ \tfrac{1}{2} J[S(S + 1) - \tfrac{3}{2}] \\ &+ \tfrac{1}{2} (J_{xx}' S_x{}^2 + J_{yy}' S_y{}^2 + J_{zz}' S_z{}^2) \end{aligned} \tag{V, 36}$$

since $S_x{}^2 + S_y{}^2 + S_z{}^2 = S(S + 1)$, where S can take the values 0 and 1 and $S_z = +S, \cdots, -S$. For *isotropic* exchange where the $J_{ii}' = 0$, the energy levels from (V, 36) are then

$$S = 1 \qquad \left.\begin{aligned} W_{+1} &= +\tfrac{1}{4} J + g_z \beta H_z \\ W_0 &= +\tfrac{1}{4} J \\ W_{-1} &= +\tfrac{1}{4} J - g_z \beta H_z \end{aligned}\right\} \text{Triplet}$$

$$S = 0 \qquad W_0 = -\tfrac{3}{4} J \qquad \text{Singlet} \tag{V, 37}$$

The anisotropic exchange term, i.e., the third term on the right-hand side of (V, 36) can be converted into a more familiar form by letting [Abragham and Bleaney (1), page 152]:

$$D = \tfrac{3}{4} J_z' \quad \text{and} \quad E = \tfrac{1}{4}(J_x' - J_y') \tag{V, 38}$$

Then it becomes

$$D[S_z^2 - \tfrac{1}{3} S(S+1)] + \tfrac{1}{2} E[S_+^2 + S_-^2] \tag{V, 39}$$

This expression is analogous in form to the familiar zero-field-splitting term arising from spin-spin interaction in states where $S \geqslant 1$ and therefore can be expected to contribute in all cases of exchange coupling. Here the origin of D and E arises from exchange interactions, but as noted above, it also includes magnetic dipole-dipole interactions between unpaired spins. Although not an exchange coupling, the ever-present magnetic dipolar interaction between the spins on neighboring atoms [see Section II-1c(γ)] has the same mathematical form as anisotropic exchange:

$$\mathcal{H}_d = D_d(3S_{1z}S_{2z} - \mathbf{S}_1 \cdot \mathbf{S}_2) \tag{V, 40}$$

If one assumes point dipoles on nonoverlapping charge distributions, this dipolar interaction can be obtained, for an isotropic $\boldsymbol{g}$ tensor, from

$$D_d = -g^2\beta^2/r^3 \tag{V, 41}$$

or more generally

$$D_d = -(\beta^2/3r^3)[2g_z^2 + \tfrac{1}{2}(g_x^2 + g_y^2)] \tag{V, 42}$$

where r is the distance between the two dipoles. Thus, within the limits of the point-dipole approximation this contribution can be reliably calculated and, in fact, in the absence of other anisotropic effects (see below) can provide a value for the interatomic distance.

When spin-orbit coupling is important, the total spin ceases to be a good quantum number, and a more general form of $\mathcal{H}_{ex}$ in (V, 34) is necessary [Stevens (425); Kanamori (255)]. However, for axial symmetry and identical atoms, it can be simplified to yield again a form like (V, 40) termed *pseudodipolar exchange* [Van Vleck (443)], which for the axial case is

$$\mathcal{H}_E = D_E(3S_{1z}S_{2z} - \boldsymbol{S}_1 \cdot \boldsymbol{S}_2) \qquad (V, 43)$$

D_E is proportional to $(\lambda/\Delta)^2 J$ or to $(\Delta g/g)^2 J$, where λ is a spin-orbit coupling constant and Δ is the energy difference between coupled ground and excited electronic states. Smith, Pilbrow, and coworkers (420) consider this pseudodipolar contribution to be negligible if $J \lesssim 30\ \text{cm}^{-1}$.

Thus the total anisotropic exchange is characterized by a measurable parameter, D_e, where

$$D_e = D_d + D_E \qquad (V, 44)$$

It appears that D_E cannot be reliably calculated at the present time so that it must then either be assumed negligible for small J, or its presence identified by deriving it from $D_E = D_e - D_d$ if the interatomic distance is known.

Antisymmetric exchange adds an additional term to $\mathcal{H}_{ex}$ of the form $\boldsymbol{C} \cdot (\boldsymbol{S}_1 \times \boldsymbol{S}_2)$ [Stevens (425); Dzyaloshinski (152); Moriya (348); Erdös (155); Owen and Harris (366)]. It tends to align the two spins $\boldsymbol{S}_1$ and $\boldsymbol{S}_2$ perpendicular to each other and to the vector $\boldsymbol{C}$. Again, spin-orbit interaction is its essential origin and $C \simeq (\lambda/\Delta) J \simeq (\Delta g/g) J$, but it vanishes for pairs with a center of inversion symmetry so that it can be neglected for most of the dimer exchange cases where the symmetry is usually high.

Other possible interactions that may be involved are electric quadrupole-quadrupole and virtual phonon interactions. The latter two are generally small and not easily distinguishable [Stevens (425); Kanamori (255); Owen and Harris (366)].

(b) Dimers, Judd-Owen Equations

Often the dominant interaction is the isotropic exchange. Then for equivalent, or similar, interacting atoms with $S_i = S_j$ the ESR spectrum can usually be characterized by just the parameters J, D_S, and E_S where the values of the anisotropic parameters will depend upon S. The total spin Hamiltonian is given by

$$\mathcal{H} = \mathcal{H}_i + \mathcal{H}_j + \mathcal{H}_{ex} \qquad (V, 45)$$

where for each ion i and j

$$\mathcal{H}_i = \beta \boldsymbol{H} \cdot \boldsymbol{g} \cdot \boldsymbol{S}_i + D_c [S_{iz}{}^2 - \tfrac{1}{3} S(S+1)] + E_c (S_{ix}{}^2 - S_{iy}{}^2) \tag{V, 46}$$

and D_c and E_c arise from the environmental crystal field.

$$\begin{aligned}\mathcal{H}_{\text{ex}} &= -J\boldsymbol{S}_i \cdot \boldsymbol{S}_j + D_e(3S_{iz}S_{jz} - \boldsymbol{S}_i \cdot \boldsymbol{S}_j) \\ &\quad + E_e(S_{ix}S_{jx} - S_{iy}S_{jy})\end{aligned} \tag{V, 47}$$

Here J = isotropic exchange constant and $D_e = D_d$ (dipolar) + D_E (pseudodipolar exchange). If the term in J is dominant in (V, 47), then $\boldsymbol{S} = \boldsymbol{S}_i + \boldsymbol{S}_j$ where $S = S_i + S_j,\ S_i + S_j - 1, \cdots, 0$ and energies of the individual spin states, $S = 0$ (singlet), $S = 1$ (triplet), etc., are given by

$$W_S = (J/2)[S(S+1) - S_i(S_i + 1) - S_j(S_i + 1)] \tag{V, 48}$$

The spacing of these levels follows a Landé interval rule, as shown in the left side of Fig. V-42, for the antiferromagnetic case (J negative) where $S = 0$ has the lowest energy. Each of these states can be treated independently so that each has its own characteristic zero-field-splitting parameters D_S and E_S which can be determined via ESR spectroscopy. Thus in a magnetic field

$$\begin{aligned}\mathcal{H} &= W_S + \beta \boldsymbol{H} \cdot \boldsymbol{g} \cdot \boldsymbol{S} + D_S [S_z{}^2 - \tfrac{1}{3} S(S+1)] \\ &\quad + E_S(S_x{}^2 - S_y{}^2)\end{aligned} \tag{V, 49}$$

where

$$\begin{aligned}D_S &= 3\alpha_S D_e + \beta_S D_c \\ E_S &= \alpha_S E_e + \beta_S E_c\end{aligned} \tag{V, 50}$$

and

$$\begin{aligned}\alpha_S &= \tfrac{1}{2} [S(S+1) + 4S_i(S_i + 1)]/(2S - 1)(2S + 3) \\ \beta_S &= [3S(S+1) - 3 - 4S_i(S_i + 1)]/(2S - 1)(2S + 3)\end{aligned} \tag{V, 51}$$

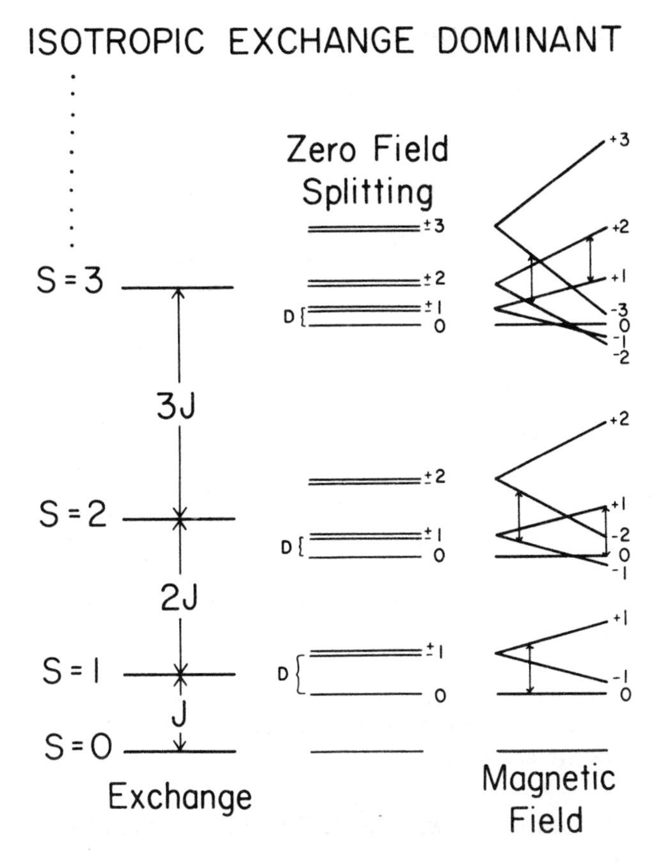

Fig. V-42. Schematic energy-level diagram for a pair of antiferromagnetically coupled atoms with $S_i = S_j > \frac{3}{2}$ and isotropic exchange dominating [adapted from Hutchings, Birgeneau, and Wolfe (237)]. Zero-field splittings and Zeeman levels are exaggerated, and only a few possible ESR transitions are indicated.

Values of α_S and β_S for $S_i = S_j = \frac{1}{2}$ to $\frac{7}{2}$ are given in Table V-6 [Owen and Harris (366)]. Equations (V, 49) to (V, 51) are usually designated as the Judd–Owen equations [Owen (365)]. Figure V-42 illustrates approximately the Landé spacing, zfs, and Zeeman splittings in this usual case of a large isotropic exchange interaction.

In practice, one can observe the intensities of the various ESR lines as a function of temperature and thereby both identify to which S value each line belongs and also determine the energy level spacings, i.e., J in Fig. V-42. For Landé spacing the intensity of the

Table V-6. Values of α_S and β_S for $S_i = S_j = \frac{1}{2}$ to $\frac{7}{2}$.[†]

$S_i = S_j =$	$\frac{1}{2}$	1	$\frac{3}{2}$	2	$\frac{5}{2}$	3	$\frac{7}{2}$
α_S							
$S = 1$	1/2	1	17/10	13/5	37/10	5	13/2
2	–	1/3	1/2	5/7	41/42	9/7	23/14
3	–	–	3/10	2/5	47/90	2/3	5/6
4	–	–	–	2/7	5/14	34/77	83/154
5	–	–	–	–	5/18	1/3	31/78
6	–	–	–	–	–	3/11	7/22
7	–	–	–	–	–	–	7/26
β_S							
$S = 1$	0	−1	−12/5	−21/5	−32/5	−9	−12
2	–	+1/3	0	−3/7	−20/21	−11/7	−16/7
3	–	–	+2/5	+1/5	−2/45	−1/3	−2/3
4	–	–	–	+3/7	+2/7	+9/77	−6/77
5	–	–	–	–	+4/9	+1/3	+8/39
6	–	–	–	–	–	+5/11	+4/11
7	–	–	–	–	–	–	+6/13

[†]From Owen and Harris (366).

transitions within a given triplet, quintet, etc., spin state will vary with temperature according to

$$I_S = \exp(-W_S/kT) / \sum_S (2S+1) \exp(-W_S/kT) \qquad (V, 52)$$

For $S_i = S_j = \frac{5}{2}$, a plot of these intensities versus kT/J is given in Fig. V-43 when $S = 0$ is the lowest energy state. D_S and E_S for individual spin states are then obtained from the identified $\Delta m_S = \pm 1$ (or other) transitions by use of the methods discussed earlier in Chapter III for $S = 1$ and in this chapter for $S > 1$.

(c) Hyperfine Interaction

The hyperfine interaction for a pair is given by [Owen and Harris (366)]

$$A_i \boldsymbol{S}_i \cdot \boldsymbol{I}_i + A_j \boldsymbol{S}_j \cdot \boldsymbol{I}_j + a_i \boldsymbol{S}_i \cdot \boldsymbol{I}_j + a_j \boldsymbol{S}_j \cdot \boldsymbol{I}_j \qquad (V, 53)$$

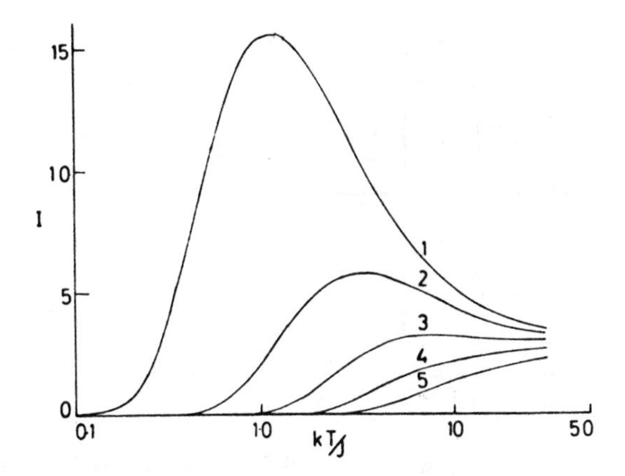

Fig. V-43. Relative intensities I of the transitions for an antiferromagnetically coupled pair of spins $S_i = S_j = \frac{5}{2}$ plotted against the reduced temperature kT/J [from Owen (365)]. Calculated from (V, 52) where the total spin states $S = 1, 2, 3, 4, 5$ are labeled in the figure.

where A_i is the coupling at nucleus i due to spins at i, and a_i is the coupling at nucleus j due to spins at i. If $S_i = S_j$, $A_i = A_j$, and $a_i = a_j$, then this becomes

$$\tfrac{1}{2}(A + a)\boldsymbol{S} \cdot (\boldsymbol{I}_i + \boldsymbol{I}_j) + \tfrac{1}{2}(A - a)(\boldsymbol{S}_i - \boldsymbol{S}_j) \cdot (\boldsymbol{I}_i - \boldsymbol{I}_j) \quad \text{(V, 54)}$$

where $\boldsymbol{S} = \boldsymbol{S}_i + \boldsymbol{S}_j$ so that S takes the values $S_i + S_j, S_i + S_j - 1, \cdots, 0$. The last term in (V, 54) can be neglected, since it connects the different spin states which are widely separated relative to A for the usual case of $J >> A$. Then the hyperfine splitting, for equivalent atoms, assuming it is isotropic, will be $\frac{1}{2}(A + a)$, but correspondingly the number of hyperfine lines will have increased to $(4I_i + 1)$. Since the transferred hyperfine constant, a, is small relative to A, the pattern for two exchange-coupled atoms with $I = \frac{5}{2}$ (such as Mn^{2+} or Mn) will be 11 hf lines with relative intensities 1:2:3:4:5:6:5:4:3:2:1 separated essentially by $A/2$, as shown in Fig. V-44. If the hfs parameter A for an isolated atom is known, then the dimer hfs will be well characterized and will identify the exchange-coupled pair.

The transferred hf constant a has apparently only been detected for pairs of V^{2+} ions in MgO [May, as referenced by Owen and Harris (366)]. This interaction and its measurement in other systems are discussed in more detail by Owen and Harris.

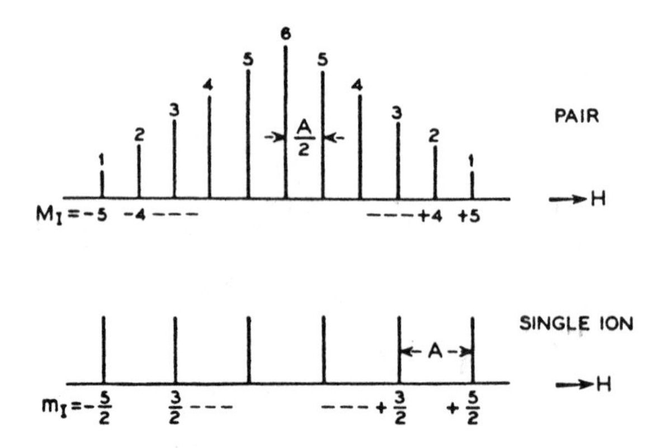

Fig. V-44. Hyperfine splitting pattern expected for an exchange-coupled pair relative to that of a single $^{55}Mn^{2+}$ ion or ^{55}Mn atom where $I_i = \frac{5}{2}$.

For an anisotropic $\boldsymbol{A}$ tensor in the dimer one can expect to observe a variation in the hfs for different lines in the ESR spectrum just as in the monomer in a similar environment. This is shown in the $S = 1$ spectrum of the exchanged-coupled Cu^{2+} pair in a polycrystalline copper adenine complex in Fig. V-45 [Kokoszka and Duerst (288)]. The two large perpendicular lines of opposite phase clearly do not exhibit the seven-line pattern expected for $I = \frac{3}{2}$, whereas the $\Delta m_S = 2$ transition at low field does. Since the latter is due to transitions in molecules at intermediate orientations ($\theta \simeq 30°$) [Kokoszka, Linzer, and Gordon (289)], it is clear that $A_{\parallel} > A_{\perp}$, as occurs in other ESR spectra of the $S = \frac{1}{2}$ Cu^{2+} ion in an axial field. This may be verified by observation of the parallel lines in the triplet spectra of dimers, but these weak transitions are often difficult to detect. In this particular case a comparative monomer spectrum can be obtained by doping the copper salt with an equimolar amount of the zinc salt [Kokoszka, Linzer, and Gordon (289)]. The classic single-crystal study of dimeric copper acetate will be considered in more detail below.

(d) Direct and Super Exchange

It is apparently still controversial whether the observed strong exchange in Cu^{2+} dimers occurs predominantly by direct metal–metal interaction or via the ligands bridging the metal ions, or if one can

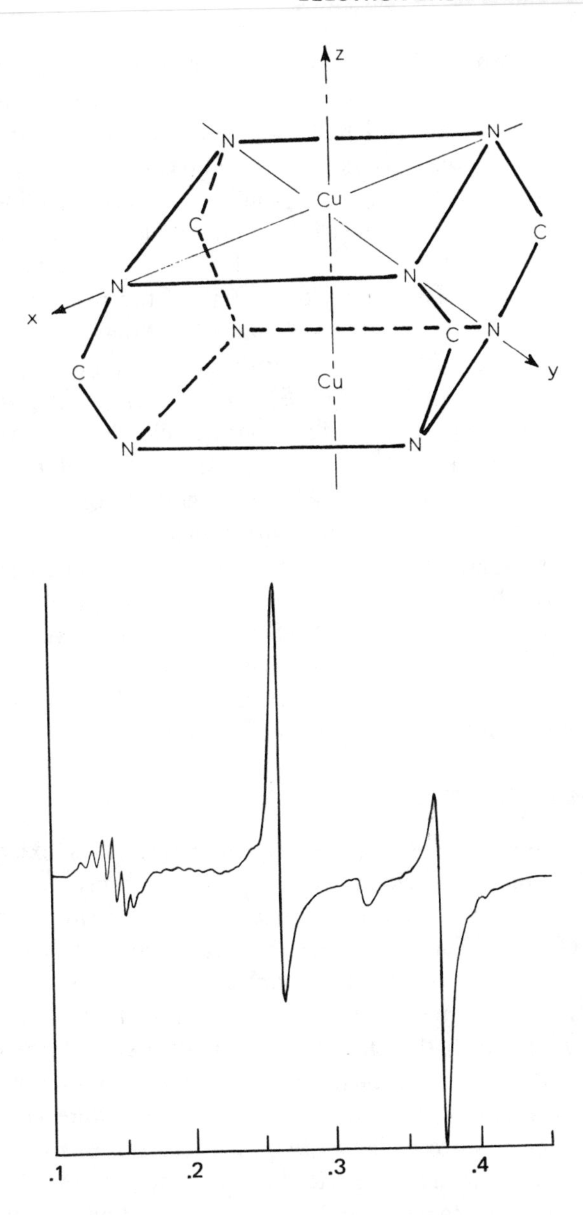

Fig. V-45. Structure of copper adenine complex and its polycrystalline ESR spectrum [from Kokoszka and Duerst (288)]. Note hyperfine structure on the low-field $\Delta m_S = 2$ transition.

make such a complete distinction. The exchange was early proposed to proceed through a "δ-bond" (overlapping $d_{x^2-y^2}$ orbitals on adjacent Cu atoms) [Figgis and Martin (164); Ross and Yates (395)], later through a σ bond [Hansen and Ballhausen, (208)], and most recently by superexchange, as originally proposed by Bleaney and Bowers (72) [Jotham and Kettle (253); Goodgame, Hill, Marsham, et al. (189); Hay, Thibeault, and Hoffmann (215); Kahn and Briat (254); Gerloch and Harding (178)]. Within the pseudonym superexchange there are also several mechanisms that may be contributing.

An *ab initio* calculation on cupric acetate dimer by de Loth, Cassoux, Daudey, and Malrieu (135), which attempts to give a definitive answer to this question, demonstrates clearly the difficulty in calculating this small singlet–triplet splitting (10^{-7} of the total energy). Their theory indicates that the magnitude of J is a result of the cancellation of several contributions and that higher-order effects, such as polarization of the ligands and $3d$ closed shells, are significant contributors.

On the other hand, an SCF-Xα-SW calculation for the Cl bridged dimer $Mo_2Cl_9{}^{3-}$, which is also strongly exchange-coupled ($J \simeq -500$ cm^{-1}) [Ginsberg (182)], concludes that the bonding is purely direct metal–metal interaction with no superexchange.

(e) Exchange-Coupled Dimers

Copper Acetate Monohydrate [$Cu_2(CH_3CO_2)_4(H_2O)_2$]. This exists in the crystal as the dimeric species in Fig. V-46, where X-ray measurements have shown that the Cu–Cu distance is 2.64 Å [van Niekerk and Schoening (439)]. The anomalous magnetic character of this crystal relative to most compounds of divalent copper was first indicated by the susceptibility (χ) measurements of Guha (201), and this behavior was finally resolved in the classic work of Bleaney and Bowers (72). They demonstrated that both the ESR and χ measurements could be explained by assuming that pairs of copper ions were coupled together by exchange forces. Thus, two identical ions of spin $S_i = \frac{1}{2}$ interact to form a singlet and an excited triplet state with $J = -310$ cm^{-1}. The individual ions are subjected to a crystal field with approximate tetragonal symmetry, and the interaction of that field and the spin-orbit coupling in the triplet state causes a shift in the g values from g_e at 300°K [Abe and Shimada (26)]:

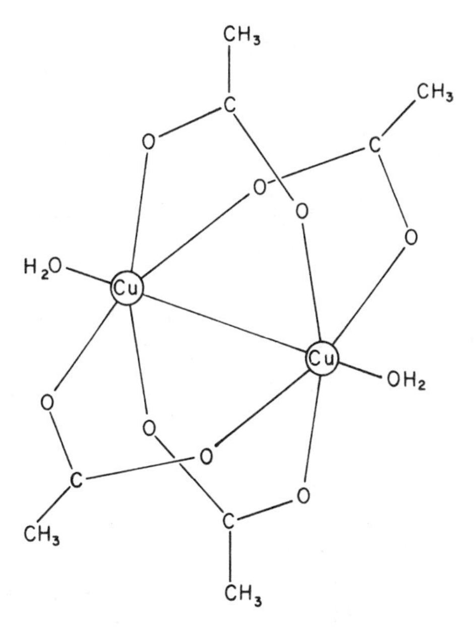

Fig. V-46. Structure of the dimer of copper acetate monohydrate $[Cu_2(CH_3CO_2)_4(H_2O)_2]$ [from van Niekirk and Schoening (439)].

$$g_z = 2.344(10) \simeq g_{\parallel} = 2(1 - 4\lambda/\Delta_1)$$

$$\left.\begin{aligned} g_x &= 2.043(5) \\ g_y &= 2.093(5) \end{aligned}\right\} \simeq g_{\perp} = 2(1 - \lambda/\Delta_2) \qquad \text{(V, 55)}$$

Here λ is the spin-orbit coupling constant ($\simeq -830\ \text{cm}^{-1}$), and Δ_1, Δ_2 are crystal-field splittings derived to be $\sim 20{,}000\ \text{cm}^{-1}$. The zero-field-splitting parameters in that state are given by $|D| = 0.345(5)$ and $E = 0.005(3)\ \text{cm}^{-1}$.

The hyperfine interaction is typical of such pairs in that the spacing is approximately one-half of that for isolated Cu^{2+} ions [obtained from zinc-doped copper acetate by Kokoszka, Allen, and Gordon (287)] and along the z axis gives a clear 1:2:3:4:3:2:1 intensity pattern due to the $I_i = I_j = \frac{3}{2}$ of the copper nuclei.

Application of (V, 50) for $S = 1$ with $\beta = 0$ gives for the experimental parameter

$$D_S = \tfrac{3}{2} D_e = \tfrac{3}{2}(D_d + D_E) \qquad \text{(V, 56)}$$

with D_d the dipolar and D_E the pseudodipolar exchange terms. For this case of an anisotropic $\boldsymbol{g}$ tensor and $r = 2.64$ Å, from (V, 42):

$$\tfrac{3}{2} D_d = -(2g_{\parallel}^2 + g_{\perp}^2)\,\beta^2/2r^3 = -0.19 \text{ cm}^{-1}$$

which indicates that pseudodipolar exchange is contributing significantly to the measured $|D| = |D_S| = 0.34$ cm^{-1}.

An accurate calculation of the pseudodipolar term from knowledge of the properties of the metal ion pair is apparently still not possible. Application of existing theory to ions in which $d_{x^2-y^2}$ is the ground state and the g components are as given in (V, 55) leads to an expression:

$$D_E = -(\tfrac{1}{12})[\tfrac{1}{4}(g_{\parallel} - 2)^2 J_{\parallel} - (g_{\perp} - 2)^2 J_{\perp}]$$

in which it was originally assumed that $J_{\parallel} = J_{\perp} = J$ [Bleaney and Bowers (72); Owen and Harris (366)]. Substituting the experimental g and J data yields $D_E = +0.93$ cm^{-1}, much too large compared with the value +0.35 calculated from (V, 56) with positive D_S. This suggests that the exchange interactions $J_{\parallel}$ and $J_{\perp}$ between the ground state of one ion and the excited state of the other are smaller than assumed.

Isolated Pairs of Magnetic Ions in Insulators. The example chosen here is the exchange interaction of nearest-neighbor Mn^{2+} ions introduced as dilute impurities in isomorphous MgO and CaO crystals. These Mn^{2+} ions are separated by $r \simeq 3.06$ Å and 3.32 Å, respectively, in the two crystals and their paired interactions can be depicted, when the surrounding oxygen ions are included, as

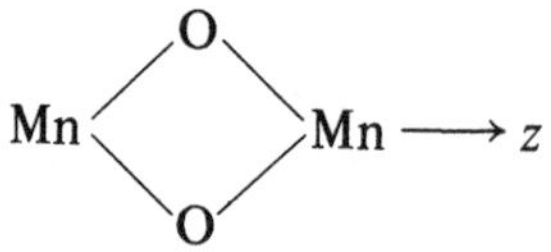

$$\mathrm{Mn} \langle {}^{\mathrm{O}}_{\mathrm{O}} \rangle \mathrm{Mn} \longrightarrow z \qquad \langle 110 \rangle \text{ crystal direction}$$

[Coles, Orton, and Owen (120); Owen (365); Harris (209); Harris and Owen (210)]. Each transition-metal ion has $S_i = \frac{5}{2}$ (and also $I_i = \frac{5}{2}$), and the pairs are found to be antiferromagnetically coupled with $-J < 20$ cm^{-1}. Thus the lowest level with $S = 0$ is diamagnetic, and successively higher spin levels $S = 1$ to $S = 4$, as in Fig. V-42 ($S = 5$ is obscured), have been observed in the ESR by raising the temperature. Each line appears split by hyperfine interaction into $(2I + 1)$ lines with relative intensities 1:2:3:4:5:6:5:4:3:2:1 and with a spacing of approximately half that for the isolated ions. Figure V-47 shows part of the spectra observed for Mn^{2+} in CaO as a function of temperature.

It was found that the Landé interval rule of (V, 48) theoretically predicted from the Judd–Owen treatment does not apply for this pair interaction in the MgO crystals even though the isotropic exchange term is dominant. From the intensity measurements as a function of temperature it was found that the successive intervals between spin levels were not strictly in the order $\Delta_{0,1} = \frac{1}{2}\,\Delta_{1,2} = \frac{1}{3}\,\Delta_{2,3} = \frac{1}{4}\,\Delta_{3,4}$ but monotonically decreased as S increased, from $\Delta_{0,1} = 20.0 \pm 1.5$ to $\frac{1}{4}\,\Delta_{3,4} = 12.0 \pm 2.0\ \text{cm}^{-1}$. This anomaly is attributed to exchange striction, although contributions from biquadratic $[j(\mathbf{S}_i \cdot \mathbf{S}_j)^2]$ or higher terms in (V, 32) have not been excluded. The determination of the individual spacings between S states as carried out for Mn^{2+} in MgO is unusual, since it is difficult to make sufficiently accurate measurements of line intensities as a function of temperature. This is particularly true where there are many S values requiring measurements over a broad temperature range, and where there is the possibility of overlapping lines.

Exchange striction is derived from a balancing of elastic and exchange forces [Kittel (273); Bean and Rodbell (53); Owen and Harris (366)], such that for antiferromagnetic coupling the ions will tend to move closer together in states with lower S. Since the isotropic exchange is a sensitive function of r, it follows that the effective J values in higher spin states will then be decreased owing to this effect. Harris (209) has been able to fit their Mn^{2+} in MgO data by assuming J to be an exponential function of the distortion $r = r_{ij} - r_0$, that is, $J = J_0 \exp(-19r/r_0)$.

Mn_2 Molecules in the Solid Rare Gases. When Mn atoms are trapped in solid krypton in a molar ratio of approximately 1:1000, the X-band ESR spectrum of the $^6S_{5/2}$ atom is observed at $g = 2.00$ with a six-line hyperfine pattern split by about 28 G. The hfs is due to spin polarization, as discussed in Chapter I. If the concentration of Mn atoms is increased, Mn_2 dimers are formed, but at 4°K their spectrum is not observed. Upon warming to 10°K and up to 50°K the

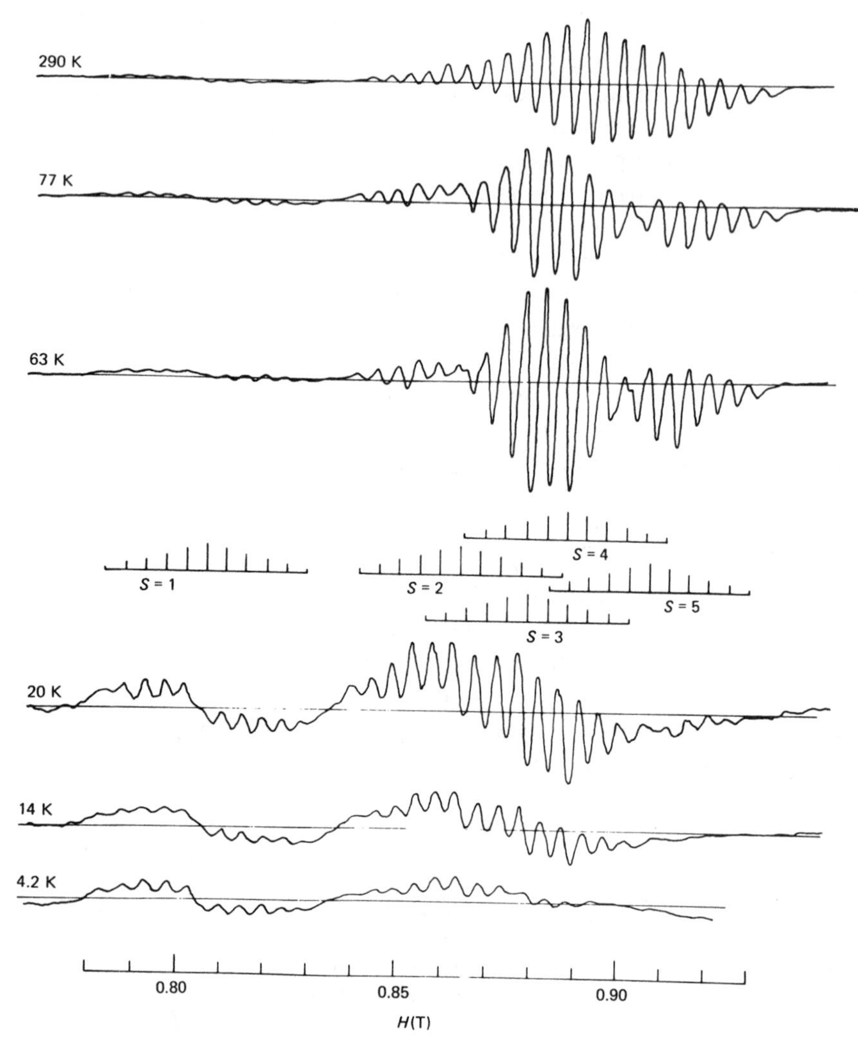

Fig. V-47. A portion of the ESR spectra of Mn^{2+} exchanged-coupled pairs in CaO as a function of temperature [from Harris (209), copyright The Institute of Physics, 1972, England]. Eleven-line hf patterns for $S = 1$ to 5 are indicated.

series of lines shown in Fig. V-48 gradually appears [Van Zee, Baumann, and Weltner (444)]. Here the hyperfine structure is clearly in accord with two exchange-coupled Mn atoms, since the 11-line pattern has the expected 1:2:3:4:5:6:5:4:3:2:1 intensity distribution

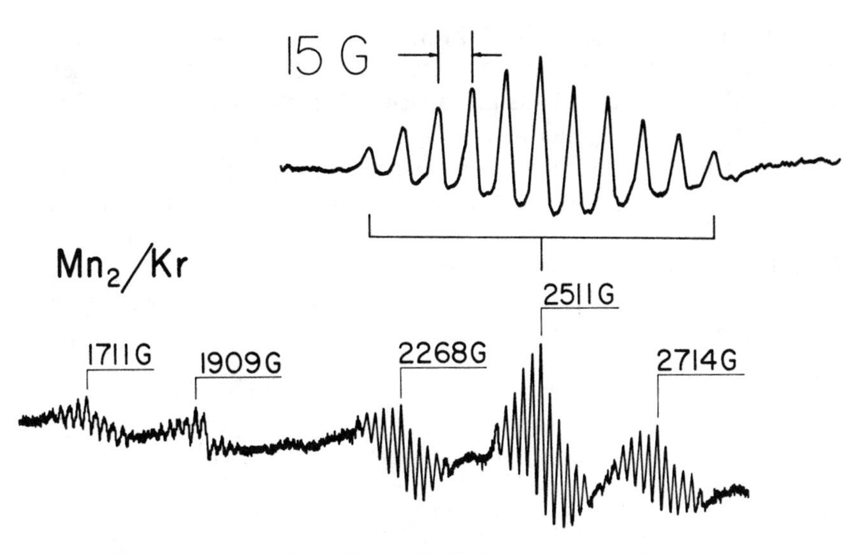

Fig. V-48. ESR spectrum of the Mn_2 molecule in a krypton matrix at a temperature of ~50° K [from Van Zee, Baumann, and Weltner (444)].

and a spacing of approximately one-half of that in the isolated atom. Except for the temperature range, the observations are similar to those described above for Mn^{2+} in MgO. Similar patterns are observed in argon matrices, although there are some site effects, and in xenon matrices, where only two eleven-line patterns are detected presumably because of matrix broadening effects.

The antiferromagnetic character of the Mn_2 molecule was predicted by Nesbet (359) in 1964 from an *ab initio* Hartree–Fock calculation. He found the molecule to have ${}^1\Sigma_g^+$ ground state with $J = -4.13\ \text{cm}^{-1}$. The bond energy was calculated to be 0.79 eV at $r = 2.88$ Å.

A total of eight 11-line patterns were detected in krypton matrices and assigned as xy lines in the $S = 1$, 2, and 3 spin states from their intensity variations with temperature. Detection of transitions in higher spin states was limited by the softening temperature of the matrix. From more detailed intensity measurements of the strong $S = 2$ line at 2511 G (see Fig. V-48), J was found to be $-9 \pm 3\ \text{cm}^{-1}$, essentially confirming Nesbet's predicted value. From analysis of the lines, D_S values were found to be 0.49, 0.140, and 0.070 cm^{-1} for $S = 1$, 2, 3 spin states, respectively. From these values one derives, via (V, 50), $D_e = -0.043(2)$ and $D_c = -0.001(4)$ cm^{-1}, where the uncertainties encompass the total variation among the three sets of derived parameters. Thus the lattice-induced anisotropy is negligible. The value of D_e can provide a measure of the interatomic distance if the only mechanism is magnetic dipole–dipole coupling. Since for small J the other contributions are negligible [Smith and Pilbrow (420)], one can validly calculate $r = 3.4$ Å from (V, 41). Although it disagrees with Nesbet's predicted r value, this bond distance is reasonable for this molecule where the experimental bond energy is only about 10 kcal/mole.

References

I. BOOKS

1. Abragham, A. and Bleaney, B., *Electron Paramagnetic Resonance of Transition Ions* (Oxford University Press, London, 1970).
2. Atkins, P. W. and Symons, M. C. R., *The Structure of Inorganic Radicals* (Elsevier, New York, 1967).
3. Ayscough, P. B., *Electron Spin Resonance in Chemistry* (Methuen, London, 1967).
4. Ballhausen, C. J., *Introduction to Ligand Field Theory* (McGraw-Hill, New York, 1962).
5. Carrington, A., *Microwave Spectroscopy of Free Radicals* (Academic Press, New York, 1974).
6. Carrington, A. and McLachlan, A. D., *Introduction to Magnetic Resonance* (Harper & Row, New York, 1967).
7. Goodenough, J. B., *Magnetism and the Chemical Bond* (Interscience, New York, 1963).
8. Gordy, W., *Theory and Applications of Electron Spin Resonance* (Wiley, New York, 1980).
9. Gordy, W., Smith, W. V., and Trambarulo, R. F., *Microwave Spectroscopy* (Wiley, New York, 1953).
10. Griffith, J. S., *The Theory of Transition-Metal Ions* (Cambridge University Press, 1961).
11. Herman, F. and Skillman, S., *Atomic Structure Calculations*, (Prentice-Hall, Englewood Cliffs, N.J., 1963).
12. Huber, K. P. and Herzberg, G., *Molecular Spectra and Molecular Structure, IV. Constants of Diatomic Molecules* (Van Nostrand Reinhold, New York, 1979).
13. Ingram, D. J. E., *Biological and Biochemical Applications of Electron Spin Resonance*, (Plenum Press, New York, 1969).
14. Kopfermann, H., *Nuclear Moments* (Academic Press, New York, 1958), Chapter I, Parts I, II, and III, pages 1–161.
15. Low, W., *Paramagnetic Resonance in Solids*, Supplement 2 to Solid State Physics Series, Ed. Seitz, F. and Turnbull, D. (Academic Press, New York, 1960).
16. McGlynn, S. P., Azumi, T., and Kinoshita, M., *Molecular Spectroscopy of the Triplet State* (Prentice Hall, Englewood Cliffs, N.J., 1969).
17. Meyer, B., *Low Temperature Spectroscopy: Optical Properties of Molecules*

in Matrices, Mixed Crystals, and Organic Glasses (Elsevier, New York, 1970).
18. Poole, C. P., Jr., *Electron Spin Resonance, A Comprehensive Treatise on Experimental Techniques* (Interscience, New York, 1967).
19. Ramsey, N. F., *Molecular Beams* (Oxford University Press, London, 1956), pages 68–88, particularly pages 76–80.
20. Slichter, C. P., *Principles of Magnetic Resonance*, Harper & Row, New York, 1963, pages 195–215.
21. Townes, C. H. and Schawlow, A. L., *Microwave Spectroscopy* (McGraw-Hill, New York, 1955).
22. Van Vleck, J. H., *The Theory of Electric and Magnetic Susceptibilities* (Oxford, 1932), pages 272–275, 287–297.
23. Wertz, J. E. and Bolton, J. R., *Electron Spin Resonance: Elementary Theory and Practical Applications* (McGraw-Hill, New York, 1972).

II. ARTICLES IN JOURNALS AND IN BOOKS

24. Aasa, R., *J. Chem. Phys. 52*, 3919 (1970).
25. Aasa, R. and Vänngard, T., *J. Magn. Reson. 19*, 308 (1975).
26. Abe, H. and Shimada, J., *Phys. Rev. 90*, 316 (1953); *J. Phys. Soc. Japan 12*, 1255 (1957).
27. Abragham, A. and Pryce, M. H. L., *Proc. Roy. Soc. (London) A205*, 135 (1951).
28. Adrian, F. J., *J. Chem. Phys. 32*, 972 (1960).
29. Adrian, F. J., *J. Chem. Phys. 36*, 1692 (1962).
30. Adrian, F. J., *Phys. Rev. 127*, 837 (1962).
31. Adrian, F. J., *J. Colloid and Interface Science 26*, 317–354 (1968).
32. Adrian, F. J. and Bowers, V. A., *Chem. Phys. Lett. 41*, 517–520 (1976).
33. Adrian, F. J., Cochran, E. L. and Bowers, V. A., *Advances in Chemistry Series* (American Chemical Society) *36*, 50 (1962).
34. Adrian, F. J., Cochran, E. L. and Bowers, V. A., *J. Chem. Phys. 36*, 1661 (1962).
35. Adrian, F. J., Cochran, E. L., and Bowers, V. A., *J. Chem. Phys. 47*, 5441 (1967).
36. Adrian, F. J., Cochran, E. L. and Bowers, V. A., *J. Chem. Phys. 59*, 56 (1973).
37. Allen, B. T., *J. Chem. Phys. 43*, 3820 (1965).
38. Amano, T. and Hirota, E., *J. Mol. Spectr. 53*, 346 (1974).
39. Ammeter, J. H. and Schlosnagle, D. C., *J. Chem. Phys. 59*, 4784 (1973).
40. Anderson, P. W., *Solid State Phys. 14*, 99–214 (1963).
41. Andrews, L., *J. Phys. Chem. 73*, 3922 (1969); *J. Chem. Phys. 50*, 4288 (1969).
42. Atkins, P. W. and Jamieson, A. M., *Mol. Phys. 14*, 425 (1967).
43. Atkins, P. W., Keen, N., and Symons, M. C. R., *J. Chem. Soc.*, 2873 (1962).
44. Bagus, P. S. and Schaefer, H. F., III, *J. Chem. Phys. 58*, 1844 (1973).

45. Barnes, C. E., Brown, J. M., Carrington, A., Pinkstone, J., Sears, T. J., and Thistlethwaite, P. J., *J. Mol. Spectr. 72*, 86 (1978).
46. Baranowski, J., Cukierda, T., Jezowska-Trzebiatowska, B., and Kozlowski, H., *J. Magn. Reson. 33*, 585 (1979).
47. Barrow, R. F. in *Donnees Spectroscopiques relatives aux Molecules Diatomiques*, Ed. Rosen, B., (Pergamon Press, Oxford, 1970), page 405.
48. Barrow, R. F., *Essays in Structural Chemistry*, Ed. Downs, A. J., Long, D. A., and Staveley, L. A. K. (Macmillan, London, 1971), Chapter 15.
49. Barrow, R. F., Bastin, M. W., and Longborough, B., *Proc. Phys. Soc. (London) 92*, 518 (1967).
50. Barrow, R. F., Du Parq, R. P., and Ricks, J. M., *J. Phys. B2*, 413 (1969).
51. Bass, I. L. and Mieher, R. L., *Phys. Rev. 175*, 421 (1968).
51a. Baumann, C. A., Van Zee, R. J., and Weltner, W., Jr., (preprint).
52. Baumann, C. A., Van Zee, R. J., Zeringue, K. J., and Weltner, W., Jr., *J. Chem. Phys. 75*, 5291 (1981).
53. Bean, C. P. and Rodbell, D. S., *Phys. Rev. 126*, 104 (1962).
54. Beltrán-Lopez, V. and Castro-Tello, J., *J. Magn. Reson. 39*, 437 (1980).
55. Bender, C. F. and Davidson, E. R., *Phys. Rev. 183*, 23 (1969).
56. Bennett, J. E., Ingram, D. J. E., Symons, M. C. R., George, P., and Griffith, J. S., *Phil. Mag. 46*, 443 (1955).
57. Berg, L.-E. and Klynning, L., *Phys. Scripta 10*, 331 (1974).
58. Berg, R. A. and Sinanoglu, O., *J. Chem. Phys. 32*, 1082 (1960).
59. Bernath, P. F., Cummins, P. G., and Field, R. W., *Chem. Phys. Lett. 70*, 618 (1980).
60. Bernheim, R. A., Bernard, H. W., Wang, P. S., Wood, L. S., and Skell, P. S., *J. Chem. Phys. 53*, 1280 (1970).
61. Bernheim, R. A., Bernard, H. W., Wang, P. S., Wood, L. S., and Skell, P. S., *J. Chem. Phys. 54*, 3223 (1971).
62. Bernheim, R. A., Kempf, R. J., Gramas, J. V., and Skell, P. S., *J. Chem. Phys. 43*, 196 (1965).
63. Bernheim, R. A., Kempf, R. J., Reichenbecher, E. F., *J. Magn. Reson. 3*, 5 (1970).
64. Bessis, N., Lefebvre-Brion, H., and Moser, C. M., *Phys. Rev. 124*, 1124 (1961).
65. Bhat, S. V. and Weltner, W., Jr., *J. Chem. Phys. 73*, 1498 (1980).
66. Bhat, S. V., Van Zee, R. J., and Weltner, W., Jr., (unpublished).
67. Bicknell, B. R., Graham, W. R. M., and Weltner, W., Jr., *J. Chem. Phys. 64*, 3319 (1976).
68. Biddles, I. and Hudson, A., *Mol. Phys. 25*, 707 (1973).
69. Bleaney, B., *Proc. Phys. Soc. (London) A63*, 407 (1950).
70. Bleaney, B., *Phil. Mag. 42*, 441 (1951).
71. Bleaney, B., *Proc. Phys. Soc. A75*, 621 (1960).
72. Bleaney, B. and Bowers, K. D., *Proc. Roy. Soc. A214*, 451 (1952).
73. Bleaney, B. and O'Brien, M. C. M., *Proc. Phys. Soc. B69*, 1216 (1956).
74. Bleaney, B. and Rubins, R. S., *Proc. Phys. Soc. (London) 77*, 103 (1961); corrigendum *78*, 778 (1961).

75. Blinder, S. M., *J. Chem. Phys. 33*, 748 (1960).
76. Bloch, F., *Phys. Rev. 102*, 104 (1956); *105*, 1206 (1957).
77. Bloembergen, N., Purcell, E. M., and Pound, R. V., *Phys. Rev. 73*, 679 (1948).
78. Blumberg, W. E., *Magnetic Resonance in Biological Systems*, Eds. Ehrenberg, A., Malstrom, B. E., and Vanngard, T. (Pergamon Press, Elmsford, N.Y., 1967), page 119.
78a. Boland, B. J., Brown, J. M., Carrington, A., and Nelson, A. C., *Proc. Roy. Soc. (London) A360*, 507 (1978).
79. Boorstein, S. A. and Gouterman, M., *J. Chem. Phys. 39*, 2443 (1963); *41*, 2776 (1964).
80. Bowers, K. D., *Proc. Phys. Soc. (London) A66*, 666 (1953).
81. Brailsford, J. R., Morton, J. R., and Vannotti, L. E., *J. Chem. Phys. 49*, 2237 (1968).
82. Brandon, R. W., Closs, G. L. Davoust, C. E., Hutchison, C. A., Jr., Kohler, B. E., and Silbey, R., *J. Chem. Phys. 43*, 2006 (1965).
83. Brandon, R. W., Gerkin, R. E., and Hutchison, C. A., Jr., *J. Chem. Phys. 41*, 3717 (1964).
84. Breit, G. and Rabi, I., *Phys. Rev. 38*, 2082 (1931).
85. Brickmann, J. and Kothe, G., *J. Chem. Phys. 59*, 2807 (1973).
86. Brith, M. and Schnepp, O., *J. Chem. Phys. 39*, 2714 (1963).
87. Brom, J. M., Jr., Durham, C. H., Jr., and Weltner, W., Jr., *J. Chem. Phys. 61*, 970 (1974).
88. Brom. J. M., Jr., Graham, W. R. M., and Weltner, W., Jr., *J. Chem. Phys. 57*, 4116 (1972).
89. Brom, J. M., Jr. and Weltner, W., Jr., *J. Chem. Phys. 57*, 3379 (1972).
90. Brom, J. M., Jr. and Weltner, W., Jr., *J. Chem. Phys. 58*, 5322 (1973).
91. Brom, J. M., Jr. and Weltner, W., Jr., *J. Chem. Phys. 64*, 3894 (1976).
92. Brown, J. M., *MTP Intern. Rev. Sci.*, Phys. Chem. Ser. One, Ed. Buckingham, A. D. Vol. 4, *Magn. Reson.* (1972).
93. Brown, J. M., Kaise, M., Kerr, C. M. L., and Milton, D. J., *Mol. Phys. 36*, 553 (1978).
94. Brown, J. M. and Milton, D. J., *Mol. Phys. 31*, 409 (1976).
95. Brown, J. M. and Sears, T. J., *Mol. Phys. 34*, 1595 (1977).
96. Brown, J. M. and Sears, T. J., *J. Mol. Spectr. 75*, 111 (1979).
97. Brown, J. M., Steimle, T. C., Coles, M. E., and Curl, R. F., Jr., *J. Chem. Phys. 74*, 3668 (1981).
98. Cade, P. E. and Huo, W. M., *J. Chem. Phys. 45*, 1063 (1966); *47*, 614 (1967).
99. Carlson, K. D. and Moser, C. M., *J. Chem. Phys. 44*, 3259 (1966); *46*, 35 (1967).
100. Carlson, K. D. and Nesbet, R. K., *J. Chem. Phys. 41*, 1051 (1964).
101. Carrington, A., Levy, D. H., and Miller, T. A., *Adv. Chem. Phys. 18*, 149 (1970).
102. Carrington, A. and Longuet-Higgins, H. C., *Mol. Phys. 5*, 447 (1962).
103. Carrington, A. and Lucas, N. J. D., *Proc. Roy. Soc. A314*, 567 (1970).

104. Castle, J. G. and Beringer, R., *Phys. Rev. 80*, 114 (1950).
105. Castner, T. G. and Känzig, W., *J. Phys. Chem. Solids 3*, 178 (1957).
106. Castner, T., Jr., Newell, G. S., Holton, W. C., and Slichter, C. P., *J. Chem. Phys. 32*, 668 (1960).
107. Chan, A. C. and Davidson, E. R., *J. Chem. Phys. 49*, 727 (1968).
108. Cheung, A. S.-C., Hansen, R. C., and Merer, A. J., *J. Mol. Spectr. 91*, 165 (1982).
109. Childs, W. J., Goodman, G. L., and Goodman, L. S., *J. Mol. Spectr. 86*, 365 (1981).
110. Childs, W. J. Goodman, L. S., and Renhorn, I., *J. Mol. Spectr. 87*, 522 (1981).
111. Christensen, H. and Veseth, L., *J. Mol. Spectr. 72*, 438 (1978).
112. Clementi, E., *J. Am. Chem. Soc. 83*, 4501 (1961).
113. Clementi, E. and Clementi, H., *J. Chem. Phys. 36*, 2824 (1962).
114. Cochran, E. L., Adrian, F. J., and Bowers, V. A., *J. Chem. Phys. 36*, 1938 (1962).
115. Cochran, E. L., Adrian, F. J., and Bowers, V. A., *J. Chem. Phys. 40*, 213 (1964).
116. Cochran, E. L., Adrian, F. J., and Bowers, V. A., *J. Chem. Phys. 44*, 4626 (1966).
117. Cochran, E. L., Bowers, V. A., Foner, S. N., and Jen, C. K., *Phys. Rev. Lett. 2*, 43 (1959).
118. Cole, T., Harding, J. T., Pellam, J. R., and Yost, D. M., *J. Chem, Phys. 27*, 593 (1957).
119. Cole, T. and McConnell, H. M., *J. Chem. Phys. 29*, 451 (1958).
120. Coles, B. A., Orton, J. W., and Owen, J., *Phys. Rev. Lett. 4*, 116 (1960).
121. Colin, R., *J. Mol. Spectr. 44*, 230 (1972).
122. Colin, R. and Jones, W. E., *Can. J. Phys. 45*, 301 (1967).
123. Cook, T. J., Zegarski, B. R., Breckenridge, W. H., and Miller, T. A., *J. Chem. Phys. 58*, 1548 (1972).
124. Coufal, H. J., Burger, M., Nagel, U., Lüscher, E., Böning, K., and Vogl, G., *Phys. Lett. 48A*, 143–144 (1974).
125. Coufal, H. and Luscher, E., *Phys. Lett. 48A*, 445–446 (1974).
126. Coulson, C. A., *Vol. Commemoratif Victor Henri, Contribution to the Study of Molecular Structure* (Maison Desoer, Liege, 1947).
127. Curl, R. F., *Mol. Phys. 9*, 585 (1965).
128. Davies, P. B., Wayne, F. D., and Stone, A. J., *Mol. Phys. 28*, 1409 (1974).
129. de Boer, E. and Mackor, E. L., *Mol. Phys. 5*, 493 (1962).
130. de Groot, M. S., Hesselmann, I. A., and van der Waals, J. H., *Mol. Phys. 16*, 45 (1969).
131. de Groot, M. S. and van der Waals, J. H., *Mol. Phys. 3*, 190 (1960).
132. de Groot, M. S. and van der Waals, J. H., *Mol. Phys. 6*, 545 (1963).
133. de Groot, M. S. and van der Waals, J. H., *Physica 29*, 1128 (1963).
134. Deile, O., *Z. Physik 106*, 405 (1937).
135. de Loth, P., Cassoux, P., Daudey, J. P., and Malrieu, J. P., *J. Am. Chem. Soc. 103*, 4007 (1981).

136. Demuynck, J. and Schaefer, H. F. III, *J. Chem. Phys. 72*, 311 (1980).
137. Denison, A. B., *Magn. Reson. Rev. 2*, 1 (1973).
138. Descleaux, J. P. and Bessis, N., *Phys. Rev. A2*, 1623 (1970).
139. Devillers, C. and Ramsay, D. A., *Can. J. Phys. 49*, 2839 (1971).
140. DeVore, T. C., Van Zee, R. J., and Weltner, W., Jr., *J. Chem. Phys. 68*, 3522 (1978).
141. DeVore, T. C., Van Zee, R. J., and Weltner, W., Jr., *Proc. Symp. on High Temp. Metal Halide Chemistry*, Eds. Hildenbrand, D. L. and Cubicciotti, D. D. (The Electrochem. Soc., Int. 1978), Proceedings Vol. 78-1, page 190.
142. DeVore, T. C. and Weltner, W., Jr., *J. Am. Chem. Soc. 99*, 4700 (1977).
143. de Wijn, H. W. and van Balderen, R. F., *J. Chem. Phys. 46*, 1381 (1967).
144. Dixon, R. N., *Can. J. Phys. 37*, 1171 (1959).
145. Dixon, T. A. and Woods, R. C., *J. Chem. Phys. 67*, 3956 (1977).
146. Doedens, R. J., *Prog. Inorg. Chem. 21*, 209 (1976).
147. Domaille, P. J., Steimle, T. C., and Harris, D. O., *J. Mol Spectr. 68*, 146 (1977).
148. Dousmanis, G. C., *Phys. Rev. 97*, 967 (1955).
149. Dowsing, R. D., *J. Magn. Reson. 2*, 332 (1970).
150. Dowsing, R. D. and Gibson, J. F., *J. Chem. Phys. 50*, 294 (1969).
151. Duerig, W. H. and Mador, I. L., *Rev. Sci. Instr. 23*, 421 (1952).
152. Dzyaloshinski, I., *J. Phys. Chem. Solids 4*, 241 (1958).
153. Easley, W. C. and Weltner, W., Jr., *J. Chem. Phys. 52*, 197 (1970).
154. Eidels-Dubovoi, S. and Beltrán-López, V., *J. Magn. Reson. 32*, 441 (1978).
155. Erdös, P., *J. Phys. Chem. Solids 27*, 1705 (1966).
156. Eshbach, J. R. and Strandberg, M. W. P., *Phys. Rev. 85*, 24 (1952).
157. Evenson, K. M. and Mizushima, M., *Phys. Rev. A6*, 2197 (1972).
158. Evenson, K. M., Radford, H. E., and Moran, M. M., *Appl. Phys. Lett. 18*, 426 (1971).
159. Evenson, K. M., Saykally, R. J., Jennings, D. A., Curl, R. F., Jr., and Brown, J. M., in *Chemical and Biochemical Applications of Lasers*, Ed. Moore, C. B., (Academic Press, New York, 1980), Vol. V, pages 95–138.
160. Ewing, G. E., Thompson, W. E., and Pimentel, G. C., *J. Chem. Phys. 32*, 927 (1960).
161. Farmer, J. B., Gerry, M. C. L., and McDowell, C. A., *Mol. Phys. 8*, 253 (1964).
162. Ferrante, R. F., Wilkerson, J. L., Graham, W. R. M., and Weltner, W., Jr., *J. Chem. Phys. 67*, 5904 (1977).
163. Fessenden, R. W. and Schuler, R. H., *J. Chem. Phys. 43*, 2704 (1965).
164. Figgis, B. N. and Martin, R. L., *J. Chem. Soc.* 3837 (1956).
165. Fischer, P. H. H., Charles, S. W., and McDowell, C. A., *J. Chem. Phys. 46*, 2162 (1967).
166. Fixman, M., *J. Chem. Phys. 48*, 223 (1968).
167. Foner, S. N., Cochran, E. L., Bowers, V. A., and Jen, C. K., *Phys. Rev. Lett. 1*, 91 (1958).
168. Foner, S. N., Cochran, E. L., Bowers, V. A. and Jen, C. K., *J. Chem. Phys. 32*, 963 (1960).

169. Foner, S. N., Jen, C. K., Cochran, E. L., and Bowers, V. A., *J. Chem. Phys. 28*, 351 (1958).
170. Freed, J. H., Bruno, G. V., and Polnaszek, C., *J. Chem. Phys. 55*, 5270 (1971).
171. Freed, J. H. and Fraenkel, G. K., *J. Chem. Phys. 39*, 326 (1963).
172. Friedman, E. and Low, W., *Phys. Rev. 120*, 408 (1960).
173. Froese, C., *J. Chem. Phys. 45*, 1417 (1966).
174. Frosch, R. A. and Foley, H. M., *Phys. Rev. 88*, 1337 (1952).
175. Fujioka, Y. and Tanaka, Y., *Sci. Pap. Inst. Phys. Chem. Res. (Tokyo) 34*, 713 (1938).
176. Fuller, G. H., *J. Phys. Chem. Ref. Data 5*, 835 (1976).
177. Fuller, G. H. and Nier, A. O., Appendix 2, "Relative Isotopic Abundances," Nucl. Data Sheets, as given in Table 8b-1 of *American Institute of Physics Handbook*, 3rd Edition (McGraw-Hill, New York, 1972).
178. Gerloch, M. and Harding, J. H., *Proc. Roy. Soc. (London) A360*, 211 (1978).
179. Gersmann, H. R. and Swalen, J. D., *J. Chem. Phys. 36*, 3221 (1962).
180. Geusic, J. E., Peter, M., and Schulz-DeBois, E. O., *Bell System Tech. J. 38*, 291 (1959).
181. Gibson, J. F., Ingram, D. J. E., and Schonland, D., *Discussions Faraday Soc. 26*, 72 (1958).
182. Ginsberg, A. P., *J. Am. Chem. Soc. 102*, 111 (1980).
183. Godfrey, M., Kern, C. W., and Karplus, M., *J. Chem. Phys. 44*, 4459 (1966).
184. Goeppert-Mayer, M., *Ann. d. Phys. (Leipzig) 9*, 273 (1931).
185. Golding, R. M., Singhasuwich, T., and Tennant, W. C., *Mol. Phys. 34*, 1343 (1977).
186. Golding, R. M. and Tennant, W. C., *Mol. Phys. 25*, 1163 (1973).
187. Golding, R. M. and Tennant, W. C., *Mol. Phys. 28*, 167 (1974).
188. Goldsborough, J. P. and Koehler, T. R., *Phys. Rev. 133*, A135 (1964).
189. Goodgame, D. M. L., Hill, N. J., Marsham, D. F., Scapoki, A. C., Smart, M. L., and Troughton, P. G. H., *J. Chem. Soc., Chem. Commun.*, 629 (1969).
190. Goodman, B. A. and Raynor, J. B., *Adv. Inorg. Chem. and Radiochem. 13*, 136 (1970).
191. Gordon, R. M. and Merer, A. J., *Can. J. Phys. 58*, 642 (1980).
192. Goudsmit, S., *Phys. Rev. 43*, 636 (1933).
193. Gouterman, M., *J. Chem. Phys. 30*, 1369 (1959).
194. Gouterman, M. and Moffitt, W., *J. Chem. Phys. 30*, 1107 (1959).
195. Graham, W. R. M., Dismuke, K. I., and Weltner, W., Jr., *J. Chem. Phys. 60*, 3817 (1974).
196. Graham, W. R. M., Dismuke, K. I., and Weltner, W., Jr., *Astrophys. J. 204*, 301 (1976).
197. Graham, W. R. M. and Weltner, W., Jr., *J. Chem. Phys. 65*, 1516 (1976).
198. Grivet, J.-Ph., *Mol. Phys. 19*, 389 (1970).
199. Grivet, J.-Ph. and Lhoste, J.-M., *Mol. Phys. 21*, 999 (1971).
200. Gruen, D. M., *Progr. Inorg. Chem. 14*, 119 (1971).

201. Guha, B. C., *Proc. Roy. Soc. A206*, 353 (1951).
202. Hagland, L., Kopp, I., and Åslund, N., *Ark. Fys. 32*, 321 (1966).
203. Hall, J. A., *J. Chem. Phys. 58*, 410 (1973).
204. Hameka, H. F., *J. Chem. Phy. 31*, 315 (1959).
205. Hameka, H. F., in *The Triplet State*, Proceedings of an International Symposium in Beirut, Lebanon, February 14–19, 1967, Ed. Zahlan, A. B. (Cambridge University Press, 1967), pages 1–27.
206. Hameka, H. F. and Oosterhoff, L. J., *Mol. Phys. 1*, 358 (1958).
207. Hand, R. W., Hunt, W. J., and Schaefer, H. F., III., *J. Am. Chem. Soc 95*, 4517 (1973).
208. Hansen, A. E. and Ballhausen, C. J., *Trans. Faraday Soc. 61*, 631 (1965).
209. Harris, E. A., *J. Phys. C5*, 338 (1972).
210. Harris, E. A. and Owen, J., *Phys. Rev. Lett. 11*, 9 (1963).
211. Harrison, J. F., *Accounts Chem. Res. 7*, 378 (1974).
211a. Harrison, J. F., *J. Chem. Phys. 54*, 5413 (1971).
212. Harvey, J. S. M., Evans, L., and Lew, H., *Can. J. Phys. 50*, 1719 (1972).
212a. Hasegawa, A., Sogabe, K., and Miura, M., *Mol. Phys. 30*, 1889 (1975).
213. Hastie, J. W., Hauge, R. H., and Margrave, J. L., *J. Chem. Phys. 51*, 2648 (1969).
214. Hatfield, W. E., *Theory and Applications of Molecular Paramagnetism*, Eds. Boudreaux, E. A. and Mulay, L. N., (John Wiley & Sons, New York, 1976), Chapter VII.
215. Hay, P. J., Thibeault, J. C., and Hoffmann, R., *J. Am. Chem. Soc. 97*, 4884 (1975).
216. Herrmann, A., Schumacher, E., and Wöste, L., *J. Chem. Phys. 68*, 2327 (1978).
217. Herzberg, G. and Johns, J. W. C., *J. Chem. Phys. 54*, 2276 (1971).
218. Herzberg, G. and Ramsay, D. A., *Proc. Roy. Soc. (London) A233*, 34 (1955).
219. Herzberg, G. and Travis, D. N., *Can. J. Phys. 42*, 1658 (1964).
220. Higuchi, J., *J. Chem. Phys. 38*, 1237 (1963).
221. Higuchi, J., Kubota, S., Kumamoto, T., and Tokue, I., *Bull. Chem. Soc. Japan 47*, 2775 (1974).
222. Hinchliffe, A., *J. Mol. Struct. 64*, 289 (1980).
223. Hirokawa, S., *J. Phys. Soc. Japan, 35*, 12 (1973).
224. Hodgson, D. J., *Progr. Inorg. Chem. 19*, 173–241, Ed. Lippard, S. J. (John Wiley, New York, 1975).
225. Hoffmann, R., *Tetrahedron 22*, 521 (1966).
226. Holmberg, R. W., *J. Chem. Phys. 51*, 3255 (1969).
227. Holuj, F., *Can. J. Phys. 44*, 503 (1966).
228. Hornig, A. W. and Hyde, J. S., *Mol. Phys. 6*, 33 (1963).
229. Hou, S. L., Summit, R. W., and Tucker, R. F., *Phys. Rev. 154*, 258 (1967).
230. Hougen, J. T., *Can. J. Phys. 40*, 598 (1962).
231. Hougen, J. T., Leroi, G. E., and James, T. C., *J. Chem. Phys. 34*, 1670 (1961).
232. Hougen, J. T., Mucha, J. A., Jennings, D. A., and Evenson, K. M., *J. Mol. Spectr. 72*, 463 (1978).

233. Huber, R. and Schwoerer, M., *Chem. Phys. Lett. 72*, 10 (1980).
234. Hudson, A. and Luckhurst, G. R., *Chem. Rev. 69*, 191 (1969).
235. Hurd, C. M. and Coodin, P., *J. Phys. Chem. Solids 28*, 523 (1967).
236. Hutchings, M. T., *Solid State Phys. 16*, 227–273 (1964).
237. Hutchings, M. T., Birgeneau, R. J., and Wolfe, W. P., *Phys. Rev. 168*, 1026 (1968).
238. Hutchison, C. A., Jr., in *The Triplet State*, Proceedings of an International Symposium in Beirut, Lebanon, February 14–19, 1967, Ed. Zahlan, A. B. (Cambridge University Press, 1967) pages 63–100.
239. Hutchison, C. A., Jr. and Mangum, B. W., *J. Chem. Phys. 29*, 952 (1958); *34*, 908 (1961).
240. Ichikawa, T., Iwasaki, M., and Kuwata, K., *J. Chem. Phys. 44*, 2979 (1966).
241. Isomoto, A., Watari, H., and Kotani, M., *J. Phys. Soc. Japan 29*, 1571 (1970).
242. Itoh, K., *Chem. Phys. Lett. 1*, 235 (1967).
243. Itzkowitz, M. S., *J. Chem. Phys. 46*, 3048 (1967).
244. Iwasaki, M., *J. Magn. Reson. 16*, 417 (1974).
245. Iwasaki, M., Toriyama, K., and Muto, H., *J. Chem. Phys. 71*, 2853 (1979).
246. Jackel, G. S. and Gordy, W., *Phys. Rev. 176*, 443 (1968).
247. Jackel, G. S., Nelson, W. H., and Gordy, W., *Phys. Rev. 176*, 453 (1968).
248. Jen, C. K., *Formation and Trapping of Free Radicals*, Ed. Bass, A. A. and Broida, H. P. (Academic Press, New York, 1960), pages 213–255.
249. Jen, C. K., Bowers, V. A., Cochran, E. L., and Foner, S. N., *Phys. Rev. 126*, 1749 (1962).
250. Jen, C. K., Foner, S. N., Cochran, E. L., and Bowers, V. A., *Phys. Rev. 104*, 846 (1956).
251. Jen, C. K., Foner, S. N., Cochran, E. L., and Bowers, V. A., *Phys. Rev. 112*, 1169 (1958).
252. Jones, W. E., *Can. J. Phys. 45*, 21 (1967).
253. Jotham, R. W. and Kettle, S. F., *Inorg. Chem. 9*, 1390 (1970).
254. Kahn, O. and Briat, B., *J. Chem. Soc., Faraday Trans.* 2, *72*, 268 (1976).
255. Kanamori, J., *Magnetism*, Ed. Rado, G. T. and Suhl, H. (Academic Press, New York, 1963), Vol. 1, page 161.
256. Känzig, W., *J. Phys. Chem. Solids 17*, 80 (1960).
257. Känzig, W. and Cohen, M. H., *Phys. Rev. Lett. 3*, 509 (1959).
258. Kasai, P. H., *J. Chem. Phys. 49*, 4979 (1968).
259. Kasai, P. H., *Phys. Rev. Lett. 2*, 67 (1968); *Accounts Chem. Res. 4*, 329 (1971).
260. Kasai, P. H., *J. Am. Chem. Soc. 94*, 5950 (1972).
261. Kasai, P. H., Hedaya, E., and Whipple, E. B., *J. Am. Chem. Soc. 91*, 4364 (1969).
262. Kasai, P. H. and McLeod, D., Jr., *J. Chem. Phys. 55*, 1566 (1971).
263. Kasai, P. H. and McLeod, D., Jr., *J. Phys. Chem. 79*, 2324 (1975); *82*, 1554 (1978).
264. Kasai, P. H., Weltner, W., Jr., and Whipple, E. B., *J. Chem. Phys. 42*, 1120 (1965).
265. Kasai, P. H. and Whipple, E. B., *Mol. Phys. 9*, 497 (1965).

266. Kasai, P. H., Whipple, E. B., and Weltner, W., Jr., *J. Chem. Phys. 44*, 2581 (1966).
267. Kaving, B. and Scullman, R., *J. Mol. Spectr. 32*, 475 (1969).
268. Kawaguchi, K., Yamada, C., and Hirota, E., *J. Chem. Phys. 71*, 3338 (1979).
269. Kayama, K., *J. Chem. Phys. 42*, 622 (1965).
270. Kayama, K. and Baird, J. C., *J. Chem. Phys. 46*, 2604 (1967).
271. Kedzie, R. W., Lyons, D. H., and Kestigan, M., *Phys. Rev. 138*, A918 (1965).
272. Kirkpatrick, E. S., Muller, K. A., and Rubins, R. S., *Phys. Rev. 135*, A86 (1964).
273. Kittel, C., *Phys. Rev. 120*, 335 (1960).
274. Kivelson, D., *J. Chem. Phys. 33*, 1094 (1960); *41*, 1904 (1964).
275. Kneubuhl, F. K., *J. Chem. Phys. 33*, 1074 (1960).
276. Knight, L. B., Jr., Brom, J. M., Jr., and Weltner, W., Jr., *J. Chem. Phys. 56*, 1152 (1972).
277. Knight, L. B., Jr., Easley, W. C., and Weltner, W., Jr., *J. Chem. Phys. 54*, 1610 (1971).
278. Knight, L. B., Easley, W. C., Weltner, W., Jr., and Wilson, M., *J. Chem. Phys. 54*, 322 (1971).
279. Knight, L. B., Jr., Mouchet, A., Beaudry, W. T., and Duncan, M., *J. Magn. Reson. 32*, 383 (1978).
280. Knight, L. B., Jr. and Weltner, W., Jr., *J. Chem. Phys. 53*, 4111 (1970).
281. Knight, L. B., Jr. and Weltner, W., Jr., *J. Mol. Spectr. 40*, 317 (1971).
282. Knight, L. B., Jr. and Weltner, W., Jr., *J. Chem. Phys. 54*, 3875 (1971).
283. Knight, L. B., Jr. and Weltner, W., Jr., *J. Chem. Phys. 55*, 2061 (1971).
284. Knight, L. B., Jr., and Weltner, W., Jr., *J. Chem. Phys. 55*, 5066 (1971).
285. Knight, L. B., Jr., Wise, M. B., Childers, A. G., Davidson, E. R., and Daasch, W. R., *J. Chem. Phys. 73*, 4198 (1980).
286. Knight, L. B., Jr., Wise, M. B., Davidson, E. R., and McMurchie, L. E., *J. Chem. Phys. 76*, 126 (1982).
287. Kokoszka, G. F., Allen, H. C., and Gordon, G., *J. Chem. Phys. 42*, 3693 (1965).
288. Kokoszka, G. F. and Duerst, R. W., *Coord. Chem. Rev. 5*, 209 (1970) and references therein.
289. Kokoszka, G. F., Linzer, M., and Gordon, G., *Inorg. Chem. 7*, 1730 (1968).
290. Kon, H., *J. Am. Chem. Soc. 95*, 1045 (1973).
291. Kopp, I., Kronekvist, M., and Guntsch, A., *Ark. Fys. 32*, 371 (1966).
292. Korst, M. N. and Khazanovitch, T. N., *J. Expt. Theor. Phys. 18*, 1049 (1964).
293. Kothe, G., Ohmes, E., Brickmann, J., and Zimmermann, H., *Angew. Chem. 83*, 1015 (1971); *Angew. Chem.* (Int. Ed. Engl.) *10*, 938 (1971).
294. Kottis, P. and Lefebvre, R., *J. Chem. Phys. 39*, 393 (1963).
295. Kottis, P. and Lefebvre, R., *J. Chem. Phys. 41*, 379 (1964).
296. Kovacs, I. and Pacher, P., *J. Phys. B8*, 796 (1975); Pacher, P., *Acta Phys. Sci. Hung. 35*, 73 (1974).
297. Kupferman, S. L. and Pipkin, F. M., *Phys. Rev. 166*, 207 (1968).

298. Kusch, P. and Hughes, V. W., *Handbuch der Physik*, Ed. Flugge, S. (Springer-Verlag, Berlin, 1959), Vol. 37, Part 1, pp. 100 and 117.
299. Langhoff, S. R., *J. Chem. Phys. 61*, 1708 (1974).
300. Langhoff, S. R., *J. Chem. Phys. 61*, 3881 (1974).
301. Langhoff, S. R. and Davidson, E. R., *Int. J. Quantum Chem. 7*, 759 (1973).
302. Langhoff, S. R. and Davidson, E. R., *J. Chem. Phys. 64*, 4699 (1976).
303. Langhoff, S. R., Davidson, E. R., and Kern, C. W., *J. Chem. Phys. 63*, 4800 (1975).
304. Langhoff, S. R., Elbert, S. T., and Davidson, E. R., *Int. J. Quantum Chem. 7*, 999 (1973).
305. Langhoff, S. R. and Kern, C. W., in *Applications of Electronic Structure Theory*, Ed. Schaefer, H. F., III, (Plenum Press, New York, 1977), Chapter 10.
306. Larsson, S., Brown, R. E., and Smith, V. H., Jr., *Phys. Rev. 6*, 1375 (1972).
307. Lembke, R. R., Ferrante, R. F., and Weltner, W., Jr., *J. Am. Chem. Soc. 99*, 416 (1977).
308. Lew, H. and Wessel, G., *Phys. Rev. 90*, 1 (1953).
309. Lin, W. C., *Mol. Phys. 25*, 247 (1973).
310. Lindsay, D. M., Herschbach, D. R., and Kwiram, A. L., *Chem. Phys. Lett. 25*, 175 (1974).
311. Lindsay, D. M., Herschbach, D. R., and Kwiram, A. L., *J. Chem. Phys. 60*, 315 (1974).
312. Lindsay, D. M., Herschbach, D. R., and Kwiram, A. L., *Mol. Phys. 32*, 1199 (1976).
313. Lindsay, D. M., Herschbach, D. R., and Kwiram, A. L., *Mol. Phys. 39*, 529 (1980).
314. Lohr, L. L., Jr. and Lipscomb, W. N., *J. Chem. Phys. 38*, 1607 (1963).
315. Lontz, R. J. and Gordy, W., *J. Chem. Phys. 37*, 1357 (1962).
316. Lounsbury, J. B., *J. Chem. Phys. 46*, 2193 (1967); *42*, 1549 (1965).
317. Ludwig, G. W. and Woodbury, H. H., *Solid State Physics*, Ed. Seitz, F. and Turnbull, D. (Academic Press, New York, 1962), Vol. 13, page 223.
318. Ludwig, G. W., Woodbury, H. H., and Ham, F. S., reference 80 in the chapter by G. W. Ludwig and H. H. Woodbury in *Solid State Physics 13*, 223 (1962).
319. Luz, Z., Reuveni, A., Holmberg, R. W., and Silver, B. L., *J. Chem. Phys. 51*, 4017 (1969).
320. Mackey, J. H. and Wood, D. E., *J. Chem. Phys. 52*, 4914 (1970).
321. Máckowiak, M., *Nuovo Cimento 29B*, 207 (1975).
322. Máckowiak, M. and Kurzyński, M., *Phys. Stat. Sol.* (b) *51*, 841 (1972).
323. Mahieu, J. M., Jacquinot, D., Schamps, J., and Hall, J. A., *J. Phys. B8*, 308 (1975).
324. Malmberg, C., Scullman, R., and Nylén, P., *Ark. Phys. 39*, 495 (1968).
325. Martin, R. W. and Merer, A. J., *Can. J. Phys. 51*, 634 (1973).
326. Martinez, J. V. and Weltner, W., Jr., *J. Chem. Phys. 65*, 4256 (1976).
327. Maruani, J., McDowell, C. A., Nakajima, H., and Raghunathan, P., *Mol. Phys. 14*, 349 (1968).
328. McCarty, M., Jr. and Robinson, G. W., *Mol. Phys. 2*, 415 (1959).

329. McClung, R. E. D., *Can. J. Phys. 46*, 2271 (1968).
330. McClure, D. S., *J. Chem. Phys. 20*, 682 (1952).
331. McConnell, H. M., *J. Chem. Phys. 25*, 709 (1956).
332. McConnell, H. M., *J. Chem. Phys. 29*, 1422 (1958).
333. McConnell, H. M., *Proc. Natl. Acad. Sci. 44*, 766 (1958).
334. McConnell, H. M., *Proc. Natl. Acad. Sci. U.S. 45*, 172 (1959).
335. McConnell, H. M. and Chesnut, D. B., *J. Chem. Phys. 28*, 107 (1958).
336. McConnell, H. M., Heller, C., Cole, T., and Fessenden, R. W., *J. Am. Chem. Soc. 82*, 766 (1960).
337. McConnell, H. M. and Strathdee, J., *Mol. Phys. 2*, 129 (1959).
337a. McDowell, C. A., Nakajima, H., and Raghunathan, P., *Can. J. Chem. 48*, 805 (1970).
338. McGarvey, B. R., *J. Chem. Phys. 41*, 3743 (1964).
339. McGarvey, B. R., in *Transition Metal Chemistry*, Vol. 3, Ed. Carlin, R. L. (Marcel Dekker, New York, 1966).
340. McIver, J. W., Jr. and Hameka, H. F., *J. Chem. Phys. 45*, 767 (1966); *46*, 825 (1967).
341. McMillan, J. A. and Smaller, B., *J. Chem. Phys. 35*, 1698 (1861).
342. McNeil, R. I., Williams, F., and Yim, M. B., *Chem. Phys. Lett. 61*, 293 (1979).
343. Meyer, H., O'Brien, M. C. M., and Van Vleck, J. H., *Proc. Roy. Soc. (London) A243*, 414 (1958).
344. Milligan, D. E. and Jacox, M. E., *J. Chem. Phys. 44*, 2850 (1966).
345. Miyagawa, I. and Gordy, W., *J. Chem. Phys. 32*, 255 (1960).
346. Mizushima, M. and Koida, M., *J. Chem. Phys. 20*, 765 (1952).
347. Morehouse, R. L., Christiansen, J. J., and Gordy, W., *J. Chem. Phys. 45*, 1747 (1966).
348. Moriya, T., *Magnetism*, Eds. Rado, G. T. and Suhl, H. (Academic Press, New York, 1963), Vol. 1, page 85.
349. Morton, J. R., *Chem. Rev. 64*, 453 (1964).
350. Morton, J. R. and Preston, K. F., *J. Magn. Reson. 30*, 577 (1978).
350a. Morton, J. R. and Preston, K. F., *Landolt-Börnstein*, Group II, Vol. 9a, Eds. Fischer, H. and Hellwege, K.-H., (Springer-Verlag, Berlin, 1977), pages 1–268.
351. Morton, J. R., Preston, K. F., Strach, S. J., Adrian, F. J., and Jette, A. N., *J. Chem. Phys. 70*, 2889 (1979).
352. Morton, J. R., Preston, K. F., Wang, J. T., and Williams, F., *Chem. Phys. Lett. 64*, 71 (1979).
353. Morton, J. R., Rowlands, J. R., and Whiffen, D. H., National Physical Laboratory (U.K.) Report No. BPR 13 (1962).
354. Mulliken, R. S. and Christy, A., *Phys. Rev. 38*, 87 (1931).
355. Mulliken, R. S., Rieke, C. A., Orloff, D., and Orloff, H., *J. Chem. Phys. 17*, 1248 (1949).
356. Muto, H. and Kispert, L. D., *J. Chem. Phys. 72*, 2300 (1980).
357. Myers, G. H., Easley, W. C., and Zilles, B. A., *J. Chem. Phys. 53*, 1181 (1970).

358. Neiman, R. and Kivelson, D., *J. Chem. Phys. 35*, 156 (1961).
359. Nesbet, R. K., *Phys. Rev. A135*, 460 (1964).
360. Nevin, T. E., Conway, M., and Cranley, M., *Proc. Phys. Soc. (London) A65*, 115 (1952); Hayes, W., McCavill, P. D., and Nevin, T. E., *Proc. Phys Soc. (London), A70*, 904 (1957).
361. Nicklin, R. C., Poole, C. P., Jr., and Farach, H. A., *J. Chem. Phys. 58*, 2579 (1973).
362. Norris, J. R. and Weissman, S. I., *J. Phys. Chem. 73*, 3119 (1969).
363. Oosterhuis, W. T., *Struct. Bonding 20*, 59 (1974).
364. Ovenall, D. W. and Whiffen, D. H., *Mol. Phys. 4*, 135 (1961).
365. Owen, J., *J. Applied Phys.*, Supplement to *32*, 2135 (1961).
366. Owen, J. and Harris, E. A., in *Electron Paramagnetic Resonance*, Ed. Geschwind, S. (Plenum Press, New York, 1972), pages 427–492.
367. Palmiere, P. and Sink, M. L., *J. Chem. Phys. 65*, 3641 (1976).
368. Peiser, H. S., *Formation and Trapping of Free Radicals*, Ed. Bass, A. M. and Broida, H. P. (Academic Press, New York, 1960), Chapter 9.
369. Pendlebury, J. M., *Proc. Phys. Soc. (London) 84*, 857 (1964).
370. Peyerimhoff, S. D. Buenker, R. J., Kammer, W. E., and Hsu, H., *Chem. Phys. Lett. 8*, 129 (1971).
371. Pick, H., *Z. Physik 159*, 69 (1960).
372. Pilbrow, J. R., *Mol. Phys. 16*, 307 (1969).
373. Pilbrow, J. R., *J. Magn. Reson. 31*, 479 (1978).
374. Pilbrow, J. R. and Lowrey, M. R., *Rep. Prog. Phys. 43*, 28 (1980).
375. Pinchemel, B. and Schamps, J., *Can J. Phys. 53*, 431 (1975); *Chem. Phys. 18*, 481 (1976).
376. Pitzer, K. S. and Clementi, E., *J. Am. Chem. Soc. 81*, 4477 (1959).
377. Pitzer, R. M. and Hameka, H. F., *J. Chem. Phys. 37*, 2725 (1962).
378. Pollack, G. L., *Rev. Mod. Phys. 36*, 748 (1964).
379. Ponte Goncalves, A. M. and Hutchison, C. A., Jr., *J. Chem. Phys. 49*, 4235 (1968).
380. Poole, C. P., Jr. and Farach, H. A., *J. Magn. Reson. 4*, 312 (1971).
381. Poole, C. P., Jr., Farach, H. A., and Jackson, W. K., *J. Chem. Phys. 61*, 2220 (1974).
382. Pople, J. A., Beveridge, D. L., and Dobosh, P. A., *J. Am. Chem. Soc. 90*, 4201 (1968).
383. Prather, J. L., *Atomic Energy Levels in Crystals*, U.S. Natl. Bur. Standards, Monograph 19, Washington, D.C. (1961).
384. Pritchard, R. H., Bender, C. F., and Kern, C. W., *Chem. Phys. Lett. 5*, 529 (1970).
385. Radford, H. E., *Phys. Rev. 122*, 114 (1961).
386. Radford, H. E., *Phys. Rev. 126*, 1035 (1962).
387. Raghunathan, P. and Shimokoshi, K., *Spectrochim. Acta 36A*, 285 (1980).
388. Raynes, W. T., *J. Chem. Phys. 44*, 2755 (1966); *41*, 3020 (1964).
389. Redfield, A. G., *Adv. Magn. Reson. 1*, 1 (1965).
390. Robbins, E. J., Leckenby, R. E., and Willis, P., *Adv. Phys. 16*, 739 (1967); Foster, P. J., Leckenby, R. E., and Robbins, E. J., *J. Phys. B2*, 478 (1969).

391. Roberts, J. and Lynden-Bell, R. M., *Mol. Phys. 21*, 689 (1971).
392. Rockenbauer, A. and Simon, P., *J. Magn. Reson. 11*, 217 (1973).
393. Rogers, M. T. and Kispert, L. D., *J. Chem. Phys. 46*, 3193 (1967).
394. Rollmann, L. D. and Chan, S. I., *J. Chem. Phys. 50*, 3416 (1969).
395. Ross, I. G. and Yates, J., *Trans. Faraday Soc. 55*, 1064 (1959).
396. Rostas, J., Cossart, D., and Bastien, J. R., *Can. J. Phys. 52*, 1274 (1974).
397. Sakaguchi, U., Arata, Y., and Fujiwara, S., *J. Magn. Reson. 9*, 118 (1973).
398. Sandars, P. G. H. and Beck, J., *Proc. Roy. Soc. A289*, 97 (1965).
399. Sands, R. H., *Phys. Rev. 99*, 1222 (1955).
400. Saunders, M. and Johnson, C. S., Jr., *J. Chem. Phys. 48*, 534 (1968).
401. Saykally, R. J. and Evenson, K. M., *Phys. Rev. Lett. 43*, 515 (1979).
402. Saykally, R. J. and Evenson, K. M., *Astrophys. J. 238*, L1 (1980).
403. Schaafsma, T. J. and Kommandeur, J., *Mol. Phys. 14*, 517 (1967).
404. Schmauss, G., Baumgärtel, H., and Zimmermann, H., *Angew Chem. 77*, 619 (1965); *Angew. Chem.* (Int. Ed. Engl.) *4*, 596 (1965).
405. Schnepp, O., *J. Phys. Chem. Solids 17*, 188 (1961).
406. Schoemaker, D., *Phys. Rev. 149*, 693 (1966).
407. Schoemaker, D., *Phys. Rev. 174*, 1060 (1968).
408. Schoemaker, D., *Phys. Rev. B7*, 786 (1973).
409. Schwoerer, M., Huber, R. A., and Hartl, W., *Chem. Phys. 55*, 97 (1981).
410. Schwoerer, M. and Wolf, H. C., in *The Triplet State*, Proceedings of an International Symposium in Beirut, Lebanon, February 14–19, 1967, Ed. Zahlan, A. B. (Cambridge University Press, 1967), pages 133–140.
411. Scott, P. R. and Richards, W. G., Chemical Society, London Specialist Periodical Reports, *Molecular Spectroscopy*, Vol. 4, pp. 40–95 (The Chemical Society, London, 1976).
412. Seidel, H., Schwoerer, M. and Schmid, D., *Z. Physik 182*, 398 (1965).
413. Shaffer, J. S., Farach, H. A., and Poole, C. P., Jr., *Phys. Rev. B13*, 1869 (1976).
414. Sillescu, H. and Kivelson, D., *J. Chem. Phys. 48*, 3493 (1968).
415. Singer, L. S., *J. Chem. Phys. 23*, 379 (1955).
416. Sinn, E., *Coord. Chem. Rev. 5*, 313 (1970).
417. Smardzewski, R. R. and Andrews, L., *J. Chem. Phys. 57*, 1327 (1972).
418. Smith, D. Y., *Phys. Rev. 131*, 2056 (1963).
419. Smith, D. Y., *Phys. Rev. 133*, A1087 (1964).
420. Smith, T. D. and Pilbrow, J. R., *Coord. Chem. Rev. 13*, 173–278 (1974).
421. Smith, G. R. and Weltner, W., Jr., *J. Chem. Phys. 62*, 4592 (1975).
422. Stephen, M. J. and Fraenkel, G. K., *J. Chem. Phys. 32*, 1435 (1960).
423. Sterne, T. E., *Proc. Roy. Soc. (London) A130*, 551 (1931).
424. Stevens, K. W. H., *Proc. Phys. Soc. (London) A65*, 209 (1952).
425. Stevens, K. W. H., *Magnetism*, Eds. Rado, G. T. and Suhl, H. (Academic Press, New York, 1963), Vol. 1, page 1.
426. Stone, A. J., *Proc. Roy. Soc. (London) A271*, 424 (1963).
427. Strickler, S. J. and Pitzer, K. S., *Molecular Orbitals in Chemistry, Physics and Biology*, Ed. Pullman, B. and Lowdin, P. O. (Academic Press, New York, 1964).

428. Sweeney, W. V., Coucouvanis, D., and Coffman, R. E., *J. Chem. Phys. 59*, 369 (1973).
428a. Takui, T. and Itoh, K., *Chem. Phys. Lett. 19*, 120 (1973).
429. Tinkham, M. and Strandberg, M. W. P., *Phys. Rev. 97*, 937, 951 (1955).
430. Tippins, H. H., *Phys. Rev. 160*, 343–345 (1967).
431. Townes, C. H., *Phys. Rev. 71*, 909L (1947); Townes, C. H. and Dailey, B. P., *J. Chem. Phys. 17*, 782 (1949); *Phys. Rev., 74*, 1245A (1949).
432. Trammell, G. T., Zeldes, H., and Livingston, R., *Phys. Rev. 110*, 630 (1958).
433. Tucker, K. D., Kutner, M. L., and Thaddeus, P., *Astrophys. J. 193*, L115 (1974).
434. Uslu, K. A., Code, R. F., and Harvey, J. S. M., *Can. J. Phys. 52*, 2135 (1974).
435. van der Waals, J. H. and de Groot, M. S., *Mol. Phys. 2*, 333 (1959).
436. van der Waals, J. H. and de Groot, M. S., in *The Triplet State*, Proceedings of an International Symposium in Beirut, Lebanon, February 14–19, 1967, Ed. Zahlan, A. B. (Cambridge University Press, 1967), pages 101–132.
437. van der Waals, J. H. and ter Maten, G., *Mol. Phys. 8*, 301 (1964).
438. van Doorn, C. Z. and Haven, Y., *Philips Res. Reports 11*, 479 (1956); *12*, 309 (1957).
439. Van Niekerk, J. N. and Schoening, F. R. L., *Acta Cryst. 6*, 227 (1953).
440. Vannotti, L. E. and Morton, J. R., *Phys. Rev. 174*, 448 (1968).
441. van Veen, G., *J. Magn. Reson. 30*, 91 (1978).
442. Van Vleck, J. H., *Phys. Rev., 33*, 467 (1929).
443. Van Vleck, J. H., *Phys. Rev. 52*, 1178 (1937).
444. Van Zee, R. J., Baumann, C. A., and Weltner, W., Jr., *J. Chem. Phys. 74*, 6977 (1981).
445. Van Zee, R. J., Baumann, C. A., and Weltner, W., Jr., unpublished.
446. Van Zee, R. J., Brown, C. M., and Weltner, W., Jr., *Chem. Phys. Lett. 64*, 325 (1979).
447. Van Zee, R. J., Brown, C. M., Zeringue, K. J., and Weltner, W., Jr., *Accounts Chem. Res. 13*, 237 (1980).
448. Van Zee, R. J., DeVore, T. C., and Weltner, W., Jr., *J. Chem. Phys. 71*, 2051 (1979).
449. Van Zee, R. J., DeVore, T. C., Wilkerson, J. L., and Weltner, W., Jr., *J. Chem. Phys. 69*, 1869 (1978).
450. Van Zee, R. J., Ferrante, R. F., Zeringue, K. J., and Weltner, W., Jr., *J. Chem. Phys. 75*, 5297 (1981).
451. Van Zee, R. J., Seely, M. L. and Weltner, W., Jr., *J. Chem. Phys. 67*, 861 (1977).
452. Van Zee, R. J. and Weltner, W., Jr., *J. Chem. Phys. 74*, 4330 (1981); erratum, ibid. *75*, 2484 (1981).
453. Veseth, L., *Phys. Scripta 12*, 125 (1975).
454. Veseth, L., *J. Mol. Spectr. 63*, 180 (1976).
455. Vinokurov, V. M., Zaripov, M. M., Stepanov, V. G., Pol'skii, Yu E., Chirkin, G. K., and Shekun, L. Ya., *Soviet Physics, Solid State 4*, 470 (1963).

456. Walsh, A. D., *J. Chem. Soc.*, 2266 (1953).
457. Wasserman, E., Barash, L., and Yager, W. A., *J. Am. Chem. Soc. 87*, 2075 (1965).
458. Wasserman, E., Hutton, R. S., Kuck, V. J., and Yager, W. A., *J. Chem. Phys. 55*, 2593 (1971).
459. Wasserman, E., Kuck, V. J., Hutton, R. S., Anderson, E. D., and Yager, W. A., *J. Chem. Phys. 54*, 4120 (1971).
459a. Wasserman, E., Schueller, K., and Yager, W. A., *Chem. Phys. Lett. 2*, 259 (1968).
460. Wasserman, E., Snyder, L. C., and Yager, W. A., *J. Chem. Phys. 41*, 1763 (1964).
461. Wasserman, E., Yager, W. A., and Kuck, V. J., *Chem. Phys. Lett. 7*, 409 (1970).
462. Watson, J. K. G., *Can. J. Phys. 46*, 1637 (1968).
463. Watson, R. E. and Freeman, A. J., *Phys. Rev. 123*, 521 (1961); *124*, 1117 (1961).
464. Wayne, F. D., *Chem. Phys. Lett. 31*, 97 (1974).
465. Wayne, F. D. and Colburn, E. A., *Mol. Phys. 34*, 1141 (1977).
466. Wayne, F. D., Davies, P. B., and Thrush, B. A., *Mol. Phys. 28*, 989 (1974).
467. Wayne, F. D. and Radford, H. E., *Mol. Phys. 32*, 1407 (1976).
468. Wei, M. S., Current, J. H., and Gendell, J., *J. Chem. Phys. 52*, 1592 (1970).
469. Weil, J. A. and Anderson, J. H., *J. Chem. Phys. 35*, 1410 (1961).
470. Weil, J. A. and Hecht, H. G., *J. Chem. Phys. 38*, 281 (1963).
471. Weissbluth, M., in *Molecular Biophysics*, Proceedings, International Summer School, Squaw Valley, California, August 17–28, 1964, Ed. Pullman, B. and Weissbluth, M. (Academic Press, New York, 1965), pages 205–238.
472. Weissman, S. I. and Kothe, G., *J. Am. Chem. Soc. 97*, 2537 (1975).
473. Weltner, W., Jr., *J. Chem. Phys. 28*, 477 (1958).
474. Weltner, W., Jr., *Adv. High Temp. Chem. 2*, 85 (1969).
475. Weltner, W., Jr., *Berichte der Bunsengesellschaft, Phys. Chem. 82*, 80 (1978).
476. Weltner, W., Jr., in *Characterization of High Temperature Vapors and Gases*, Proceedings of the 10th Materials Research Symposium at the National Bureau of Standards (U.S.), Gaithersburg Maryland, September 18–22, 1978, Ed. Hastie, J. W. (NBS Special Publication 561, issued October 1979), Vol. 1, page 587.
477. Weltner, W., Jr., McLeod, D., Jr., and Kasai, P., *J. Chem. Phys. 46*, 3172 (1967).
478. Wessel, G. and Lew, H., *Phys. Rev. 92*, 641 (1953).
479. Wickman, H. H., Klein, M. P., and Shirley, D. A., *J. Chem. Phys. 42*, 2113 (1965).
480. Woodgate, G. K., *Proc. Roy. Soc. (London) A293*, 117 (1966).
481. Wyard, S. J., Smith, R. C., and Adrian, F. J., *J. Chem. Phys. 49*, 2780 (1968).
482. Yates, J. H. and Pitzer, R. M., *J. Chem. Phys. 70*, 4049 (1979).
483. Zeldes, H. and Livingston, R., *J. Chem. Phys. 35*, 563 (1961).

484. Zeldes, H., Trammell, G. T., Livingston, R., and Holmberg, R. W., *J. Chem. Phys. 32*, 618 (1960).
485. Zhitnikov, R. A. and Kolesnikov, N. V., *Soviet Physics JETP* (Engl. Transl.) *19*, 65 (1964).
486. Zhitnikov, R. A. and Kolesnikov, N. V., *Soviet Physics, Solid State* (Engl. Transl.) *6*, 2645 (1965).
487. Zhitnikov, R. A. and Kolesnikov, N. V., *Soviet Physics, Solid State* (Engl. Transl.) *7*, 927 (1965).
488. Zhitnikov, R. A., Kolesnikov, N. V., and Kosyakov, V. I., *Soviet Physics JETP* (Engl. Transl.) *16*, 839 (1963).
489. Zhitnikov, R. A., Kolesnikov, N. V., and Kosyakov, V. I., *Soviet Physics JETP* (Engl. Transl.) *17*, 815 (1963).

Appendix A

Stable Magnetic Nuclei: Their Abundances, Spins, Moments, and Magnetogyric Ratios[a]

Nucleus	Natural abundance (%)	I ($\hbar$)	μ^b (nm)	Q^c (b)	γ_n^d (10^4 rad/G sec)
^{1}H	99.985	$\frac{1}{2}$	+2.79278		+2.67520
^{2}H	0.015	1	+0.85742	+0.0028	+0.41066
^{3}He	0.00013	$\frac{1}{2}$	−2.1276	–	−2.0380
^{6}Li	7.42	1	+0.82203	−0.0008	+0.39371
^{7}Li	92.58	$\frac{3}{2}$	+3.25636	−0.04	+1.03975
^{9}Be	100	$\frac{3}{2}$	−1.17745	+0.05	−0.375960
^{10}B	19.78	3	+1.8006	+0.085	+0.28746
^{11}B	80.22	$\frac{3}{2}$	+2.6885	+0.041	+0.85841
^{13}C	1.108	$\frac{1}{2}$	+0.7024	–	+0.6728
^{14}N	99.63	1	+0.40375	+0.01	+0.193375
^{15}N	0.37	$\frac{1}{2}$	−0.2831	–	−0.2712
^{17}O	0.037	$\frac{5}{2}$	−1.8937	−0.026	−0.36279
^{19}F	100	$\frac{1}{2}$	+2.6288	–	+2.5181
^{21}Ne	0.257	$\frac{3}{2}$	−0.66176	+0.09	−0.21157
^{23}Na	100	$\frac{3}{2}$	+2.21740	+0.10	+0.708015
^{25}Mg	10.13	$\frac{5}{2}$	−0.8554	+0.22	−0.16388
^{27}Al	100	$\frac{5}{2}$	+3.6413	+0.15	+0.69760
^{29}Si	4.70	$\frac{1}{2}$	−0.55526	–	−0.53188
^{31}P	100	$\frac{1}{2}$	+1.1317	–	+1.08405
^{33}S	0.76	$\frac{3}{2}$	+0.6435	−0.055	+0.20547
^{35}Cl	75.53	$\frac{3}{2}$	+0.82181	−0.10	+0.26240

Nucleus	Natural abundance (%)	I ($\hbar$)	μ^b (nm)	Q^c (b)	γ_n^d (10^4 rad/G sec)
^{37}Cl	24.47	$\frac{3}{2}$	+0.68407	−0.079	+0.21842
^{39}K	93.10	$\frac{3}{2}$	+0.39143	+0.049	+0.124982
^{41}K	6.88	$\frac{3}{2}$	+0.21487	+0.060	+0.068609
^{43}Ca	0.145	$\frac{7}{2}$	−1.3172	<±0.2	−0.18025
^{45}Sc	100	$\frac{7}{2}$	+4.7559	−0.22	+0.65081
^{47}Ti	7.28	$\frac{5}{2}$	−0.78846	+0.29	−0.151053
^{49}Ti	5.51	$\frac{7}{2}$	−1.10414	+0.24	−0.151093
^{51}V	99.76	$\frac{7}{2}$	+5.1485	−0.05	+0.70453
^{53}Cr	9.55	$\frac{3}{2}$	−0.4735	−0.15(5)	−0.15119
^{55}Mn	100	$\frac{5}{2}$	+3.449	+0.4	+0.6608
^{57}Fe	2.19	$\frac{1}{2}$	+0.09042	–	+0.086613
^{59}Co	100	$\frac{7}{2}$	+4.616	+0.38	+0.6317
^{61}Ni	1.19	$\frac{3}{2}$	−0.7498	+0.16	−0.23941
^{63}Cu	69.09	$\frac{3}{2}$	+2.2228	−0.211	+0.709739
^{65}Cu	30.91	$\frac{3}{2}$	+2.3812	−0.195	+0.76032
^{67}Zn	4.11	$\frac{5}{2}$	+0.87524	+0.16	+0.167678
^{69}Ga	60.4	$\frac{3}{2}$	+2.0145	+0.19	+0.64323
^{71}Ga	39.6	$\frac{3}{2}$	+2.5597	+0.18	+0.81731
^{73}Ge	7.76	$\frac{9}{2}$	−0.87918	−0.18	−0.093574
^{75}As	100	$\frac{3}{2}$	+1.439	+0.29	+0.4595
^{77}Se	7.58	$\frac{1}{2}$	+0.534	–	+0.5115
^{79}Br	50.54	$\frac{3}{2}$	+2.1055	+0.37	+0.67229
^{81}Br	49.46	$\frac{3}{2}$	+2.2696	+0.31	+0.72468
^{83}Kr	11.55	$\frac{9}{2}$	−0.9703	+0.26	−0.10327
^{85}Rb	72.15	$\frac{5}{2}$	+1.3524	+0.26	+0.25909
^{87}Rb*	27.85	$\frac{3}{2}$	+2.7500	+0.13	+0.87807
^{87}Sr	7.02	$\frac{9}{2}$	−1.093	+0.3	−0.1163
^{89}Y	100	$\frac{1}{2}$	−0.13733	–	−0.13155
^{91}Zr	11.23	$\frac{5}{2}$	−1.3028	–	−0.24959
^{93}Nb	100	$\frac{9}{2}$	+6.167	−0.22	+0.65635
^{95}Mo	15.72	$\frac{5}{2}$	−0.9135	±0.12	−0.17501
^{97}Mo	9.46	$\frac{5}{2}$	−0.9327	±1.1	−0.17869
^{99}Ru	12.72	$\frac{5}{2}$	−0.62	–	−0.119
^{101}Ru	17.1	$\frac{5}{2}$	−0.68	—	−0.130

Nucleus	Natural abundance (%)	I ($\hbar$)	μ^b (nm)	Q^c (b)	γ_n^d (10^4 rad/G sec)
^{103}Rh	100	$\frac{1}{2}$	−0.0883	–	−0.0846
^{105}Pd	22.23	$\frac{5}{2}$	−0.642	+0.8	−0.1230
^{107}Ag	51.82	$\frac{1}{2}$	−0.1135	–	−0.1087
^{109}Ag	48.18	$\frac{1}{2}$	−0.1305	–	−0.1250
^{111}Cd	12.75	$\frac{1}{2}$	−0.59428	–	−0.56926
^{113}Cd	12.26	$\frac{1}{2}$	−0.62167	–	−0.59550
^{113}In	4.28	$\frac{9}{2}$	+5.5229	+0.82	+0.58782
^{115}In	95.72	$\frac{9}{2}$	+5.5348	+0.83	+0.58908
^{115}Sn	0.35	$\frac{1}{2}$	−0.9178	–	−0.87916
^{117}Sn	7.61	$\frac{1}{2}$	−0.9999	–	−0.9578
^{119}Sn	8.58	$\frac{1}{2}$	−1.0461	–	−1.0021
^{121}Sb	57.25	$\frac{5}{2}$	+3.3592	−0.28	+0.35753
^{123}Sb	42.75	$\frac{7}{2}$	+2.5466	−0.36	+0.34848
^{125}Te	6.99	$\frac{1}{2}$	−0.8872	–	−0.8498
^{127}I	100	$\frac{5}{2}$	+2.8091	−0.79	+0.53817
^{129}Xe	26.44	$\frac{1}{2}$	−0.7768	–	−0.7441
^{131}Xe	21.18	$\frac{3}{2}$	+0.6908	−0.12	+0.22057
^{133}Cs	100	$\frac{7}{2}$	+2.5779	−0.003	+0.35276
^{135}Ba	5.59	$\frac{3}{2}$	+0.8365	+0.18	+0.26709
^{137}Ba	11.32	$\frac{3}{2}$	+0.9357	+0.28	+0.29877
^{138}La*	0.089	5	+3.704	+0.51	+0.3548
^{139}La	99.911	$\frac{7}{2}$	+2.778	+0.22	+0.3801
^{141}Pr	100	$\frac{5}{2}$	+4.16	−0.058	+0.797
^{143}Nd	12.17	$\frac{7}{2}$	−1.063	−0.48	−0.1455
^{145}Nd	8.30	$\frac{7}{2}$	−0.654	−0.25	−0.895
^{147}Sm*	14.97	$\frac{7}{2}$	−0.813	−0.18	−0.1113
^{149}Sm	13.83	$\frac{7}{2}$	−0.670	+0.052	−0.0917
^{151}Eu	47.82	$\frac{5}{2}$	+3.4631	+1.1	+0.66346
^{153}Eu	52.18	$\frac{5}{2}$	+1.530	+2.8	+0.2931
^{155}Gd	14.73	$\frac{3}{2}$	−0.2584	+1.6	−0.08251
^{157}Gd	15.68	$\frac{3}{2}$	−0.3388	+1.7	−0.10818
^{159}Tb	100	$\frac{3}{2}$	±2.008	+1.3	±0.6412
^{161}Dy	18.88	$\frac{5}{2}$	−0.482	+2.4	−0.0923

Nucleus	Natural abundance (%)	I ($\hbar$)	μ^b (nm)	Q^c (b)	γ_n^d (10^4 rad/G sec)
^{163}Dy	24.97	$\frac{5}{2}$	+0.676	+2.5	+0.1295
^{165}Ho	100	$\frac{7}{2}$	+4.12	+2.7	+0.564
^{167}Er	22.94	$\frac{7}{2}$	−0.564	+2.83	−0.0772
^{169}Tm	100	$\frac{1}{2}$	−0.231	–	−0.221
^{171}Yb	14.31	$\frac{1}{2}$	+0.4919	–	+0.4712
^{173}Yb	16.13	$\frac{5}{2}$	−0.6776	+3.0	−0.12981
^{175}Lu	97.41	$\frac{7}{2}$	+2.230	+5.6	+0.3051
^{176}Lu*	2.59	7	+3.18	+8.0	+0.2174
^{177}Hf	18.50	$\frac{7}{2}$	+0.7902	+4.5	+0.10813
^{179}Hf	13.75	$\frac{9}{2}$	−0.638	+5.1	−0.0679
^{181}Ta	99.988	$\frac{7}{2}$	+2.35	+3	+0.321
^{183}W	14.40	$\frac{1}{2}$	+0.1169	–	+0.1120
^{185}Re	37.07	$\frac{5}{2}$	+3.172	+2.3	+0.6078
^{187}Re*	62.93	$\frac{5}{2}$	+3.204	+2.2	+0.6138
^{187}Os	1.64	$\frac{1}{2}$	+0.0643		+0.0616
^{189}Os	16.1	$\frac{3}{2}$	+0.6565	+0.8	+0.20964
^{191}Ir	37.3	$\frac{3}{2}$	+0.1454	+1.1	+0.04642
^{193}Ir	62.7	$\frac{3}{2}$	+0.1583	+1.0	+0.05054
^{195}Pt	33.8	$\frac{1}{2}$	+0.6022	–	+0.5768
^{197}Au	100	$\frac{3}{2}$	+0.14486	+0.59	+0.046254
^{199}Hg	16.84	$\frac{1}{2}$	+0.50271	–	+0.48154
^{201}Hg	13.22	$\frac{3}{2}$	−0.55671	+0.44	−0.177757
^{203}Tl	29.50	$\frac{1}{2}$	+1.6115	–	+1.5437
^{205}Tl	70.50	$\frac{1}{2}$	+1.6274	–	+1.5589
^{207}Pb	22.6	$\frac{1}{2}$	+0.5783	–	+0.55395
^{209}Bi	100	$\frac{9}{2}$	+4.080	−0.38	+0.4343
^{235}U	0.72	$\frac{7}{2}$	−0.43	+4.9	−0.059

[a] From Fuller (176), except for natural abundances which were from Fuller and Nier (177).
[b] Magnetic dipole moment in n m; 1 n m = 1 nuclear magneton = β_n = 5.050951 × 10^{-24} erg/gauss.
[c] Electric quadrupole moment in units of b = barn = 10^{-24} cm^2.
[d] Magnetogyric ratio in radians/gauss second derived from columns 3 and 4 as $\gamma_n = \mu\beta_n/I\hbar$ [where $\hbar$ = 1.0545919 × 10^{-27} ergs/(radian/sec)]. Multiply γ_n in column 6 by 1.591695 × 10^{-3} to obtain it in units of MHz/Tesla.
*Halflife $T_{1/2} > 10^{10}$ years.

Appendix B

Atomic Hyperfine Parameters for Use with ESR Data

There have been several tabulations of the atomic hyperfine parameters often used for estimation of spin densities in molecules, according to the simple relationships given in (II, 42) and (II, 43). Although such relationships are very approximate, since they are based on an elementary LCAO-MO description of the wavefunction of the molecular state, they are nevertheless very convenient and almost universally applied to interpret ESR data. The atomic data are generally obtained from computed values of $\langle r^{-3} \rangle$ and $|\Psi(0)|^2$ utilizing the best available wavefunctions, and are therefore subject to constant improvement. The first such tabulation was that of Morton, Rowlands, and Whiffen (353) utilizing Hartree-Fock wavefunctions [Watson and Freeman (463); Froese (173)]. These are also tabulated by Goodman and Raynor (190). Later work by Hurd and Coodin (235) utilizing the Hartree-Fock-Slater atomic orbitals of Herman and Skillman (11) has been extended and made more precise by Morton and Preston (350). The latter authors have also included a relativistic correction to the isotropic hf parameter of essentially the form suggested by Mackey and Wood (320). Morton and Preston's tabulation is given here in Table B1.

A_{dip} is obtained from P in column 7 of Table B1 by multiplying it by the appropriate angular factor $\bar{\alpha}$, tabulated in Table B2. $P = g_e \beta_e g_I \beta_n \langle r^{-3} \rangle$, and $\bar{\alpha}$ is calculated as described in section II-1c(β). A discussion of $\bar{\alpha}$ may be found in Morton and Preston (350), Ayscough [(3), page 82] and Goodman and Raynor [(190), pages 150-154]. (Note that A_{aniso} tabulated in Goodman and Raynor is $2A_{\text{dip}}$ as defined here.) In the usual applications $\bar{\alpha}$ is $\bar{\alpha}_{zz}/2$, so that for p electrons, $\bar{\alpha} = \frac{2}{5}$; for d_{z^2}, $\bar{\alpha} = \frac{2}{7}$; for $d_{xz} = d_{yz}$, $\bar{\alpha} = \frac{1}{7}$; for $d_{x^2-y^2,\, xy}$, $\bar{\alpha} = -\frac{2}{7}$.

Table B1. Atomic Parameters Obtained from Herman and Skillman's Wavefunction[a]

Mass	Element	Code[b]	$\Psi^2(0)$ (a.u.$^{-3}$)	A_{iso}[c] (MHz)	$\langle r^{-3}\rangle$ (a.u.$^{-3}$)	pd (MHz)
1	Hydrogen	1s	0.318	1420	–	–
3	Helium	1s	1.867	-6357	–	–
7	Lithium	2s	0.2101	364.9	–	–
9	Beryllium	2s	0.7188	–451.6	–	–
11	Boron	2s2p	1.775	2547	0.9293	159.1
13	Carbon	2s2p	3.358	3777	2.002	268.5
14	Nitrogen	2s2p	5.599	1811	3.599	138.8
17	Oxygen	2s2p	8.669	–5263	5.820	–421.0
19	Fluorine	2s2p	12.53	52870	8.766	4400
21	Neon	2s2p	17.48	–6194	12.53	–528.0
23	Sodium	3s	0.7797	927.1	–	–
25	Magnesium	3s	1.763	–485.9	–	–
27	Aluminum	3s3p	3.327	3911	1.493	207.7
29	Silicon	3s3p	5.115	–4594	2.691	–285.5
31	Phosphorus	3s3p	7.252	13306	4.242	917.0
33	Sulfur	3s3p	9.930	3463	6.131	251.3
35	Chlorine	3s3p	12.81	5723	8.389	439.0
40	Argon	3s3p	16.34	–	11.07	–
39	Potassium	4s	1.066	228.5	–	–
43	Calcium	4s	2.063	–640.7	–	–
45	Scandium	4s3d	2.506	2823	1.851	240.3
47	Titanium	4s3d	2.975	–782.0	2.444	–73.64
51	Vanadium	4s3d	3.378	4165	3.114	437.6
53	Chromium	4s3d	2.811	–748.2	3.414	–103.0
55	Manganese	4s3d	4.300	5036	4.721	622.1
57	Iron	4s3d	4.832	747.2	5.659	97.75
59	Cobalt	4s3d	5.233	5947	6.710	845.3
61	Nickel	4s3d	5.755	–2499	7.864	–375.5
63	Copper	4s3d	4.617	5995	8.455	1197
67	Zinc	4s3d	6.739	2087	10.52	351.8
69	Gallium	4s4p	10.18	12210	3.973	509.6
73	Germanium	4s4p	13.40	–2363	6.439	–120.2
75	Arsenic	4s4p	16.75	14660	9.102	834.1
77	Selenium	4s4p	20.41	20120	12.05	1229
79	Bromine	4s4p	24.47	32070	15.25	2044
83	Krypton	4s4p	29.12	–5937	18.76	–386.4
85	Rubidium	5s	2.000	1037	–	–
87	Strontium	5s	3.617	–853.6	–	–
89	Yttrium	5s4d	4.616	–1250	2.373	–62.25
91	Zirconium	5s4d	5.283	–2753	3.126	–155.6
93	Niobium	5s4d	4.736	6590	3.494	457.3
95	Molybdenum	5s4d	5.264	–1984	4.318	–150.7
99	Technetium	5s4d	7.136	9443	5.690	686.2
101	Ruthenium	5s4d	6.085	–1764	6.145	–159.7

Table B1. (Continued)

Mass	Element	Code[b]	$\Psi^2(0)$ (a.u.$^{-3}$)	A_{iso}[c] (MHz)	$\langle r^{-3}\rangle$ (a.u.$^{-3}$)	pd (MHz)
103	Rhodium	5s4d	6.414	−1229	7.179	−121.1
105	Palladium	4d	–	–	7.666	−188.0
107	Silver	5s4d	7.170	−1831	9.451	−204.9
111	Cadmium	5s4d	10.03	−13650	11.37	−1290
115	Indium	5s5p	14.06	20180	6.051	710.9
119	Tin	5s5p	17.64	−43920	9.160	−1831
121	Antimony	5s5p	21.51	35100	12.25	1572
125	Tellurium	5s5p	25.29	−55590	15.47	−2622
127	Iodine	5s5p	29.27	41600	18.92	2031
129	Xenon	5s5p	33.79	−67790	22.57	−3350
133	Cesium	6s	2.538	2467	–	–
137	Barium	6s	4.722	3971	–	–
139	Lanthanum	6s5d	5.492	6007	3.127	237.1
140	Cerium	6s4f	4.950	–	4.953	–
141	Praseodymium	6s4f	5.208	12490	5.565	884.6
143	Neodymium	6s4f	5.309	−2377	6.201	−179.9
147	Samarium	6s4f	5.618	−2014	7.546	−167.4
153	Europium	6s4f	5.952	5722	8.261	482.9
157	Gadolinium	6s5d	6.970	−2546	3.993	−86.14
159	Terbium	6s4f	6.142	13630	9.783	1251
163	Dysprosium	6s4f	6.459	2963	10.59	273.6
165	Holmium	6s4f	6.624	13560	11.43	1285
167	Erbium	6s4f	6.739	−1934	12.31	−189.5
169	Thulium	6s4f	6.919	−5835	13.26	−585.4
173	Ytterbium	6s4f	7.245	−3670	14.19	−367.5
175	Lutetium	6s5d	8.700	10630	4.588	279.2
177	Hafnium	6s5d	9.942	4410	5.756	124.1
181	Tantalum	6s5d	11.11	15020	6.945	445.4
183	Tungsten	6s5d	11.97	5777	8.174	182.5
187	Rhenium	6s5d	13.09	35490	9.454	1157
189	Osmium	6s5d	13.90	13200	10.79	451.0
193	Iridium	6s5d	14.87	3492	12.19	122.9
195	Platinum	6s5d	12.53	34410	12.81	1474
197	Gold	6s5d	12.86	2876	14.31	132.0
199	Mercury	6s5d	17.37	41880	16.78	1611
205	Thallium	6s6p	22.97	183800	10.13	3150
207	Lead	6s6p	27.96	81510	14.72	1626
209	Bismuth	6s6p	33.09	77530	19.15	16.59

[a] Reference (11).
[b] Atomic orbitals for which $\Psi^2(0)$ and $\langle r^{-3}\rangle$ are given in columns 4 and 6, respectively.
[c] Includes Mackey-Wood correction (see text).
[d] Must be used in conjunction with angular factors ($\overline{\alpha}$) given in Table B2.

Table B2. Angular Factors $\bar{\alpha}$ for Real Atomic Orbitals

Orbital	Principal Values		
	$\bar{\alpha}_{xx}$	$\bar{\alpha}_{yy}$	$\bar{\alpha}_{zz}$
p_z	−2/5	−2/5	4/5
p_y	−2/5	4/5	−2/5
p_x	4/5	−2/5	−2/5
d_{z^2}	−2/7	−2/7	4/7
d_{xz}	2/7	−4/7	2/7
d_{yz}	−4/7	2/7	2/7
d_{xy}	2/7	2/7	−4/7
$d_{x^2-y^2}$	2/7	2/7	−4/7
f_{z^3}	−4/15	−4/15	8/15
f_{xz^2}	1/5	−3/5	2/5
f_{yz^2}	−3/5	1/5	2/5
f_{xyz}	0	0	0
$f_{z(x^2-y^2)}$	0	0	0
$f_{x(x^2-3y^2)}$	1/3	1/3	−2/3
$f_{y(3x^2-y^2)}$	1/3	1/3	−2/3

Appendix C

Magnetic Parameters and Spin Densities of Small Molecules Studied by ESR in Matrices and Crystals

This list includes diatomic, triatomic, and some tetra-atomic molecules. Molecules with doublet ground states are listed first, then triplets, etc., in each diatomic, triatomic, or tetra-atomic category. Molecules in each category have been grouped according to similarity, such as hydrides, halides, etc., and/or according to their properties; e.g., isoelectronic σ or π radicals are tabulated together. Many of the doublet radicals have been discussed by Atkins and Symons (2), and the recent tabulation of them by Morton and Preston (350a) has been very helpful. The data for each table are taken from the references cited at the end of that table. The references may not be exhaustive, and for doublet molecules the reader is often referred to Morton and Preston (350a) for a complete list. Generally an effort has been made to examine the data critically, and parameters obtained in several matrices are sometimes given to indicate the environmental effects.

In the tables:

- Column 1 gives the molecule (and sometimes its electronic state) with the isotopic nuclei indicated for which hyperfine structure was observed.
- Column 2 indicates the matrix (solid N_2, Ar, etc.), microcrystalline solid or powder (P), or single crystal (S.C.), in which the molecule was observed.
- Columns 3 to 6 give the experimentally determined $\boldsymbol{g}$ and $\boldsymbol{A}$ tensor components in MHz. [If desired, the hfs components in gauss may be obtained from $A_i\,(\mathrm{G}) = (2.0023/g_i)(0.356828)\,A_i\,(\mathrm{MHz})$.]

The remaining columns vary depending upon whether the tabulated molecules are axial and whether $S \geqslant 1$, i.e., whether they have zero-field-splitting parameters D and E.

For doublet molecules:

- Column 7 gives A_{iso} (MHz) calculated from A_{iso} (MHz) = $(A_x + A_y + A_z)$ (MHz)/3. Even for rather extreme $\boldsymbol{g}$ tensor anisotropy (as occurs for HgH) this yields essentially the same result as A_{iso} (MHz) = $g_{iso} \beta A_{iso}$ (G)/$10^6 h$, where $g_{iso} = (g_x + g_y + g_z)/3$ and A_{iso} (G) = $(A_x + A_y + A_z)$ (G)/3.
- Column 8 (for axial molecules) or columns 8 to 10 (for nonaxial molecules) are obtained from A_{dip} (MHz) = $(A_{\parallel} - A_{\perp})$ (MHz)/3 or T_i (MHz) = $(A_i - A_{iso})$ (MHz), respectively. Again, the difference from anisotropic terms calculated from g_i, A_i (G), and A_{iso} (G) are small except for the few cases where the g anisotropy is large.
- Columns 9 and 10 (axial molecules) or columns 11 to 13 (nonaxial molecules) are spin populations taken from the published work if included there, or if not included then they are often calculated from (II, 42) and (II, 43) using Table B1.
- Column 11 (axial molecules) or column 14 (nonaxial molecules) contains comments concerning preparation procedures, matrix sites, etc., and gives the temperature at which the ESR measurements were made.

For triplet and high-spin molecules the tabulation is the same as for axial molecules except that a column containing the zero-field-splitting parameter D (cm^{-1}) has been inserted after column 6.

The titles of publications have been abbreviated in the following way in the references for each table:

AJ	*Astrophysical Journal*
AP	*Annalen der Physik*
APL	*Applied Physics Letters*
BBPC	*Berichte der Bunsen-Gesellschaft für physikalische Chemie*
CC	*Chemical Communications*
CJP	*Canadial Journal of Physics*
CL	*Chemistry Letters*
CP	*Chemical Physics*
CPL	*Chemical Physics Letters*
CR	*Compte Rendus Hebdomadaires des Seances de l'Academie des Sciences, Paris*
CRe	*Chemical Reviews*
DANS	*Doklady Akademii Nauk SSSR*
FDCS	*Faraday Discussions of the Chemical Society*
FTT	*Fizika Tverdogo Tela*
HPA	*Helvetica Physica Acta*
IAN	*Izvestiya Akademii Nauk SSSR, Seriya Khim*

IC	*Inorganic Chemistry*
IJRPC	*International Journal of Radiation Physics and Chemistry*
JCP	*Journal of Chemical Physics*
JCS	*Journal of the Chemical Society*
JCS/DT	*Journal of the Chemical Society, Dalton Transactions*
JCS/FTI	*Journal of the Chemical Society, Faraday Transactions I*
JCS/FTII	*Journal of the Chemical Society, Faraday Transactions II*
JFC	*Journal of Fluorine Chemistry*
JMR	*Journal of Magnetic Resonance*
JMS	*Journal of Molecular Spectroscopy*
JP	*Journal of Physics*
JPC	*Journal of Physical Chemistry*
JPCS	*Journal of Physics and Chemistry of Solids*
JPP	*Journal de Physique (Paris)*
JPSJ	*Journal of the Physical Society of Japan*
LB	*Landolt–Bornstein*, New Series, Group II, Vol. 9, Part a, Inorganic Radicals, pages 5 to 268 (Springer-Verlag, Berlin, 1977)
MC	*Molecular Crystals*
MP	*Molecular Physics*
PCS	*Proceedings of the Chemical Society*
PL	*Physics Letters*
PM	*Philosophical Magazine*
PNAS	*Proceedings of the National Academy of Sciences (U.S.A.)*
PR	*Physical Review*
PRL	*Physical Review Letters*
PRS	*Proceedings of the Royal Society (London)*
PSS	*Physica Status Solidi*
S	*Science*
SA	*Spectrochimica Acta*
SL	*Spectroscopy Letters*
SS	*Surface Science*
T	*Tetrahedron*
TFS	*Transactions of the Faraday Society*
ZN	*Zeitschrift für Naturforschung*
ZNK	*Zhurnal Neorganicheskoi Khimii*
ZP	*Zeitschrift für Physik*
ZSK	*Zhurnal Strukturnoi Khimii*

Table C-1. $^2\Sigma$ Molecules: Diatomic Metal Hydrides and Carbides

MX	Matrix	Experimental Data				A_{iso} (MHz)	A_{dip} (MHz)	Spin Density		Comments
		$g_\parallel$	$g_\perp$	$A_\parallel$ (MHz)	$A_\perp$ (MHz)			a_s^2	$a_{p,d}^2$	
$^9Be^1H$	Ar	2.0022(2)	2.0021(3)	(Be) 208(1)	194.8(3)	199(1)	4(1)	0.32		using $A_{iso}(Be^+) = 618$
				(H) 201(1)	190.8	194(1)	3(1)	0.14		metal vapor + H atoms, 4°K
$^{25}Mg^1H$	Ar	2.0002(4)	2.0020(4)	(Mg) 226(3)	218(1)	220(2)	3(1)	0.45		metal vapor + H atoms, 4°K
				(H) 298(1)	264(1)	296(1)	0.9(2)	0.19		
Ca^1H	Ar	2.0013(4)	1.9966(4)	138(1)	134(1)	135(1)	1.4(3)	0.095		metal vapor + H atoms, 4°K
Sr^1H	Ar	2.0004(4)	1.9865(4)	123(2)	121(2)	122(2)	0.6(5)	0.086		metal vapor + H atoms, 4°K
Ba^1H	Ar	1.9984(4)	1.9746(4)	47(2)	46(2)	47(2)	0.0(4)	0.032		metal vapor + H atoms, 4°K
Zn^1H	Ar	2.0003(3)	1.9852(6)	485(2)	487(2)	486(2)	1(4)	0.35		metal vapor + H atoms, 4°K
$^{111}Cd^1H$	Ar	1.9970(4)	1.9524(4)	(Cd) 4358(35)	3966(3)	4097(13)	131(13)	0.33	(0.6)	aver. of sites
				(H) 515(1)	515(1)	515(1)	0(2)	0.36		metal vapor + H atoms, 4°K
$^{199}Hg^1H$	Ar	1.976(2)	1.8280(5)	(Hg) 7790(70)	6608(10)	7002(30)	394(27)	0.20	(0.5)	aver. of sites
				(H) 707(6)	711(2)	710(5)	1(3)	0.50		metal vapor + H atoms, 4°K
$^{105}Pd^1H$	Ar	1.965(2)	2.2932(4)	(Pd) 867(10)	801(2)	823(5)	22(4)	0.35	0.76 (*d*)	site α
				(H) 103(6)	106(1)		1(2)	0.075		metal vapor + H atoms, 4°K
$^{171}Yb^1H$	Ar	1.9953(4)	1.9402(2)	(Yb) 5724(20)	5266(5)			0.56	0.68 (*p*)	metal vapor + H atoms, 4°K
				(H) 226(2)	224(2)	225(2)	1(1)	0.16		
$^{27}Al^1H^+$	Ne	2.002(1)	2.000(1)	(Al) 1009(3)	877(3)	921(3)	44(3)	0.33	0.59	vapor. of $Al_2O_3(s)$ (+H_2O), 4°K
				(H) 282(3)	283(3)	283(3)	~0	0.20		
$^{23}Na^1H^+$	H_2O	2.002		(Na) 59.4	39.2	45.9	6.7	0.05		γ irrad. of aqu. NaCl, 77°K
	P			(H) 1437.7	1431.0	1433.2	2.2	1.0		
$^{39}K^1H^+$	H_2O	2.0016		(K)	11			~0.05		γ irrad. of aqu.
	P			(H)		1400		~1.0		$BaCl_2 + Na_2SO_4$, 77°K
$^{109}Ag^1H^+$	P	2.002	2.046	(Ag) −365.6	−254.4	−292.6	−36.5	0.14	0.3 (d_z^2)	γ irrad. $H_2SO_4 \cdot Ag^+$
			2.036		−257.8					
				(H) 865.0	843.4	852.7	6.5	0.55		
					849.0					

Table C-1. *(Continued)*

MX	Matrix	Experimental Data					A_{iso} (MHz)	A_{dip} (MHz)	Spin Density		Comments
		$g_\parallel$	$g_\perp$		$A_\parallel$ (MHz)	$A_\perp$ (MHz)			a_s^2	$a_{p,d}^2$	
$^{199}Hg^1H^{2+}$	P	(2.0)		(Hg)			12,610(150)		0.30		γ irrad. of $Hg^{2+} \cdot HClO_4$, 77°K
				(H)			840(30)		0.59		
$^{119}Sn^1H^{2+}$	P	(2.0)		(Sn)			12,050		0.27		γ irrad. of $SnSO_4$, 77°K
				(H)			~730		0.51		
$^{207}Pb^1H^{2+}$	P	~2.00		(Pb)			13,870		0.17		γ irrad. of $PbAc_2$, 77°K
				(H)			~450		0.32		
$^{103}Rh^{13}C$	Ne	2.0039(6)	2.0541(6)	(Rh)	1113(1)	1089(1)	1097(1)	8.(1)	0.69	0.25	Rh + C vapor, 4°K
				(C)	89(1)	55(1)	66(1)	11(1)	0.02	0.12	

BeH: Knight, Brom, and Weltner, *JCP 56*, 1152 (1972).
MgH, CaH, SrH, BaH: Knight and Weltner, *JCP 54*, 3875 (1971).
ZnH, CdH, HgH: Knight and Weltner, *JCP 55*, 2061 (1971).
PdH: Knight and Weltner, *JMS 40*, 317 (1971).
YbH: Van Zee, Seely, and Weltner, *JCP 67*, 861 (1977); Morton and Preston, *JMR 30*, 577 (1978).
AlH^+: Knight, Martin, and Davidson, *JCP 71*, 3991 (1979).
NaH^+: Claxton and McWilliams, *TFS 66*, 513 (1970); Claxton and Smith, *TFS 67*, 1859 (1971); Bloom, Eachus, and Symons, *JCS* A833 (1971); Ginns and Symons, *CC* 949 (1971); Ginns and Symons, *JCS/DT* 143 (1972).
KH^+: Bloom, Eachus, and Symons, *JCS* A833 (1971).
AgH^+: Eachus and Symons, *CC* 285 (1969); *JCS* A1336 (1970).
HgH^{2+}: Booth, Starkie, and Symons, *JCS* A3198 (1971).
SnH^{2+}: Booth, Starkie, and Symons, *JCS* A3198 (1971); Starkie and Symons, *JCS/DT* 731 (1974).
PbH^{2+}: Roberts and Eachus, *JCP 57*, 3022 (1972); Roberts and Eachus, *JCP 59*, 5251 (1973); Starkie and Symons, *JCS/DT* 731 (1974).
RhC: Brom, Graham, and Weltner, *JCP 57*, 4116 (1972).

Table C-2. $^2\Sigma$ Molecules: Diatomic Fluorides and Chlorides

MX	Matrix	Experimental Data				A_{iso} (MHz)	A_{dip} (MHz)	Spin Density		Comments
		$g_{\parallel}$	$g_{\perp}$	$A_{\parallel}$ (MHz)	$A_{\perp}$ (MHz)			a_s^2	$a_{p,d}^2$	
$^9Be^{19}F$	Ar	2.0014(5)	2.0014(5)	(Be) 303	297	294	(2)	0.48		probably rotating, assume
				(F) 241	227	229	(5)	0.004	(0.003)	$A_{iso}(Be^+) = 618$
$Mg^{19}F$	Ne	2.0020(5)	2.0010(5)	331(3)	143(3)	206(3)	63(3)	0.045	0.04	metal vapor + F_2 or
$Ca^{19}F$	Ne	2.0020(5)	2.000(1)	149(3)	106(3)	120(3)	14(3)	0.012	0.008	vaporiz. ɔf MF_2 + solid Al or B
$^{87}Sr^{19}F$	Ne	2.0020(5)	1.9970(5)	(Sr) 591(3)	570(3)	576(3)	7(3)	0.58	0.38, 0.04	″
				(F) 126(3)	95(3)	105(3)	10(3)	0.009	0.006	
$^{137}Ba^{19}F$	Ne	2.0010(5)	1.9950(5)	(Ba) 2453(9)	2401(6)	2415(7)	17(7)	0.60	0.17, 0.24	″
				(F) 67(3)	59(3)	62(3)	3(3)	0.003	0.002	
$Zn^{19}F$	Ne	2.002(1)	1.996(1)	673(2)	143(2)	320(2)	177(2)	0.0061	0.10	″
$Cd^{19}F$	Ne	2.001(3)	1.985(2)	670(7)	64(2)	266(4)	202(4)	0.0050	0.11	″
$^{199}Hg^{19}F$	Ar	1.993(1)	1.961(1)	(Hg) 22,622(10)	21,880(8)	22,163(12)	223(8)	0.54	0.33	″
				(F) 1344(10)	195(5)	670(8)	320(5)	0.013	0.18	
$^{171}Yb^{19}F$	Ar	1.9975(5)	1.9954(5)	(Yb) 7822(5)	7513(5)	7617(5)	102(3)	0.78		″
				(F) 220(2)	134(2)	163(2)	28(1)	0.003	0.02	
$Kr^{19}F$	KrF_4	2.000	2.068			600	330	0.04	0.61	γ irrad. of S.C., 77° K
$^{129}Xe^{19}F$	XeF_4	1.9740	2.1264	(Xe) 2368	1224	(Xe) 1605	382	0.05	0.36	γ irrad. of S.C., 77° K
				(F) 2637	526	(F) 1243	703	0.03	0.47	
$^9Be^{35}Cl$	Ar	2.001(1)	1.998(1)	(Be) 282(2)	268(2)	273(2)	5(1)	0.44		Be vapor + Cl_2, 12° K
				(Cl) 53(2)	24(2)	34(2)	10(1)	0.0072	0.07	
$Ca^{35}Cl$	Ar	1.998(1)				28		0.006		Ca vapor + Cl_2, 4° K
$^{199,201}HgCl$	P	(2.00)				29,692		0.71		γ irrad. C_2H_5HgCl
$^{129}Xe^{35}Cl$	Ar	1.962	2.303	(Xe) 582	~280	~381	~101	0.006	0.08	uv irrad. of $Xe/Cl_2/Ar$, 4° K
				(Cl) 288	~0	~96	~96	0.02	0.55	

BeF: Knight, Wise, Childers, Davidson, and Daasch, *JCP 73*, 4198 (1980).
MgF, CaF, SrF, BaF: Knight, Easley, Weltner, and Wilson, *JCP 54*, 322 (1971).
ZnF, CdF: Knight, Mouchet, Beaudry, and Duncan, *JMR 32*, 383 (1978); Knight, Fisher, and Wise, *JCP 74*, 6009 (1981).
HgF: Knight, Fisher and Wise, *JCP 74*, 6009 (1981).
YbF: Van Zee, Seely, DeVore, and Weltner, *JCP 82*, 1192 (1978).
KrF: Falconer, Morton, and Streng, *JCP 41*, 902 (1964).
XeF: Morton and Falconer, *PCS* 95 (1963); *JCP 39*, 427 (1963); Eachus and Symons, *JCS* A304 (1971).
BeCl: Knight, Wise, Childers, Daasch, and Davidson, *JCP 74*, 4256 (1981).
CaCl: Martinez and Weltner, *JCP 65*, 4256 (1976).
HgCl: Fullam and Symons, *JCS/DT* 1086 (1974).
XeCl: Adrian and Bowers, *JCP 65*, 4316 (1976).

Table C-3. $^2\Sigma$ Molecules: Diatomic Metal Oxides and Sulfides, and Triatomic Metal Cyanides and Hydroxides

MX	Matrix	Experimental Data				A_{iso} (MHz)	A_{dip} (MHz)	Spin Density		Comments
		$g_\parallel$	$g_\perp$	$A_\parallel$ (MHz)	$A_\perp$ (MHz)			a_s^2	$a_{p,d}^2$	
^{87}RbO	N_2	2.001	2.081	573	428	477	48	0.46		Rb vapor + N_2O, 4° K
^{133}CsO	N_2	2.001	1.982	465	348	387	39	0.16		Cs vapor + N_2O, 4° K
^{11}BO	Ne	2.0015(3)	2.0012(3)	1018(1)	1034(1)	1028(2)	−5(1)	0.51		vapor. of B(s) + MO(s), 4° K
^{27}AlO	Ne	2.0015(3)	2.0004(3)	872(1)	713(1)	766(2)	53.0(7)	(0.15)	(0.43)	vapor. of Al_2O_3(s), 4° K
^{45}ScO	Ar	2.00(1)	2.00(1)	2063(10)	1990(10)	2010(10)	25(10)	0.71	0.26	vapor. of M_2O_3(s), 4° K
^{89}YO	Ar	2.003(2)	2.003(2)			803(3)		0.64		vapor. of M_2O_3(s), 4° K
^{139}LaO	Ar	2.01(1)	2.01(1)			3890(10)		0.65		vapor. of M_2O_3(s), 4° K
^{11}B^{33}S	Ne	2.0016(1)	1.9942(1)	(B) 853.3(3) (S)	766.7(3) 16(2)	795.6(3)	28.9(3)	0.31	0.45	vapor. of solid B + ZnS, 4° K
^{45}ScS	Ar	1.9975(10)	1.9915(5)	1937(20)	1821(16)	1860(20)	39(10)	0.41	0.66	vapor. of ZnS + Sc, 4° K
^{89}YS	Ne	2.000(1)	1.9918(5)	733(2)	720(1)	724(2)	4(1)	0.43	0.2	vapor. of Y_2S_3, 4° K
^{111}CdCN	Ar		1.9681(2)	7890(40)	7558(4)	7680(20)	104(10)	0.53	~0.5	CN radicals + M vapor, 4° K
^{199}HgCN	Ar		1.8789(3)	16,770(30)	15,390(10)	15,960(20)	396(15)	0.39	~0.6	
^{109}Ag(C^{14}N)$^-$	K[Ag(CN)$_2$] P	2.002 (bent structure?)			(Ag) 434 (N) 48.5			0.24 0.03		γ irrad., 77° K
^{9}BeO^{1}H	Ar	2.0009(4)	2.0009(4)	(Be) 271(1) (H) 7.3(5)	260(1) <3.9	264(1) <5	4(1) ~1	0.43 0.003	0.57	metal vapor + H_2O_2, 4° K
^{25}MgO^{1}H	Ar	2.0013(1)	2.0020(1)	(Mg) 321.4(3) (H) 13.7(5)	308.4(3) 8.5(1)	312.7(3) 10.2(3)	4.3(3) 1.7(3)	0.52 0.008	0.49	metal vapor + H_2O_2, 4° K

RbO, CsO: Lindsay, Herschbach, and Kwiram, *JCP 60*, 315 (1974).
BO: Knight, Easley, and Weltner, *JCP 54*, 1610 (1971);
Knight, Wise, Davidson, and McMurchie, *JCP 76*, 126 (1982).
AlO: Knight and Weltner, *JCP 55*, 5066 (1971); see also Ammeter and Schlosnagle, *JCP 59*, 4784 (1973).
ScO, YO, LaO: Weltner, McLeod, and Kasai, *JCP 46*, 3172 (1967).
BS: Brom and Weltner, *JCP 57*, 3379 (1972).
ScS, YS: McIntyre, Lin, and Weltner, *JCP 56*, 5576 (1972).
CdCN, HgCN: Knight and Lin, *JCP 56*, 6044 (1972).
AgCN$^-$: Symons and Zimmerman, *JCS/DT*, 1970 (1976).
BeOH: Brom and Weltner, *JCP 64*, 3894 (1976).
MgOH: Brom and Weltner, *JCP 58*, 5322 (1973).

Table C-4. $^2\Sigma$ Diatomic Metal Molecules and Ions

MX	Matrix	Experimental Data $g_\parallel$	$g_\perp$	$A_\parallel$ (MHz)	$A_\perp$ (MHz)	A_{iso} (MHz)	A_{dip} (MHz)	Spin Density a_s^2	$a_{p,d}^2$	Comments
$^{107}AgMg$	Ar	2.0001(2)		837				0.46		metal vapors, 4° K
$^{107}AgCa$	Ar	1.9982(2)		435				0.24		metal vapors, 4° K
$^{107}AgSr$	Ar		1.9823(2)		372					metal vapors, 4° K
$^{107}AgBa$	Ar		1.9538(2)		218					metal vapors, 4° K
$^{107}AgZn$	Ar	2.0025(2)	1.9905(2)	1324(3)	1324(3)	1324(3)	~0	0.73	~0	metal vapors, 4° K
$^{107}Ag^{111}Cd$	Ar	2.0014(2)	1.9711(2)	(Ag) 1327(3)	1327(3)	1327(3)	~0	0.73	~0	metal vapors, 4° K
				(Cd) 2180(250)	1990(30)	2050	60	0.16	~0.1	
$^{107}Ag^{199}Hg$	Ar	1.9958(2)	1.9136(2)	(Ag) 1562(3)	1562(3)	1562(3)	~0	0.86	~0	metal vapors, 4° K
				(Hg) 3130(200)	2520(30)	2720	200	0.07	0.31	
$^{109}Ag_2^{+}$	P	2.001	1.974	−877	−834	−848	−14	0.45	0.17	γ irrad., $CH_3OH \cdot Ag^+$
$^{109}Ag^{111,113}Cd^{2+}$	P	2.001	1.975	(Ag) (−)924	(−)865	(−)885	20	0.48	0.24	γ irrad., 5 M $H_2SO_4 \cdot Ag^+ \cdot Cd^{2+}$
				(Cd) (−)4537	(−)4368	4424	56	0.32	0.11	
$^{107,109}Ag^{199}Hg^{2+}$	P	1.997		(Ag)		724		0.40		γ irrad., $HClO_4 \cdot Hg^{2+} \cdot Ag^{2+}$
				(Hg)		24,570		0.59		
$^{111,113}Cd_2^{3+}$	P	1.988				2540		0.19		γ irrad., 5 M $H_2SO_4 \cdot Cd^{2+}$
$^{199}Hg_2^{3+}$	P	1.9677				9378		0.22		γ irrad., 5 M $H_2SO_4 \cdot Hg^{2+}$
$^{27}Al_2^{5+}$	Al_2O_3	2.0023		Al_1: 975(5)	840(5)	885(5)	45	0.23	0.54	γ irrad., 77° K, 2 sites
(or Al_2^{3+})	S.C.			Al_2: 423(5)	335(5)	364(5)	29	0.09	0.35	site ‖ to *c* axis

AgMg, AgCa, AgSr, AgBa, AgZn, AgCd, AgHg: Kasai and McLeod, *JPC 79*, 2325 (1975); *82*, 1554 (1978).

Ag_2^+: McMillan and Smaller, *JCP 35*, 1698 (1961);
Shields, *TFS 62*, 1042 (1966);
Shields and Symons, *MP 11*, 57 (1966);
Zhitnikov, Baranov, and Melnikov, *PSS B59*, K111 (1973);
Forbes and Symons, *MP 27*, 467 (1974);
Zhitnikov and Peregud, *FTT 17*, 1655 (1975);
Alenko, Zhitnikov, Krasikov, and Peregud, *FTT 18*, 1552 (1976);
Brown, Findlay, and Symons *JCS/FTI 72*, 1792 (1976).

$AgCd^{2+}$: Eachus and Symons, *JCS* A3080 (1970).

$AgHg^{2+}$: Symons and Marov, *ZNK 17*, 2601 (1972).

Cd_2^{3+}: Eachus and Symons, *JCS* A3080 (1970);
Aleksandrov, Ershov, Pikaev, and Spitsyn, *IAN* 249 (1976).

Hg_2^{3+}: Booth, Starkie, and Symons, *JCS* A3198 (1971);
Aleksandrov, Ershov, Pikaev, and Spitsyn, *IAN* 249 (1976);
Morton and Preston, *LB*, page 261.

Al_2^{5+} (or Al_2^{3+}): Cox and Hervé, *CR 261*, 5080 (1965).

Table C-5. $^2\Sigma$ Diatomic Molecules and Anions

MX	Matrix	Experimental Data				A_{iso} (MHz)	A_{dip} (MHz)	Spin Density		Comments
		$g_\parallel$	$g_\perp$	$A_\parallel$ (MHz)	$A_\perp$ (MHz)			a_s^2	$a_{p,d}^2$	
$^{13}C_2^{-}\ ^{23}Na^+$	Ar	2.0124(30)	2.0021(6)*	(Na) −12.4(6) (C)	−11.4(6)* ~233(20)	−11.8				A(Na) assumed minus, * = aver. of x and y
$^{13}C^{14}N$	Ar	2.0015(5)	2.0003(5)	(C) 678(1) (N) 18.2(4)	543.1(5) −28.0(3)	588.1(3) −12.6(3)	45(3) 15.4(3)	0.19 −0.008	0.48 0.32	Photolysis of ICN
$^{13}CO^+$	Ne	2.0004(7)	1.9996(5)	1665(2)	1527(2)	1573(2)	46(1)	0.42	0.43	uv irrad. of CO + Cl_2 in Ne, 4° K
	Zeo	(x) 2.0005	(y) 2.0005 (z) 2.0045	(x) 843	(y) 714 (z) 772	776	(x) 43 (z) 19	0.25	$2p\sigma$ 0.47 $2p_z$ 0.21	γ irrad., 77° K polarized by Zeolite
$^{13}CO^-$	MgO Surf.	2.0021(3)	2.0055(3)	0	(x) 76 (y) 31	36	(x) 25 (y) 11	0.01	$2p_x$ 0.28 $2p_y$ 0.12	CO adsorbed on MgO, z = CO axis

C_2^-: Graham, Dismuke, and Weltner, *JCP 61*, 4793 (1974).
CN: Cochran, Adrian, and Bowers, *JCP 36*, 1938 (1962);
Easley and Weltner, *JCP 52*, 197 (1970);
Adrian and Bowers, *CPL 41*, 517 (1976).
CO^+: Vedrine and Naccache, *CPL 18*, 190 (1973);
Knight and Steadman, *JCP 77*, 1750 (1982).
CO^-: Lunsford and Jayne, *JCP 44*, 1492 (1966).

Table C-6. $^2\Sigma$ Diatomic Anions and $^2A'$ Triatomic Halogen-Hydroxide Anions

MX	Matrix	Experimental Data				A_{iso} (MHz)	A_{dip} (MHz)	Spin Density		Comments
		$g_\parallel$	$g_\perp$	$A_\parallel$ (MHz)	$A_\perp$ (MHz)			a_s^2	a_p^2	
$^{19}F_2^-$	NaF	2.0017(3)	2.0229(3)*	2516.1(15)	−133(12)	750	883	0.01	0.50	V_k center[a]
$^{35}Cl_2^-$	Ar	2.0006	2.0037	282.0	22	109	87	0.02	0.50	
$^{35}Cl_2^-$	NaCl	2.0013(1)	2.0458(2)*	276.5(15)	25.88	109.4	83.5	0.02	0.48	V_k center[a]
$^{81}Br_2^-$	NaBr	1.9791(1)	2.1741(5)*	1193.9(3)	211.2	538.8	327.6	0.02	0.40	V_k center[a]
$^{127}I_2^-$	NaI	1.8584(1)	2.31*	895.3(6)	452.6(30)	600.2	147.6	0.01	0.18	V_k center[a]
$^{19}F^{35}Cl^-$	KCl	2.0018(5)	2.030(2)	(F) 2257(3) (Cl) 353(3)	−250(15) 51(15)	586 152	837 101	0.01 0.03	0.56 0.70	F^- and Pb^{2+} doped[a]
$^{19}F^{81}Br^-$	KCl	1.9891(5)	2.125(2)	(F) 2047(3) (Br) 1673(3)	−149(60) 366(30)	583 802	732 436	0.009 0.03	0.54 0.72	Br^-, F^-, Pb^{2+} doped[a]
$^{19}F^{127}I^-$	KCl	1.9363	[2.26]	(F) ~1724 (I) ~1520	~0 661(50)	~575 947	~575 286	0.006 0.035	0.50 0.70	I^-, F^-, Pb^{2+} doped[a]
$^{35}Cl^{81}Br^-$	KCl	1.9839(1)	2.1340(5)*	(Cl) 248.8(3) (Br) 1345.6(3)	22.4(3) 345.0(60)	97.9 678.5	75.5 333.5	0.024 0.018	0.55 0.82	Br^-, Pb^{2+} doped
$^{35}Cl^{127}I^-$	KCl	1.887(7)	2.385(8)*	(Cl) 161.6(50) (I) 1041.9(5)	<13 1001(48)	60 1015(30)	50 14	0.01 0.024	0.3 0.02	I^-, Pb^{2+} doped
$^{17}O_2^{3-}$	CaF_2	2.0037(8)	2.0118(10) 2.0264(8)	(O) 203.9(30) (Y) 11.53(12)	17(7) 11.64(12)*	79 11.60	62 −0.04	0.02 0.14	0.43 ~0	$^{89}Y^{3+}$ counterion X irrad., 293° K
$^{77}Se_2^{3-}$	AgBr	2.005(2)	2.116(2) 2.086(3)	385(35)	0(30) 60(20)	~168(30)	~120	0.008	~0.2	vis. irrad., 180° K
$^{125}Te_2^{3-}$	AgBr	1.887(1)	2.248(2) 2.062(3)	780(50)	40(40) 160(50)	327(50)	~230	0.006	~0.2	vis. irrad., 180° K
$^{19}F^1H^-$	H_2O Glass	2.1	2.009	(F) ±21 (H) ~0	93, ±19.7 −70, −112	45	−24, 0	0.001	0.014	γ irrad. of KF, 77° K

Table C-6. *(Continued)*

MX	Matrix	Experimental Data $g_{\parallel}$	$g_{\perp}$		$A_{\parallel}$ (MHz)	$A_{\perp}$ (MHz)	A_{iso} (MHz)	A_{dip} (MHz)	Spin Density a_s^2	$a_{p,d}^2$	Comments
$^{35}ClO^1H^-$	H_2O	2.0054	2.0174	(Cl)	165	−46.3	24.1	70	0.005	0.48	γ irrad. of KCl, 77° K
	Glass			(H)	70.0	69.5	70	~0	0.05		
$^{81}BrO^1H^-$	H_2O	2.00	2.08	(Br)	1190	146	494	348	0.017	0.57	γ irrad. of NaBr, 77° K
	Glass			(H)	~56	58	57		0.04		
$^{127}IO^1H^-$	H_2O	1.980(4)	2.13	(I)	1330	209	583	374	0.021	0.70	γ irrad. of NaI, 77° K
	Glass			(H)	~42	~45	44		0.03		

*Aver. of x and y.
[a]Several crystals and counterions.

F_2^-, Cl_2^-, Br_2^-, I_2^-: Känzig, *PR 99*, 1890 (1955); Cohen, *PR 101*, 1432 (1956); Castner and Känzig, *JPCS 3*, 178 (1957); Woodruff and Känzig, *JPCS 5*, 268 (1958); Delbecq, Smaller, and Yuster, *PR 111*, 1235 (1958); Känzig, *JPCS 17*, 80 (1960); Symons, *JCS 570* (1963); Schoemaker, *PR 174*, 1060 (1968) [I_2^-]; *B7*, 786 (1973); Bass and Micher, *PR 175*, 421 (1968).

Cl_2^- in Ar: Martinez and Weltner, *JCP 65*, 4256 (1976).

FCl^-, FBr^-, FI^-, $ClBr^-$, ClI^-: Wilkins and Gabriel, *PR 132*, 1950 (1963); Schoemaker, *PR 149*, 693 (1966); Goldberg and Meistrich, *PR 172*, 877 (1968); Delbecq, Schoemaker, and Yuster, *PR B3*, 473 (1971); *B7*, 3933 (1973); Schoemaker and Shirkey, *PR B6*, 1562 (1972); Van Puymbroeck, Lagendijk, and Schoemaker, *PSS 59*, 585 (1980). [See Morton and Preston (350a) for more complete references.]

O_2^{3-}: Bill, *HPA 42*, 771 (1969); Wagner and Murphy, *PR B6*, 1638 (1972); Lagendijk, Glasbeck, and van Voorst, *CPL 20*, 92 (1973).

Se_2^{3-}: Hohne and Stasiw, *PSS 20*, 657, 667 (1967); *24*, 591 (1967).

Te_2^{3-}: Hohne and Stasiw, *PSS 20*, 657, 667 (1967).

FOH^-, $ClOH^-$: Ginns and Symons, *CC* 949 (1971); *JCS/DT* 143 (1972).

$ClOH^-$: Ginns and Symons, *JCS/DT* 143 (1972); Catton and Symons, *CC* 1472 (1968); *JCS* A446 (1969).

$BrOH^-$, IOH^-: Marov and Symons, *JCS* A201 (1971).

Table C-7. Quenched $^2\Pi$ Diatomic Molecules and Ions

MX	Matrix	Experimental Data				A_{iso} (MHz)	A_{dip} (MHz)	Spin Density		Comments
		$g_{\parallel}$	$g_{\perp}$	$A_{\parallel}$ (MHz)	$A_{\perp}$ (MHz)			a_s^2	$a_{p,d}^2$	
O_2^-	KCl	2.4359	1.9551							salt heated in O_2
			1.9512							
O_2^-	NaO_2	2.175(5)	2.002(5)							
$^{23}Na^+O_2^-$	Ar	2.1112	2.0063	−10.0	−5.1	−8.3		0.009		metal vapor + O_2, 4° K
			2.0029		−9.8					
$^{23}Na^+O_2^-$	Kr	2.1106	2.0075	±10.6	±~6.2	±~9.1		0.01		metal vapor + O_2, 4° K
			2.0022		±~10.6					
$K^+O_2^-$	Kr	2.1184	2.0068							metal vapor + O_2, 4° K
			2.0007							
$^{87}Rb^+O_2^-$	Kr	2.1227	2.0069	±24.1	±~18.3	±21.6		0.02		metal vapor + O_2, 4° K
			1.9996		±~22.4					
$^{133}Cs^+O_2^-$	Kr	2.1067	2.0069	±14.5	±13.6	±14.3		0.006		metal vapor + O_2, 4° K
			2.0013		±14.8					
O_2^+	AsF_6^-	1.743	2.000							O_2AsF_6 powd., 5–300° K
			1.973							
^1HO	H_2O	2.0597	2.0028	9.5	−80.2	−65.2				site I, X-irrad., 77° K
	S.C.		2.0089		−124.8					
	S.C.	2.0618	2.0034	10.1	−75.7	−63.7				γ irrad., 77° K
			2.0081		−125.6					NaH maleate trihydrate
$^1HO^{2-}$	CaO	2.0042(2)	2.0088(2)	236(3)	121(3)	163(3)				n irrad., 77° K
			2.0074(2)		132(3)					
$^{19}F^{17}O^{2-}$	MgO	2.0022(2)	2.0085(2)	(F) 1275(3)	484(3)	748(3)	264(3)	0.014	0.15	n irrad., 77° K
	S.C.			(O) 331(3)	42(9)	138	96	0.026	0.57	
^{35}ClO	$KClO_4$	2.0100	2.0035	(+)50.3	−5.7	(+)10.6		0.002		γ irrad. + vis. light, 93° K
	S.C.		1.9965		−12.9					
$^{35}ClO^{2-}$	$Ca_5(PO_4)_3Cl$	2.0031(2)	2.0255(2)	84.9(15)	38.3(3)	53.8	15.5	0.009	0.09	x irrad., 300° K
$^{33}S_2^-$	KBr	3.5037	0.8434	145.2						heat KBr in S
	S.C.		0.8388							

Table C-7. *(Continued)*

MX	Matrix	Experimental Data				A_{iso} (MHz)	A_{dip} (MHz)	Spin Density		Comments
		$g_{\parallel}$	$g_{\perp}$	$A_{\parallel}$ (MHz)	$A_{\perp}$ (MHz)			a_s^2	$a_{p,d}^2$	
$^{33}S_2^+$	Si: S	2.0008				115		0.03		heated to 1300°C, 300°K
$^{77}Se^{33}S^-$	KI	3.6290	0.9532	(Se) 750(5)	120(5)	~290		0.01		heat KI doped with Se and S, 4°K
	S.C.		0.9681		<20					
				(S) 140(5)	<20	~70		0.02		
					<60					
$^{77}Se_2^-$	KI	3.7079	0.7698	740(5)	70(10)	285		0.01		heat KI + Se, 4°K
	S.C.		0.7824		45(15)					

O_2^-: Känzig and Cohen, *PRL 3*, 509 (1959);
Bennett, Ingram, Symons, George, and Griffith, *PM 46*, 443 (1955);
Atkins and Symons (2), page 111.
$Na^+O_2^-$: Adrian, Cochran, and Bowers, *JCP 59*, 56 (1973).
$Na^+O_2^-$, $K^+O_2^-$, $Rb^+O_2^-$, $Cs^+O_2^-$: Lindsay, Herschbach, and Kwiram, *CPL 25*, 175 (1974).
O_2^+: Jaccard, *PR 124*, 60 (1961);
Goldberg, Christie, and Wilson, *IC 14*, 152 (1975);
DiSalvo, Falconer, Hutton, Rodriguez and Waszcak, *JCP 62*, 2575 (1975);
see also Atkins and Symons (2), page 111.
HO: McMillan, Matheson, and Smaller, *JCP 33*, 609 (1960);
Wigen and Cowen, *JPCS 17*, 26 (1960);
Ovenall, *JPCS 21*, 309 (1961);
Symons, *JCS* 570 (1963);
Brivati, Symons, Tinling, Wardale, and Williams, *CC* 402 (1965);
Box, Budzinski, Lilga, and Freund, *JCP 50*, 5422 (1969);
see also Atkins and Symons (2), page 104.
HO: Gunter, *JCP 46*, 3818 (1967);
Torijama and Iwasaki, *JCP 55*, 1890 (1971).
HO^{2-}: McGeehin, Henderson, Boas, and Hall, *JP 8*, 1718 (1975).
FO^{2-}: Bill and Lacroix, *PL 21*, 257 (1966);
Bill and Lacroix, *JPP Colloq. 28*, C4-138 (1967);
Schulze and Gaebler, *AP 28*, 37 (1972);
Ruis, Cox, Freund, and Owen, *JP 7*, 581 (1974).
ClO: Atkins, Brivati, Keen, Symons, and Trevalion, *JCS* 4785 (1962);
Byberg, *JCP 47*, 861 (1967).
ClO^{2-}: Knottnerus, den Hartog, and van der Lugt, *PSS A13*, 505 (1972);
Roufosse, Stapelbrock, Bartram, and Gilliam, *PR B9*, 855 (1974);
Knottnerus and den Hartog, *PSS A29*, 183 (1975).
S_2^-: Morton, *JCP 43*, 3418 (1965); *JPC 71*, 89 (1967);
Vannotti and Morton, *PR 161*, 282 (1967).
S_2^+: Ludwig, *PR A137*, 1520 (1965).
SeS^-, Se_2^-: Vannotti and Morton, *PL A24*, 520 (1967); *JCP 47*, 4210 (1967).

Table C-8. Triplet and High-Spin ($S = \frac{3}{2}$ to 4) Diatomic Molecules

Molecule	Matrix	Experimental Data							Spin Density		Comments
		$g_\parallel$	$g_\perp$	$A_\parallel$ (MHz)	$A_\perp$ (MHz)	D (cm^{-1})	A_{iso} (MHz)	A_{dip} (MHz)	a_s^2	a_d^2	
$^3\Sigma$:											
$^{17}O_2$	N_2	2.00				~+4	110(3)				torsionally oscil., <10° K
$^4\Sigma$:											
^{51}VO	Ar	2.0023(10)	1.9804(10)	714.7(5)	837.1(5)	>>0.3	796	−41	0.57	0.49	vapor. of V_2O_5 + V solids, 4° K
^{93}NbO	Ne	2.0023(10)	1.9577(10)	1587(1)	1647(1)	>>0.3	1627	−24	0.74	0.20	vapor. of Nb_2O_5 + Nb solids, 4° K
^{95}MoN	Ne		1.969(1)	480(8)	506.3(6)	>>0.3	498	−9			sputtered Mo with N_2 + Ne, 4° K
$^5\Sigma$:											
$^{45}Sc_2$	Ne		~2.00	308(30)	238(6)	0.11(1)	261	27	0.25/2	0.67/2	vapor. of Sc (s), 4° K
$^6\Sigma$:											
$^{53}Cr^1H$	Ar	1.995(5)	2.005(5)	(Cr)	±53(5)	±0.34					Cr vapor + H atoms, 4° K
				(H)			49(5)		0.03		
$^{53}Cr^{19}F$	Ar	(2.002)	1.989	(Cr)	~±36	±0.56					superheat CrF_2 vapor, 4° K
				(F)	<~20						
$Cr^{107}Ag$	Ar	(2.0023)	1.950(3)	~47(20)	47(1)	⩾2.0	~47	~0(10)	0.03		Cr + Ag vapors, 4° K
$Cr^{197}Au$	Ar	(2.0023)	2.001(1)	~39(40)	39(6)	⩾2.0	~39	~0(20)	0.01		Cr + Au vapors, 4° K
^{55}MnO	Ar	(2.002)	1.995(5)	479(12)	433(1)	±1 to 2.3	448	15			Mn vapor + O_2, 4° K
^{55}MnS	Ar	(2.002)	2.009(5)	6(5)	141(6)	⩾±1.9	96	−45			Mn vapor + CS_2, 4° K
$^7\Sigma$:											
$^{55}Mn^1H$	Ar	(2.002)	2.001(1)	(Mn) 322(6)	299(2)	−0.002	306		0.4		Mn vapor + H atoms, 4° K
				(H)			⩽20				
$^{55}Mn^{19}F$	Ne	(2.002)	2.001(1)	(Mn) 504	429	−0.008	454	25	0.58		Photolysis of Mn/F_2/Ne, 4° K
				(F) 85(2)	60(1)		68		0.00		
$^{55}MnCl$	Ar	(2.002)	1.995(3)	308(28)	411(3)	±0.070	376	−34	0.49		superheat. MnX_2 vapor or uv irrad. of Mn + X_2, 4° K
				^{35}Cl hfs unresolved							
$^{55}MnBr$	Ar	(2.002)	1.993(3)	415(18)	430(6)	±0.311	425	−5	0.55		"
				^{79}Br hfs unresolved							
^{55}MnI	Ar	(2.002)	(1.993)	221(18)	404(6)	~±0.50	343	−61	0.45		"
				^{127}I hfs unresolved							
$^{55}Mn^{107}Ag$	Kr	(2.002)	1.998(2)	(Mn) 275(60)	213(6)	0.181	233	20			Mn + Ag vapors, 4° K
				(Ag) 213(80)	112(15)		146	34			

Table C-8. *(Continued)*

Molecule	Matrix	Experimental Data							Spin Density		Comments
		$g_{\parallel}$	$g_{\perp}$	$A_{\parallel}$ (MHz)	$A_{\perp}$ (MHz)	D (cm^{-1})	A_{iso} (MHz)	A_{dip} (MHz)	a_s^2	a_d^2	
CrZn	Ar	1.999 (4)	2.0046 (9)			0.0836 (2)					Cr + Zn vapors, 4° K
$^4\Sigma$:											
$Cr^{63}Cu$	Kr	(2.0023)	2.0030 (10)	631 (60)	836 (9)	−0.007 (5)	781	−75	0.91		Cr + Cu vapors, 4° K
				$\lvert eqQ(^{63}Cu)\rvert$ = 310 (120) MHz							
$^9\Sigma$:											
GdO	Ar	1.986 (5)	1.996 (5)			0.2078 (3)					vaporiz. of solid Gd_2O_3, 4° K
				$\lvert b_4^0\rvert = 0.00040(5)\ cm^{-1}$							
Exchange Coupled Diatomic:											
$^{55}Mn_2$		$J = -9(3)\ cm^{-1}$									Mn vapor, 4° to 50° K
$^3\Sigma$	Kr	(2.002)	2.00		43 (2)	0.49					
$^5\Sigma$	Kr	(2.002)	1.999 (1)		43 (2)	0.14					
$^7\Sigma$	Kr	(2.002)	1.999 (1)		43 (2)	0.07					

O_2: Graham, Harvey, and Kiefte, *JCP 52*, 2235 (1970);
Simoneau, Harvey, and Graham, *JCP 54*, 4819 (1971);
Kon, *JACS 95*, 1045 (1973);
Hirokawa, *JPSJ 35*, 12 (1973); *37* 897 (1974).
VO: Kasai, *JCP 49*, 4979 (1968);
Cheung, Hansen, and Merer, *JMS 91*, 165 (1982).
NbO: Brom, Durham, and Weltner, *JCP 61*, 970 (1974).
MoN: Knight and Steadman, *JCP 76*, 3378 (1981).
Sc_2: Knight, Van Zee, and Weltner CPL (in press)
CrH: Van Zee, DeVore, and Weltner, *JCP 71*, 2051 (1979).
CrF: Van Zee, DeVore, and Weltner, *Proc. Symp. High Temp. Metal Halide Chem.*, Vol. 78-1, page 187 (Hildenbrand and Cubicciotti, eds., Electrochem. Soc., Princeton, N.J., 1978).
CrAg, CrAu: Baumann, Van Zee, and Weltner (preprint).
MnO: Ferrante, Wilkerson, Graham, and Weltner, *JCP 67*, 5904 (1977);
Baumann, Van Zee, and Weltner JPC (in press)
MnS: Baumann, Van Zee, and Weltner JPC (in press)
MnH: Van Zee, DeVore, Wilkerson, and Weltner, *JCP 69*, 1869 (1978).
MnF: DeVore, Van Zee, and Weltner, *JCP 68*, 3522 (1978).
MnCl, MnBr, MnL: Baumann, Van Zee, and Weltner JPC (in press)
MnAg, CrZn: Baumann, Van Zee, and Weltner (preprint).
CrCu: Van Zee and Weltner, *JCP 74*, 4330 (1981); erratum *75*, 2484 (1981); corrected to $^4\Sigma$ in Baumann, Van Zee, and Weltner, *JCP 79*, 5272 (1983).
GdO: Van Zee, Ferrante, Zeringue, and Weltner, *JCP 75*, 5297 (1981).
Mn_2: Van Zee, Baumann, and Weltner, *JCP 74*, 6977 (1981).

Table C-9. Doublet Triatomic σ Radicals

Molecule	Matrix	Experimental Data								Spin Density			Comments
		g_x, g_y, g_z	A_x (MHz)	A_y (MHz)	A_z (MHz)	A_{iso} (MHz)	T_x (MHz)	T_y (MHz)	T_z (MHz)	a_s^2	a_p^2	a_d^2	
$^{1}H^{13}C_\alpha{}^{13}C_\beta$	Ar	2.0029	(H) 41(1)		51(1)	45.4(3)	A_{dip} = 3(1)			0.03			uv irrad. of C_2H_2 in Ar,
($^2\Sigma$)		2.0029	(α) 863(1)		980(1)	902(1)	A_{dip} = 39(1)			0.29	0.43		4° K
		2.0025	(β) 139(1)		191(1)	156(1)	A_{dip} = 17(1)			0.05	0.19		
$^{11}B^{13}C_\alpha{}^{13}C_\beta$	Ne	2.0002	(B) 752.4		815.3	773.4	A_{dip} = 21.0			0.40	0.41		site (II-III)
($^2\Sigma$)		2.0002	(α) 33.0		42.0	36.0	A_{dip} = 3.0			0.01	0.06		vapor. of solid carbide,
		2.0012	(β) ~0		~0	~0	A_{dip} = ~0			0.00	0.00		4° K
$^{45}Sc^{1}H_2$	Ar	1.987(1)	(Sc) −128(2)		−212(2)	−156(2)	+28	+28	−56	0.06			linear?
($^2\Sigma$?)			(H) 12(1)		12(3)	12(2)	~0	~0	~0	0.08			vaporization of hydride
		1.980(1)						or A_{dip} = −28					powder, 12° K
$^{89}Y^{1}H_2$	Ar	1.991(1)	(Y) −40(10)										linear?
($^2\Sigma$?)			(H) 13(2)										vapor. of hydride, 12° K
$^{45}Sc^{19}F_2$	Ar	–	(Sc) –	220(3)	–					~0.04			vapor. of ScF_2 + Sc
		1.995(1)	(F)	28(2)				probably bent		<0.01			solids, 12° K
		–											
$^{47}Ti^{19}F_2^+$	CH_3OH	1.9465(4)			(Ti)	50.0(9)							HF + $TiCl_3$ in CH_3OH,
	liq.				(F)	21.1(3)							228° K
$^{89}Y^{19}F_2$	Ar	1.997(1)	(Y) 758(3)	757(3)	728(3)	748(3)	10	9	−20	0.39		0.67	vaporization of YF_3 + Y
		1.989	(F) 98(2)	89(2)	75(2)	87(2)	11	2	−12	0.002	0.005		solids, 12° K
		1.985						probably bent					
$^{89}Y(CN)_2$	Ar	1.973(1)				672							identification uncertain
$^{63}Cu^{19}F_2$	Ar	2.601(1)	(Cu) 2090(5)		1933(5)	2038(5)	A_{dip} = −52(4)			0.25		0.75	vapor. of solid fluoride,
($^2\Sigma_g$)		2.601(1)	(F) 206(1)		308(1)	240(1)	A_{dip} = 34(1)						4° K
		1.913(1)											

Table C-9. *(Continued)*

Molecule	Matrix	g_x, g_y, g_z	A_x (MHz)	A_y (MHz)	A_z (MHz)	A_{iso} (MHz)	T_x (MHz)	T_y (MHz)	T_z (MHz)	a_s^2	a_p^2	a_d^2	Comments
		Experimental Data								Spin Density			
$^{109}Ag(CN)_2$	K[Ag(CN)$_2$]	2.078	84	84	−182	−5	A_{dip} = −89					~1.0	γ irrad., 77° K
	P	2.078			σ* radical								
		2.000			Linear, bent by lattice below 77° K								
$^{109}Ag(CN)_2^{2-}$	K[Ag(CN)$_2$]	2.005	84	84	168	112	A_{dip} = −28			0.06		≤0.45	γ irrad., 77° K
	P	2.006											
		2.000			Likely linear (z axis), electron in σ* orbital								
$^{19}F^{133}Cs^{19}F$	CsF	2.189(10)	(Cs) 31(30)	31(30)	646.7(6)	236	A_{dip} = −205				0.48		γ irrad., 77° K
($^2\Sigma_u$)	S.C.	2.189(10)	(F) 214(150)	214(150)	1764.4(15)	731	A_{dip} = −520			0.006	0.59/2		H_1 center
		1.935(2)											

HC_2: Adrian and Bowers, *CPL 41*, 517 (1976);
Graham, Dismuke, and Weltner, *JCP 60*, 3817 (1974);
Cochran, Adrian, and Bowers, *JCP 40*, 213 (1964);
Jinguji, McDowell, and Raghunathan, *JCP 61*, 1489 (1974).

BC_2: Easley and Weltner, *JCP 52*, 1489 (1970).

ScH_2, YH_2: Knight, Wise, Fisher, and Steadman, *JCP 74*, 6636 (1981).

ScF_2: Knight and Wise, *JCP 71*, 1578 (1979).

TiF_2^+: Waters and Maki, *PR 125*, 233 (1962).

YF_2, $Y(CN)_2$: Knight and Wise, *JCP 73*, 4946 (1980).

CuF_2: Kasai, Whipple, and Weltner, *JCP 44*, 2581 (1966).

$Ag(CN)_2$, $Ag(CN)_2^{2-}$: Symons and Zimmerman, *JCS/DT*, 1970 (1976).

FCsF: Kato, *JPSJ 38*, 1691 (1975);
Kazumata, Kato, Ueda, and Nishi, *JPSJ 38*, 190 (1975).

Table C-10. Doublet Triatomic σ Radicals

Molecule	Matrix	Experimental Data: g_x, g_y, g_z		A_x (MHz)	A_y (MHz)	A_z (MHz)	A_{iso} (MHz)	T_x (MHz)	T_y (MHz)	T_z (MHz)	Spin Density: a_s^2	$a_{p\sigma}^2$	a^2	References and Comments
$^1H^{11}BO^-$	KBH_4	2.006	(H)	253	285	253	264	−11	21	−11	0.18			γ irrad., 20° K
	P	1.997	(B)	267	316	267	283	−16	33	−16	0.14	>0.32		
		2.006												
$^1H^{13}C^{14}N^-$	KCl	2.0039	(H)	370	384	398	384	−14	0	+14	0.27			uv or γ irrad. CN^-
	P	2.0022	(C)	194.4	255.9	182.8	211	−17	45	−28	0.07	0.24		doped, 4° K
		2.0005	(N)	0	59	0	20	−20	39	−20	0.01			
						$A(^{13}C)$ rotated 34° in plane								
$^{19}F^{13}C^{14}N^-$	KCl		(F)	936	986	2163	1362	−426	−376	801	0.025	0.23		X-irrad., 4° K
	CN^-	(2.0023)	(C)	610.4	626.6	704.5	647.1	−36.7	−20.5	57.4	0.17	0.27		(also 77°, 200° K)
	S.C.	assumed	(N)	5.0	3.1	45.0	17.9	−12.9	−14.8	27.1	0.01			(axes undet'd.?)
$^{35}ClC^{14}N^-$	KCl	2.0094	(Cl)	84	84	367	178	−94	−94	189				uv irrad., 77° K
		2.0085	(N)	5.1	10.7	20.3	12.0	−6.9	−1.3	8.3				
		1.9887						z axis chosen in analogy with ClCO						
$^{81}BrCN^-$	BrCN	(⊥) 2.065(5)		(⊥) 540		(∥) 1587	889	−349		698	0.04	0.53		γ irrad., 77° K
	S.C.	(∥) 2.00						linear σ^* radical						
$^1H^{13}CO$	H_2CO_2	2.0037	(H)	337	346	378	354	−17	−8	25	0.25			γ irrad., 77° K
	S.C.	2.0023	(C)	341	438	317	365	−24	72	−48	0.13			
		1.9948												
	Ar	2.0041	(H)	366.9	378.9	396.0	380.6	−13.7	−1.7	15.4	0.27			uv irrad. of HI, CH_2O,
		2.0027	(C)	365.7	427.9	338.9	377.5	−11.8	50.4	−38.6		0.45		or CH_3OH in Ar, 4° K
		1.9960			Bent at $\theta \cong 125°$				A_{yz} (H) = ±16.2					
$^{19}F^{13}CO$	Ar	(⊥) 2.0030	(F)		(⊥) 638.2	(∥) 1438.4	904.9	(⊥) −266.7		(∥) 533.5	0.02	0.17	(0.28)	∥ along C-F
			(C)		(⊥) 801.9	(∥) 803.8	802.5	(−24.8)*	(+25.1)*	(−0.3)*	0.26	(0.27)	on O	*see Cochran, et al.
		(∥) 2.0019			Bent at $\theta \cong 110°$			$g + A(^{13}C)$ are relative to $A(^{19}F)$						
$^{35}ClCO$	CO	2.0061		58	50	227	112	−54	−62	115	0.02	0.42		z along C–Cl
		2.0003												x ⊥ to plane
		1.9980				$e^2Qq = 76, \eta = 0.25$								

HBO⁻: Catton, Symons, and Wardale, *JCS* A2622 (1969).
HCN⁻: Root, Symons, and Weatherly, *MP 11*, 161 (1966);
Cochran, Weatherly, Bowers and Adrian, *Bull. Am. J. Phys. 13*, 357 (1968), Abstract AD7.
FCN⁻, ClCN⁻: Othmer, Ph.D. thesis, Cornell University, Ithaca, N.Y., 1970 (as given by Morton and Preston, *LB*, pages 24, 30).
BrCN⁻: Mishra, Neilson, and Symons, *JCS/FTII/70*, 1280 (1974).
HCO: Adrian, Cochran, and Bowers, *JCP 36*, 1661 (1962);
Cochran, Adrian, and Bowers, *JCP 44*, 4626 (1966);
Holmberg, *JCP 51*, 3255 (1969).
FCO: Adrian, Cochran, and Bowers, *JCP 43*, 462 (1965);
Cochran, Adrian, and Bowers, *JCP 44*, 4626 (1966).
ClCO: Adrian, Cochran, and Bowers, *JCP 56*, 6251 (1972).

Table C-11. Doublet Triatomic σ Radicals

Molecule	Matrix	Experimental Data								Spin Density			References and Comments
		g_x, g_y, g_z	A_x (MHz)	A_y (MHz)	A_z (MHz)	A_{iso} (MHz)	T_x (MHz)	T_y (MHz)	T_z (MHz)	a_s^2	$a_{p_y}^2$	$a_{p_z}^2$	
$^{11}B^{19}F_2$	Xe	2.0012 (3)			(B) 826.2 (F) 532	$\theta \cong 112°$				0.41 0.02/2			γ irrad. BF_3 in Xe at 4° K
$^{14}N_3^{2-}$	KN_3 S.C.	2.0010 2.0046 1.9983	(Mid. N) 46.2 (Out. N) −18.8	56.1 −21.0	86.7 39.2	$\theta \cong 139°$				0.04		0.26	γ irrad., 77° K
$^{14}N^{14}NO^-$	CS_2 P	2.008 2.004 2.004	(Mid. N) 87 (9) (Out. N) 20 (9)	87 (9) 20 (9)	132 (3) 42 (6)								γ irrad. of N_2O in CS_2 at 77° K
$^{14}N^{17}O_2$	$NaNO_2$ S.C.	2.0061 1.9908 2.0013	(N) 138.5 (O) −12.0	130.5 ±4.85*	190.5 152.1*	153.2 47 (* Princ. A comp.)	−15.2 −59	−21.9 −47	+37.2 106	0.09 0.01	 0.01	0.4 0.7	γ irrad., 77° K
$^{14}NO_2$	Ne	2.0056 1.9920 2.0030	146.0	128.3	178.0	150.8	−4.8	−22.5	+27.2				4° K
$^{31}PO_2$		Identification uncertain											
$^{29}SiO_2^-$	$ZrSiO_4$	2.0050 1.9960 1.9920	452	469	563								77°, 300° K
$^{23}Na^{+13}CO_2^-$	$NaHCO_2$ S.C.	2.0032 1.9975 2.0014	(Na) 21.0 (C) 436	21.0 422	26.1 546	22.7 468 $\theta = 134°$	−1.7 −32	−1.7 −46	+3.4 +78	0.02 0.14		 0.66 0.22/2	γ irrad., room temp. on O
$^{23}Na^+CO_2^-$	CO_2	2.0029 1.9974 2.0012	50.8	49.2	54.3	51.4	−0.6	−2.2	+2.9				Na (or K) vapor plus CO_2 at 77° K
OCS^-	OCS MgO surf.	2.0049 1.986 2.002											uv irrad., 77° K

Table C-11. Doublet Triatomic σ Radicals *(Continued)*

Molecule	Matrix	Experimental Data								Spin Density			References and Comments
		g_x, g_y, g_z	A_x (MHz)	A_y (MHz)	A_z (MHz)	A_{iso} (MHz)	T_x (MHz)	T_y (MHz)	T_z (MHz)	a_s^2	$a_{p_y}^2$	$a_{p_z}^2$	
$^{23}Na^{+13}CS_2^-$	CS_2	2.0079	(Na) 47.8	47.9	48.7	48.1	~0	~0	~0	0.06			Na (or K) vapor in CS_2 at 77°K
	P	1.9661	(C) 205.4	184.4	340.3	243.4	−38	−59	+96.9	0.08		0.56	
		1.9993				θ = 141°							
$^{14}N^{13}CO^{2-}$	KNCO	2.0015	(N)			~14							γ irrad., 77°K
	P	1.997	(C) 499	349	350								
		2.002											
$^{14}N^{13}CS^{2-}$	KNCS	1.997	(N) ~0	~0	41.9					0.009		0.29	γ irrad., 77°K
	P		(C) 224	224	335					0.008		0.40	

BF_2: Nelson and Gordy, *JCP 51*, 4710 (1969).

N_3^{2-}: Marinkas, *JCP 52*, 5144 (1970);
Neilson and Symons, *JCS/FTII 68*, 1772 (1972);
Adams and Owens, *JCP 58*, 3532 (1973);
Claxton, Overill, and Symons, *MP 26*, 75 (1973).

NNO^-: Mishra and Symons, *CC* 510 (1972).

NO_2: Jen, Foner, Cochran, and Bowers, *PR 112*, 1169 (1958);
Zeldes and Livingston, *JCP 35*, 563 (1961);
Adrian, *JCP 36* 1692 (1962);
Atkins, Keen, and Symons, *JCS* 2873 (1962);
Kasai, Weltner, and Whipple, *JCP 42*, 1120 (1965);
Luz, Reuveni, Holmberg, and Silver, *JCP 51*, 4017 (1969);
Myers, Easley, and Zilles, *JCP 53*, 1181 (1970);
McDowell, Nakajima, and Raghunathan, *CJC 48*, 805 (1970);
Morton and Preston *LB*, pages 56–63.

PO_2: Identification uncertain, Morton and Preston, *LB*, page 141;
Bershov, Samoilovich, Lushnikov, and Tarashchan, *ZSK 9*, 309 (1968);
Bershov, *Teor. i Eksperim. Khim., Akad. Nauk Ukr. SSR 6*, 397 (1970).

SiO_2^-: Solntsev, Shcherbakova, and Dvornikov, *ZSK 15*, 217 (1974).

CO_2^-: Ovenall and Whiffen, *MP 4*, 135 (1961);
Brivati, Keen, Symons, and Trevalian, *PCS* 66 (1961);
Atkins, Keen, and Symons, *JCS* 2873 (1962);
Marshall, Reinberg, Serway, and Hodges, *MP 8*, 225 (1964);
Bennett, Mile, and Thomas, *TFS 61*, 2357 (1965).

OCS^-: Lin, Johnson, and Lunsford, *CPL 15*, 412 (1972).

CS_2^-: Bennett, Mile, and Thomas, *TFS* 63, 262 (1967);
Lin, Johnson, and Lunsford, *CPL 15*, 412 (1972);
Behar and Fessenden, *JPC 76*, 1706 (1972).

NCO^{2-}, NCS^{2-}: Ginns and Symons, *JCS/DT* 3 (1973).

Table C-12. Doublet Triatomic π_x Radicals

Molecule	Matrix	Experimental Data g_x, g_y, g_z	A_x (MHz)	A_y (MHz)	A_z (MHz)	A_{iso} (MHz)	T_x (MHz)	T_y (MHz)	T_z (MHz)	Spin Density a_s^2	a_p^2	a^2	Comments
$^{14}N^1H_2$	Ar	2.0038		(rotating) Parameters dependent upon trapping medium		(N) 29.1 (H) −67.0				0.02 0.05			(also Kr, Xe) uv irrad. of HN_3, 4°K
$^1H_2O^+$	Beryl	2.0065				7.19							γ irrad., 77°K
$^{14}N^{19}F_2$	Ar	2.0022 2.0085 2.0051	(N) 138.2 (F) 594.1	2.8 −46.4	2.8 −46.3	47.9 167.1	90.3 427	−45.1 −213.5	−45.1 −213.4	0.03 0.00	0.80 0.24/2		(also Ne, Kr)
$^{14}N^{35}Cl_2$	Ar	2.006 2.026 2.023	(N) 112 (Cl) 67	±14 −48	±14 −48	47 (or 28) −9.7	(65) (77)	(−33) (−38)	(−33) (−38)	0.03 0.00	0.57 0.44/2		NCl_3 in Ar,
$^{31}P^1H_2$	Kr	2.0087				(P) 244 (H) 50				0.02 0.04			γ irrad. of PH_3 in Kr, 4°K
$^{31}PH_2$	PH_3	2.002	771			224	576			0.02			γ irrad., 77°K
$^{31}P^{19}F_2$	Xe	2.0027 2.0016 2.0016	(P) 863.6 (F) 351.8	−76.2 −37.8	−76.2 −37.8	237.1 92.1		A_{dip} = 313.3 A_{dip} = 129.9		0.02 0.00	0.83 0.16/2		γ irrad. of PF_3 in Xe, 4°K
$^{35}Cl^{31}P^{35}Cl$	PCl_3 P	1.9953 2.0187 2.0187	(P) 761.0 (Cl) 46.1	−63.6 ~0	−63.6 ~0	211.3 15.4		A_{dip} = 274.9 A_{dip} = 15.4					γ irrad., 77°K

Table C-12. Doublet Triatomic π_x Radicals *(Continued)*

Molecule	Matrix	Experimental Data								Spin Density			Comments
		g_x, g_y, g_z	A_x (MHz)	A_y (MHz)	A_z (MHz)	A_{iso} (MHz)	T_x (MHz)	T_y (MHz)	T_z (MHz)	a_s^2	a_p^2	a^2	
$^{31}P^{35}Cl^{35}Cl$	Ar	2.002	(P) 767	−49.0	−36	227	540	−276	−263		0.71		Photolysis of PCl_3 in Ar, 20° K
		1.9995	(Cl) 25.2	−49.0	−49.0	−24.3	49.5	−24.7	−24.7		0.14		
		1.9995	(Cl) 25.2	~0	~0	8	17	−8	−8		0.05		
$^{119}Sn^{35}Cl_2^-$	KCl	1.6494	(Sn) 2154		2196	2170		A_{dip} = 14		0.05			γ irrad., 24° K, 2 other unsym. sites
	$(SnCl_2)$	1.6494	(Cl) 33.0		52.8	40		A_{dip} = 7		0.01	0.04		
	S.C.	1.8952			linear molecule in center with tetragonal symmetry								

NH_2: Parameters from solid state ESR are very dependent upon the trapping medium, particularly because of hydrogen bonding.
Morton and Smith, *CJC 44*, 1951 (1966);
Rao and Symons, *JCS* A2163 (1971);
Michaut, Roncin, and Marx, *CPL 36*, 599 (1975);
Foner, Cochran, Bowers, and Jen, *PRL 1*, 91 (1958);
Coope, Farmer, Gardner, and McDowell, *JCP 42*, 2628 (1965);
Fischer, Charles, and McDowell, *JCP 46*, 2162 (1967).

H_2O^+: Samoilovich and Novozhilov, *ZNK 15*, 84 (1970).

NF_2: Doorenbos and Loy, *JCP 39*, 2393 (1963);
Colburn, Ettinger, and Johnson, *IC 3*, 455 (1964);
Farmer, Gerry, and McDowell, *MP 8*, 253 (1964);
Kasai and Whipple, *MP 9*, 497 (1965);
Piette, Johnson, Booman, and Colburn, *JCP 35*, 1481 (1961);
Adrian, Cochran, and Bowers, *Advanc. in Chem. Series 36*, 50 (1962);
McDowell, Nakajima, and Raghunathan, *CJC 48*, 805 (1970).

NCl_2: Wei, Current, and Gendell, *JCP 57*, 2431 (1972);
Fullam, Mishra, and Symons *JCS/DT* 2145 (1974).

PH_2: Morehouse, Christiansen, and Gordy *JCP 45*, 1747 (1966);
Jackel and Gordy, *PR 176*, 443 (1968);
Fullam, Mishra, and Symons, *JCS/DT* 2145 (1974).

PF_2: Nelson, Jackel, and Gordy, *JCP 52*, 4572 (1970);
Wei, Current, and Gendell, *JCP 52*, 1592 (1970); *57*, 2431 (1972);
Fullam, Mishra, and Symons, *JCS/DT* 2145 (1974).

ClPCl: Kokoszka and Brinckman, *CC* 349 (1968); *JACS 92*, 1199 (1970);
Begum and Symons, *JCS* A2065 (1971);
Wei, Current, and Gendell, *JCP 57*, 2431 (1972);
Fullam, Mishra, and Symons, *JCS/DT* 2145 (1974).

PClCl: Wei, Current, and Gendell, *JCP 57*, 2431 (1972).

$SnCl_2^-$: Delbecq, Hartford, Schoemaker, and Yuster, *PSS* B*71*, K81 (1975).

Table C-13. Doublet Triatomic π_x Radicals

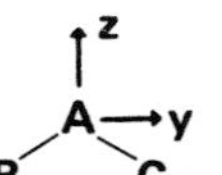

Molecule	Matrix	Experimental Data g_x, g_y, g_z		A_x (MHz)	A_y (MHz)	A_z (MHz)	A_{iso} (MHz)	T_x (MHz)	T_y (MHz)	T_z (MHz)	Spin Density a_s^2	$a_{p_x}^2$	a^2	References and Comments
$^{17}O^{35}Cl^{17}O$	$KClO_3$ S.C.	2.0018 2.0167 2.0111	(O) (Cl)	129 223.9	−23 −37.8	−23 −35.2	28 50.3	101 173.6	−51 −88.1	−51 −85.5	0.01/2	0.53/2 0.49		X irrad., 150°K
$O^{35}ClO$	Ne	2.0022 2.0157 2.0125		197.4	−32.62	−28.63 $e^2Q\,(V_{xx}-V_{yy})\,(^{35}Cl) = 95,\ e^2Qq_z = 6$	45.4	152.0	−78.0	−74.0	0.01	0.43		condensation of gas mix., 4°K
$O^{81}BrO$	$KBrO_4$ S.C.	2.0017 2.0080 2.0076		101	−38	−48	5	96	−33	−43	0.00	0.05		X irrad., 26°K
^1HOO	Ar	$\perp$ = 2.0044 $\parallel$ = 2.0393		−40.9	−7.0	−37.8 ($x \perp$ to plane, z = O–O direction)	−28.6	−12.3	+21.6	−9.2 O (end) (cent.)	0.02	 0.71 0.29		uv irrad. of $HI + O_2$ in Ar, 4°K
^{19}FOO	Ar	2.0008 2.0022 2.0080		141.1	39.2	−288.4 (z = O–F bond)	−108.1	249.2	147.3	−180.3 O (end) (cent.)	0.002	 0.60 0.40		uv irrad. of $F_2 + O_2$ in Ar, 4°K
$^{35}ClOO$	Ar	1.9987 2.0100 1.9915		−9	−50	+16 (z = O–O direction) $e^2Qq\,(^{35}Cl) = 105$	14	−23	−64	+2	0.00			uv irrad. of $Cl_2 + O_2$ in Ar, 4°K
$^{19}F^{35}ClO^+$	SbF_5	2.002 2.008 2.008	(F) (Cl)	270.1 185.8	−49.5 −54.8	−49.5 −54.8 (^{17}O)	57.0 25.4 50.5	213 160	−107 −80	−107 −80	0.00 0.01 0.01	0.06 0.64		$Cl_2 + H_2O$, 77°K uv irrad. $ClF_3:H_2O$
$^{35}ClO^{35}Cl^+$	SbF_5	2.003 1.991 2.001		23.8	0.0	0.0 (^{17}O)	7.9 56.6	15.9	−7.9	−7.9	0.00 0.01	0.08/2		$Cl_2 + H_2O$, 77°K
$^{35}ClSO$	Powd.	2.000 2.017 2.008		33.0	−18.6	−21.9	−2.5	35.5	−21.1	−24.4		0.13		γ irrad., 77°K

Table C-13. Doublet Triatomic π_x Radicals *(Continued)*

Molecule	Matrix	Experimental Data								Spin Density			References and Comments
		g_x, g_y, g_z	A_x (MHz)	A_y (MHz)	A_z (MHz)	A_{iso} (MHz)	T_x (MHz)	T_y (MHz)	T_z (MHz)	a_s^2	$a_{p_x}^2$	a^2	
$^{81}BrSO$	Powd.	2.006 2.052 2.002	225	−52	−115	19	206	−71	−134		0.15		γ irrad., 77°K
$^{35}Cl^{77}SeO$	Powd.	1.994 2.039 1.995	(Cl) 31.0 (Se) 798	−7.4 −391	−10.3 −355	4.4 17	26.6 781	−11.8 −408	−14.7 −372		0.01 1.04		γ irrad., 77°K
$^{35}ClSS$	Ar	2.0019 2.0225 2.0384	17.7	−3.9	−10.5 (z = Cl–S bond) e^2Qq_z (^{35}Cl) = 68(2)	1.1	16.6	−5.0	−11.5	0.0	0.04 −0.01 (p_z)		uv irrad. of S_2Cl_2 in Ar, 4°K

OClO: Cole, *PNAS 46*, 506 (1960);
Byberg, Jensen, and Muus, *CP 46*, 131 (1967);
Eachus, Edwards, Subramanian, and Symons, *JCS A1704* (1968);
Fayet, Pariset, and Thieblemont, *CR B268*, 78 (1969);
Schlick and Silver, *MP 26*, 177 (1973);
McDowell, Raghunathan, and Tait, *JCP 59*, 5858 (1973);
Morton and Preston, *LB*, pages 187–192.

OBrO: Collins, Cosgrove, and Betteridge, *JP B3*, L48 (1970);
Rao and Symons, *JCS/DT* 147 (1972);
Byberg and Linderberg, *CPL 33*, 612 (1975).

HOO: Livingston, Ghormley, and Zeldes, *JCP 24*, 483 (1956);
Atkins and Symons [2], page 155;
Bennett, Mile, and Thomas, *Proc. 11th Symp. (Int.) on Combustion*, Berkeley, Calif., U.S.A., 853 (1967);
Adrian, Cochran, and Bowers, *JCP 47*, 5441 (1967);
Wyard, Smith, and Adrian, *JCP 49*, 2780 (1968);
Milhelcic, Ehhalt, Klomfass, Kulessa, Schmidt, and Trainer, *BBPC 82*, 16 (1978).

FOO: Kasai and Kirschenbaum, *JACS 87*, 3069 (1965);
Fessenden and Schuler, *JCP 44*, 434 (1966);
Adrian, *JCP 46*, 1543 (1967);
Morton and Preston, *IC 13*, 1786 (1974).

ClOO: Atkins, Brivati, Keen, Symons, and Trevalion, *JCS* 4785 (1962);
Eachus, Edwards, Subramanian, and Symons, *CC* 1036 (1967);
Byberg, *JCP 47*, 861 (1967);
Eachus, Edwards, Subramanian, and Symons, *JCS* A1704 (1968);
Adrian, Cochran, and Bowers, *JCP 56*, 6251 (1972).

$FClO^+$: Morton and Preston, *IC 13*, 1786 (1974);
Eachus and Symons, *JCS/DT* 431 (1976).

$ClOCl^+$: Morton and Preston, *IC 13*, 1786 (1974);
Eachus and Symons, *JCS/DT* 431 (1976).

ClSO, BrSO, ClSeO: Nishikida and Williams, *JMR 14*, 348 (1974).

ClSS: Herring, McDowell, and Tait, *JCP 57*, 4564 (1972).

Table C-14. Doublet Triatomic π_x Radicals

Molecule	Matrix	Experimental Data								Spin Density			References and Comments
		g_x, g_y, g_z	A_x (MHz)	A_y (MHz)	A_z (MHz)	A_{iso} (MHz)	T_x (MHz)	T_y (MHz)	T_z (MHz)	a_s^2	$a_{p_x}^2$	a_d^2	
$^{14}N^{17}O_2^{2-}$	KN_3 S.C.	2.0029 2.0078 2.0071	(N) 103.2 (O) 74.8		−11.8 −8.4	26.5 19.3	76.7 55.5		−38.3 −63.9		0.80 0.38/2		X irrad., 300° K and up
PO_2^{2-}			Identification uncertain										
$^{75}AsO_2^{2-}$	$CaCO_3$ S.C.	1.9991 1.9910 2.0150	614 (1)	−173 (2)	−152 (2)	96	518	−248	−269	0.01	0.96	0.03	γ irrad., 77° K *T* varied: 4° K to 400° K
					$\theta \cong 90°$, $e^2qQ = 16.2(1)$								
$^{17}O_3^-$	$KClO_3$ S.C.	2.0035 2.0187 2.0123	(Mid. O) 231.6 (Out. O) 122.3	−23 −14	−23 −14	61.9 31.4	169.7 90.9	−84.9 −45.4	−84.9 −45.4		0.59 0.64/2		γ irrad., 77° K
$^{23}Na^+O_3^-$	Ar	2.0022 2.0165 2.0121	−3.87*	−5.98*	−3.29								Na vapor + O_3, 4° K
					*Not princ. axis Nonplanar ion pair, Q' (^{23}Na) observed								
$^{33}S^{17}O_2^-$	$K_2S_2O_5$ S.C.	2.0019 2.012 2.0057	(S) 162.5 (O) 84.1		−11 −8	47 23	116 61		−58 −31		0.74 0.43/2		γ irrad., 300° K
$^{23}Na^+SO_2^-$	Ar	2.0024 2.0101 2.0060	−4.48	−5.20	−3.79	−4.27							Na vapor + SO_2, 4° K
					Nonplanar ion pair, A_{xz} (^{23}Na) = +1.6								
$^{33}S^{33}SO^-$	$Na_2S_2O_3$ S.C.	2.0035 2.0287 2.0106	(Mid. S) 136.3 (Out. S) 91.4	<3 <3	<3 <3	45 30	91 61	−45 −30	−45 −30		(0.54) (0.37)		γ irrad., 300° K hfs from Lin and Lunsford (MgO surf)

Table C-14. Doublet Triatomic π_x Radicals *(Continued)*

		Experimental Data								Spin Density			
Molecule	Matrix	g_x, g_y, g_z	A_x (MHz)	A_y (MHz)	A_z (MHz)	A_{iso} (MHz)	T_x (MHz)	T_y (MHz)	T_z (MHz)	a_s^2	$a_{p_x}^2$	a_d^2	References and Comments
$^{33}S_3^-$	KCl	2.0026	(Mid. S) 145	29	6	60	85	−31	−54	0.02	0.65		S doped, 300° K
	S.C.	2.0499	(Out. S) 54	4	~0	19	35	−15	−19	0.01	0.44/2		
		2.0319				*A* (outer S) at 20 ± 10° to S–S bond in plane							
$^{77}SeO_2^-$	NaH	1.9967(10)	670	270(20)	240(20)	390	280	−120	−140	0.03	0.41		γ irrad., 300° K
	SeO_3	2.0268(10)											(aver. of two sites)
	S.C.	2.0062(10)											
$^{77}Se_3^-$	KCl	1.9885	(Mid. S) 694	135	166	332	362	−197	−166	0.02	0.4		Se doped, 100° K
	S.C.	2.2205	(Out. S) 203	48	~0	84				0.01/2			
		2.1545				*A* (outer Se) at 20 ± 10° to Se–Se bond in plane							

NO_2^{2-}: Reuveni, Poupko, and Luz, *JMR 18*, 358 (1975).

PO_2^{2-}: Identification uncertain, Morton and Preston, *LB*, page 141;
Bershov, Samoilovich, Lushnikov, and Tarashchan, *ZSK 9*, 309 (1968);
Bershov, *Teor. i Eksperim. Khim., Akad. Nauk Ukr. SSR 6*, 397 (1970).

AsO_2^{2-}: Marshall and Serway, *JCP 50*, 435 (1969);
Chan, Marshall, and Serway, *PSS B59*, 687 (1973).

O_3^-: Schlick, *JCP 56*, 654 (1972);
Adrian, Bowers, and Cochran, *JCP 61*, 5463 (1974);
Morton and Preston, *LB*, pages 113–116

SO_2^-: Schneider, Dischler, and Rauber, *PSS 13*, 141 (1966);
Reuveni, Luz, and Silver, *JCP 53*, 4619 (1970);
Schoonheydt and Lunsford, *JPC 76*, 323 (1972);
Adrian, Cochran, and Bowers, *JCP 59*, 56 (1973).

SSO^-; Morton, *JCP 71*, 89 (1967);
Aki, *JPSJ 26*, 939 (1969);
McLaughlan and Marshall, *JPC 74*, 1359 (1970);
Lin and Lundsford, *JPC 80*, 635 (1976).

S_3^-: Schneider, Dischler, and Rauber, *PSS 13*, 141 (1966);
Suwalski and Seidel, *PSS 13*, 159 (1966);
Lunsford and Johnson, *JCP 58*, 2079 (1973).

SeO_2^-: Cook, Rowlands, and Whiffen, *MP 8*, 195 (1964);
Atkins, Symons, and Wardale, *JCS* 5215 (1964);
Bline, Poberaj, Schara, and Stepisnik, *JPCS 27*, 1391 (1966);
Schneider, Dischler, and Rauber, *PSS 13*, 141 (1966).

Se_3^-: Schneider, Dischler, and Rauber, *PSS 13*, 141 (1966).

Table C-15. Doublet Triatomic Metal Molecules

Molecule	Matrix	Experimental Data: g_x, g_y, g_z	A_x (MHz)	A_y (MHz)	A_z (MHz)	A_{iso} (MHz)	T_x (MHz)	T_y (MHz)	T_z (MHz)	Spin Density: a_s^2	a_p^2	a^2	Comments
$^{23}Na_3$	Ar	2.0012(12)			(mid. Na)	-60(25)				-0.07			site A (sites B & C also obs.), ≥4° K
($^2B_2$ or $^2\Sigma_u^+$)					(out. Na)	424(40)				0.94/2			
($^2E'$)	Ar					261.0(6)							pseudorotating, Na vapor in Ar
$^{39}K_3$	Ar	1.9985(5)			(mid. K)	-13.1(6)				-0.06			K vapor in Ar, ≥4° K
($^2B_2$ or $^2\Sigma_u^+$)					(out. K)	109.1(9)				0.94/2			
($^2E'$)	Ar	1.9990(5)				68.5(20)							pseudorotating
$^{107}Ag_3$	Benz	1.960	g_{iso} = 1.9622(1)		(mid. Ag)	(-) 105.7(9)				(-) 0.06			Ag vapor in benz., 77° K
		1.962			(out. Ag)	810.2(9)				0.88/2			
		1.966											
$^{109}Ag_3{}^{2+}$	Toluene Ag^+, H_2O	1.973(2)				561(6)				0.31			γ irrad., 77° K

Na_3: Lindsay, Herschbach, and Kwiram, *MP 32*, 1199 (1976); *39*, 529 (1980).
K_3: Thompson and Lindsay, *JCP 74*, 959 (1981).
Thompson, Tischler, Garland, and Lindsay, *SS 106*, 408 (1981).
Lindsay, Garland, Tischler, and Thompson, *ACS Symp. Series* No. 179, page 69 (J. L. Gole and W. C. Stwalley, eds., 1982).
Ag_3: Howard, Preston, and Mile, *JACS 103*, 6226 (1981).
$Ag_3{}^{2+}$: Forbes and Symons, *MP 27*, 467 (1974).

Table C-16. Doublet Triatomic Miscellaneous Ions

		Experimental Data								Spin Density			
Molecule	Matrix	g_x, g_y, g_z	A_x (MHz)	A_y (MHz)	A_z (MHz)	A_{iso} (MHz)	T_x (MHz)	T_y (MHz)	T_z (MHz)	a_s^2	$a_{p_y}^2$	a_d^2	Comments
$^{27}Al_2O^-$	Al_2O_3 P	2.0036				28							γ irrad., 77° K
O_3^{3-}	O_2, TiO_2 surf.	2.008 2.001 2.001											uv irrad., 77° K
$^{19}F_3^{2-}$	LiF S.C.	(2.002)	(mid. F) 751 (out. F) 224	3146 1060	594 398	1497 561	-746 -337	1649 499	-903 -163	0.03 0.01	0.56 0.36/2		X irrad., 77° K
			A (out. F) = 17° to $F_1 \cdots F_3$ (linear in gas, bent in crystal), $y = F_1 \cdots F_3$ direction										
$K^{+1}H_2{}^{33}S^-$	H_2S P	2.0023 2.0164 2.0164				(H) 21.7 (S) 169				0.02 0.05			K (or Na) vapor + H_2S, 77° K
$^1H_2S^-$	KCl SH^-	1.9865 2.2055 2.2055	92.6	60.8	60.8	71				0.10/2			uv irrad., 21° K
$^{109}Ag^{14}NO^+$	Zeol	1.934 2.000 2.000	(N) 3 (Ag) 195	34 182	34 182	24 186	-21 9	+10 -4	+10 -4	0.02 0.11		0.13	NO added to Ag^+ exchanged Zeol., 77° K
$^{63}Cu^{14}NO^+$	Zeol.	1.89 2.009 2.009	(N) (Cu) 640	~70 530	~70 530	570	A_{dip} = 40			0.10		0.11	NO added to Cu^{2+} exchanged Zeol., 77° K

Al_2O^-: Shirikov, Anufrienko, and Sazonov, *DANS 208*, 902 (1973).
O_3^{3-}: Meriaudeau and Vedrine, *JCS/FTII 72*, 472 (1976).
F_3^{2-}: Cohen, Kanzig, and Woodruff, *JPCS 11*, 120 (1959);
Symons, *JCS* 570 (1963);
Morton, *CRe 64*, 453 (1964);
Atkins and Symons [2], page 158.
H_2S^-: Grundig, Hausmann, and Sander, *Kurznachr. Akad. Wiss. Goettingen, Sammelh.* 9, 47 (1965);
Hausmann, *ZP 192*, 313 (1966);
Bennett, Mile, and Thomas, *CC* 182 (1966).
$AgNO^+$: Chao and Lunsford, *JPC 78*, 1174 (1974).
$CuNO^+$: Chao and Lunsford, *JPC 76*, 1546 (1972).

Table C-17. Triplet Triatomic Molecules

Molecule	Matrix	Experimental Data							Spin Density		Comments
		$g_\parallel$	$g_\perp$	$A_\parallel$ (MHz)	$A_\perp$ (MHz)	D (cm^{-1})	A_{iso} (MHz)	A_{dip} (MHz)	a_s^2	a^2	
CH_2 ($^3B_1$)		Torsionally oscillating in all matrices									uv irrad. of CH_2N_2, 4° and 77°K
	Xe					0.69	(E = 0.003)	D (rigid) = 0.93			
	Ne					0.601	(E = 0.0069)	D (rigid) = 0.76			
	C_4F_8							D (rigid) = 0.775			
$^{13}CD_2$	Xe	2.00	Derived θ = 137°	196	277	0.759 (E = 0.0044)	249		0.075		
NCN ($^3\Sigma$)	Fluoro-carbons					1.544	(E < 0.002)				uv irrad. of NCN_3, 77 or 4°K
$^{13}C^{14}N_\alpha{}^{14}N_\beta$ ($^3\Sigma$)	Ne	(2.0023)	(2.0023)	(C) (N_α) (N_β)	50(5) 35(5) 19(5)	1.159					C vapor + N_2, 4°K, (Tors. oscil. in matrix)
$^{13}C_\alpha{}^{13}C_\beta O$ ($^3\Sigma$)	Ne	(2.0023)	2.0029	(C_α) 17(3) (C_β) −32(3)	57(3) −26(3)	0.739	44(3) −30	−13 −2	0.01 0.01	0.12 0.02	C vapor + CO, 4°K, (Tors. oscil.)
$^{29}Si^{14}N_2$ ($^3\Sigma$)	Ar		(2.0023)	(Si) (N)	95(3) 17(3)	2.206		Bent in N_2 and Ar matrices			Si vapor + N_2, 4°K
$^{29}Si^{13}CO$ ($^3\Sigma$)	Ar		(2.0023)	(Si) (C)	84(3) <14	2.237		Bent in CO and Ar matrices			Si vapor + CO, 4°K
$Ti^{19}F_2$ ($^3B_1$)	Ne	1.9880(3)	1.9149(3) 1.9229(3)	<60(40)	<30(20)	±0.078 E = ±0.002		∥ = F----F direction			vapor. of TiF_3 + Ti solids, 4°K

CH_2: Wasserman, Yager, and Kuck, *CPL 7*, 409 (1970);
Wasserman, Kuck, Hutton, and Yager, *JACS 92*, 7491 (1970);
Wasserman, Kuck, Hutton, Anderson, and Yager, *JCP 54*, 4120 (1971);
Bernheim, Bernard, Wang, Wood, and Skell, *JCP 53*, 1280 (1970); *54*, 3223 (1971);
Wasserman, Hutton, Kuck, and Yager, *JCP 55*, 2593 (1971);
Bicknell, Graham, and Weltner, *JCP 64*, 3319 (1976).

NCN: Wasserman, Barash, and Yager, *JACS 87*, 2075 (1965).

CNN: Wasserman, Barash, and Yager, *JACS 87*, 2075 (1965);
Smith and Weltner, *JCP 62*, 4592 (1975).

CCO: Smith and Weltner, *JCP 62*, 4592 (1975).

SiCO, SiN_2: Lembke, Ferrante, and Weltner, *JACS 99*, 416 (1977).

TiF_2: DeVore and Weltner, *JACS 99*, 4700 (1977).

Table C-18. High-Spin ($S = \frac{3}{2}$ to 4) Triatomic Molecules

Molecule	Matrix	Experimental Data							Spin Density		Comments
		$g_{\parallel}$	$g_{\perp}$	$A_{\parallel}$ (MHz)	$A_{\perp}$ (MHz)	D (cm^{-1})	A_{iso} (MHz)	A_{dip} (MHz)	a_s^2	a_d^2	
Quartet ($^4\Sigma$):											
$^{51}V^{19}F_2$	Ar	(2.002)		(V) 205 (20)	581 (3)	≥0.48	456	125	0.11	~1	vapor. VF_3(s), 4° K
				(F)	28.6 (9)				~0.0		
$^{55}MnO_2$	Ar	(2.002)		723(11)	727(3)	1.45	726(6)	−1(5)			Mn atoms + O_2, 4° K
Sextet ($^6\Sigma$):											
$^{55}Mn^1H_2$	Ar	(2.002)		(Mn)	73	0.26					Mn vapor + H atoms, 4° K
($^6\Sigma$ or $^6A_1$)				(H)	36						
$^{55}Mn^{19}F_2$	Ne	(2.0023)	1.994 (4)	(Mn) 153 (6)	124 (1)	0.37	134	10			vapor of MnF_2 solid, 4° K
				(F)	19 (1)						
$MnCl_2$	Ne	2.002 (1)	2.001 (1)			0.44					vapor of solid salt, 4° K
$MnBr_2$	Ne	(2.0023)	2.005 (2)			1.25					vapor of solid salt, 4° K
MnI_2	Ne	(2.0023)	1.996 (2)			≥1.9					vapor of solid salt, 4° K
Nonet:											
GdF_2?	Ne		(2.0023)			0.20					vapor of solid GdF_3, 4° K

VF_2: DeVore, Van Zee, and Weltner, *Proc. Symp. High Temp. Metal Halide Chem.*, Vol. 78-1, page 187 (Hildenbrand and Cubicciotti, eds., Electrochem. Soc., Princeton, N.J., 1978).

MnO_2: Ferrante, Wilkerson, Graham, and Weltner, *JCP 67*, 5904 (1977); corrected (unpublished).

MnH_2: Van Zee, DeVore, Wilkerson, and Weltner, *JCP 69*, 1869 (1978);
Van Zee, Brown, and Weltner, *CPL 64*, 325 (1979);
Demuynck and Schaefer, *JCP 72*, 311 (1980).

MnF_2: DeVore, Van Zee, and Weltner, *JCP 68*, 3522 (1978);
Van Zee, Brown, and Weltner, *CPL 64*, 325 (1979).

$MnCl_2$, $MnBr_2$, MnI_2: Baumann, Van Zee, and Weltner, *JPC* (in press).

GdF_2: Baumann, Van Zee, Zeringue, and Weltner, *JCP 75*, 5291 (1981).

Table C-19. Doublet Tetra-Atomic (Plus CH_2OH) π Radicals. (All are Pyramidal Except the Three Indicated)

Molecule	Matrix	Experimental Data $g_\parallel$	$g_\perp$	$A_\parallel$ (MHz)	$A_\perp$ (MHz)	A_{iso} (MHz)	A_{dip} (MHz)	Spin Density a_s^2	$a_{p,d}^2$	Comments
$^{13}C^1H_3$ (planar)	CH_4 Liq.	(2.002)			(C) (H)	107.43 −64.56				e-irrad., 96° K also deut. derivatives
$^{13}C^1H_3$	Kr	2.0026			(C) (H)	108.0 −64.46		0.03		e-irrad. of CH_4, 85° K
$^{13}C^1H_3$	NaAc · $3H_2O$ S.C.	(2.002)		(C) 231.7(14) (H) −61.1(8)	43.4(14) 42.0(14) −63.0(8)	105.6 −62.5	63			e-irrad., 77° K rotating about 3-fold axis
$^{13}C^1H_2{}^{19}F$	Xe	2.0045			(H) (C) (F)	59.2 153.7 180.4		0.04		e-irrad. of CH_3F, 85° K
$^{13}C^1H^{19}F_2$	Xe	2.0041			(H) (C) (F)	62.3 417.4 236.2		0.11		e-irrad. of CH_2F_2, 85° K
$^{13}C^{19}F_3$	C_2F_6 Liq.	2.0031			(F)	405.82				e-irrad., 178° K
$^{13}C^{19}F_3$	Xe	2.0031			(C) (F)	761.5 399.2		0.21 0.01		e-irrad. of CHF_3, 85° K
$C^{19}F_3$	Kr	2.0024	2.0042	(z) 738	(x) 244 (y) 224					uv irrad. of CF_3I, 4° K
		z along C–F bond, x ⊥ to C–F bond in direction of molecular axis.								
$^{13}C^1H_2O^1H$	H_2O Liq.	2.00322			(H on C) (H on O) (C)	49.26 unresolved 132.81				e-irrad. of CH_3OH, 300° K
$^{29}Si^1H_3$	Kr	2.0012	2.0014		(H) (Si)	22 745		0.22		γ irrad. of SiH_4, 4° K
$Si^1H_2{}^{19}F$	SF_6 P	2.0024			(H) (F)	97.0 154.7				e-irrad. of SiH_nF_m deriv., 98° K
$Si^1H^{19}F_2$	SF_6 P	2.0013			(H) (F)	251.8 217.9				e-irrad. of SiH_nF_m deriv., 98° K

Table C-19. Doublet Tetra-Atomic (Plus CH_2OH) π Radicals. (All are Pyramidal Except the Three Indicated) *(Continued)*

Molecule	Matrix	Experimental Data						Spin Density		Comments
		$g_\parallel$	$g_\perp$	$A_\parallel$ (MHz)	$A_\perp$ (MHz)	A_{iso} (MHz)	A_{dip} (MHz)	a_s^2	$a_{p,d}^2$	
$^{29}Si^{19}F_3$	SF_6		2.0003		(Si)	1394.2		0.31		e-irrad. of SiH_nF_m deriv., 98°K
	P				(F)	382.4				
	SiF_4		2.004	(Si) 1575.2	1414.8	1468.1	53.6	0.43	0.38	γ irrad., 77°K
	S.C.			(F) 603.6	298.7	376.1	125.9	0.01		
					226.1		24.1			
$^{29}Si^{35}Cl_3$	$SiCl_4$		2.0035(10)		(Si)	1234				γ irrad., 180°K
	P		near tetrahedral		(Cl)	37.6				
$^{73}Ge^1H_3$	Kr	2.003	2.017		(H)	42(6)				γ irrad. of GeH_4 (also in Xe), 4°K
					(Ge)	211(8)		0.14		
$^{119}Sn^1H_3$	Kr	2.003	2.025		(H)	73(12)				γ irrad. of SnH_4 (also in Xe), 4°K
					(Sn)	(1073)		0.12		
$^{11}B^1H_3^-$ (planar)	$(CH_3)_4NBH_4$		2.0023(2)		(B)	58.0(3)		0.02		γ irrad., 300°K
	P				(H)	−42.3(3)		0.03		
$^{11}B^1H_3^-$	H_2O		2.0015	(B) 101	42	62	20	0.02	0.31	γ irrad. of BH_4^-, 77°K
	P			(H) −42	−39	−40	1	0.03		
$^{27}Al^1H_3^-$	$LiAlH_4$ or $NaAlH_4$		2.001	(Al) 714(14)	588(14)	630	42	0.23	~0.7	γ irrad., 77°K, 15°K
	P			(H)		<56				
$^{14}N^1H_3^+$	NH_4ClO_4	2.0027	2.0030	(H) −78.5	−70.1	−72.4		0.05		γ irrad., 77°K
	S.C.				−68.7					
(planar)				(N) 101.7	35.6	54.6		0.03		(77 ≤ T ≤ 300°K)
					26.4					
		Rapid rotation about 3-fold axis, also some axial motion.								
		Magnetic parameters are temperature-dependent owing to radical motion.								
				(H) −83	−77					at 4°K
				(N) 180	71					
$^{11}B^{19}F_3^-$	TMS		2.0021(3)		(B)	429(3)		0.21		γ irrad. of BF_3, 77–105°K
					(F)	499(3)				
$^{14}N^{19}F_3^+$	NF_4^+ Salts	2.003	2.009	(N) 322.4	253	276	23	0.18	0.505	γ irrad., 77°K
	P			(F) 841.0	281	468	187			
$^{19}F_2{}^{14}N^{17}O$	SF_6		2.00592(7)		(N)	262.9		0.15		γ irrad., of NO, 110°K
					(F)	402.3				
					(O)	39.6				

Table C-19. *(Continued)*

Molecule	Matrix	Experimental Data				A_{iso} (MHz)	A_{dip} (MHz)	Spin Density		Comments
		$g_{\parallel}$	$g_{\perp}$	$A_{\parallel}$ (MHz)	$A_{\perp}$ (MHz)			a_s^2	$a_{p,d}^2$	
$^1H_2{}^{14}NO$	H_2O Liq.		2.0057(1)		(N)	35.9				photolysis of $NH_2OH + S_2O_8{}^{2-}$, 300° K
					(H)	−35.9				
			low inversion barrier							
$^1H_2{}^{13}CO^-$	H_2O Liq.		2.00367		(C)	105.7				e-irrad. of CH_3OH, 300° K
					(H)	−40				

CH_3: Jen, Foner, Cochran, and Bowers, *PR 112*, 1169 (1958);
McConnell, *JCP 29*, 1422 (1958);
Cole, Pritchard, Davidson, and McConnell, *MP 1*, 406 (1958);
Karplus, *JCP 30*, 15 (1959);
Symons, *Ann. Rept. Prog. Chem. 57*, 68 (1960);
Fessenden and Schuler, *JCP 43*. 2704 (1965);
Morehouse, Christiansen, and Gordy, *JCP 45*, 1751 (1966);
Turkevich and Fujita, *S 152*, 1619 (1966);
Rogers and Kispert, *JCP 46*, 221 (1967);
Jackel, Christiansen, and Gordy, *JCP 47*, 4274 (1967);
Fessenden, *JPC 71*, 74 (1967);
Jackel and Gordy, *PR 176*, 443 (1968);
Janecka, Vyas, and Fujimoto, *JCP 54*, 3229 (1971);
Morton and Preston, *LB*, page 32.

CH_2F, CHF_2: Fessenden and Schuler, *JCP 43*, 2704 (1965).

CF_3: Adrian, Cochran, and Bowers, *Free Radicals in Inorganic Chemistry*, page 50, (American Chemical Society, Washington, D.C., 1962);
Fessenden and Schuler, *JCP 43*, 2704 (1965);
Jackel, Christiansen, and Gordy, *JCP 47*, 4274 (1967);
Rogers and Kispert, *JCP 46*, 3193 (1967);
Maruani, McDowell, Nakajima, and Raghunathan, *MP 14*, 349 (1968) [see, however, Hasegawa, Sogabe and Miura, *MP 30*, 1889 (1975)];
Fessenden, *JMR 1*, 277 (1969).

CH_2OH: Dixon and Norman, *JCS* 3119 (1963);
Livingston and Zeldes, *JCP 44*. 1245 (1966);
Fessenden, *JPC 71*, 74 (1967);
Dobbs, Gilbert, and Norman, *JCS* A124 (1971);
Laroff and Fessenden, *JCP 57*, 5614 (1972).

SiH_3: Cochran, *4th Int. Symp. Free Radical Stabilization* G-I-1. (Washington, D.C., 1959);
Morehouse, Christiansen, and Gordy, *JCP 45*, 1751 (1966);
Jackel and Gordy, *PR 176*, 443 (1968);
Wong, Hutchinson, and Wan, *SL 6*, 665 (1973);
Katsu, Yatsurugi, Sato, and Fujita, *CL 4*, 343 (1975);
Raghunathan and Shimokoshi, *SA 36A* 285 (1980).

SiH_2F, $SiHF_2$: Merritt and Fessenden, *JCP 56*, 2353 (1972).

SiF_3: Merritt and Fessenden, *JCP 56*, 2353 (1972);
Hasegawa, Sogabe, and Miura, *MP 30*, 1889 (1975).

$SiCl_3$: Leray and Roncin, *JCP 42*, 800 (1965);
Roncin, *MC 3*, 117 (1967);
Hesse, Leray, and Roncin, *JCP 57*, 749 (1972).

GeH_3: Morehouse, Christiansen, and Gordy, *JCP 45*, 1751 (1966);
Jackel, Christiansen, and Gordy, *JCP 47*, 4274 (1967);
Jackel and Gordy, *PR 176*, 443 (1968).

SnH_3: Morehouse, Christiansen, and Gordy, *JCP 45*, 1751 (1966);
Jackel and Gordy, *PR 176*, 443 (1968).

$BH_3{}^-$: Catton, Symons, and Wardale, *JCS* A2622 (1969);
Sprague and Williams, *MP 20*, 375 (1971);
Symons, *IJRPC 8*, 381 (1976).

$AlH_3{}^-$: Catton and Symons, *JCS* A2001 (1969).

$NH_3{}^+$: Hyde and Freeman, *JPC 65*, 1636 (1961);
Cole, *JCP 35*, 1169 (1961);
Fujimoto and Morton *CJC 43*, 1012 (1965);
Janecka and Fujimoto, *JMR 4*, 47 (1971);
Rao and Symons, *JCS* A2163 (1971);
Morton and Preston, *LB*, page 65.

$BF_3{}^-$: Patten, *CPL 18*, 112 (1973) (see, however, Hudson and Williams);
Hudson and Williams, *JCP 65*, 3381 (1976).

$NF_3{}^+$: Mishra, Symons, Christe, Wilson, and Wagner, *IC 14*, 1103 (1975);
Hudson and Williams, *JCP 65*, 3381 (1976).

F_2NO: Nishikida and Williams, *JACS 97*, 7166 (1975);
Vanderkooi, Mackenzie, and Fox, *JFC 7*, 415 (1976);
Morton and Preston, *LB*, page 75.

H_2NO: Gutch and Waters, *JCS* 751 (1965);
Adams, Nicksie, and Thomas, *JCP 45*, 654 (1966).
Chawla and Fesssenden, *JPC 79*, 2693 (1975);
Ellinger, Subra, Rassat, Douady and Berthier, *JACS 97*, 476 (1975);
Briere, Claxton, Ellinger, Rey, and Laugier, *JACS 104*, 34 (1982).

H_2CO^-: Eiben and Fessenden, *JPC 75*, 1186 (1971);
Laroff and Fessenden, *JCP 57*, 5614 (1972); *JPC 77*, 1283 (1973);
Ellinger, Subra, Rassat, Douady, and Berthier, *JACS 97*, 476 (1975).

Table C-20. Doublet Tetra-Atomic Pyramidal Molecules (ϕ = O–X–O Angle)

Molecule	Matrix	Experimental Data g_x, g_y, g_z	A_x (MHz)	A_y (MHz)	A_z (MHz)	A_{iso} (MHz)	T_x (MHz)	T_y (MHz)	T_z (MHz)	Spin Density a_s^2	a_p^2	a^2	Comments
$^{14}N^{17}O_3^{2-}$	KN_3 S.C.	2.0060(3) 2.0060(2) 2.0018(3)	(N) 88.7(6) (O) 12.4(6)	88.7(6) 12.4(6)	170.3(6) 92.5(6) φ = 114°	116 39	−27 −27	−27 −27	54 53	0.08 0.02/3	0.57 0.48/3		γ irrad. of KNO_3 doped, 90°, 300°K
$^{31}P^{17}O_3^{2-}$	a S.C.	2.001 2.001 1.999	(P) 1538 (O) 39.8	1540 39.8	1958 92.9 φ = 110°	1680 58	−142 −18	−140 −18	278 35	0.16 0.03/3	0.53 0.32/3		X irrad., 300°K
$^{75}As^{17}O_3^{2-}$	b S.C.	2.0052 2.0051 2.0043	(As) 1586 (O) 11(9)	1574 39(9)	2031 90(11) φ = 110°	1730 47	−154 −36	−154 −8	309 43	0.18 0.03/3	0.59 0.45/3		X irrad., 300°K
$^{33}SO_3^-$	K_2SO_4 S.C.	2.0033 2.0033 2.0023	269	275	314 φ = 111°	359	−34	−34	70	0.13	0.49		X irrad., 300°K
$^{77}SeO_3^-$	K_2SeO_4 S.C.	2.0145 2.0156 2.0031	1210.7	1211.4	1728.1 φ = 112°	1383	−172	−172	345	0.10	0.47		γ irrad., 300°K
TeO_3^-	K_2TeO_4 P	2.0165 2.0165 2.0047											γ irrad., 77°K
$^{35}ClO_3$	$NaClO_3$ S.C.	2.009 2.009 2.002	348	348	515 φ = 112°	404 f_{zz} = −0.38 = field gradient at Cl nucleus	−56	−56	111	0.09	0.40		X irrad., 26°K
$^{79}BrO_3$	KNO_3 S.C.	2.024 2.024 2.004	1625	1625	2294 φ = 111°	1848 f_{zz} = −0.29 = field gradient at Br nucleus	−223	−223	446	0.09	0.35		γ irrad. of $KBrO_3$ doped, 77°K

[a] $Na_2DPO_3 \cdot 5D_2O$

[b] $Na_2HAsO_4 \cdot 7H_2O$

NO_3^{2-}: Symons, *JCS* A1998 (1970);
Reuveni and Luz, *JMR 23*, 265 (1976);
Morton and Preston, *LB*, page 72.

PO_3^{2-}: Horsfield, Morton, and Whiffen, *MP 4*, 475 (1961);
Symons, *JCS*, A1998 (1970);
Schlick, Silver, and Luz, *JCP 52*, 1232 (1970);
Koksal and Yuksel, *ZN 30a*, 1044 (1975);
Morton and Preston, *LB*, page 143.

AsO_3^{2-}: Lin and McDowell, *MP 7*, 223 (1964).
Atkins and Symons [2], page 174;
Garbutt and Gesser, *CJC 48*, 2685 (1970);
Reuveni and Luz, *JMR 16*, 339 (1974);
Morton and Preston, *LB*, page 209.

SO_3^-: Chantry, Horsfield, Morton, Rowlands, and Whiffen, *MP 5*, 233 (1962);
Gromov and Morton, *CJC 44*, 527 (1966);
Suzuki and Abe, *JPSJ 30*, 586 (1971);
Lind and Kewley, *CJC 50*, 43 (1972);
Morton and Preston, *LB*, page 167.

SeO_3^-: Atkins, Symons, and Wardale, *JCS* 5215 (1964);
Aiki, *JPSJ 29*, 379 (1970);
Morton and Preston, *LB*, page 218.

TeO_3^-: Constantinescu, Pascaru, and Constantinescu, *Rev. Roumaine Phys. 12*, 223 (1967);
Morton and Preston, *LB*, page 248.

ClO_3: Eachus and Symons, *JCS* 2433 (1968);
Vinther, *JCP 57*, 183 (1972);
Byberg, *CPL 23*, 414 (1973);
Suryanarayana and Sobhanadri, *MP 28*, 1127 (1974);
Morton and Preston, *LB*, page 192.

BrO_3: Byberg and Kirkegaard, *JCP 60*, 2594 (1974).

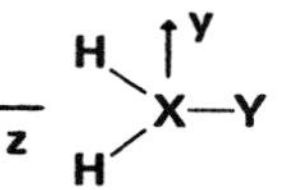

Table C-21. Doublet Tetra-Atomic Planar σ Radicals

Molecule	Matrix	g_x, g_y, g_z		A_x (MHz)	A_y (MHz)	A_z (MHz)	A_{iso} (MHz)	T_x (MHz)	T_y (MHz)	T_z (MHz)	a_s^2	a_p^2	a^2	Comments
		Experimental Data									Spin Density			
$^1H_2C^{14}N$	Ar	–	(N)		96.4									uv irrad. of HI + HCN, 4°K
		2.00	(H)		244.9									
		–												
	KCl	2.0031	(N)				26.6				0.01			γ irrad. of CN^- doped, 77°, 300°K
		(aver.)	(H)				255.7				0.36/2			
							freely rotating							
	KCl	2.0022 (2)	(N)	-9.0	90.2	-9.0	24.1		A_{dip} = 33.1		0.01	0.60		e-irrad. of $K_3Ir(CN)_6$ doped, 77°, 300°, 373°K
	S.C.		(H)		241 to 248		245				0.34/2			
		2.0037 (2)				rotating about C–N bond (z-axis)								
	$HCONH_2$	2.003	(N)	~0	93	~0	31		A_{dip} = 31		0.02	0.64		γ irrad., 77°K
			(H)				255				0.36/2			
	P													
$^1H_2{}^{13}C_2^-$	MgO	(2.002)	(H)				163				0.24/2			C_2H_4 + O^- center, 77°K
	surf.		(C)	~0		~0								
			(C)	42	216	42	100		A_{dip} = 58		0.03	0.63		
$^1H_2CO^+$	H_2SO_4 sol'n	2.000		260	241	260	254				0.36/2			γ irrad. of H_2CO, 77°K, (see Vugman et. al. on H_2CN)
		2.025				rotating about C–O bond (z-axis)?								
$^1H_2{}^{11}BO$	Ar	(2.002)	(B)		87 (3)									B vapor (+H_2O), 4°K
			(H)		372 (1)						0.52/2			
		2.0082 (6)				rotating about B–O bond (z-axis)?								

H_2CN: Cochran, Adrian, and Bowers, *JCP 36*, 1936 (1962);
Brivati, Root, Symons, and Tinling, *JCS* A1942 (1969);
Rao and Symons, *JCS* A2163 (1971);
Banks and Gordy, *MP 26*, 1555 (1973);
Symons, *T 29*, 615 (1973);
Vugman, Elia, and Muniz, *MP 30*, 1813 (1975).

$H_2C_2^-$: Ben Taant, Symons, and Tench, *JCS/FTI*, 1149 (1977).

H_2CO^+: Mishra and Symons, *CC*, 909 (1975).

H_2BO: Graham and Weltner, *JCP 65*, 1516 (1976).

Table C-22. Doublet Tetra-Atomic σ Radicals (C_s Symmetry, Nonplanar) ($y \perp$ to Plane, Therefore Coincident Axis of g and A)

		Experimental Data									Spin Density			
Molecule	Matrix	g_x, g_y, g_z		A_x (MHz)	A_y (MHz)	A_z (MHz)	A_{iso} (MHz)	T_x (MHz)	T_y (MHz)	T_z (MHz)	a_s^2	a_p^2	a^2	Comments
$^1H^{14}NO_2^-$	KCl	2.0020	(N)	109.8	33.7	33.7	59	51	−25	−25	0.03	0.54		uv irrad. of OH^- and
	S.C.	2.0074	(H)	−14.0	−35.1	36.5	−13	−1	−22	50	0.01			CN^- doped, 4°K
		2.0074					x axis (unpaired orbital) ≅ 100° from any bond							
$^1H^{31}PO_2^-$	NH_4^+	2.0019 (8)	(P)	1698 (5)	1228 (5)	1228 (5)	1385	313	−157	−157	0.14	0.55		γ irrad., 300°K
	$H_2PO_2^-$	2.0035 (8)	(H)	238 (3)	227 (3)	224 (3)	231	7	−4	−7	0.16			
	S.C.	2.0037 (8)												
$^1HSO_2$	Ar	2.0028 (2)		311 (2)	306 (3)	325 (5)	314	−3	−8	+11	0.22			uv irrad. of SO_2 + HI,
		2.0068 (5)												4°K
		2.0086 (5)												
$^{19}F^{31}PO_2^-$	KPO_2F_2	1.9920	(P)	2250	~1740	~1770	1920	330	−180	−150	0.19	0.57		γ irrad., 77°, 300°K
	P	2.0095	(F)	493	582	340	472	21	110	−132	0.01			
		2.0062												
$^{19}FSO_2$	Ar	2.0014		239	199	750	396	−157	−197	354	0.01			uv irrad. of F_2SO_2 or
		2.0016												of SO_2 + CF_3OF,
		2.0039												4°K

HNO_2^-: Adrian, Cochran, and Bowers, *JCP 51*, 1018 (1969);
Symons and Zimmerman, *JCS/FTI 72*, 409 (1976).
HPO_2^-: Morton, *MP 5*, 217 (1962);
Köksal and Yüksel, *ZN 30a*, 1044 (1975).
HSO_2: McDowell, Herring, and Tait, *JCP 63*, 3278 (1975).
FPO_2^-: Begum, Subramanian, and Symons, *JCS* A700 (1971).
FSO_2: Morton and Preston, *JCP 58*, 2657 (1973); *JMR 41*, 127 (1980);
McDowell, Herring, and Tait, *JCP 63*, 3278 (1975).

Table C-23. Doublet Tetra-Atomic Molecules

Molecule	Matrix	Experimental Data								Spin Density			Comments
		g_x, g_y, g_z	A_x (MHz)	A_y (MHz)	A_z (MHz)	A_{iso} (MHz)	T_x (MHz)	T_y (MHz)	T_z (MHz)	a_s^2	a_p^2	a^2	
$^{14}N_4^-$	KN_3	2.0016 (4)	25.8	−7.9	−7.8	3.4	22.4	−11.2	−11.2	0.01/4	0.80/4		γ irrad., 77° K
	S.C.	2.0051 (4)				x axis ⊥ to distorted square planar structure							
		1.9876 (4)											
$^{13}CO_3^-$	$CaCO_3$	2.0031	312.12	312.12	479.60	368	−56	−56	112	0.10	0.52		X, γ irrad., 77° K
	S.C.	2.0031											
		2.0013				C_{3v} symmetry, rotation about C_3 axis down to 4° K							
$C^{17}O_3^-$	$KHCO_3$	2.0065	(O′) 63.08	240.10	40.62	45.6	−108.6	195.15	−86.39	0.01	0.64		γ irrad., 100° K
	S.C.	2.0083	(O″) 20.02	136.38	32.65	28.0	−47.9	108.74	−60.75	0.01	0.39		
		2.0198				~C_{2v} sym., third unique O carries minor spin density. Crystal-field effects may lead to distortion from Y shape.							
$^{14}NO_3$	$NaNO_3$	2.0221 (4)	9.93	9.93	11.52	10.5	−0.5	−0.5	1.0	0.01	0.01		e-irrad., 20° K
	S.C.	2.0221 (4)											
		2.0025 (4)				Parameters very dependent upon irradiated salt							
$N^{17}O_3$	$NaNO_3$	2.0231	8	8	110.0	42	−4	−4	68	0.02/3	0.60/3		γ irrad., 77° K
	S.C.	2.0231				Rotating about C_3 axis down to 10° K or C_{2v} symmetry.							
		2.0047											

N_4^-: Shuskus, Young, Gilliam, and Levy, *JCP 33*, 622 (1960);
Horst, Anderson, and Milligan, *JPCS 23*, 157 (1962);
Symons, *JCS* 570 (1963);
Neilson and Symons, *JCS/FTII 68*, 1772 (1972);
Guha, Bogan, and Gillian, *PR B7*, 4047 (1973).

CO_3^-: Chantry, Horsfield, Morton, and Whiffen, *MP 5*, 589 (1962);
Serway and Marshall, *JCP 46*, 1949 (1967);
Olsen and Burnelle, *JACS 92*, 3659 (1970);
Holmberg, *JCP 55*, 1730 (1971);
Top, Raziel, Luz, and Silver, *JMR 12*, 102 (1973);
Marshall, McMillan, and Nistor, *JMR 14*, 20 (1974);
Morton and Preston, *LB*, page 42.

NO_3: Livingston and Zeldes, *JCP 41*, 4011 (1964);
Zdansky and Sroubek, *PSS 12*, K85 (1965);
Adde, *CR C264*, 1905 (1967);
Atkins and Symons [2], page 166;
Reuveni and Luz, *JMR 23*, 271 (1976);
Morton and Preston, *LB*, page 70.

Table C-24. Doublet Tetra-Atomic Transition Metal Molecules

Molecule	Matrix	Experimental Data					A_{iso} (MHz)	A_{dip} (MHz)	Spin Density		Comments
		$g_{\parallel}$	$g_{\perp}$		$A_{\parallel}$ (MHz)	$A_{\perp}$ (MHz)			a_s^2	$a_{p,d}^2$	
$^{47,49}Ti^{19}F_3$ ($^2A_1'$)	Ne	1.9912(2)	1.8910(2)	(Ti)	197.9(2)	178.2(2)	185	7	0.24	0.33	TiF_3 vapor, 4° K (also in Ar)
				(F)	47.9(2)	11.5(2)	24	12	0.00	0.00	
					probably planar (D_{3h})						
$^{55}MnO_3$ ($^2A_1'$)	Ne	2.0036(8)	2.0084(8)		1772(3)	1532(3)	1613	81	0.32	0.46	Mn vapor + O_3, 4° K (also in Ar)
					probably planar (D_{3h})						

TiF_3: DeVore and Weltner, *JACS 99*, 4700 (1977).
MnO_3: Ferrante, Wilkerson, Graham, and Weltner, *JCP 67*, 5904 (1977).

Table C-25. Triplet Tetra-Atomic Molecules

Molecule	Matrix	Experimental Data					A_{iso} (MHz)	A_{dip} (MHz)	Spin Density		Comments
		$g_\parallel$	$g_\perp$	$A_\parallel$ (MHz)	$A_\perp$ (MHz)	D (cm^{-1})			a_s^2	a^2	
HCCN	PE[a]	2.189	2.013			0.849	($E = 0 \pm 0.003$)				uv irrad. of diazo compd., 77° K
$^{13}C_4$	Ar		2.0040(5)	(mid. C) (out. C)	<4 29.6(5)	0.2253(3)					graphite vapor, 4° K

[a] Polychlorotrifluoroethylene

HCCN: Bernheim, Kempf, Gramas, and Skell, *JCP 43*, 196 (1965);
Bernheim, Kempf, and Reichenbecher, *JMR 3*, 5 (1970).
C_4: Graham, Dismuke, and Weltner, *AJ 204*, 301 (1976).

Table C-26. High-Spin ($S = \frac{3}{2}$ to $\frac{7}{2}$) Tetra-Atomic Molecules

		Experimental Data							Spin Density		
Molecule	Matrix	$g_{\parallel}$	$g_{\perp}$	$A_{\parallel}$ (MHz)	$A_{\perp}$ (MHz)	D (cm^{-1})	A_{iso} (MHz)	A_{dip} (MHz)	a_s^2	a^2	Comments
Quartet ($S = \frac{3}{2}$):											
$Cr^{19}F_3$ (^4A)	Ar		(2.002)		35	$\geqslant$0.59					
Sextet ($S = \frac{5}{2}$):											
$Fe^{19}F_3$ ($^4A_1{}'$)	Ar		(2.002)		73	$\geqslant$0.84					vapor of solid salt, 4° K
Octet ($S = \frac{7}{2}$):											
Gd^1H_3 ($^8A_1{}'$?)	Ar				<30	0.7 to 1.3					Gd vapor + H atoms, 4° K
$Gd^{19}F_3$ ($^8A_1$)	Ne	1.995	1.990		<30	0.43					vapor of solid salt, 4° K
$GdCl_3$	Ar		(1.99)			0.36?					Gd vapor + Cl_2 + uv rad., 4° K

CrF_3, FeF_3: DeVore, Van Zee, and Weltner, *Proc. Symp. High Temp. Metal Halide Chem.*, Vol. 78-1, page 187 (Hildenbrand and Cubicciotti, eds., Electrochem. Soc., Princeton, N.J. 1978).

GdH_3, GdF_3, $GdCl_3$: Baumann, Van Zee, Zeringue, and Weltner, *JCP* 75, 5291 (1981).

Appendix D

Magnetic Resonance Parameters for Gas-Phase Molecules

These tables include all molecules studied by magnetic resonance techniques in the gas phase as of June 1982. The symbolism is that used in Chapter IV. The *g* values of a few molecules and their analyses are also discussed in that chapter.

The abbreviations for the titles of journals are the same as those listed and used in Appendix C.

Table D-1. Gas-Phase Parameters of Doublet ($S = \frac{1}{2}$) Diatomic Molecules

Molecule	Electr. State	Rot. Const. B_e (cm^{-1})	Spin-Orb. A (cm^{-1})	Magnetic Hyperfine Parameters (MHz)					eQq (MHz)	Spin-Rot. Const. γ_0 (10^{-3} cm^{-1})	Dipole Moment μ_e (D)
				a	b	c	d	h^*			
O^1H	$X^2\Pi_i$	18.9108	−139.21	86.01	−118.08	132.12	56.62				1.6676
S^1H	$X^2\Pi_i$	9.4611	−376.96	32.58	−63.44	32.44	27.36				0.7580
SeH	$X^2\Pi_i$	7.77	−1815								0.49
Te^1H	$X^2\Pi_i$	5.56	−2250								
$S^{19}F$	$X^2\Pi_i$	0.552174	−387					428.60			0.794
$Se^{19}F$	$X^2\Pi_i$	0.3625	−560					325.6			1.52
^{19}FO	$X^2\Pi_i$	1.05956	−177					696			
^{35}ClO	$X^2\Pi_i$	0.62347	−321.8	136.230	47.2	−94.7	173.021		−88.05		1.239
	$C^2\Sigma$	0.695206								−12.9	
^{79}BrO	$X^2\Pi_i$	0.429598	−900					504.68	650.6		1.765
^{127}IO	$X^2\Pi_i$	0.34026	−446					583.92	−1891.06		2.45
^7LiO	$X^2\Pi_i$		−112.0		−29.58			3.950	0.4439		6.84
^{14}NO	$X^2\Pi_r$	1.67195	123.16	84.195	41.79	−58.66	112.600		−1.852		0.15872
^{14}NS	$X^2\Pi_r$	0.769602	222.94	61.90			87.03	56.35	−2.62		1.81
C^1H	$X^2\Pi_r$	14.457	27.95	52	−74	52	43.6				
$C^{19}F$	$X^2\Pi_r$	1.4172	77.12		190			662.9			0.65
$C^{14}N$	$X^2\Sigma^+$	1.89974			−33.966	60.329			−1.307	7.25455	1.45
	$A^2\Pi_i$	1.7151	−52.64	64.2	457.5	103.2	−90		1.5		0.56
	$B^2\Sigma^+$	1.973			470	55			−5	15.65	−1.15

Table D-1. *(Continued)*

Molecule	Electr. State	Rot. Const. B_e (cm^{-1})	Spin-Orb. A (cm^{-1})	Magnetic Hyperfine Parameters (MHz) a	b	c	d	h*	eQq (MHz)	Spin-Rot. Const. γ_0 (10^{-3} cm^{-1})	Dipole Moment μ_e (D)
KAr	$X^2\Sigma$									($\bar{\gamma}$) 0.24	
$^1H_2^+$	$X^2\Sigma_g$	30.21			804.065	98.034				1.0886	
CO^+	$X^2\Sigma^+$	1.97720			1480.8	92.1				9.1050	
	$A^2\Pi_i$	1.58940	−117.5					67.8			
$H^{79}Br^+$	$X^2\Pi_i$							1460			
$H^{35}Cl^+$	$X^2\Pi_i$	9.78804 (B_0)	−643					365	8	−297	
$Ca^{19}F$	$X^2\Sigma^+$	0.34370			108.476	40.647				1.3175	
$Ca^{79}Br$	$X^2\Sigma^+$	0.094466			95.329	77.62			20.02	3.0262	
	$B^2\Sigma^+$	0.096515			7				<3	−69	
CaI	$X^2\Sigma^+$	0.06933			116					1.87	
	$B^2\Sigma^+$	0.07179			18				−66	−49.7	
$Sr^{19}F$	$X^2\Sigma^+$	0.2505			97.083	30.27				2.4949	

*$h = a + (b + c)/2$

OH: Radford, *PR 122*, 114 (1961); *PR 126*, 1035 (1962); Chung and Levy, *AJ 162*, L161 (1970); Carrington and Lucas, *PRS A314*, 567 (1970); Clough, Curran, and Thrush, *PRS A323*, 541 (1971); Lee, Tam, Larouche, and Woonton, *CJP 49*, 2207 (1971); Hinkley, Walker, and Richards, *PRS A331*, 553 (1973); Lee and Tam, *CP 4*, 434 (1974); Veseth, *JMS 63*, 180 (1976); Brown, Kaise, Kerr, and Milton, *MP 36*, 553 (1978).

SH: Radford and Linzer, *PRL 10*, 443 (1963); McDonald, *JCP 39*, 2587 (1963); Uehara and Morino, *JMS 36*, 158 (1970); Miller, *JCP 54*, 1658 (1971); Byfleet, Carrington, and Russell, *MP 20*, 271 (1971); Brown and Thistlethwaite, *MP 23*, 635 (1972); Tanimoto and Uehara, *MP 25*, 1193 (1973); Meerts and Dymanus, *AJ 187*, L45 (1974); *CJP 53*, 2123 (1975); Veseth, *JMS 63*, 180 (1976).

Table D-1. Gas-Phase Parameters of Doublet ($S = \frac{1}{2}$) Diatomic Molecules *(Continued)*

SeH:	Radford, *JCP 40*, 2732 (1964); Carrington, Currie, and Lucas, *PRS A315*, 355 (1970).
TeH:	Radford, *JCP 40*, 2732 (1964).
SF, SeF:	Carrington, Currie, Miller, and Levy, *JCP 50*, 2726 (1969); Byfleet, Carrington, and Russell, *MP 20*, 271 (1971); Brown, Blyfleet, Howard, and Russell, *MP 23*, 457 (1972); Amano and Hirota, *JMS 45*, 417 (1973).
FO:	McKellar, *CJP 57*, 2106 (1979).
ClO:	Carrington, Dyer, and Levy, *JCP 47*, 1756 (1967); Carrington, Levy, and Miller, *JCP 47*, 3801 (1967); Amano, Saito, Hirota, Morino, Johnson, and Powell, *JMS 30*, 275 (1969); Uehara, Tanimoto, and Morino, *MP 22*, 799 (1971); Amano and Hirota, *JMS 66*, 185 (1977); Kakar, Cohen, and Geller, *JMS 70*, 243 (1978); Coxon, *CJP 57*, 1538 (1979).
BrO, IO:	Powell and Johnson, *JCP 50*, 4596 (1969). Carrington, Dyer, and Levy, *JCP 52*, 309 (1970); Byfleet, Carrington, and Russell, *MP 20*, 271 (1971); Amano, Yoshimaga, and Hirota, *JMS 44*, 594 (1972); Saito, *JMS 48*, 530 (1973).
LiO:	Freund, Herbst, Mariella, and Klemperer, *JCP 56*, 1467 (1972); Veseth, *JMS 59*, 51 (1976).
NO:	Brown and Radford, *PR 147*, 6 (1966); Neumann, *AJ 161*, 779 (1970); Meerts and Dymanus, *JMS 44*, 320 (1972); Ashford, Jarke, and Solomon, *JCP 57*, 3867 (1972); *64*, 3097 (1976); Meerts, *CP 14*, 421 (1976); Meerts and Veseth, *JMS 82*, 202 (1980).
NS:	Carrington and Levy, *JCP 44*, 1298 (1966); Carrington, Howard, Levy, and Robertson, *MP 15*, 187 (1968); Uehara and Morino, *MP 17*, 239 (1969); Amano, Saito, Hirota, and Morino, *JMS 32*, 97 (1969); O'Hare, *JCP 52*, 2992 (1970); Byfleet, Carrington, and Russell, *MP 20*, 271 (1971).
CH:	Evenson, Radford, and Moran, *APL 18*, 426 (1971); Hougen, Mucha, Jennings, and Evenson, *JMS 72*, 463 (1978).
CF:	Carrington and Howard, *MP 18*, 225 (1970).
CN:	Evenson, Dunn, and Broida, *PR 136*, A1566 (1964); Radford, *PR 136* A1571 (1964); Thomson and Dalby, *CJP 46*, 2815 (1968); Meakin and Harris, *JMS 44*, 219 (1972); Cook and Levy, *JCP 58*, 3547 (1973); *59*, 2387 (1973); Penzias, Wilson, and Jefferts, *PRL 32*, 701 (1974); Turner and Gammon, *AJ 198*, 71 (1975); Dixon and Woods, *JCP 67*, 3956 (1977).
KAr:	Mattison, Pritchard, and Kleppner, *PRL*, *32*, 507 (1974).
H_2^+:	Jefferts, *PRL 23*, 1476 (1969).
CO^+:	Dixon and Woods, *PRL 34*, 61 (1975); Gagnaire and Goure, *CJP 54*, 2111 (1976); Carrington, Milverton, and Sarre, *MP 35*, 1505 (1978).
HBr^+:	Saykally and Evenson, *PRL 43*, 515 (1979).
HCl^+:	Brown and Watson, *JMS 65*, 65 (1977); Brown, Colbourn, Watson, and Wayne, *JMS 74*, 294 (1979); Ray, Lubic, and Saykally, *MP 46*, 217 (1982).
CaF:	Field, Harris, and Tanaka, *JMS 57*, 107 (1975); Nakagawa, Domaille, Steimle, and Harris, *JMS 70*, 374 (1978); Bernath, Cummins, and Field, *CPL 70*, 618 (1980); Childs and Goodman, *PRL 44*, 316 (1980); Childs and Goodman, *PR A21*, 1216 (1980); Dulick, Bernath, and Field, *CJP 58*, 703 (1980); Childs, Goodman, and Goodman, *JMS 86*, 365 (1981).
CaBr:	Bernath, Field, Pinchemel, Lefebvre, and Schamps, *JMS 88*, 175 (1981); Bernath, Pinchemel, and Field, *JCP 74*, 5508 (1981); Bernath, Pinchemel, Field, Moller, and Torring, *JMS 88*, 420 (1981); Childs, Cok, Goodman, and Goodman, *JCP 75*, 501 (1981).
CaI:	Bernath, Pinchemel, and Field, *JCP 74*, 5508 (1981); Reisner, Bernath, and Field, *JMS 89*, 107 (1981).
SrF:	Brown, Milton, and Steimle, *FDCS* No. 71, 151 (1981); Childs, Goodman, and Renhorn, *JMS 87*, 522 (1981).

Table D-2. Gas-Phase Parameters for Triplet ($S = 1$) Diatomic Molecules (and Some $^1\Delta$ Excited States)

Molecule	Electr. State	Rot. Const. B_e (cm^{-1})	Spin-Orb. A (cm^{-1})	Magn. Hyper. Parameters (MHz) a	b	c	d	eQq (MHz)	Spin-Rot. Const. γ_0 (10^{-3} cm^{-1})	Spin-Spin Const. λ_0 (cm^{-1})	Dipole Moment μ_e (D)	Comments
$^{16}O_2$	$X^3\Sigma_g^-$	1.43768							−8.42536	1.9847511		
$^{17}O^{16}O$	$a^1\Delta_g$				−100.92	139.3		96	−8.17989	1.984889		
SO	$X^3\Sigma^-$	0.7208171			51.04^a	-96.65^a		-15.40^a	−5.6153	5.2787981	1.55	^{a} ^{33}S
	$a^1\Delta$			145.1^a							1.32	
S_2	$X^3\Sigma_g^-$	0.2955							−8.57	11.774		
SeO	$X^3\Sigma^-$	0.4655			$h = \pm 360$(^{77}Se)				−30	86.4		
	$a^1\Delta$	0.461									2.01	
$^{14}N_2$	$A^3\Sigma_u^+$	1.4546			25.805	−37.980		−1.8	−2.518	−1.285		
$^{14}N^1H$	$X^3\Sigma^-$	16.6668			(N) 42.8	−68			−54.66	0.9198		
					(H) −100	89						
$^{31}P^1H$	$X^3\Sigma^-$	8.5377			(P) 277	−456			−76.48	2.207		
					(H) −54	15						
$^{14}N^{19}F$	$a^1\Delta$	1.2225		109(^{14}N)							0.37	
				753(^{19}F)								
^{13}CO	$a^3\Pi_r$	1.69124	41.41	162.5	1284.8	15.8	213.0				1.374	
$^1H_2$ (ortho)	$c^3\Pi_u$	31.07	−0.1248	26.6	450.5^b	$c - 3d = 104.177$			0.80	0.318970^c		para also
	$d^2\Pi_u$	30.364	−0.02719	6.7	443.4^b	$c - 3d = 73.0$			0.87	0.07475^c		para also

$^b a_F = b + c/3$
$^c B_0 - \sqrt{6}\, B_2$

O_2: Falick, Mahan, and Myers, *JCP 42*, 1837 (1965);
Arrington, Falick, and Myers, *JCP 55*, 909 (1971);
Miller, *JCP 54*, 330 (1971);
Mizushima, Wells, Evenson, and Welch, *PRL 29*, 831 (1972);
Welch and Mizushima, *PR A5*, 2692 (1972);
Evenson and Mizushima, *PR A6*, 2197 (1972);
Gerber, *HPA 45*, 655 (1972);
Johns and Lepard, *JMS 55*, 374 (1975);
Steinbach and Gordy, *PR A8*, 1753; *A11*, 729 (1975);

Table D-2. Gas-Phase Parameters for Triplet ($S = 1$) Diatomic Molecules (and Some $^1\Delta$ Excited States) *(Continued)*

	Tomuta, Mizushima, Howard, and Evenson, *PR A12*, 974 (1975);
	Christensen and Veseth, *JMS 72*, 438 (1978);
	Scalabrin, Saykally, Evenson, Radford, and Mizushima, *JMS 89*, 344 (1981).
SO:	Daniels and Dorain, *JCP 45*, 26 (1966);
	Carrington, Levy, and Miller, *PRS A293*, 108 (1966);
	Carrington, Levy, and Miller, *MP 13*, 401 (1967); *PRS A298*, 340 (1967);
	Uehara, *MP 21*, 407 (1971);
	Brown and Uehara, *MP 24*, 1169 (1972);
	Davies, Wayne, and Stone, *MP 28*, 1409 (1974);
	Clark and DeLucia, *JMS 60*, 332 (1976);
	Christensen and Veseth, *JMS 72*, 438 (1978);
	Kawaguchi, Yamada, and Hirota, *JCP 71*, 3338 (1979).
S_2:	Channappa, Pendlebury, and Smith, in *La Structure Hyperfine Magnétique des Atomes et des Molécules*, page 73 (Moser and Lefebvre, eds., éditions du CNRS, Paris, 1967);
	Wayne, Davies, and Thrush, *MP 28*, 989 (1974).
SeO:	Carrington, Currie, Levy, and Miller, *MP 12*, 535 (1969).
N_2:	Miller, *JCP 43*, 1695 (1965);
	Freund, Miller, DeSantis, and Lurio, *JCP 53*, 2290 (1970);
	Bullock and Hause, *JMS 39*, 519 (1971);
	DeSantis, Lurio, Miller, and Freund, *JCP 58*, 4625 (1973).
NH:	Wayne and Radford, *MP 32*, 1407 (1976).
PH:	Davies, Russell, Smith, and Thrush, *CJP 57*, 522 (1979).
NF:	Curran, MacDonald, Stone, and Thrush, *CPL 8*, 451 (1971).
CO:	Freund and Klemperer, *JCP 43*, 2422 (1965);
	Stern, Gammon, Lesk, Freund, and Klemperer, *JCP 52*, 3467 (1970);
	Gammon, Stern, Lesk, Wicke, and Klemperer, *JCP 54*, 2136 (1971);
	Gammon, Stern, and Klemperer, *JCP 54*, 2151 (1971);
	Wicke, Field, and Klemperer, *JCP 56*, 5758 (1972);
	Wicke, Klemperer, and Field, *JCP 62*, 3544 (1975).
H_2:	Lichten, *PR 120*, 848 (1960); *126*, 1020 (1962); *A3*, 594 (1971);
	Chiu, *PR 137*, A384 (1965); *145*, 1 (1966); *159*, 190 (1967);
	Jette and Cahill, *PR 160*, 35 (1967);
	Brooks, Lichten, and Reno, *PR A4*, 2217 (1971);
	Jost, Marechal, and Lombardi, *PR A5*, 740 (1972);
	Miller and Freund, *JCP 56*, 3165 (1972); *58*, 2345 (1973);
	Freund and Miller, *JCP 58*, 3565 (1973); *59*, 4093 (1973);
	Miller, *JCP 59*, 4078 (1973);
	Jette, *JCP 61*, 816 (1974); *CPL 25*, 590 (1974);
	Jette and Miller, *CPL 29*, 547 (1974).

Table D-3. Gas-Phase Parameters of Doublet ($S = \frac{1}{2}$) Linear Triatomic Molecules

Molecule	Electr. State	Rot. Const. B (cm^{-1})	Spin-Orb. A (cm^{-1})	Magnetic Hyperfine Parameters (MHz) a	$b + c$	d	h	eQq (MHz)	Dipole Moment μ_e (D)	Comments
CC^1H	$X^2\Sigma$	1.45683			40.39 + 12.32					$\gamma = -62.62$ MHz
^{14}NCO	$X^2\Pi_{3/2}$	0.38940	−95.86	64.31	−21.20	90.96		−1.85	0.742	$v_2 = 0^b$
	$^2\Delta_{5/2}$	0.39046	(−95.9)	60.47	−17.38			−2.11		$v_2 = 1$; $^2\Phi_{7/2}$ $v_2 = 2$ obs.
^{14}NCS	$X^2\Pi_{3/2}$	0.2036	−319.9				25.4	3.3	2.5	$v_2 = 0^c$; $^2\Delta_{5/2}$ also

[b,c] Renner parameters = −0.146 and −0.16, respectively.

CCH: Tucker, Kutner, and Thaddeus, *AJ 193*, L115 (1974);
Tucker and Kutner, *AJ 222*, 859 (1978);
Wootten, Bozyan, Garrett, Loren, and Snell, *AJ 239*, 844 (1980);
Ziurys, Saykally, Plambeck, and Erickson, *AJ 254*, 94 (1982).

NCO: Dixon, *PTRS A252*, 165 (1960);
Hougen, *JCP 36*, 519 (1962);
Saito and Amano, *JMS 34*, 383 (1970);
Carrington, Fabris, Howard, and Lucas, *MP 20*, 961 (1971);
Amano and Hirota, *JCP 57*, 5608 (1972);
Carrington [5], page 225 ff;
Barnes, Brown, Fackerell, and Sears, *JMS 92*, 485 (1982).

NCS: Dixon and Ramsay, *CJP 46*, 2619 (1968);
Carrington, Fabris, and Lucas, *MP 16*, 195 (1969);
Fabris (1970) Ph.D. thesis, Univ. of Southampton [as quoted by J. M. Brown, *MTP Int. Rev. of Sci., Phys. Chem., Series One, 4*, 235 (1972)].

Table D-4. Gas-Phase Parameters of Doublet ($S = \frac{1}{2}$) Bent Triatomic Molecules

Molecule	Electr. State	Rot. Const. (cm^{-1})	Magn. Hf. Parameters (MHz) a_F	T_{aa}	T_{bb}	T_{cc}	Electr. Quadru. (MHz)	Spin-Rot. Const. (cm^{-1})	Comments
^{1}HCO	$X\,^2A'$	$A = 24.32690$ $B = 1.490057$ $C = 1.402563$	372.2		$T_{bb} - T_{cc} = 19.2$			$aa = 0.387768$ $bb = 0.6278 \times 10^{-3}$ $cc = -0.006872$	
$^{1}HO_2$	$X\,^2A''$	$A = 20.356557$ $B = 1.1179011$ $C = 1.0564528$	−27.48	−8.34	19.68	−11.34		$aa = -1.65352$ $bb = -0.0140955$ $cc = 0.0002876$	spin rot: $\frac{1}{2}\|ab + ba\| =$ 0.647×10^{-5}
^{1}HSO	$X\,^2A''$	$A = 9.989732$ $B = 0.683959$ $C = 0.638239$	−36.37	−11.96 $\|T_{ab}\| = -7.8$	10.44			$aa = -0.34577$ $bb = -0.014232$ $cc = 0.0000334$	spin rot: $\frac{1}{2}\|ab + ba\| =$ 0.01261
^{19}FSO	$X\,^2A''$	$A = 1.290832$ $B = 0.311575$ $C = 0.250342$	67.23	−118.2 $\|T_{ab}\| = 10.3$	−117.1			$aa = -0.0113259$ $bb = -0.0011640$ $cc = 0.0000621$	spin rot: $\frac{1}{2}\|ab + ba\| =$ 0.0069365
$^{14}N^{1}H_2$	$X\,^2B_1$	$A = 23.6932$ $B = 12.9516$ $C = 8.17275$	(^{14}N) 28.5 (^{1}H) −67.4	−44.8 21.4	−43.5 −12.9	88.3 −8.5	$aa = 0.3$ $bb = -1.9$ $cc = 1.6$	$aa = -0.30912$ $bb = -0.04516$ $cc = 0.0004023$	obs. also in $v_2 = 1$ state
PH_2	$X\,^2B_1$	$A = 9.13233$ $B = 8.08349$ $C = 4.21434$						$aa = -0.28107$ $bb = -0.08171$ $cc = 0.00017$	
$^{14}NO_2$	$X\,^2A_1$	$A = 8.00115$ $B = 0.433637$ $C = 0.410398$	147.23	−22.16	39.85	−17.69	$aa = 0.45$ $bb = -1.71$ $cc = 1.26$	$aa = 0.1805304$ $bb = 2.636 \times 10^{-4}$ $cc = -3.1863 \times 10^{-3}$	

Table D-4. *(Continued)*

Molecule	Electr. State	Rot. Const. (cm^{-1})	Magn. Hf. Parameters (MHz) a_F	T_{aa}	T_{bb}	T_{cc}	Electr. Quadru. (MHz)	Spin-Rot. Const. (cm^{-1})	Comments
$^{35}ClO_2$	$X\,^2B_1$	$A = 1.737243$ $B = 0.3319849$ $C = 0.277976$	46.23	−77.72	−83.22	160.94	$aa = -8.61$ $bb = 0.43$ $cc = 8.28$	$aa = -0.0463050$ $bb = -0.0072375$ $cc = 1.536 \times 10^{-4}$	5 vib. levels obs. with excitation of all 3 modes. Equil. rot. constants determined.

HCO: Bowater, Brown, and Carrington, *JCP 54*, 4957 (1971); *PRS A333*, 265 (1973);
Saito, *AJ 178*, L95 (1972);
Bolman, Brown, Carrington, and Lycett, *PRS A335*, 113 (1973);
Austin, Levy, Gottlieb, and Radford, *JCP 60*, 207 (1974);
Cook, Evenson, Howard, and Curl, *JCP 64*, 1381 (1976);
Johns, McKellar, and Riggin, *JCP 67*, 2427 (1977);
Brown, Buttenshaw, Carrington, and Parent, *MP 33*, 589 (1977);
Boland, Brown, Carrington, and Nelson, *PRS A360*, 507 (1978);
Brown, Buttenshaw, Carrington, Dumper, and Parent, *JMS 79*, 47 (1980).

HO_2: Radford, Evenson, and Howard, *JCP 60*, 3178 (1974);
Beers and Howard, *JCP 63*, 4212 (1975);
Hougen, *JMS 54*, 447 (1975);
Hougen, Radford, Evenson, and Howard, *JMS 56*, 210 (1975);
Johns, McKellar, and Riggin, *JCP 68*, 3957 (1978);
Barnes, Brown, Carrington, Pinkstone, Sears, and Thistlethwaite, *JMS 72*, 86 (1978);
McKellar, *JCP 71*, 81 (1979);
Brown and Sears, *JMS 75*, 111 (1979);
Barnes, Brown, and Radford, *JMS 84*, 179 (1980).

HSO: Kakimoto, Saito, and Hirota, *JMS 80*, 334 (1980);
Ohashi, Kakimoto, Saito, and Hirota, *JMS 84*, 204 (1980);
Endo, Saito, and Hirota, *JCP 75*, 4379 (1981).

FSO: Endo, Saito, and Hirota, *JCP 74*, 1568 (1981).

NH_2: Kroll, *JCP 63*, 319 (1975);
Davies, Russell, Thrush, and Wayne, *JCP 62*, 3739 (1975);
Hills, Cook, Curl, and Tittel, *JCP 65*, 823 (1976);
Cook, Hills, and Curl, *AJ 207*, L139 (1976); *JCP 67*, 1450 (1977);
Davies, Russell, Thrush, and Radford, *PRS A353*, 299 (1977);
Brown, Buttenshaw, Carrington, and Parent, *MP 33*, 589 (1977);
Hills, Lowe, Cook, and Curl, *JCP 68*, 4073 (1978);
Hills and McKellar, *JMS 74*, 224 (1979); *JCP 71*, 3330 (1979);
Kawaguchi, Yamada, Hirota, Brown, Buttenshaw, Parent, and Sears, *JMS 81*, 60 (1980).

PH_2: Davies, Russell, and Thrush, *CPL 37*, 43 (1976);
Hills and McKellar, *JCP 71*, 1141 (1979).

NO_2: Bird, Baird, Jache, Hodgeson, Curl, Kunkle, Bransford, Rastrup-Andersen, and Rosenthal, *JCP 40*, 3378 (1964);
Lees, Curl, and Baker, *JCP 45*, 2037 (1966);
Foster, Hodgeson, and Curl, *JCP 45*, 3760 (1966);
Schaarsma, *CPL 1*, 16 (1967);
Curl, Evenson, and Wells, *JCP 56*, 5143 (1972);
Solarz, Levy, Abe, and Curl, *JCP 60*, 1158 (1974);
Freund, Hougen, and Lafferty, *CJP 53*, 1929 (1975);
Schmiedl, Bonilla, Paech, and Demtroder, *JMS 68*, 236 (1977);
Solarz and Levy, *JMS 73*, 374 (1978).

ClO_2: Curl, Kinsey, Baker, Baird, Bird, Heidelberg, Sugden, Jenkins, and Kenney, *PR 121*, 1119 (1961);
Curl, Heidelberg, and Kinsey, *PR 125*, 1993 (1962);
Jones and Brown, *JMS 90*, 222 (1981).

Table D-5. Gas-Phase Parameters of Doublet ($S = \frac{1}{2}$) Polyatomic Molecules

Molecule	Electr. State	Rot. Constants (cm^{-1})	Magn. Hf. Parameters (MHz)				Electr. Quadr. (MHz)	Spin-Rot. Constants (cm^{-1})	Comments
			a_F	T_{aa}	T_{bb}	T_{cc}			
CH_3O	X^2E	$A = 5.3280$ $B = 0.931776$	65					$\gamma = -0.0545$	spin-orbit $A = -142.8$ cm^{-1}
CH_2OH			$(^1H\text{–}C) \cong 16$ G $(^1H\text{–}O) \cong 2$ G						

CH_3O: Radford and Russell, *JCP 66*, 2222 (1977);
Russell and Radford, *JCP 72*, 2750 (1980).
CH_2OH: Radford, Evenson, and Jennings, *CPL 78*, 589 (1981).

Table D-6. Gas-Phase Parameters of Triplet ($S = 1$) Triatomic Molecules

Molecule	Electr. State	Rot. Constants (cm^{-1})	Magn. Hf. Parameters (MHz)				Spin-Spin Constants (cm^{-1})	Spin-Rot. Constants (cm^{-1})
			a_F	T_{aa}	T_{bb}	T_{cc}		
C^1H_2	X^3B_1	$A = 73.0578$ $B = 8.41517$ $C = 7.21927$	(−20.2)	(39)	(−20)		$D = 0.7784$ $E = 0.0399$	$aa = 0.00045$ $bb = -0.00515$ $cc = -0.00411$

CH_2: Mucha, Evenson, Jennings, Ellison, and Howard, *CPL 66*, 244 (1979).
Sears, Bunker, McKellar, Evenson, Jennings, and Brown (preprint).

Appendix E

Fundamental Physical Constants; Units and Conversion Factors

FUNDAMENTAL PHYSICAL CONSTANTS*

Speed of light in vacuum: $c = 2.99792458 \times 10^{10}$ cm sec^{-1}
Planck constant: $h = 6.6260755(40) \times 10^{-27}$ erg sec
$\hbar = h/2\pi = 1.05457266(63) \times 10^{-27}$ erg sec
Boltzmann constant: $k = 1.380658(12) \times 10^{-16}$ erg K^{-1}
Avogadro constant: $N_A = 6.0221367(36) \times 10^{23}$ mol^{-1}
Rydberg constant: $R_\infty = 13.6056981(40)$ eV
Electronic charge: $e = 4.8032068(15) \times 10^{-10}$ esu
Electron rest mass: $m_e = 9.1093897(54) \times 10^{-28}$ g
Proton rest mass: $m_p = 1.6726231(10) \times 10^{-24}$ g
Bohr magneton: $\beta_e = e\hbar/2m_e = 9.2740154(31) \times 10^{-21}$ erg G^{-1}
Electron g-factor: $g_e = 2.002319304386\ (20)$
Electron magnetic moment: $\mu_e = g_e\beta_e/2 = 9.2847701(31) \times 10^{-21}$ erg G^{-1}
Nuclear magneton: $\beta_n = e\hbar/2m_p = 5.0507866(17) \times 10^{-24}$ erg G^{-1}
Proton magnetic moment: $\mu_p = 1.41060761(47) \times 10^{-23}$ erg G^{-1}

*E. R. Cohen and B. N. Taylor, *Physics Today 41* (8) Part 2 (August), 9–13 (1988).

UNITS AND CONVERSION FACTORS

10^7 ergs = 1 joule (J)
1 electron volt (eV) = $1.60217733(49) \times 10^{-19}$ J
1 Tesla (T) = 10^4 gauss (G)
(gauss in cgs units = g$^{1/2}$ cm$^{-1/2}$ sec^{-1})
1 atomic unit (a.u.) of length: 1 bohr = $0.529177249(24) \times 10^{-8}$ cm
1 atomic unit (a.u.) of energy: 1 hartree = 27.2113961(81) eV

Magnetic field, H (gauss), to electron resonant frequency, ν:

$$\nu(\text{MHz}) = (g_e\beta_e/h)\left(\frac{g}{g_e}\right)H = 2.802494\left(\frac{g}{g_e}\right)H(\text{G})$$

$$\nu(\text{cm}^{-1}) = \nu(\text{MHz}) \times 0.3335641 \times 10^{-4}$$

g-factor:

$$g = h\nu/\beta_e H = 0.7144775\ \nu(\text{MHz})/H(\text{G})$$

Hyperfine coupling parameters:

$$A(\text{MHz}) = (g_e\beta_e/h)\left(\frac{g}{g_e}\right)A(\text{G}) = 2.802494\left(\frac{g}{g_e}\right)A(\text{G})$$

$$A_{\text{iso}}(\text{MHz}) = 800.2375\,\frac{\mu_n(\text{n.m.})}{I}\,|\psi(0)|^2(\text{a.u.})$$

$$A_{\text{dip}}(\text{MHz}) = 47.760658\,\frac{\mu_n(\text{n.m.})}{I}\,\langle(3\cos^2\theta - 1)/r^3\rangle(\text{a.u.})$$

or

$$\langle(3\cos^2\theta - 1)/r^3\rangle(\text{a.u.}) = 0.02093774\,\frac{I}{\mu_n(\text{n.m.})}\,A_{\text{dip}}\ (\text{MHz})$$

$$|\psi(0)|^2_{\text{a.u.}} = 0.001249629\,\frac{I}{\mu_n(\text{n.m.})}\,A_{\text{iso}}\ (\text{MHz})$$

Author Index

Aasa, R., 92, 120, 279, 280
Abe, H., 316
Abragham, A., 12, 13, 16, 17, 41, 44, 63, 65, 74, 165, 168, 174, 205, 235, 256, 265, 266, 292, 306, 308
Adrian, F. J., 21, 23–29, 52, 53, 79, 106, 114, 122, 123, 137, 138, 144–147, 150, 152, 232, 233
Allen, B. T., 274, 276
Allen, H. C., 317
Amano, T., 199, 201
Ammeter, J. H., 16, 18, 19, 21, 30, 31, 33, 36
Anderson, E. D., 202
Anderson, J. H., 75
Anderson, P. W., 304
Andrews, L., 149, 150
Arata, Y., 111
Åslund, N., 52
Atkins, P. W., 43, 48, 52, 70, 106, 108, 118, 126
Azumi, T., 202, 206

Bagus, P. S., 288, 294
Baird, J. C., 158, 199
Ballhausen, C. J., 316
Baranowski, J., 258, 259, 263
Barash, L., 211, 213
Barnes, C. E., 231, 233
Barrow, R. F., 52, 57, 201, 234
Bass, I. L., 86
Bastin, M. W., 52
Baumann, C. A., 254, 287, 295–299, 305, 320, 321
Baumgärtel, H., 252
Bean, C. P., 319
Beaudry, W. T., 100
Beck, J., 17
Beltrán-López, V., 275, 276
Bender, C. F., 127, 129, 199
Bennett, J. E., 79
Berg, L.-E., 52, 206
Beringer, R., 105
Bernath, P. F., 131
Bernard, H. W., 196, 200–203
Bernheim, R. A., 196, 200–203, 206–208
Bessis, N., 18
Beveridge, D. L., 57
Bhat, S. V., 28–30, 36
Bicknell, B. R., 197, 200, 201
Biddles, I., 125
Birgeneau, R. J., 311
Bleaney, B., 12, 13, 17, 44, 62, 65, 68, 73, 74, 89, 111, 119, 165, 168, 174, 205, 235, 246, 256, 265, 266, 276, 292, 306, 308, 316, 318
Blinder, S. M., 69
Bloch, F., 193
Bloembergen, N., 193
Blumberg, W. E., 278
Boland, B. J., 220
Bolton, J. R., 42, 44, 70, 81, 125, 159, 193, 266
Boorstein, S. A., 201, 202
Bowers, K. D., 316, 318
Bowers, V. A., 19–21, 23, 28, 36, 52, 53, 79, 87, 114, 121–123, 137, 138, 144–147, 150, 152, 232
Brailsford, J. R., 38
Brandon, R. W., 182, 208
Breckenridge, W. H., 201
Breit, G., 6, 7, 10
Briat, B., 316
Brickmann, J., 237, 241, 252, 253
Brith, M., 38
Brom, J. M., 52, 118, 119, 127–129, 132, 135, 250, 251
Brown, C. M., 235, 238, 240, 242, 259, 265, 283–285
Brown, J. M., 220, 221, 225–227, 231–234
Brown, R. E., 14, 18
Bruno, G. V., 193, 194
Buenker, R. J., 201

Cade, P. E., 127, 141
Carlson, K. D., 206
Carrington, A., 42, 46, 108, 123, 125, 159, 164, 182, 205, 220, 226, 231, 233
Cassoux, P., 316
Castner, T. G., 81-86
Castner, T., Jr., 277
Castro-Tello, J., 276
Chan, A. C., 127, 129
Chan, S. I., 94
Charles, S. W., 23
Chestnut, D. B., 61, 142
Cheung, A. S. C., 52, 234, 248
Childers, A. G., 131, 132, 134, 135
Childs, W. J., 52, 131
Chirkin, G. K., 236
Christensen, H., 227, 228
Christiansen, J. J., 120, 121
Christy, A., 50, 52, 231
Clementi, E., 209
Clementi, H., 209
Closs, G. L., 208
Cochran, E. L., 19-21, 23, 27, 52, 79, 87, 114, 121-123, 144-147, 150, 152, 232
Code, R. F., 17
Coffman, R. E., 279
Cohen, M. H., 79, 152
Colburn, E. A., 200, 201
Cole, T., 28, 75
Coles, B. A., 318
Coles, M. E., 232, 233
Colin, R., 201
Cook, T. J., 201
Coucouvanis, D., 279
Coulson, C. A., 108
Cukierda, T., 258, 259, 263
Cummins, P. G., 131
Curl, R. F., 50-52, 129, 133, 138, 176, 211, 221, 226, 228, 231-234
Current, J. H., 115, 122

Daasch, W. R., 131, 132, 134, 135
Daudey, J. P., 316
Davidson, E. R., 100, 101, 127, 129, 131, 132, 134, 135, 200, 201
Davies, P. B., 201
Davoust, C. E., 208
de Boer, E., 126
de Groot, M. S., 164, 167, 171, 174, 175, 182, 184, 186, 187, 192, 201, 202, 204
Deile, O., 52
de Loth, P., 316
Demuynck, J., 286
Denison, A. B., 202, 206
Descleaux, J. P., 18
Devillers, C., 52, 197, 212, 213
DeVore, T. C., 206, 217, 218, 250, 253, 261, 270, 281-285, 288, 291-293, 303
de Wijn, H. W., 73, 276
Dismuke, K. I., 52, 75, 209-211
Dixon, R. N., 200, 201
Dixon, T. A., 52
Dobosh, P. A., 57
Doedens, R. J., 305
Domaille, P. J., 52
Dousmanis, G. C., 59
Dowsing, R. D., 259, 278
Duerig, W. H., 87
Duerst, R. W., 305, 314
Duncan, M., 100
Du Parq, R. P., 201
Durham, C. H., Jr., 250, 251
Dzyaloshinski, I., 309

Easley, W. C., 49, 52, 53, 62, 93, 99, 100, 106, 122, 123, 130-134, 137-139, 158, 233
Eidels-Dubovoi, S., 275, 276
Elbert, S. T., 201
Erdös, P., 309
Eshbach, J. R., 229, 230
Evans, L., 17
Evenson, K. M., 221-224
Ewing, G. E., 145

Farach, H. A., 77, 184, 277, 281
Farmer, J. B., 87, 122
Ferrante, R. F., 213, 216, 252, 286, 287, 301-303
Fessenden, R. W., 75, 142-144
Field, R. W., 131
Figgis, B. N., 316
Fischer, P. H. H., 23
Fixman, M., 193
Foley, H. M., 53, 131, 225, 226
Foner, S. N., 19, 20, 21, 23, 27, 87, 121, 122
Foster, P. J., 147
Fraenkel, G. K., 123, 125
Freed, J. H., 123, 125, 193, 194
Friedman, E., 73, 276
Frosch, R. A., 53, 131, 225, 226
Fujioka, Y., 52
Fujiwara, S., 111

Gendell, J., 115, 122
George, P., 79
Gerkin, R. E., 182
Gerloch, M., 316
Gerry, M. C. L., 87, 122
Gersmann, H. R., 102
Geusic, J. E., 241
Gibson, J. F., 278, 281

Ginsberg, A. P., 316
Godfrey, M., 201
Goeppert-Mayer, M., 192
Golding, R. M., 73, 111, 112, 114, 117, 192, 281
Goldsborough, J. P., 25
Goodenough, J. B., 304
Goodgame, D. M. L., 316
Goodman, B. A., 61, 70
Goodman, G. L., 52, 131
Goodman, L. S., 52, 131
Gordon, G., 314, 317
Gordon, R. M., 287
Gordy, W., 23, 25, 26, 60–62, 75, 81, 120, 121, 125–127, 132, 144, 145, 219
Goudsmit, S., 15
Gouterman, M., 199, 201, 202
Graham, W. R. M., 30, 34, 35, 52, 75, 197, 200, 201, 209–211, 252, 286, 287, 303
Gramas, J. V., 206
Griffith, J. S., 79, 204, 205, 256
Grivet, J. -Ph., 182, 183
Gruen, D. M., 205, 206
Guha, B. C., 316
Guntsch, A., 52

Hagland, L., 52
Hall, J. A., 52, 199
Ham, F. S., 254
Hameka, H. F., 158, 164, 198–202
Hand, R. W., 282
Hansen, A. E., 316
Hansen, R. C., 52, 234, 248
Harding, J. H., 316
Harris, E. A., 305, 309, 311–313, 318-320
Harris, D. O., 52
Harrison, J. F., 200, 201
Hartl, W., 258, 259, 265
Harvey, J. S. M., 17
Hasegawa, A., 144
Hastie, J. W., 206, 218
Hatfield, W. E., 305
Hauge, R. H., 206, 218
Haven, Y., 256
Hay, P. J., 316
Hecht, H. G., 89
Hedaya, E., 115
Heller, C., 75
Herrmann, A., 147
Herschbach, D. R., 25, 79, 147, 150–155
Herzberg, G., 52, 145, 200, 206, 213
Hesselmann, I. A., 201, 202, 204
Higuchi, J., 125, 265
Hill, N. J., 316
Hinchliffe, A., 136, 137
Hirokawa, S., 195
Hirota, E., 199, 201, 227
Hodgson, D. J., 305
Hoffmann, R., 209, 316
Holmberg, R. W., 69, 106, 109, 114, 145
Holton, W. C., 277
Holuj, F., 119
Hornig, A. W., 201
Hou, S. L., 236
Hougen, J. T., 205, 221–224, 234
Hsu, H., 201
Huber, K. P., 52, 206
Huber, R. A., 258, 259, 265
Hudson, A., 62, 125
Hughes, V. W., 19
Hunt, W. J., 282
Huo, W. M., 127
Hutchings, M. T., 256, 257, 311
Hutchison, C. A., Jr., 182–184, 201, 202
Hutton, R. S., 197, 200–202
Hyde, J. S., 201

Ichikawa, T., 232, 233
Ingram, D. J. E., 79, 278, 281
Isomoto, A., 119
Itoh, K., 264, 294
Itzkowitz, M. S., 193
Iwasaki, M., 29, 36, 111, 192, 232, 233

Jackel, G. S., 23, 25, 26, 120, 144, 145
Jackson, W. K., 184
Jacox, M. E., 213
Jacquinot, D., 52
James, T. C., 205
Jamieson, A. M., 43, 48
Jen, C. K., 19–21, 23, 28, 29, 87, 120–122
Jennings, D. A., 221–224, 232
Jette, A. N., 26, 27
Jezowska-Trzebiatowska, B., 258, 259, 263
Johns, J. W. C., 200
Johnson, C. S., Jr., 193
Jones, W. E., 201
Jotham, R. W., 316

Kahn, O., 316
Kaise, M., 227, 231, 232
Kammer, W. E., 201
Kanamori, J., 308, 309
Kanzig, W., 79, 81–86, 152
Karplus, M., 201
Kasai, P. H., 8, 10, 11, 21, 23, 52, 61, 76, 77, 88, 95, 97, 98, 100, 103, 106, 115, 122, 234, 246–250, 254, 293, 303, 304
Kaving, B., 52
Kawaguchi, K., 227
Kayama, K., 158, 199
Kedzie, R. W., 278

Keen, N., 108
Kempf, R. J., 206-208
Kern, C. W., 158, 199, 201, 202, 204
Kerr, C. M. L., 227, 231, 232
Kestigan, M., 278
Kettle, S. F., 316
Khazanovitch, T. N., 193
Kinoshita, M., 202, 206
Kirkpatrick, E. S., 241, 271
Kispert, L. D., 29, 80, 142-144
Kittel, C., 319
Kivelson, D., 102, 123, 125, 193
Klein, M. P., 277, 278
Klynning, L., 52
Kneubuhl, F. K., 117
Knight, L. B., Jr., 30, 48, 50, 52, 93, 99-101, 127-132, 134, 135, 265
Koehler, T. R., 25
Kohler, B. E., 208
Koide, M., 202
Kokoszka, G. F., 305, 314, 317
Kommandeur, J., 122
Kon, H., 195, 197
Kopfermann, H., 7, 12, 14, 15
Kopp, I., 52
Korst, M. N., 193
Kotani, M., 119
Kothe, G., 237, 241, 243, 252, 253
Kottis, P., 171, 172, 174, 187-190
Kovacs, I., 288
Kozlowski, H., 258, 259, 263
Kronekvist, M., 52
Kubota, S., 125
Kuck, V. J., 195-197, 200-202
Kumamoto, T., 125
Kupferman, S. L., 21
Kurzyński, M., 258, 259
Kusch, P., 19
Kutner, M. L., 51, 52
Kuwata, K., 232, 233
Kwiram, A. L., 25, 79, 147, 150-155

Langhoff, S. R., 158, 199-202, 204
Larsson, S., 14, 18
Leckenby, R. E., 147
Lefebvre, R., 171, 172, 174, 187-190
Lefebvre-Brion, H., 18
Lembke, R. R., 213, 216
Leroi, G. E., 205
Levy, D. H., 220, 226
Lew, H., 15, 17, 19
Lhoste, J. -M., 182, 183
Lindsay, D. M., 25, 79, 147, 150-155
Lin, W. C., 111, 112, 117, 192
Linzer, M., 314
Lipscomb, W. N., 235
Livingston, R., 69, 77, 106, 107, 233
Lohr, L. L., Jr., 235
Longborough, B., 52
Longuet-Higgins, H. C., 123, 125
Lounsbury, J. B., 200, 201
Low, W., 68, 73, 276, 292
Lowrey, M. R., 111
Lucas, N. J. D., 231
Luckhurst, G. R., 62
Ludwig, G. W., 241, 246, 254
Luz, Z., 106, 109
Lynden-Bell, R. M., 193
Lyons, D. H., 278

Mackor, E. L., 126
Máckowiak, M., 258, 259, 265
Mador, I. L., 87
Mahieu, J. M., 52
Malmberg, C., 52
Malrieu, J. P., 316
Mangum, B. W., 183, 184, 201, 202
Margrave, J. L., 206, 218
Marsham, D. F., 316
Martin, R. L., 316
Martin, R. W., 234
Martinez, J. V., 63
Maruani, J., 80, 111, 143, 144
Merer, A. J., 52, 234, 248, 287
Meyer, H., 195, 196
Meyer, B., 20
Mieher, R. L., 86
Miller, T. A., 201, 220, 226
Milligan, D. E., 213
Milton, D. J., 227, 231, 232, 234
Miura, M., 144
Miyagawa, I., 75
Mizushima, M., 202, 227
Moffitt, W., 199, 201, 202
Moran, M. M., 221
Morehouse, R. L., 120, 121
Moriya, T., 309
Morton, J. R., 18, 26-28, 38, 80, 81, 126, 299
Moser, C. M., 18, 206
Mouchet, A., 100
Mucha, J. A., 221-224
Muller, K. A., 241, 271
Mulliken, R. S., 50, 52, 141, 231
Muto, H., 29, 36
Myers, G. H., 106, 122, 123, 233

McCarty, M., Jr., 38
McClung, R. E. D., 105
McClure, D. S., 202
McConnell, H. M., 28, 61, 75, 120, 121, 123-125, 142, 158

McDowell, C. A., 23, 80, 87, 98, 106, 111, 122, 143
McGarvey, B. R., 158, 160, 235
McGlynn, S. P., 202, 206
McIver, J. W., Jr., 200, 201
McLachlan, A. D., 42, 46, 108, 125, 126, 159, 164, 182, 205
McLeod, D., Jr., 8, 10, 11, 21, 23, 254
McMillan, J. A., 36, 37
McMurchie, L. E., 100, 101
McNeil, R. I., 100

Nakajima, H., 80, 98, 106, 111, 143
Neiman, R., 102
Nelson, W. H., 23, 25, 26
Nelson, A. C., 220
Nesbet, R. K., 206, 321, 322
Nevin, T. E., 288
Newell, G. S., 277
Nicklin, R. C., 281
Norris, J. R., 193
Nylén, P., 52

O'Brien, M. C. M., 111, 195
Ohmes, E., 252
Oosterhoff, L. J., 164
Oosterhuis, W. T., 281
Orloff, D., 141
Orloff, H., 141
Orton, J. W., 318
Ovenall, D. W., 108
Owen, J., 305, 309, 311–313, 318, 319

Pacher, P., 288
Palmiere, P., 201
Peiser, H. S., 21
Pendlebury, J. M., 25
Peter, M., 241
Peyerimhoff, S. D., 201
Pick, H., 256
Pilbrow, J. R., 92, 111, 119, 236, 241, 271, 305, 309, 322
Pimentel, G. C., 145
Pinchemel, B., 287
Pinkstone, J., 231, 233
Pipkin, F. M., 21
Pitzer, K. S., 209
Pitzer, R. M., 202, 205, 282
Pollack, G. L., 20
Polnaszek, C., 193, 194
Pol'skii, Yu E., 236
Ponte Goncalves, A. M., 202
Poole, C. P., Jr., 70, 77, 184, 277, 281
Pople, J. A., 57
Pound, R. V., 193
Prather, J. L., 257
Preston, K. F., 26, 27, 126, 299
Pritchard, R. H., 199
Pryce, M. H. L., 16, 41, 63
Purcell, E. M., 193

Rabi, I., 6, 7, 10
Radford, H. E., 200, 201, 221, 225, 226
Raghunathan, P., 80, 98, 106, 111, 124, 143, 145
Ramsay, D. A., 52, 145, 197, 212, 213
Ramsey, N. F., 7, 14
Raynes, W. T., 201
Raynor, J. B., 61, 70
Redfield, A. G., 193
Reichenbecher, E. F., 206–208
Renhorn, I., 52, 131
Reuveni, A., 106, 109
Richards, W. G., 206
Ricks, J. M., 201
Rieke, C. A., 141
Robbins, E. J., 147
Roberts, J., 193
Robinson, G. W., 38
Rockenbauer, A., 111
Rodbell, D. S., 319
Rogers, M. T., 80, 142–144
Rollmann, L. D., 94
Ross, I. G., 316
Rubins, R. S., 73, 241, 271, 276

Sakaguchi, U., 111
Sandars, P. G. H., 17
Sands, R. H., 89
Saunders, M., 193
Saykally, R. J., 221, 222, 232
Scapoki, A. C., 316
Schaafsma, T. J., 122
Schaefer, H. F., III, 282, 286, 290, 294
Schamps, J., 52, 287
Schawlow, A. L., 57, 60, 64, 65, 78, 219, 225, 229, 230
Schlosnagle, D. C., 16, 18, 19, 21, 30, 31, 33, 36
Schmauss, G., 252
Schmid, D., 237, 256
Schnepp, O., 38
Schoemaker, D., 63, 84–86
Schoening, F. R. L., 316, 317
Schonland, D., 281
Schueller, K., 294
Schuler, R. H., 142–144
Schulz-DeBois, E. O., 241
Schumacher, E., 147
Schwoerer, M., 183, 237, 256, 258, 259, 265
Scott, P. R., 206
Scullman, R., 52

Sears, T. J., 231, 233, 234
Seely, M. L., 52
Seidel, H., 237, 256
Shaffer, J. S., 277
Shekun, L. Ya., 236
Shimada, J., 316
Shimokoshi, K., 124, 145
Shirley, D. A., 277, 278
Silbey, R., 208
Silver, B. L., 109
Simon, P., 111
Singhasuwich, T., 281
Sillescu, H., 193
Sinanoglu, O., 206
Singer, L. S., 239
Sink, M. L., 201
Sinn, E., 305
Skell, P. S., 196, 200-203, 206
Slichter, C. P., 81, 277
Smaller, B., 36, 37
Smardzewski, R. R., 149
Smart, M. L., 316
Smith, D. Y., 25
Smith, G. R., 52, 195, 197, 212, 213, 215
Smith, R. C., 232
Smith, T. D., 305, 309, 322
Smith, V. H., Jr., 14, 18
Smith, W. V., 219
Snyder, L. C., 168, 170, 174, 177-179, 182, 185, 191, 208
Sogabe, K., 144
Steimle, T. C., 52, 231, 233
Stepanov, V. G., 236
Stephen, M. J., 123, 125
Sterne, T. E., 196
Stevens, K. W. H., 256, 308, 309
Stone, A. J., 43, 47, 48, 201
Strach, S. J., 26, 27
Strandberg, M. W. P., 199, 227, 229, 230
Strathdee, J., 61
Strickler, S. J., 209
Summit, R. W., 236
Swalen, J. D., 102
Sweeney, W. V., 279
Symons, M. C. R., 70, 79, 106, 108, 118, 126

Takui, T., 294
Tanaka, Y., 52
Tennant, W. C., 73, 111, 112, 114, 117, 192, 281
ter Maten, G., 201
Thaddeus, P., 51, 52
Thibeault, J. C., 316
Thistlethwait, P. J., 231, 233
Thompson, W. E., 145
Thrush, B. A., 201
Tinkham, M., 199, 227
Tippins, H. H., 43
Tokue, I., 125
Toriyama, K., 29, 36
Townes, C. H., 57, 60, 64, 65, 78, 219, 225, 229, 230
Trambarulo, R. F., 219
Trammell, G. T., 69, 77
Travis, D. N., 213
Troughton, P. G. H., 316
Tucker, K. D., 51, 52
Tucker, R. F., 236

Uslu, K. A., 17

van Balderen, R. F., 73, 276
van der Waals, J. H., 164, 167, 171, 174, 175, 182, 184, 186, 187, 192, 201, 202, 204
van Doorn, C. Z., 256
Vänngard, T., 92, 120
van Nickerk, J. N., 316, 317
Vannotti, L. E., 18, 38
van Veen, G., 92
van Vleck, J. H., 3, 30, 50, 51, 195, 306, 309
Van Zee, R. J., 28, 52, 235, 238, 240, 242, 250, 253, 254, 259, 261, 265, 267, 270, 272, 281-285, 287, 288, 291-293, 295-303, 305, 320, 321
Veseth, L., 227, 228, 232, 248
Vinokurov, V. M., 236

Walsh, A. D., 107, 108
Wang, P. S., 196, 200-203
Wasserman, E., 168, 170, 174, 177-179, 182, 185, 191, 195-197, 200-202, 208, 211, 213, 294
Watari, H., 119
Watson, J. K. G., 201
Wayne, F. D., 200, 201
Wei, M. S., 115, 122
Weil, J. A., 75, 89
Weissbluth, M., 188
Weissman, S. I., 193, 243
Weltner, W., Jr., 8, 10, 11, 28-30, 34-36, 48-53, 61-63, 75, 88, 91, 93, 95, 97-100, 103, 106, 118, 119, 122, 123, 127-132, 134, 135, 137-139, 158, 195, 197, 200, 201, 206, 209-212, 213, 215, 216-218, 230, 235, 238, 240, 242, 250-254, 259, 261, 265, 270, 272, 281-288, 291-293, 295-303, 305, 320, 321
Wertz, J. E., 42, 44, 70, 81, 125, 159, 193, 266
Wessel, G., 15, 19
Whiffen, D. H., 108
Whipple, E. B., 61, 88, 95, 97, 98, 103, 106, 115, 122

Wickman, H. H., 277, 278
Wilkerson, J. L., 252, 285–288, 292, 293, 303
Williams, F., 100
Willis, P., 147
Wilson, M., 52, 93, 100, 130–132, 134
Wise, M. B., 100, 101, 131, 132, 134, 135
Wolf, H. C., 183
Wolfe, W. P., 311
Wood, L. S., 196, 200–203
Woodbury, H. H., 241, 246, 254
Woodgate, G. K., 17
Woods, R. C., 52
Wöste, L., 147
Wyard, S. J., 232

Yager, W. A., 168, 170, 174, 177–179, 182, 185, 191, 195–197, 200, 201, 208, 211, 213, 294
Yamada, C., 227
Yates, J., 316
Yates, J. H., 205, 282
Yim, M. B., 100

Zaripov, M. M., 236
Zegarski, B. R., 201
Zeldes, H., 69, 77, 106, 107, 233
Zeringue, K. J., 235, 238, 240, 242, 259, 265, 295–299, 301, 302
Zilles, B. A., 106, 122, 123, 233
Zimmermann, H., 252

Subject Index

A_1, A_2, hyperfine parameters (for gas-phase molecules), 226
A_{dip} (atom), dipolar or anisotropic hyperfine constant, 13 ff.
  alteration from central field approximation, 16, 17
  correction for ionic charge, 59, 60
  free atom values, 18, 59–61
    angular factor, $\overline{\alpha}$, 59
    tables of (Appendix B), 345–348
  for hydrogenic atoms, 15
  matrix crystal-field effects, 33–38
A_{iso} (atom), isotropic hyperfine constant, 13 ff.
  alteration from central field approximation, 16, 17
  correction for ionic charge, 59, 60
  free atom values, 18, 59–61
    table of (Appendix B), 345–347
  for hydrogenic atoms, 15
  magnitude relative to A_{dip}, 62
  matrix crystal-field effects, 33–38
Ag^{2+} ions, quenched, 36, 37
  ESR spectrum, 37
AgM, where M = Mg, Ca, Zn, Hg, 254
a_F, isotropic Fermi hyperfine constant (gas-phase molecules) [*see also* A_{iso} (molecule)], 224
Al atoms, quenched, 30–35, 62
Alkaline-earth fluoride molecules, 129, 135
  comparison with corresponding hydrides, 132
Alkaline-earth hydride molecules, 126–129, 132
  comparison with corresponding fluorides, 132
Alkaline-earth hydroxide molecules, 135–137
Alkali superoxide molecules (*see* O_2^-):
  anomaly in CsO_2, 155
  calculated hf and *g* tensor components, 153–155
  ESR spectra, 150, 151
  infrared and Raman studies, 149
  magnetic parameters, 152
  metal hf parameters, 152, 153
  molecular orbitals, 153–155
AlO, 51
Alternant methylenes, 206–209
  table of parameters, 208
  ESR of $HC-(C{\equiv}C)_2-CH_3$, 207
A_{dip} (molecule), dipolar or anisotropic hyperfine constant, 54
  contribution from interatomic dipole-dipole interaction, 61, 62
  relation to $A_{\parallel}, A_{\perp}$ in axial case, 55
  relation to Frosch and Foley's parameters, 54–56
  relation to $\langle(3\cos^2\alpha - 1)/2r^3\rangle$, 54
  sign of, 62, 63
A_{iso} (molecule), isotropic hyperfine constant (designated as a_F for gas phase molecules), 54
  relation to $A_{\parallel}, A_{\perp}$ in axial case, 55
  relation to Frosch and Foley's parameters, 54–56
  relation to $|\Psi(0)|^2$, 54
  sign of, 62, 63
Anisotropic exchange, 306–309
Anisotropic hyperfine tensor [*see* Nuclear hyperfine interaction (molecules)]:
  axial molecule, 55
  nonaxial molecule (*T*), 104
Antiferromagnetic coupling, 305, 311, 318–321
Antisymmetric exchange, 309
A tensor [*see* Nuclear hyperfine interaction (molecules)]
Atoms and atomic ions in matrices, 19 ff.
Atoms in a crystal field, 28–38
Atoms, orbitally degenerate:
  crystal-field analysis of quenching, 30–38
  $^2D_{5/2}(Ag^{2+})$, quenched, 36, 37
  $^2P_{1/2}$(Al, B, Ga), quenched, 30–35, 62, 95

Atoms, orbitally degenerate (*Cont.*):
  $^2P_{3/2}$(Br, I), quenched, 29–33, 36, 37
  $^2P_{3/2}(O^-, S^-)$, quenched, 38
  quenching of orbital angular momentum, 29–38
Atoms, *S*-state, 19–29
  half-filled shells, 25
  high-spin, crystal field effects, 28, 29
  matrix effects upon g_J and A, 19–25
  spin polarization, 25
  theory of matrix shifts, 24, 25
A value, atomic hf constant, 5–12
Axial molecule, defined, 39

BaF, 51, 93, 129, 132, 133
  comparison of spin density on F among fluorides, 132
  comparison with BaH, 132
BaH, 51, 126–129, 132
  comparison with BaF, 132
B atoms, quenched, 30–35, 62, 95
  ESR spectrum, 35
BeF, 129, 132–135
  ESR spectrum, 134
BeH, 61, 126–129, 132
  comparison of $|\Psi(0)|^2$ with theory, 129
  ESR spectrum of BeD, 128
Benzene (in $^3B_{1u}$ state), 201, 202, 204
Benzene-1, 3, 5-tris-phenylmethylene, 294, 295
BeOH, 135–137
  comparison of isotropic hf data with MgOH, 137
B_n^m, spin operator coefficients (zfs parameters), 256, 257
BO, 99–101
Bohr magneton (β), 2
Br_2^-, 84
Br atoms, quenched, 29–33, 36, 37
  ESR spectrum, 35
Breit-Rabi equation, 6–12
BS, 51
t-Butylpentadiynylene, 208

C_4, 52, 176, 183, 209–211
  ESR spectrum, 210
CaF, 51, 129, 131–133
  comparison with CaH, 132
  comparison of ^{19}F hf in neon matrices and gas phase, 131
  comparison of spin density on F among fluorides, 132
CaH, 51, 126–129, 132
  comparison with CaH, 132
CCO, 52, 176, 195, 197, 211–215
  ESR spectra, 214, 215
  isotope effects in matrices, 195–198
CdH, 51
CF_3, 80, 115, 142–144
  comparison with MH_3 and CH_xF_y radicals, 144
  ESR spectrum, 143
  g and A (^{19}F) axes not coincident, 115, 143, 144
$C_2F_4^-$, oriented in matrices, 101
CH, laser-magnetic-resonance of, 221–224
CH_2, 195–204, 209
  ESR spectrum (and CD_2), 203
  isotope effects in matrices, 195–197
  matrix effects, 197, 198
  theoretical calculation of D and E, 199–201
CH_3, 80, 120, 121, 141, 142, 144
  comparison with other MH_3 and CH_xF_y radicals, 144
  ESR spectrum, 121
  ^1H hyperfine splitting, 142
  rotational effects, 120, 121
CHF_2, 144
  comparison with MH_3 and CH_xF_y radicals, 144
CH_2F, 144
  comparison with MH_3 and CH_xF_y radicals, 144
C_2H, 51, 75–77
Cl_2^-, 63, 81–86, 153
ClO, 79
CN, 49, 51, 53, 56, 62, 123, 135–141
  calculation of $\Delta g_\perp$, 140, 141
  comparison of matrix and gas phase data, 138
  ESR spectra, 138, 139
  rotation in matrices, 123
    ESR spectra, 123
    linewidth effect, 123
C_3N_2, 136, 213
  bending induced by matrix, 213
CNN, 168, 195, 197, 211–213
  isotope effects in matrices, 195–198
$^{13}CO_2^-$, 104, 107, 108
CO_3^-, 80
Copper acetate monohydrate, 304, 314, 316–318
Copper adenine complex (exchange coupling), 314, 315
  ESR in $S = 1$ state, 315
Copper phthalocyanine, 102
Correlation effects (atoms), 16
Cr^{3+} ion, 235
  in emerald, 241
CrAu:
  in solid silicon, 254
  in solid argon, 254

CrCu, 297–301
ESR spectrum, 300
CrH_2, 265
CrO, 265
Crystal-field effects in matrices (*see* Matrix effects)
CsO_2 (*see* alkali superoxide molecules), 149, 151
anomalous parameters, 155
ESR spectrum, 151
CuF_2, 61, 95, 98
$Cu(NO_3)_2$, 79, 95–98, 102, 103
Curl equation (*see* Spin-rotation interaction), 50–52, 129, 133, 138, 176, 211, 226, 228, 231–234
Cyanomethylene, 208, 209

D (zero-field splitting) versus H_r (resonant magnetic fields):
$^3\Sigma$ molecules, 170, 185
$^4\Sigma$ molecules, 240
$^5\Sigma$ molecules, 261
$^6\Sigma$ molecules, 270
$^7\Sigma$ molecules, 290
$^8\Sigma$ molecules, 297
$^9\Sigma$ molecules, 301
Diacetylene derivatives (quintet), 265
Diphenylmethylene radical, 206, 208
Dipole moment (μ) (*see* Electric-dipole transitions), 219
Direct and super exchange, 314, 316
Doublet ($S = \frac{1}{2}$) molecules:
axial nonlinear, 79
axial orbitally-degenerate, 77–79
free and hindered rotation of trapped radicals, 120–126
nonaxial, 103–120
transition probabilities, 119–120
$^2\Pi$, quenched, molecules, 77–79
table of magnetic properties of diatomic, 360, 361
randomly oriented nonaxial, 117–119
representative radicals, 126–155
$^2\Sigma$, 39
energy levels in a magnetic field, 65–70
g tensor, 42
nuclear hyperfine interaction, 52
nuclear Zeeman and nuclear quadrupole interactions, 63
spin Hamiltonian, 40
spin rotation interaction in (*see* Spin-rotation interaction)
tabulated magnetic properties:
from ESR of molecules in matrices and crystals;
diatomic fluorides and chlorides, 354
diatomic metal hydrides and carbides, 352, 353
diatomic metal molecules and ions, 356
diatomic metal oxides, sulfides, 355
diatomic molecules and anions, 357, 358
triatomic metal cyanides and hydroxides, 355
triatomic radicals, 364, 365
from gas-phase data;
diatomic, 390–392
polyatomic, 398
triatomic, linear, 395
triatomic, bent, 396, 397
tetra-atomic, tabulated magnetic properties:
in gas-phase, 398
in matrices and crystals, 385
π radicals, 379–381
pyramidal, 382
σ radicals, 383, 384
transition-metal molecules, 386
triatomic, tabulated magnetic properties:
in gas-phase, 395–397
in matrices and crystals
metal molecules, 375
miscellaneous ions, 376
π_x radicals, 369–374
D tensor (zero-field tensor), 45, 156–160
D_{SO}, spin-orbit contribution, 157, 158, 198
low-lying electronic states, 158
D_{SS}, spin-spin contribution, 158–160, 198
relative magnitudes of D_{SO} and D_{SS}, 159, 199
motional and matrix effects upon, 193
isotope effects (torsional oscillation), 193, 195–197
electronic effects, 197, 198
principal eigenvalues of, 156, 157, 160
traceless property, 156
for representative triplet molecules:
alternant methylenes, 206–209
C_4, 209–210
CNN, NCN, CCO, 211–215
SiCO, SiN_2, 213, 216
TiF_2, 217, 218
$S \geqslant 2$ molecules:
higher terms in S_i
$B_n{}^m$, coefficients, 256, 257
$O_n{}^m$, equivalent operators, 256, 257
theoretical calculation of D and E, 160, 198–206
aromatic molecules, 201–204
benzene, triplet conformations, 203, 204

D tensor (zero-field tensor) (*Cont.*)
theoretical calculation of *D* and *E* (*Cont.*)
CH_2, 200, 201
H_2CO, 201
NH, PH, NX, PF, 200, 201
O_2, SO, S_2, 199, 201
transition-metal molecules, 204–206

Electric-dipole transitions, 220, 221
allowing Stark modulation, 220
in gas-phase EPR, 220
intensities relative to magnetic-dipole transitions, 220
Electric field gradient at the nucleus (q) (*see* Nuclear quadrupole interaction):
calculation of, from central charge approximation, 64, 65
in completely ionic molecules, 65
defined, 13, 64
Electronic exchange coupling (*see* exchange coupling)
Electron paramagnetic resonance (EPR) in the gas phase (*see* Gas phase molecules), 219–221
electric dipole transitions, 220, 221
preparation of radicals, 220, 221
spin decoupling, 220, 221
in $^2\Pi_{1/2}$ states, 225
in $^2\Sigma$ molecules, 220
in nonlinear doublet molecules (HCO), 220, 221
in $^3\Sigma$ molecules, 220
spin-rotation interaction, 220, 221, 228, 229, 231–234
Stark vs. Zeeman modulation, 220, 221
Electron spin resonance spectroscopy (*see also* Atoms; Doublet, Triplet, etc., molecules):
atoms, 3, 4
hyperfine structure, 4–12
magnetic-dipole transitions, 7–12
electron exchange coupling, 311–314
hf structure for dimers, 314
ESR spectrum of copper adenine complex, 315
ESR spectrum of Mn^{2+} pairs in CaO, 320
ESR spectrum of Mn_2 in solid Kr, 321
ESR cavity, 70, 71
frequency-swept spectrometer, 92
gas-phase molecules (*see* Electron paramagnetic resonance)
matrix-isolated molecules, 86–89
cryostat apparatus, 87–89
single crystals, 81
V centers, 81–86
NO_2, CO_2^-, HCO, 105–109, 114–115
$S = \frac{1}{2}$, axial molecules, 70–77
randomly oriented molecules, 89, 118
resonant magnetic fields
allowed transitions, 72–74
forbidden transitions, 74–77
selection rules and transition probabilities, 70–72, 119, 120
$S = 1$, axial molecules, 163, 166, 169, 170
$\Delta m_S = 2$ transitions, 175, 176
randomly-oriented, 176–183
relative intensities, 174
$S = 1$, nonaxial molecules, 185–189
ESR spectra of CH_2, CD_2, 203
ESR spectra of alternant methylenes, 207
ESR spectrum of C_4, 210
motional effects, 193–198
randomly oriented, 190–193
two quantum transition, 192, 193
$S = \frac{3}{2}$ molecules, 238–256
$\Delta m_S = 2$ and 3 transitions, 241, 242
ESR spectrum of NbO, 250
ESR spectrum of VF_2, 253
ESR spectrum of VO, 248, 249
randomly oriented, 241–245
$S = 2$ molecules, 260–265
ESR spectrum of *m*-phenylene-bis-phenyl-methylene, 264
randomly oriented, 262–264
$S = \frac{5}{2}$ molecules, 268–287
$g' = 4.3$ signal, 277–281
ESR spectrum of FeF_3, 282
ESR spectrum of MnF_2, 283, 284
ESR spectrum of MnH_2, 285
ESR spectrum of MnO, 286
randomly oriented, 271–274
spectral effects for $D \simeq A$, 273–277
$S = 3$ molecules, 287–295
ESR of MnF, 293
randomly oriented, 288–292
$S = \frac{7}{2}$ molecules
ESR spectrum of CrCu, 300
ESR spectra of GdF_3, 295, 298, 299
$S = 4$ molecules
ESR spectrum of GdO, 302
Exchange-coupled dimers (*see* exchange coupling):
copper acetate monohydrate, 316–318
copper adenine complex, 314, 315
Mn_2 molecules in matrices, 319–322
Mn^{2+} pairs in crystalline MgO and CaO, 318–321
$Mo_2Cl_9^{3-}$, 316
Exchange coupling (of electrons), 235, 236, 304, 305
anisotropic exchange, 306–309
magnetic dipolar interaction, 308, 309

antiferromagnetic, 305
antisymmetric exchange, 309
biquadratic coupling, 306, 319
in dimers, 304, 305
direct and superexchange, 314, 316
ab initio calculations, 316
ferromagnetic, 305
hyperfine interaction, 304, 312–314
transferred hf constant (a), 313
isotropic (Heisenberg) coupling, 305–307
J dominant, 309
Judd–Owen equations, 309–312
crystal-field effects (D_c, E_c), 310
dipolar effects (D_d, E_d),
D_S, E_S, definitions, 310, 311
α, β coefficients, 310–312
intensities of ESR lines, 311–313
Landé-type spacing, 311
pseudodipolar effects (D_E, E_E), 309, 310
pseudodipolar exchange, 308, 309
representative dimers, 316–322
copper acetate monohydrate, 316–318
Mn_2 molecules in matrices, 319–322
Mn^{2+} pairs doped into insulators, 318–320
exchange striction, 319
Exchange striction, 319
"Extra" lines (*see* Off-principal-axis-transitions)

F_2^-, 56, 84
Fe^{3+}, in biological molecules and complexes, 277–279
calculated line positions for $E/D = 0.32$, 280
ESR spectrum in $NH_4CoEDTA$, 279
FeB, in solid silicon, 254
FeF_2, 265
FeF_3, 281, 282
ESR spectrum, 282
Fermi contact interaction (*see* Isotropic hyperfine tensor), 14, 16
Ferrichrome-A, 277, 278
Ferrihemoglobin, 281
Ferromagnetic coupling, 305
Fine structure:
defined, 156
quartet molecules, 236, 250
sextet molecules, 266, 272, 278, 279
triplet molecules, 156, 167, 180
Frosch and Foley's hyperfine parameters, 53–57, 224, 225

Ga atoms, quenched, 30–35, 62
Gas phase molecules (*see* Electron paramagnetic resonance):
electric dipole transitions, 220
EPR, 219, 220
g factors (*see* g factors of gas phase molecules, Rotational g factors), 226–234
HO_2, NO_2, 231–234
O_2 and SO, 228–230
OH, 231, 232
NH_2, 229, 234
hyperfine parameters, 225
spin and orbital contributions, 225, 226
laser magnetic resonance, 219
spin decoupling, 220, 221
$^2\Pi_{1/2}$ states, 225
spin-rotation interactions, Curl equation, 228, 229, 231–234
for high-spin molecules, 234
tabulated magnetic properties, 224, 389–399:
in $^1\Delta$ excited states, 393, 394
doublet molecules:
diatomic, 390–392
triatomic, bent, 396, 397
triatomic, linear, 395
polyatomic, 398
triplet molecules
diatomic, 393, 394
triatomic, 399
Gd^{3+} (half-filled f shell), zero-field tensor, 235
GdF_3, 235, 295, 297, 298
ESR spectrum, 298, 299
GdO, 302, 303
ESR spectrum, 302
g^e, effective g value:
$^4\Sigma$ molecules, 239–241
VO, 250
VF_2, 251, 252
$^6\Sigma$ molecules, 271
FeF_3, 282
MnO, 286
$^8\Sigma$ molecules, 295
GeH_3, 144, 145
comparison with other MH_3 and CH_xF_y radicals, 144
g factors of gas phase molecules (*see* Spin-rotation interaction, Rotational g factors):
g_l^e, g_L', g_l, electron-orbit interactions, 227, 228, 231–233
$g_l', g_r^{e'}$, Λ-type doubling effects, 231, 232
g_r, g_r^e, g_r^n, rotation interactions (*see* Rotational g factors), 227–234
g_s^e, g_s, electron spin interactions, 227, 228, 231–233

g factors of gas phase molecules (*Cont.*)
relation to spin-rotation constants, 228, 229, 231–234
g_1 factor (of microwave magnetic field H_1), 70–72, 92
for orthorhombic g tensor, 119, 120
in triplet molecules, 172
$g' = 4.3$ signal (nonaxial sextet molecules), 277–281
Fe^{3+} in $NH_4CoEDTA$ powder, 279
Mn^{2+} in As-Te-I glass, 281
g_I factor (nuclear Zeeman effect), 5
$g^{(I)}$ tensor (at a magnetic nucleus) (*see* Nuclear Zeeman interaction), 63
approximated as isotropic, 63
if axial, 63
g_J factor (atomic Zeeman effect), 3
g tensor, 39, 42–50
components for triplet alternant methylenes, 208
diagonal form for axial molecules, 45
$\Delta g_\perp$, 46–50
calculation of, 47–50
inclusion of overlap, 49
effect of isotopic substitution, 50
$^3\Sigma$ molecules, Curl equation, 176
significance of sign of, 47
table of values from matrix spectra, 51, 52
g and A tensor axes coincident, 104–108
g and A tensor axes noncoincident, 80, 104, 109, 117
$g_\parallel, g_\perp$, 39, 43, 45–50
for orbitally-degenerate molecules, if quenched, 78, 79
fom perturbation theory, 42–50
if trigonal symmetry axis, 79
single crystals, 81
V_K center, 81–86

$h_{3/2}, h_{1/2}$ (hyperfine parameters), 224, 226
H_2, 221, 227
Hamiltonian (*see also* Spin Hamiltonian):
for electron exchange coupling, 305–307
anisotropic exchange, 308
for dimer interaction, 309, 310
isotropic exchange, 305
pseudodipolar exchange, 308
for nuclear electric quadrupole interaction
atoms, 12, 13
molecules, 64, 103
for nuclear hyperfine interaction:
atoms, 5
including polarization, etc., effects, 16, 17
molecules, 40, 53–57
anisotropic (dipolar) interaction, 53
Frosch and Foley's expression (and parameters) for, 53–57
isotropic (Fermi contact) terms, 53
$L \cdot I$ terms, 54
linear molecule, 54
more than one nucleus, 56
electron-exchange coupling, 312, 313
for nuclear Zeeman interaction:
atoms, 5
molecules, $g^{(I)}$ tensor, 63, 103
for spin-orbit interaction, 40–42
for spin-spin interaction, 156, 157, 159, 160, 184, 236, 256
contribution of spin-orbit interaction, 157
for $S \geqslant 2$ molecules, 256, 266, 302
Zeeman
atoms, 1, 5
molecules, 40–46, 65, 75, 103, 110, 184, 246, 247
in gas phase, 227, 231
in $S \geqslant 2$ molecules, 256, 266
including spin-spin interaction, 160, 184, 236
HBr^+ (LMR), 223
HCO, 110, 113–115, 145–147, 220
ESR spectrum in matrix, 146
gas phase spectroscopy, 145, Appendix D
$^1H^{13}CO$ (g and A tensors not coincident), 110–115
A (^{1}H) and A (^{13}C) components, 114
angles between A and g tensor components, 114, 115
anisotropic hf components, 115
single crystal ESR, 114, 145
H_2CO (in $^3A_2$ state), 201
HgH, 50, 51, 56
High spin molecules, 235 ff.
defined, 235
tabulated magnetic properties:
diatomics, 362, 363
triatomics, 378
tetra-atomics, 388
theoretical calculations of zfs, 235
HO_2, 104, 231–234
g tensor components, 231–234
g_S^{ii} from gas-phase, solid state, and Curl equation, 233
$(g_r^e)_{gg}$, from expt. and Curl equation, 233, 234
single crystal parameters, 232, 233
spin-rotational constants, 233
Hydrogen atom, 5, 12, 77
matrix effect, sites, 24
superhyperfine interaction, 27, 28
Hyperfine interaction (*see* Nuclear hyperfine interaction)

Hyperfine parameters, gas-phase:
a, b, c, d, 225
$h_{3/2}, h_{1/2}, A_1, A_2$, 226

I atoms, quenched, 29–33, 37
Interatomic dipole-dipole interaction, 61
contribution to anisotropic hyperfine interaction, 61, 62
I, nuclear spin, 5, 341, 344
Isotropic exchange coupling (*see* Exchange coupling), 305, 307
coupling constant J, 305
J dominant (Judd-Owen equations), 309-312
intensities of ESR signals, 311–313
Landé interval rule, 310, 311
table of α_S and β_S, 312
zfs parameters, D_S, E_S, 310
Isotropic hyperfine tensor [*see* Nuclear hyperfine interaction (molecules)]:
axial molecule, 55
nonaxial molecule, 104

Judd-Owen equations, 309–312, 319

KO_2 (*see* alkali superoxide molecules), 149, 151
ESR spectrum, 151
Kramers' doublet:
$^4\Sigma$ molecules, 239, 241–243, 248, 252
$^6\Sigma$ and sextet axial molecules, 270, 271, 277, 281
$g' = 4.3$ signal, 277
$^8\Sigma$ molecules, 295

Λ doubling parameters, p and q:
relation to g_l' and $g_r^{e'}$, 231
Λ matrix, 44–46, 158
λ, spin-spin coupling constant (gas phase) (*see* Zero-field splitting), 219
for C_4 molecule, 211
Landé splitting factor (g_J), 3
LaO, 10
Laser magnetic resonance (LMR), 219, 221–224
atoms studied, 222
CH spectra and observed transitions, 223, 224
molecular ions, 223
molecules studied, 223
sensitivity, 221
spectrometer, 222
LiO, 155, 221

Magnetic dipole (dipolar) interaction:
electron-electron, 158, 159, 308, 310, 322
interatomic distance from, 308, 322
electron-nuclear
atoms, 14–19, 59
interatomic, 61
molecules, 53–55
Magnetic dipole moment:
orbital electronic contribution, 1
spin contribution, 2
table of, for stable nuclei, 341–344
Magnetic dipole transition, 7
Magnetogyric ratio (γ):
defined, 1, 344
table of, for stable nuclei, 341–344
Matrix effects (atoms), 19–38
on hf in atoms with half-filled shells, 25, 26
on H atoms, 21–24, 26–28
multiple sites, 19–21
on orbitally-degenerate atoms (quenching), 29–38
theory of crystal-field effects, 33–38
shifts in g_J and A, 21–26
table, 22, 23
on spin-spin interaction, 28, 29
superhyperfine interaction, 26–28
theory of shifts, 24, 25
Matrix effects (molecules) (*see* motional effects), 19–25
bending of linear molecule, 118, 136
nonrandom orientation, 95–102
on zero-field splittings in triplet molecules, 193–198
electronic interactions, 197–198
torsional oscillations; isotope effects, 193–197
Matrix-isolated atoms and molecules:
apparatus for preparation, 87–89
atoms and atomic ions, 19–38
matrix and crystal-field effects, 19–29
orbitally-degenerate, quenching, 29–38
molecules;
exchange-coupling in Mn_2, 319–322
motional and matrix effects upon zfs, 193–198
NO_2 and HO_2, comparison with gas data, 232, 233
nonrandom orientation in matrices, 95–102
rotation, likelihood of, 86
rotation, free and hindered, 120–126
with $S = \frac{1}{2}$
axial, 10, 89–103
nonaxial, 106, 110, 113, 120
off-principal-axis absorption, 102
representative molecules, 126–155
with $S = 1$, 175–194
representative molecules, 206–218
with $S = \frac{3}{2}$, 241–247
representative molecules, 247–255

Matrix-isolated atoms and molecules (*Cont.*):
  molecules (*Cont.*)
    with $S = 2$, 262–264, 265
    with $S = \frac{5}{2}$, 271–274
      representative molecules, 281–287
    with $S = 3$, 288–295
    with $S = \frac{7}{2}$, 295–300
    with $S = 4$ (GdO), 302, 303
    Σ molecules, matrix versus gas, 220
    spin-rotation constants for, 51, 52, 234
Mesitylene (triplet state), 204
Methylpentyldiynylene, 206–209
  ESR spectrum, 207
Methylpropargylene, 208
MgF, 100, 129, 130, 132, 135, 136
  comparison with MgH and MgOH, 132, 136
  comparison of spin density on F among fluorides, 132
  ESR spectrum, 130
MgH, 51, 61, 126–129, 132, 136
  comparison with MgF and MgOH, 132, 136
MgOH, 118, 119, 135–137
  annealing effect upon matrix spectrum, 118, 119, 135
  comparison of isotropic hf data with BeOH, 137
  comparison with MgF and MgH, 136
  ESR spectrum, 119, 135
Microwave spectroscopy, 219
Mn atoms, 319
Mn^{2+}, half-filled d shell:
  in As-Te-I glass, $g' = 4.3$ signal, 281
  in axial crystal field, 266
  doped into MgO (exchange), 305
  doped into MgO and CaO (exchange), 318–320, 321
    ESR spectra, 320
  doped into polycrystalline CaO, 272–277
    ESR spectrum, 275
  zero-field splitting, 235
Mn_2, 304, 305, 319–322
  ESR spectrum, 321
$MnAu^+$ in solid silicon, 254
MnB^+ in solid silicon, 254
MnF, 288, 291–294, 303, 304
  ESR spectrum, 293
MnF_2, 235, 271, 282, 283, 297
  ESR spectrum, 283, 284
MnH, 288, 290–294
  gas phase, 288
MnH_2 (MnD_2), 235, 282, 282, 285
  ESR spectrum, 285
MnO, 235, 266, 277, 282, 286, 287, 304
  D value, 287
  ESR spectrum, 286
  hyperfine analysis, 287
MnO_2, 250, 252
MnPt in solid silicon, 254
  ESR spectrum, 255
Motional effects in matrices, 86, 120–126
  effect on linewidths of hf structure, 123–126
  effect of nuclear spin state, 120, 121
  effect on zfs in triplet molecules, 193–197
    rotational diffusion, 193, 194
    torsional oscillation; isotope effects, 193
  molecular rotation about one axis, 121, 122
  rotational freedom, 86
  theory of spin-relaxation, 125, 126

N_2^-, 79
Na atoms, 19, 20, 22
  matrix effects, 19, 20
Na_3, 147–149
  ESR spectrum, 147
  magnetic parameters in two sites in solid argon, 149
NaO_2 (*see* O_2^- and Alkali superoxide molecules), 149
  ESR spectrum, 150, 151
Napthalene in its triplet state, 184, 186–190, 199
  theoretical calculation of D and E, 201, 202
Natural abundances of nuclei, 341–344
NbO, 52, 102, 234, 247, 250–252
  ESR spectrum, 251
NCl, theoretical calculation of D, 200, 201
NCN, 211–213
NF, theoretical calculation of D, 200, 201
NF_2, 98, 122
  rotation in matrices, 122
NH, 199–201
  theoretical calculation of D, 199, 200
NH_2, 121, 229, 234
  effect of nuclear spin state, 121
$NiCl_2$, 205
NMR transitions (atoms), 8, 9–11
NO, 78
NO_2, 51, 98, 104–108, 109, 110, 121, 122, 152, 231–233
  spin rotation constants, 231
  g tensor components, 232, 233
$^{14}NO_2$, g and A tensor axes coincident, 104–108, 115
  anisotropic hf parameters, 107
  ESR single crystal data, 107, 232, 233
  gas phase data, 105, 231–233
  g_S^{ii} values
    from Curl's equation, 233
    from gas phase, 231–233
    from solid state, 232, 233

rotation in matrices, 121, 122
ESR spectrum, 122
s, p hybridization and bond angle, 108
spin densities, 108
$N^{17}O_2$, g and A tensor axes not coincident, 109–110, 111
"bent" NO bonds, 109
NO_3, 79
Nonet ($S = 4$) molecules:
$^9\Sigma$ molecules, 295, 301–303
GdO, 302, 303
contribution of $B_4{}^0$ term, 302, 303
ESR spectrum, 302
off-principal-axis transitions, 302
resonant-magnetic fields as a function of D, 301, 302
Non-random molecular orientation in matrices, 95–102
BO, 99–101
crystallite orientation, 100
ESR spectrum, 99, 101
CuF_2, 95, 98
$Cu(NO_3)_2$, 95–103
ESR spectra, 97
orientation distribution function, 98
in matrices other than rare gas solids, 100, 101
NF_2, 98
NO_2, 98
other observations of orientation, 100
Nuclear hyperfine interaction, 4–19
A_{dip}, anisotropic dipolar hf constant, 13–19, 59, 345–348
A_{iso}, isotropic hf constant, 13–19, 59, 345–347
Hamiltonian, 5
magnetic dipole transitions, 7
NMR transitions, 8–11
polarization, correlation and relativistic effects, 16, 17
$|\Psi(0)|^2$, 14, 16–19
$\langle r^{-3}\rangle$, 14–19, 345–347
$\langle r_l^{-3}\rangle$, 16, 17
$\langle r_{sC}{}^{-3}\rangle$, 16, 17
Zeeman energy levels (Breit-Rabi equation), 6–12
Nuclear hyperfine interaction (molecules) (*see* individual molecules), 40–42, 53–57
A_{iso}, A_{dip} (*see* above in index), 54, 55, 61
A tensor
axes relative to g tensor axes, 110
diagonal form in axial case ($A_{\parallel}, A_{\perp}$), 55
Frosch and Foley's parameters, 54–56
$L \cdot I$ term, 55, 56
not true tensor, footnote, 42
relation to A_{iso} and A_{dip}, 55
in exchange-coupled atoms, 304, 312–314
in gas-phase molecules, 225
spin and orbital contributions, 225, 226
Hamiltonian, 53, 54
in terms of Frosch and Foley's parameters, 53
in high-spin ($S \geqslant 2$) molecules:
for equivalent electrons, 247, 302–304
for exchanging pairs and clusters, 304
$^4\Sigma$ molecules, 243–247
D and A small, 243
D large, 246
impurity pairs in silicon, 254
relation to electron configuration, 247
$^6\Sigma$ molecules, 266–268, 272–277
$D \simeq A$, ESR effects, 273–277
(1) intensity variations, 274
(2) peak to peak splitting, 276
(3) forbidden transitions, 276
eigenvalue matrix for $I = \frac{5}{2}$, 267, 268
interatomic dipole-dipole contribution, 61, 62
more than one magnetic nucleus, 56, 57
for equivalent nuclei, 56, 57
nonaxial doublet molecules, 103
anisotropic hf tensor, 104
A and g tensors coincident, 104–108
A and g tensors noncoincident, 80, 104, 109–112
isotropic hf tensor, 104
perturbation theory, A relatively large (Na_3), 148, 149
rotating molecules in matrices, 120–126, 142
signs of A components, 62, 63
single crystals, 82–84, 106–116
in $^3\Sigma$ molecules, 179–183
hfs of $\Delta m_S = 2$ lines, 182
hfs relative to $^2\Sigma$ molecule, 182, 183
hfs at zero field, 182
transferred hf constant (in exchange), 313
Nuclear magneton (β_n), 5
Nuclear quadrupole interaction:
in atoms, 12, 13, 64
Hamiltonian, 12, 13
in molecules, 41, 63–65, 103, 152, 219, 246, 250
in CrCu molecules, 300, 301
energy levels, 68, 69
in forbidden transitions, 74–77
resonant magnetic fields, 72, 73
quadrupole coupling constant (Q'), 13, 64

Nuclear quadrupole interaction (*Cont.*):
  quadrupole coupling constant (Q') (*Cont.*)
    calculation of q, 64, 65
    in ionic molecules, 65
Nuclear Zeeman interaction:
  atoms, Hamiltonian and g_I, 5, 6
  molecules, 41, 63–65
    Hamiltonian and $g^{(I)}$ tensor, 63–65
    in producing forbidden transitions
      high-spin (Mn^{2+}), 276, 277
      $S = \frac{1}{2}$, 74–77

O^- ion, quenched, 38
O_2, 158, 195, 197, 201, 205, 209, 227–229
  gas phase, 227, 228
    γ (spin-rotation constant), 228, 230
    g_l^e, g_S^e, 228
    g_r^e, g_r^n, 229, 230
  hyperfine splitting, 304
  theoretical calculation of D, 158, 199, 200
  torsional oscillation in matrices
    hindering barrier, 197
    isotope effects on D, 197
O_2^-, (*see* Alkali superoxide molecules), 79, 81, 149–155
  ESR of alkali superoxide molecules in solid krypton, 151
  ESR of NaO_2 molecule in solid argon, 150
Octafluorocyclobutane matrix, 197
Octet ($S = \frac{7}{2}$) molecules:
  $^8\Sigma$ and axial, 295–301
    CrCu (or $^6\Sigma$), 297–301
      ESR spectrum, 300
      nuclear quadrupole coupling, 300, 301
      table of magnetic parameters, 300
    GdF_3 (axial), 295, 298, 299
      ESR spectrum, observed and simulated, 298, 299
      off-principal-axis transitions, 295, 299
    resonant magnetic fields as a function of D, 295, 297
    Zeeman levels and magnetic dipole transitions, 295, 296
Off-principal-axis-transitions ("extra" lines):
  among fine structure transitions
    $^4\Sigma$ molecules, 243–245
    $^5\Sigma$ molecules, 263, 264
    $^6\Sigma$ molecules, 271, 273, 274, 284–286
      nonaxial, $g' = 4.3$, 279
    $^7\Sigma$ molecules, 288, 292
    Octet axial molecules (GdF_3), 295, 299
    $^9\Sigma$ molecules (GdO), 302
  among hyperfine structure transitions
    copper phthalocyanine, 102
    $Cu(NO_3)_2$, 102, 103
    NbO, 102, 250–251
    VO, 102, 248–250
OH, 79, 225, 227, 231, 232
  gas-phase, 225, 227, 231, 232
    g values in $X\,^2\Pi$ state, 232
Optical double resonance techniques, 219
Orbital angular momentum, 1, 3
  quenching of, 3
Orbital angular momentum operator, 48, 49
Orbitally-degenerate ($^2\Pi$, etc.) molecules and quenching, 77–79
  g tensor components, 78, 79
    if quenched, 79
  O_2^-, 149–155
O_n^m, spin equivalent operators, 256, 257, 266, 278, 287, 295, 301, 302

PdH, PdD, 51
PH, theoretical calculation of D, 200, 201
PF, theoretical calculation of D, 200, 201
PF_2, 115, 122
  g and A (^{19}F) axes not coincident, 115
  possible rotation in matrices, 122, 123
m-Phenylene-bis-phenyl-methylene (quintet), 264
Phenylpentadiynlylene, 208
Phenylpropargylene, 208
Phenyl radical, g and A(^1H) axes not coincident, 115
Polarization effects (atoms), 16
Polychlorotrifluoroethylene matrix, 207, 208
Propargylene, 206, 208, 209
Pseudodipolar exchange, 308–310

q (*see* Electric field gradient at the nucleus)
Q, nuclear electric quadrupole moment, 12, 13, 341–344
Quadrupole coupling (*see* Nuclear quadrupole interaction)
Quartet ($S = \frac{3}{2}$) molecules:
  $^4\Sigma$ molecules, 236–252
    absorption and ESR spectra, 242, 244, 245
    $\Delta m_S = 2$ and 3, 241–243
    hyperfine interaction, 243–247
      for $D \gg A$, 246, 247
    magnetic dipole transitions, 238
    randomly oriented, 241–245
    resonant magnetic fields as a function of D, 238, 240
    spin Hamiltonian, 236
    spin wavefunctions, 236
    VF_2 and MnO_2, 250–253
      ESR spectrum of VF_2, 253
    VO and NbO, 247–252
      comparison of parameters, 252
      ESR spectra, 248, 249, 251

Zeeman levels, 236–239
exact solution, 236
impurity pairs in solid silicon, 254, 255
ESR spectrum of MnPt, 255
organic radicals, 252, 253
R-center in KCl crystals, 256
Quenching (of the orbital angular momentum), 3
O_2^-, 79, 81, 149–155
Quintet (S = 2) molecules:
ions in crystals and inorganic molecules, 265
Sc_2, 265
organic radicals, 264, 265
diacetylene derivatives, 265
m-phenylene-bis-phenyl-methylene, 264
$^5\Sigma$ molecules, 259–264
magnetic dipole transitions, 259–261
resonant magnetic field as a function of D, 259–261
randomly oriented, 262–264
absorption and ESR spectra, 262, 263
off-principal-axis transition, 263
spin operators ($O_n{}^m$), 256, 257
spin wavefunctions, 257–259
theory, 256, 257
Zeeman levels, 256–261
zfs coefficients ($B_n{}^m$), 256, 257

Randomly oriented molecules:
g_1 factor, 72, 92, 94
ESR of matrix-isolated, 86–92
$S = \frac{1}{2}$, axial molecules, 77, 78, 89–95, 102, 103, 126–145
nonaxial molecules, 110, 113–120, 145–155
S = 1, axial molecules, 176–183, 209–216
nonaxial molecules, 190–193, 206–208, 217, 218
$S = \frac{3}{2}$, axial molecules, 241–255
S = 2, axial molecules, 262–264, 265
organic nonaxial molecules, 264, 265
$S = \frac{5}{2}$, axial molecules, 271–277, 281–287
nonaxial molecules, 277–281
S = 3, axial molecules, 288–294
organic radical, 294–295
$S = \frac{7}{2}$, axial molecules, 295–301
S = 4, axial molecules, 301, 302
RbO_2 (*see* Alkali superoxide molecules), 149, 151
ESR spectrum, 151
R-center, 256
Relativistic effects (atoms), 16
Resonant magnetic fields:
for axial $S = \frac{1}{2}$ molecules
allowed transitions, 72, 73
forbidden transitions, 74
example, C_2H, 75, 77
theory, 75–77
with g and A axes noncoincident, 111–113
with orthorhombic g and A tensors, 105
for $^3\Sigma$ molecules, as a function of θ and D, 167, 169, 170
hyperfine interaction, 181, 182
for triplet molecule, $E \neq 0$, 184, 185
$\Delta m_S = 2$ transitions for naphthalene, 187–190
two quantum transition, 192, 193
for $^4\Sigma$ molecules, as a function of D, 240
for $^5\Sigma$ molecules, as a function of D, 261
for $^6\Sigma$ molecules, as a function of D, 270
for $^7\Sigma$ molecules, as a function of D, 290
for $^8\Sigma$ molecules, as a function of D, 297
RhC, 51
Rotation of trapped radicals (*see* motional effects in matrices)
Rotational angular momentum (N), 50
Rotational constant, B or B_e, 219
Rotational g factors of molecules:
g_r, 227, 229–233
$g_r{}^e$, electron contribution, 228–233
to O_2 and SO, 228–230
to OH, 231, 232
to HO_2, NO_2, 231–234
to NH_2, 229, 234
from perturbation theory, 229
relation to spin rotation constant, 229
application to HO_2, 233
$g_r{}^{e'}$, relation to Λ doubling, 231, 232
$g_r{}^n$, nuclei contribution, 229, 230, 232, 233
calculation from rigid charge approximation, 229, 230
Rotational magnetic moments of molecules (*see* rotational g factors)
Russell-Saunders coupling, 3

S^- ion, quenched, 38
S_2, theoretical calculation of D, 199–201
Satellite lines in ESR spectrum of atomic H, 77
Sc_2, 265
ScO, 10
SeH, gas phase, 225
Septet (S = 3) molecules:
organic radicals, 294, 295
$^7\Sigma$, 287–295
MnH (MnD) and MnF, 288, 289–294
comparison with theory for MnH, 294
ESR of MnF, 293
table of magnetic parameters, 293
off-principal-axis transitions, 288, 292

Septet ($S = 3$) molecules (*Cont.*):
$^7\Sigma$ (*Cont.*)
randomly oriented, 288, 291, 292
absorption and ESR spectrum, 291
resonant magnetic fields as a function of D, 288, 290
Zeeman levels and magnetic dipole transitions, 288–290
zero-field splittings, 288
Sextet ($S = \frac{5}{2}$) molecules:
axial and $^6\Sigma$, 266–277
$D \simeq A$, ESR effects, 273–277
(1) intensity variation, 274
(2) peak-to-peak splitting, 276
(3) forbidden transitions, 276
effective g values, 271
eigenvalue matrix for $I = \frac{5}{2}$, 267, 268
FeF_3, 281, 282
ESR spectrum, 282
hyperfine interaction, 266–268, 272–277
magnetic dipole transitions, 268–271
MnF_2, MnH_2, MnD_2, 282–287
ESR spectra, 283–285
MnO, 286, 287
ESR spectrum, 286
hyperfine analysis, 287
off-principal-axis transitions, 271, 273, 274
randomly oriented, 271, 272
resonant magnetic fields as a function of D, 268–271
zero-field splitting, 268
nonaxial
ferrihemoglobin, 281
$g' = 4.3$ signal, 277–281
Fe^{3+} in powders, 277–280
ferrichrome-A, 277, 278
Mn^{2+} in glass, 281
spin Hamiltonian, 266
SH, gas phase, 225
SiCO, 136, 179, 213, 216
bending induced by matrix, 213, 216
ESR spectra, 216
SiH_3, 79, 123–125, 144, 145
rotation in matrices, 123–125
ESR spectrum, showing linewidth effects, 124, 145
comparison with other MH_3 and CH_xF_y radicals, 144
SiF_3, g and A (^{19}F) axes noncoincident, 144
$^2\Sigma$ molecules (*see* Doublet molecules), 39 ff.
$^3\Sigma$ molecules (*see* Triplet molecules), 160 ff.
$^4\Sigma$ molecules (*see* Quartet molecules), 236 ff.
$^5\Sigma$ molecules (*see* Quintet molecules), 256 ff.
$^6\Sigma$ molecules (*see* Sextet molecules), 266 ff.
$^7\Sigma$ molecules (*see* Septet molecules), 287 ff.
$^8\Sigma$ molecules (*see* Octet molecules), 295 ff.
$^9\Sigma$ molecules (*see* Nonet molecules), 301 ff.
Single crystals, ESR, 81
CF_3, 144
CO_2^-, 108
HCO, 114, 115
HO_2, 232–234
NO_2, 105–109, 232, 233
O_2^-, 152
V centers, 81–86
SiNN, bending induced by matrix, 213, 216
Sites in matrices (*see also* Matrix effects):
atoms, 19–21
molecules
CH_2, 197
CN, 137
MgOH, 118, 136
Na_3, 149
m-phenylene-bis-phenyl-methylene, 264
SnH_3, 144, 145
SO, 199, 200, 201, 227–229
γ (spin-rotation constant), 228, 230
gas phase, 227, 228
g_l^e, g_S^e, 228
g_r^e, g_r^n, 229, 230
Spin decoupling in gaseous radicals (*see also* Electron paramagnetic resonance), 220, 221, 225
Spin density (at a nucleus):
defined for s and p, d electrons, 57
in high spin molecules, 303
from measured molecular A_{iso} and A_{dip} values, 57, 58
in ionic molecules, 60
relation to atomic A_{iso} and A_{dip} values, 58
tabulated values in molecules, 349–388
Spin Hamiltonian:
definition (for $S = \frac{1}{2}$ molecules), 40–42
electron exchange coupling contribution, 305–310
g tensor for, 42–46
nuclear quadrupole interaction contribution, 64, 103
nuclear Zeeman interaction contribution, 63, 103
spin-spin interaction contribution:
for $S \geqslant 2$ molecules, 256, 257, 266, 302

for triplet molecules, 156–160
for quartet molecules, 236
Spin operators, 42, 43, 67, 161, 164, 172–174, 181
operator equivalents (O_n^m), 256, 257, 266, 278, 287, 295, 301, 302
Spin-orbit coupling, 40–46, 133, 204, 308
coupling constant, 41
Hamiltonian, 41, 42, 44–46
Spin polarization, 16, 17, 25, 26, 142
Spin relaxation, 125, 126
m_I dependence for rotating radicals, 125, 126
Spin-rotation interaction and tensor:
relation to spin decoupling in gas molecules, 220, 221
constant γ (linear molecules)
calculated from Curl equation, 50–52
relation to g_r^e, 229
HO_2, $(g_r^e)_{gg}$ calculated, 233
table, comparison of observed and calculated, 51, 52
for doublet molecules, 50–52, 129, 133, 138, 234
g_l for OH (gas), 228, 232
for high-spin molecules, 51, 52, 234
for O_2, SO, gas phase, 228
for triplet molecules, 51, 52, 176, 211, 234
tensor components ϵ_{ii}, 231
g_S^{ii} from Curl equation, 231–233
HO_2 and NO_2, 233
NH_2, 234
Spin wavefunctions,
doublet ($S = \frac{1}{2}$) molecules, 42–46
triplet ($S = 1$) molecules, 161, 164, 170–172
quartet ($S = \frac{3}{2}$) molecules, 236
quintet ($S = 2$) molecules, 257–259
SrF, 51, 129, 131–134
angular charge distribution, 134
comparison of ^{19}F hf in neon matrices and gas phase, 131
comparison of spin density on F among fluorides, 132
comparison with SrH, 132
SrH, 51, 126–129, 132
comparison with SrF, 132
Stable magnetic nuclei, 341–344
Superexchange, 304, 314, 316
Superhyperfine interaction (atoms), 26–28

TeH, gas phase, 225
Tetramethylsilane (TMS) matrix, molecular orientation in, 101
TiF_2, 204, 206, 217, 218, 297
ESR spectrum, 217
TiO, 206
Transferred hyperfine constant, 313
Tranferrin, 277, 280
Transition-metal molecules,
doublet, 51, 95
electron-exchange coupling in, 305, 313–322
high-spin, 235
hyperfine interaction in, 302–304
nonet, 301, 302
octet, 295, 297–301
septet, 288, 291–294
sextet, 266, 271, 275–287
triplet, 204–206, 217, 218
quartet, 52, 247–255
quintet, 265
Transition probabilities:
H_1 perpendicular to H
anisotropic axial g tensor, 71, 72
for randomly-oriented molecules, 72
isotropic g tensor, 70
orthorhombic g tensor, 119
g_1 factor, 119, 120
for sextet molecules, $g' = 4.2$ signal, 279, 280
for triplet molecules, 172–174
variation with angle between static field (H) and microwave field (H_1), 70
Trinitrene (septet molecule), 294
Triphenylene, 168, 178, 179, 193, 194
motional effects upon ESR spectrum, 194
Triplet ($S = 1$) molecules
axial and $^3\Sigma$, 160–183
choice of xy line positions, 178
$\Delta m_S = 2$ transitions, 163, 166–168, 170, 172–176, 182
eigenfunctions, 170–172
ESR spectra, 177–180
g tensor, Curl's equation, 176, 234
H_{min}, 170, 175, 176
hyperfine interaction, 179–183
intensities of ESR transitions, 174
magnetic-dipole transitions, 163, 165–170
randomly oriented, 176–179
resonant magnetic fields as a function of D and θ, 166–170
Zeeman levels, 160–165
perturbation theory solution, 165
motional and matrix effects upon zfs, 193–198
nonlinear, 183
D, E, and their signs, 183, 184

Triplet ($S = 3$) molecules (*Cont.*):
  nonlinear (*Cont.*)
    $\Delta m_S = 2$ transitions (H_{min}), 184, 186, 188–191
    ESR transitions, 185–190
    g and D axes noncollinear, 192
    randomly oriented, 190–192
    two quantum transition, 192, 193
    Zeeman levels, 185–189
  representative triplet molecules
    C_4, 209–211
    CNN, NCN, CCO, 211–213, 214, 215
    propargylene and alternant methylenes, 206–209
    SiCO and SiNN, 213, 216
    TiF_2, 217, 218
  spin-spin interaction, 156
  tabulated magnetic properties,
    in gas-phase
      diatomic molecules, 393, 394
      triatomics, 399
    in matrices and crystals,
      diatomic molecules, 362, 363
      triatomics, 52, 377
      tetra-atomic, 52, 387
  theoretical calculation of zfs, 198–206
  zero-field splitting: X, Y, Z, D, E, 156–160

V^{2+} ions doped into MgO, transferred hf between pairs, 313
V^{3+} ion in octahedral symmetry, 205
V_K center, 81–86, 153
VF, 250, 251, 253
VF_3, 205
VO, 52, 100, 102, 234, 246–250, 252, 303, 304
  ESR spectrum, 248, 249

YbH, 51

Zeeman energy levels
  atoms, 4, 6–9
  $S = \frac{1}{2}$, $^2\Sigma$ molecules including hyperfine interaction, 68, 69
    perturbation theory expression including hf, nuclear Zeeman and quadrupole terms, 69
    effect of $H_{applied}$ vs. H_{hf}, 69, 70
  $S = 1$, $^3\Sigma$ molecules, 162–165, 169, 182
    nonlinear triplet molecules, 185–189
      naphthalene, 186
  $S = \frac{3}{2}$, $^4\Sigma$ molecules, 236–239
  $S = 2$, quintet molecules, 258–261
  $S = \frac{5}{2}$, sextet molecules, 267–270
    eigenvalue matrix for $I = \frac{5}{2}$, 267, 268
  $S = 3$, septet molecules, 288–290
  $S = \frac{7}{2}$, octet molecules, 295, 296
Zero-field splitting (*see* D tensor, λ, Triplet molecules):
  defined, 156
  D, E, zfs parameters, 156, 157
  in electron exchanged-coupled dimers
    due to anisotropic exchange, 308–312
  motional and matrix effects upon, 193–198
  in $S = 1$, triplet molecules
    $^3\Sigma$ molecules, 156, 157, 162, 163, 169
      from ESR spectra, 177–179
    nonlinear, 183, 184
      from ESR spectra, 191
  in $S > \frac{3}{2}$, high spin molecules, 235–322
  in $S = \frac{3}{2}$, quartet molecules, 237–239
    from ESR spectra, 241–243
  in $S > 2$ molecules, contribution of higher terms in S_i, 256, 257
  in $S = 2$, quintet molecules
    $^5\Sigma$ molecules, 259, 260
  in $S = \frac{5}{2}$, sextet molecules
    $^6\Sigma$ molecules, 268, 269
      from ESR spectra, 272
    nonaxial $g' = 4.3$ signal, 277–281
  in $S = 3$, septet molecules
    $^7\Sigma$ molecules, 289
      from ESR spectra, 291
  in $S = \frac{7}{2}$, octet molecules, 296
  in $S = 4$, nonet molecules, 301, 302
  tabulated zfs parameters of small molecules, Appendices C and D
  theoretical calculation of zfs parameters
    spin-orbit and spin-spin contributions, 156–160, 198–206
Zero-field tensor (*see* D tensor)
ZnF, 100
ZnH, 51

A CATALOG OF SELECTED

DOVER BOOKS

IN ALL FIELDS OF INTEREST

THE BOOK OF BEASTS: Being a Translation from a Latin Bestiary of the Twelfth Century, T. H. White. Wonderful catalog real and fanciful beasts: manticore, griffin, phoenix, amphivius, jaculus, many more. White's witty erudite commentary on scientific, historical aspects. Fascinating glimpse of medieval mind. Illustrated. 296pp. 5⅜ × 8¼. (Available in U.S. only) 24609-4 Pa. $5.95

FRANK LLOYD WRIGHT: ARCHITECTURE AND NATURE With 160 Illustrations, Donald Hoffmann. Profusely illustrated study of influence of nature—especially prairie—on Wright's designs for Fallingwater, Robie House, Guggenheim Museum, other masterpieces. 96pp. 9¼ × 10¾. 25098-9 Pa. $7.95

FRANK LLOYD WRIGHT'S FALLINGWATER, Donald Hoffmann. Wright's famous waterfall house: planning and construction of organic idea. History of site, owners, Wright's personal involvement. Photographs of various stages of building. Preface by Edgar Kaufmann, Jr. 100 illustrations. 112pp. 9¼ × 10.
23671-4 Pa. $7.95

YEARS WITH FRANK LLOYD WRIGHT: Apprentice to Genius, Edgar Tafel. Insightful memoir by a former apprentice presents a revealing portrait of Wright the man, the inspired teacher, the greatest American architect. 372 black-and-white illustrations. Preface. Index. vi + 228pp. 8¼ × 11. 24801-1 Pa. $9.95

THE STORY OF KING ARTHUR AND HIS KNIGHTS, Howard Pyle. Enchanting version of King Arthur fable has delighted generations with imaginative narratives of exciting adventures and unforgettable illustrations by the author. 41 illustrations. xviii + 313pp. 6⅛ × 9¼. 21445-1 Pa. $5.95

THE GODS OF THE EGYPTIANS, E. A. Wallis Budge. Thorough coverage of numerous gods of ancient Egypt by foremost Egyptologist. Information on evolution of cults, rites and gods; the cult of Osiris; the Book of the Dead and its rites; the sacred animals and birds; Heaven and Hell; and more. 956pp. 6⅛ × 9¼.
22055-9, 22056-7 Pa., Two-vol. set $21.90

A THEOLOGICO-POLITICAL TREATISE, Benedict Spinoza. Also contains unfinished *Political Treatise*. Great classic on religious liberty, theory of government on common consent. R. Elwes translation. Total of 421pp. 5⅜ × 8½.
20249-6 Pa. $6.95

INCIDENTS OF TRAVEL IN CENTRAL AMERICA, CHIAPAS, AND YUCATAN, John L. Stephens. Almost single-handed discovery of Maya culture; exploration of ruined cities, monuments, temples; customs of Indians. 115 drawings. 892pp. 5⅜ × 8½. 22404-X, 22405-8 Pa., Two-vol. set $15.90

LOS CAPRICHOS, Francisco Goya. 80 plates of wild, grotesque monsters and caricatures. Prado manuscript included. 183pp. 6⅝ × 9⅝. 22384-1 Pa. $4.95

AUTOBIOGRAPHY: The Story of My Experiments with Truth, Mohandas K. Gandhi. Not hagiography, but Gandhi in his own words. Boyhood, legal studies, purification, the growth of the Satyagraha (nonviolent protest) movement. Critical, inspiring work of the man who freed India. 480pp. 5⅜ × 8½. (Available in U.S. only)
24593-4 Pa. $6.95

ILLUSTRATED DICTIONARY OF HISTORIC ARCHITECTURE, edited by Cyril M. Harris. Extraordinary compendium of clear, concise definitions for over 5,000 important architectural terms complemented by over 2,000 line drawings. Covers full spectrum of architecture from ancient ruins to 20th-century Modernism. Preface. 592pp. 7½ × 9⅝. 24444-X Pa. $14.95

THE NIGHT BEFORE CHRISTMAS, Clement Moore. Full text, and woodcuts from original 1848 book. Also critical, historical material. 19 illustrations. 40pp. 4⅝ × 6. 22797-9 Pa. $2.50

THE LESSON OF JAPANESE ARCHITECTURE: 165 Photographs, Jiro Harada. Memorable gallery of 165 photographs taken in the 1930's of exquisite Japanese homes of the well-to-do and historic buildings. 13 line diagrams. 192pp. 8⅜ × 11¼. 24778-3 Pa. $8.95

THE AUTOBIOGRAPHY OF CHARLES DARWIN AND SELECTED LETTERS, edited by Francis Darwin. The fascinating life of eccentric genius composed of an intimate memoir by Darwin (intended for his children); commentary by his son, Francis; hundreds of fragments from notebooks, journals, papers; and letters to and from Lyell, Hooker, Huxley, Wallace and Henslow. xi + 365pp. 5⅜ × 8. 20479-0 Pa. $5.95

WONDERS OF THE SKY: Observing Rainbows, Comets, Eclipses, the Stars and Other Phenomena, Fred Schaaf. Charming, easy-to-read poetic guide to all manner of celestial events visible to the naked eye. Mock suns, glories, Belt of Venus, more. Illustrated. 299pp. 5¼ × 8¼. 24402-4 Pa. $7.95

BURNHAM'S CELESTIAL HANDBOOK, Robert Burnham, Jr. Thorough guide to the stars beyond our solar system. Exhaustive treatment. Alphabetical by constellation: Andromeda to Cetus in Vol. 1; Chamaeleon to Orion in Vol. 2; and Pavo to Vulpecula in Vol. 3. Hundreds of illustrations. Index in Vol. 3. 2,000pp. 6⅛ × 9¼. 23567-X, 23568-8, 23673-0 Pa., Three-vol. set $37.85

STAR NAMES: Their Lore and Meaning, Richard Hinckley Allen. Fascinating history of names various cultures have given to constellations and literary and folkloristic uses that have been made of stars. Indexes to subjects. Arabic and Greek names. Biblical references. Bibliography. 563pp. 5⅜ × 8½. 21079-0 Pa. $7.95

THIRTY YEARS THAT SHOOK PHYSICS: The Story of Quantum Theory, George Gamow. Lucid, accessible introduction to influential theory of energy and matter. Careful explanations of Dirac's anti-particles, Bohr's model of the atom, much more. 12 plates. Numerous drawings. 240pp. 5⅜ × 8½. 24895-X Pa. $4.95

CHINESE DOMESTIC FURNITURE IN PHOTOGRAPHS AND MEASURED DRAWINGS, Gustav Ecke. A rare volume, now affordably priced for antique collectors, furniture buffs and art historians. Detailed review of styles ranging from early Shang to late Ming. Unabridged republication. 161 black-and-white drawings, photos. Total of 224pp. 8⅜ × 11¼. (Available in U.S. only) 25171-3 Pa. $12.95

VINCENT VAN GOGH: A Biography, Julius Meier-Graefe. Dynamic, penetrating study of artist's life, relationship with brother, Theo, painting techniques, travels, more. Readable, engrossing. 160pp. 5⅜ × 8½. (Available in U.S. only) 25253-1 Pa. $3.95